INTRODUCTION TO PROGRAMMABLE LOGIC CONTROLLERS

Third Edition

by

Gary Dunning

DELMAR
CENGAGE Learning™

Australia   Canada   Mexico   Singapore   Spain   United Kingdom   United States

Introduction to Programmable Logic Controllers, Third Edition

Gary Dunning

Vice President, Technology and Trades SBU:

Alar Elken

Editorial Director:

Sandy Clark

Senior Acquisitions Editor:

Stephen Helba

Developmental Editor:

Sharon Chambliss

Marketing Director:

David Garza

Channel Manager:

Dennis Williams

Marketing Coordinator:

Stacey Wiktorek

Production Director:

Mary Ellen Black

Production Editor:

Barbara L. Diaz

Editorial Assistant:

Dawn Daugherty

For product information and technology assistance, contact us at
Cengage Learning Customer & Sales Support, 1-800-354-9706

For permission to use material from this text or product, submit all requests online at **cengage.com/permissions**
Further permissions questions can be emailed to
permissionrequest@cengage.com

Library of Congress Control Number: 2005028293

ISBN-13: 978-1-4018-8426-0

ISBN-10: 1-4018-8426-1

Delmar Cengage Learning
5 Maxwell Drive
Clifton Park, NY 12065-2919
USA

Cengage Learning products are represented in Canada by Nelson Education, Ltd.

For your lifelong learning solutions, visit **delmar.cengage.com**

Visit our corporate website at **www.cengage.com**

Notice to the Reader

Publisher does not warrant or guarantee any of the products described herein or perform any independent analysis in connection with any of the product information contained herein. Publisher does not assume, and expressly disclaims, any obligation to obtain and include information other than that provided to it by the manufacturer. The reader is expressly warned to consider and adopt all safety precautions that might be indicated by the activities described herein and to avoid all potential hazards. By following the instructions contained herein, the reader willingly assumes all risks in connection with such instructions. The publisher makes no representations or warranties of any kind, including but not limited to, the warranties of fitness for particular purpose or merchantability, nor are any such representations implied with respect to the material set forth herein, and the publisher takes no responsibility with respect to such material. The publisher shall not be liable for any special, consequential, or exemplary damages resulting, in whole or part, from the readers' use of, or reliance upon, this material.

Printed in the United States of America
2 3 4 5 6 7 11 10 09 08

TABLE OF CONTENTS

Part I　Introduction to PLCs

Part II   PLC Instructions

PREFACE

Welcome to the world of programmable logic controllers. Since their development in the early 1970s, programmable logic controllers (PLCs) have literally taken control of practically every modern manufacturing process.

The third edition text has been brought up to date, providing the latest possible information on and introduction to PLC products. We have added PLC information and pictures from AEG Schneider Automation/Square D-Modicon and Mitsubishi Electric Automation, Inc. Chapter 2 introduces the newer and smaller micro and pico PLCs. The accompanying lab manual has been updated from Rockwell Automation's older APS programming software to the current RSLogix 500 Windows programming software. All chapters and lab exercises were developed to be an easy-to-understand introduction to the world of PLCs.

Developed as a first quarter or semester introductory textbook, *Introduction to Programmable Logic Controllers, Third Edition,* discusses the top players in today's PLC market and their products, including Rockwell Automation/Allen-Bradley, General Electric, Mitsubishi, Omron, Schneider Automation, and Siemens. The text was developed to be as generic as possible. Although separate, the lab manual's hands-on programming exercises were developed for the Allen-Bradley SLC 500 and MicroLogix family of PLCs using RSLogix 500 Windows software.

This book was developed for electrical technicians, maintenance personnel, machine design engineers, and individuals selling PLCs in today's ever-changing manufacturing environment. Students will quickly learn the basics of today's modern PLCs. Because the PLC is an industrial computer, many electricians and technicians are apprehensive about diving into the world of computers and the associated, seemingly foreign terminology. This text is developed for these individuals. These new terms are explained in plain English so those new to the world of PLCs will feel comfortable. For additional help in understanding PLC terminology, there is a comprehensive glossary near the end of the book.

Topics covered include exploring what a PLC is, operation, usage, instructions, applications, hardware selection and configuration, introductory programming examples and exercises, and some troubleshooting hints.

The accompanying lab manual gives the student hands-on programming and hookup exercises. Students will configure different PLCs, set up the RSLinx drivers to communicate from a personal computer to the PLC, select proper modules, and calculate power supply loading. A lab exercise will walk students through a motor starter interface to an SLC 500 PLC along with issues in converting conventional ladder start-stop into PLC logic, separation of I/O, and proper instruction selection. Programming exercises provide hands-on experience developing simple programs that incorporate basic instructions up to and including timers, counters, and sequencers.

This textbook is organized in the following manner:

Chapter	Chapter Title	Chapter Description
1	Welcome to the World of Programmable Logic Controllers	Provides an introduction to what programmable controllers are and why and where they are used.
2	Micro Programmable Logic Controllers	Introduces the student to the world of the new and smaller micro and pico PLCs.
3	Programming a Programmable Controller	Describes the available options for programming a programmable controller.
4	Number Systems	Covers the basics of numbers and number systems used with PLCs.
5	Introduction to Digital and Analog PLC Interface	Introduces the basic operating principles of the PLC.
6	Introduction to Logic	Describes the basic logic functions and how they relate to programmable controllers.
7	Input Modules	Describes the basic input modules available, selecting the correct module and basic module interface.
8	Output Modules	Describes the basic output modules available, selecting the correct module and basic module interface.
9	Putting Together a Modular PLC	Introduces the basic considerations when putting together a modular PLC. The chapter reviews power calculations for configuring rack I/O modules and power supply selection.
10	PLC Processors	Introduces the PLC processor and its capabilities, features, and basic operation.
11	Introduction to ControlNet and DeviceNet	Introduces the ControlNet and DeviceNet networks.
12	Processor Data Organization	Introduces program and data organization inside a PLC processor.
13	The Basic Relay Instructions	Describes the basic instructions used in developing PLC ladder programs.
14	Understanding Relay Instructions and the Programmable Controller Input Modules	Describes the use of relay instructions and considerations when interfacing to real-world input devices.
15	Documenting Your PLC System	Introduces documentation features available when using PLC ladder program development software.
16	Timer and Counter Instructions	Describes timer and counter instructions. Specifically introduces the Allen-Bradley SLC 500 timer and counter instructions.
17	Comparison and Data-Handling Instructions	Introduces data handling and comparison instructions. Specifically covers the SLC 500 instructions.
18	Sequencer Instructions	Introduces the sequencer instructions. Specifically covers the SLC 500 instructions.
19	Program Flow Instructions	Introduces program flow instructions such as JMP, LBL, JSR, SBR, RET, IIM, ICM, MCR, and REF.
	Appendix A	A quick reference to the SLC 500 instruction set.
	Appendix B	A quick reference to the SLC 500 status file.
	Appendix C	Reviews how hexadecimal numbers are used in masking applications.
	Glossary	A comprehensive list of defined PLC terms.

ACKNOWLEDGMENTS

I wish to thank my wife, Jean, for her endless hours of typing, proofreading, spell-checking, figure development, and art.

Developing any major technical textbook requires a substantial amount of technical information. We are greatly thankful to the following corporations who provided literature, manuals, and art along with the technical support necessary to complete this project.

AEG Schneider Automation/Square D-Modicon

Allen-Bradley, a Rockwell Automation business

ASAP Inc., Chagrin Falls, Ohio

GE Fanuc Automation

Mitsubishi Electric Automation, Inc.

Omron Electronics, Inc.

Siemens Energy & Automation, Inc.

Along with Thomson Delmar Learning, I would like to express appreciation to the following reviewers for their encouragement and suggestions during the preparation of this manuscript:

Keith Elliott
Rockingham Community College
Wentworth, NC

William Shepherd
Owens Community College
Perrysburg, OH

Ron Meyer
Central Community College
Doniphan, NE

Daniel Lewis
James Rumsey Technical College
Bunker Hill, WV

Paul F. Owens
San Juan College
Aztec, NM

Michael Brumbach
York Technical College
Rock Hill, SC

IF PASSWORD IS CORRECT GO TO DRIVE SET UP SCREEN

B3:10/0

0022
```
    ┌── EQU ──────────┐         ┌─ OSR ─┐                                              B3:1/15
    │ Equal           │         [  OSR  ]                                              ─( )─
    │ Source A   N12:21│                                  RESET PASSWORD
    │              7777<│                                 ATTEMPTS COUNTER
    │ Source B    32767 │                                 C5:5
    │             32767<│                                 ─( RES )─
    └─────────────────┘
```

0023
```
    ┌── GRT ──────────┐    ┌── NEQ ──────────┐      B3:10/1
    │ Greater Than (A>B)│   │ Not Equal       │      [  OSR  ]
    │ Source A   N12:21 │   │ Source A   N12:21│
    │              7777<│   │              7777<│
    │ Source B      0   │   │ Source B    32767 │
    │               0<  │   │             32767<│
    └─────────────────┘    └─────────────────┘
```

```
                                    ┌── CTU ──────────────┐
                                    │ Count Up            │─( CU )─
                                    │ Counter      C5:5   │
                                    │ Preset        3<    │─( DN )─
                                    │ Accum         2<    │
                                    └─────────────────────┘

                                    ┌── MOV ──────────────┐
                                    │ Move                │
                                    │ Source          −1  │
                                    │                 −1< │
                                    │ Dest        N12:21  │
                                    │             7777<   │
                                    └─────────────────────┘
```

NOTIFICATION   HANDSHAKE

B3:0/7   B3:0/8
0024   ─] [─   ─( )─

Normal state of password display. This will display enter password and press enter when then password entry attempts counter is equal to zero.

0025
```
    ┌── EQU ──────────┐         B3:10/2                ┌── MOV ──────────────┐
    │ Equal           │         [  OSR  ]              │ Move                │
    │ Source A  C5:5.ACC│                              │ Source          0   │
    │              2<   │                              │                 0<  │
    │ Source B      0   │                              │ Dest        N12:6   │
    │               0<  │                              │              1<     │
    └─────────────────┘                               └─────────────────────┘
```

The first time the incorrect password is entered send a 1 to the multi state indicator on the panel View so as to display "Password does not match - REENTER."

0026
```
    ┌── EQU ──────────┐         B3:10/2                ┌── MOV ──────────────┐
    │ Equal           │         [  OSR  ]              │ Move                │
    │ Source A  C5:5.ACC│                              │ Source          1   │
    │              2<   │                              │                 1<  │
    │ Source B      1   │                              │ Dest        N12:6   │
    │               1<  │                              │              1<     │
    └─────────────────┘                               └─────────────────────┘
```

The second time the incorrect password is entered send a 1 to the multi state indicator on the PanelView so as to display "Password does not match - REENTER."

0027
```
    ┌── EQU ──────────┐         B3:10/3                ┌── MOV ──────────────┐
    │ Equal           │         [  OSR  ]              │ Move                │
    │ Source A  C5:5.ACC│                              │ Source          1   │
    │              2<   │                              │                 1<  │
    │ Source B      2   │                              │ Dest        N12:6   │
    │               2<  │                              │              1<     │
    └─────────────────┘                               └─────────────────────┘
```

PART

I

Introduction to PLCs

There are many programmable logic controllers (PLCs) available from various manufacturers in the marketplace, and even though they may look different, they basically operate in a similar fashion. Part I of this book will introduce the basic operating principles of the modern PLC. We will be introduced to a number of popular PLCs in use today as we learn about the basic operating principles and features available. Chapter 2 introduces the group of smaller, more powerful, micro PLCs. Chapter 3 explores programming options. Number systems as they are used in the PLC world are introduced in Chapter 4. Chapter 5 looks into the inner workings of PLCs, and Chapter 6 introduces logic and how it is used to solve the ladder program. The next four chapters, Chapters 7 through 10, introduce PLC hardware. The new Chapter 11 provides an introduction to two popular industrial networks, ControlNet and DeviceNet. Understanding input modules and output modules, putting together a modular PLC along with processor operation, and an introduction to industrial networks completes the first section. There are review questions at the end of each chapter, and lab exercises, including programming exercises with RSLogix 500 software in the accompanying Lab Manual.

Part two, beginning with Chapter 12, will introduce PLC instructions to prepare you for the Lab Manual's programming exercises using Rockwell RSLogix 500 software.

IF PASSWORD IS CORRECT GO TO DRIVE SET UP SCREEN

0022
```
      ─── EQU ───
      Equal
      Source A    N12:21
                   7777<
      Source B    32767
                  32767<
```
B3:10/0
[OSR]

B3:1/15
( )
RESET PASSWORD
ATTEMPTS COUNTER
C5:5
(RES)

0023
```
      ─── GRT ───
      Greater Than (A>B)
      Source A    N12:21
                   7777<
      Source B       0
                      0<
```
```
      ─── NEQ ───
      Not Equal
      Source A    N12:21
                   7777<
      Source B    32767
                  32767<
```
B3:10/1
[OSR]

CTU
(CU)

(DN)

−1
−1<
N12:21
7777<

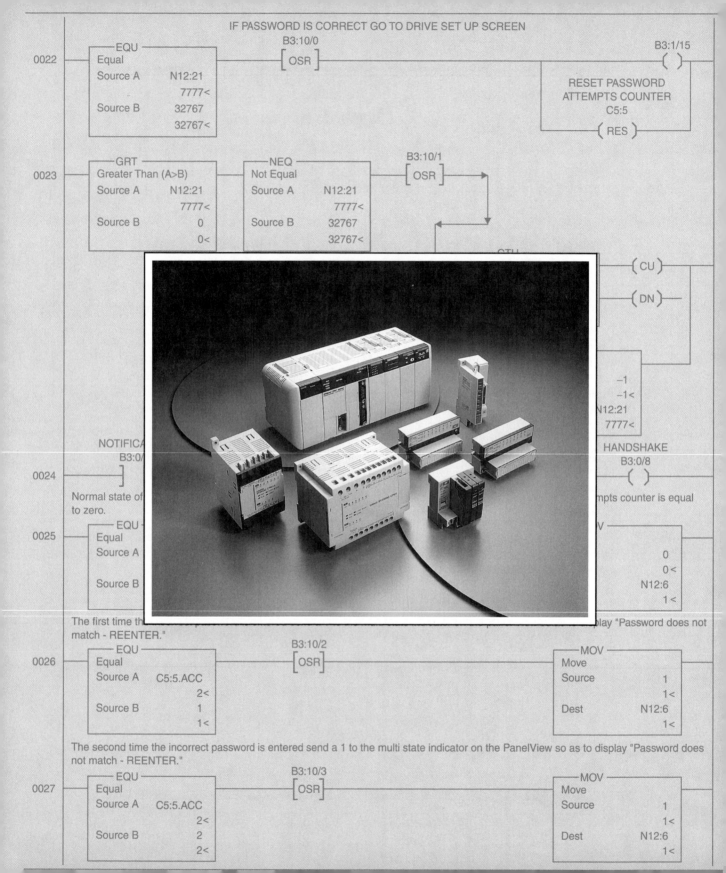

NOTIFICA...
B3:0/...
0024 ┤ ├

HANDSHAKE
B3:0/8
( )

Normal state of ...
to zero.

...mpts counter is equal

0025
```
      ─── EQU ───
      Equal
      Source A
      Source B
```
...V
0
0<
N12:6
1<

The first time the ... play "Password does not match - REENTER."

0026
```
      ─── EQU ───
      Equal
      Source A    C5:5.ACC
                      2<
      Source B       1
                      1<
```
B3:10/2
[OSR]
```
      ─── MOV ───
      Move
      Source         1
                     1<
      Dest       N12:6
                     1<
```

The second time the incorrect password is entered send a 1 to the multi state indicator on the PanelView so as to display "Password does not match - REENTER."

0027
```
      ─── EQU ───
      Equal
      Source A    C5:5.ACC
                      2<
      Source B       2
                      2<
```
B3:10/3
[OSR]
```
      ─── MOV ───
      Move
      Source         1
                     1<
      Dest       N12:6
                     1<
```

CHAPTER

1

Welcome to the World of Programmable Logic Controllers

OBJECTIVES

After completing this chapter, you should be able to:

- define "PLC"
- explain where PLCs came from
- explain why their use is valuable
- explain where they are used
- detail what PLCs can do
- explain how PLCs know what they are supposed to do

INTRODUCTION

A **programmable logic controller,** which is usually called a PLC or, more commonly, simply a programmable controller, is a solid-state, digital, industrial computer. Figure 1-1 illustrates the family of Allen-Bradley MicroLogix PLCs. The top three units are PLCs of different sizes within the Allen-Bradley MicroLogix 1000 family. A handheld programming terminal is pictured at the bottom of the figure.

Upon first glance, a programmable controller may seem to be no more than a black box with wires bringing signals in and other wires sending signals out. It might also appear there is some magic being done inside that somehow decides when field devices should be turned on. In actuality, there is no magic. The PLC is a computer, and someone had to tell it what to do. The PLC knows what to do through a program that was developed and then entered into its memory. The PLC is a computer; however, without a set of instructions telling it what to do, it is nothing more than a box full of electronic components. Without

Figure 1-1   Allen-Bradley MicroLogix 1000 PLCs and a handheld programmer. (Used with permission of Rockwell Automation, Inc.)

instructions, the black box that we call a PLC can do nothing. The user program is the list of instructions that tells the PLC what to do.

Computers such as PLCs can be wonderful tools; however, although it might appear otherwise, they only do exactly what the human programmer told them to do.

WHAT IS A PLC?

The PLC, or programmable controller, can be classified as a solid-state member of the computer family. A programmable controller is an industrial computer in which control devices such as limit switches, push buttons, proximity or photoelectric sensors, float switches, or pressure switches, to name a few, provide incoming control signals into the unit. An incoming control signal is called an **input.**

Inputs interact with instructions specified in the user ladder program, which tells the PLC how to react to the incoming signals. The user program also directs the PLC on how to control field devices like motor starters, pilot lights, and solenoids. A signal going out of the PLC to control a field device is called an **output.** Figure 1-2 gives an overview of the interaction between SW 1 (the systems input), the PLC and its ladder program, and the pilot light output.

A formal definition of a PLC comes from the National Electrical Manufacturers Association (NEMA):

> [A programmable controller is] a digitally operated electronic system, designed for use in an industrial environment, which uses a programmable memory for the internal

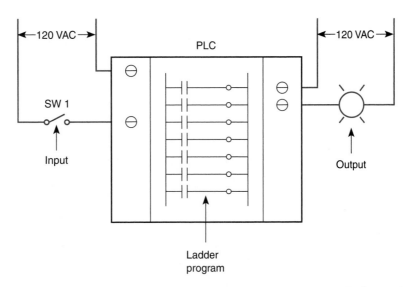

Figure 1-2   A general overview of a PLC identifying SW 1, the input, the PLC, and the output device.

storage of user-orientated instructions for implementing specific functions such as logic, sequencing, timing, counting, and arithmetic to control, through digital or analog inputs and outputs, various types of machines or processes. Both the PLC and its associated peripherals are designed so that they can be easily integrated into an industrial control system and easily used in all their intended functions.[1]

Figure 1-3 illustrates two Siemens PLCs in a control application. Notice the wires connected to the PLCs: these wires are the input and ouput connections.

WHERE DID THE PLC COME FROM?

In the 1960s and 1970s, electromechanical relays, timers, counters, and sequencers were the standard. Many control panels contained hundreds of these electromechanical devices and, in some cases, a mile or more of wire. The primary negative aspect of mechanical control was that reliability was low, in contrast to the maintenance costs associated with keeping these panels operating, which were extremely high. The auto industry complained that the real cost to purchase and replace a single relay could run as high as $50. A second major factor was the time, expense, and labor required when a change in control needs dictated a control panel modification. In fact, during the model year changeovers, the auto industry discarded entire control panels and replaced them with new ones as the quickest and cheapest solution.

The programmable controller is a solid-state electronic device designed in the early 1970s to replace electromechanical relays, mechanical timers, counters, and sequencers.

---

1. National Electrical Manufacturers Association, NEMA IA 2.1 standard, adopted from the International Electrotechnical Commission, IEC 1131 standard, part 1, section 2.50.

Figure 1-3   PLC and its input and output wiring. (Courtesy of Siemens Energy &
Automation, Inc.)

The Hydramatic division of General Motors was the first to see the need for a device that
would become what we know as the programmable logic controller. High-speed manufacturing, as in the auto industry, required reliable control devices that were smaller, consumed less power, featured fast switching, and were quickly and easily changeable. These
devices must also be able to withstand the harsh industrial environment. Keep in mind
that the first PLCs were little more than relay replacers.

Figure 1-4 shows the General Electric Fanuc Family of Programmable Controllers.
The rear PLC is the Series 90-70, the center row from left to right shows Series 90-30 and
VersaMax, and the front row left to right shows two Series 90 Micro PLCs with the Genius
I/O at the far right.

WHY USE A PLC?

The question, "Why use a PLC?" really should be rephrased to, "Why automate?" The
PLC is the tool that provides the control for an automated process. What will automating
a process do for a company? Automation will help a manufacturing facility to:

1. gain complete control of the manufacturing process
2. achieve consistency in manufacturing
3. improve quality and accuracy

Figure 1-4   General Electric's family of PLCs. (Courtesy of GE Fanuc Automation)

4. work in difficult or hazardous environments
5. increase productivity
6. shorten the time to market
7. lower the cost of quality, scrap, and rework
8. offer greater product variety
9. quickly change over from one product to another
10. control inventory

Why automate? Let us look at a story from *Fortune* magazine entitled, "Manufacturing the Right Way." A small part of this article, quoted here, refers to the Caterpillar factory in East Peoria, Illinois, and its factory redesign. The story describes the presence of old technology side-by-side with new, automated technology.

[The Caterpillar plant] . . . turns out 120 types of transmissions for the whole catalogue of Caterpillar machines. In the area where transmission cases are prepared for assembly, a long line of 35 machine tools, each requiring its own operator, stretches down one side of the aisle. On the other side are four (of an eventual 32) flexible cellular systems. Both do the same work of milling, drilling, boring, tapping, deburring, and reaming the crude steel from the foundry.

Because the old machines can handle only one kind of case at a time, the area around them is jammed with bins of cases waiting to go through in batches. When a batch is finished, the operator spends anywhere from four hours to two days setting up for the next batch. Once he is ready, he might need two or three tries to get the tool running right.

With adjustments being made at 35 stations, many $1,000 cases end up in the scrap heap before the next batch runs through smoothly.

On the other side of the aisle, the four cellular units are cutting metal. Since they are programmed to handle any case the plant makes, setup time is reduced to seconds; the system selects the right tools from a rotating belt and inserts them into spindles. Because the tool does the job right the first time, there is little scrap.[2]

This short story is one of hundreds illustrating the advantages of automating a manufacturing process. If you go back to the questions we asked at the beginning of this section, you should be able to address each one from this story about Caterpillar.

Now that we know why we would want to automate a manufacturing environment, let us look at what makes programmable controllers work.

WHAT MAKES A PLC WORK?

The heart of any computer is the microprocessor. The computer's microprocessor, also called the **processor** or central processing unit (**CPU**), supervises system control through the user program. The microprocessor reads input signals and follows the instructions that a programmer has stored in the PLC's **memory.** As a result of the solved program, the PLC turns outputs, or field-controlled devices, on or off. When the PLC is running and following the programmer's instructions, this is called solving the user program.

As we begin to consider what makes up a PLC, Figure 1-5 provides a look at a typical small, or micro, PLC. Figure 1-5 is a picture of a Mitsubishi FX2N SuperMicro PLC.

Figure 1-5 The Mitsubishi FX2N-80MR SuperMicro PLC. (Courtesy of Mitsubishi Electric Automation, Inc.)

HOW DO PLCs KNOW WHAT THEY ARE SUPPOSED TO DO?

A PLC simply follows the instructions stored in memory. To retain user program instructions, they are stored in the memory of the PLC for future use and reference. Think of an instruction as a sentence. Each instruction that is entered will be placed in memory in ascending order. The list of instructions is called the user ladder program. See Figure 1-6 for a sample list of instructions that could be turned into a program.

---

2. "Manufacturing the Right Way," *Fortune*, May 21, 1990, p. 60.

Instruction number	Instruction
0000	IF input number one is ON
0001	AND input number two is ON
0002	THEN turn ON output number one
0003	END

Figure 1-6   A representation of a PLC program where Switch 1 and Switch 2 are in series with Light 1. If both switches or inputs are ON, they turn on output one, which is our pilot light. See Figure 1-7.

Figure 1-7 illustrates what a schematic representation of the instruction list from Figure 1-6 would look like. Figure 1-8 shows what the two previous figures would look like represented as a PLC ladder logic rung.

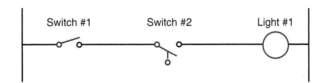

Figure 1-7   Circuit representation of the program illustrated in Figure 1-6.

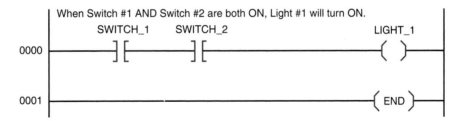

Figure 1-8   PLC ladder program rung representing Figure 1-7's schematic.

HOW DO INSTRUCTIONS GET INTO THE PLC'S MEMORY?

The instructions you wish your PLC to carry out are transferred to the memory of the controller from either a handheld programmer or a personal computer. The user ladder program is created by the operator pushing the correct sequence of buttons on a handheld programmer. The handheld programmer illustrated in Figure 1-9 can be used to program the General Electric Series 90-20 or 90-30 PLC.

A PLC user program can also be created using a personal or industrial computer. When using a personal or industrial computer to develop the user ladder program, a PLC ladder development software package is used. The primary difference between a personal computer and an industrial computer is that the industrial computer has been hardened

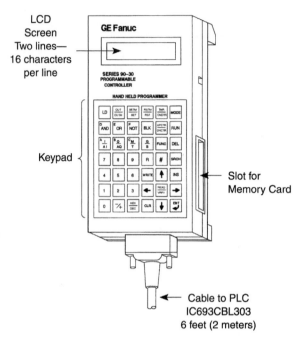

Figure 1-9   Series 90-30 and 90-20 handheld programmer. (Courtesy of GE Fanuc Automation)

to withstand the factory environment. For now we will use the term "personal computer" for both. In Chapter 3 the differences and advantages of industrial computer programming terminals will be discussed.

When the entire user ladder program has been developed, entered, and verified for correctness, the next step is to **download** the program into the processor's memory. Transferring the PLC program from a personal computer's memory to PLC memory is called downloading the program. Figure 1-10 illustrates a personal computer interfaced to an Omron CQM1 PLC.

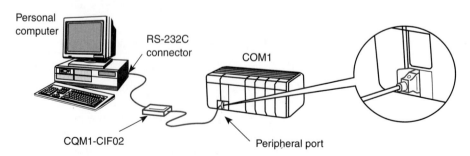

Figure 1-10   Interfacing a personal computer to an Omron CQM 1 PLC. (Courtesy of Omron Electronics, Inc.)

Before downloading a user program, the processor must be in **program mode.** After downloading the program, if all input and output signals are wired to the correct screw terminals, the processor can be put in **run mode.** In run mode, the program will continuously run and solve the programmed instructions. Solving the programmed instructions is sometimes called solving the logic. This continual running of the program in a PLC is called *scanning*. As part of the processor's problem-solving routine, the PLC will look at the incoming signals, follow the preprogrammed instructions associated with each input signal, and control the programmed output field devices. Refer to Figure 1-2.

Now that we know what a PLC is and what it does, let us explore innovations that PLCs brought with them.

INNOVATIONS INCORPORATED INTO THE PROGRAMMABLE CONTROLLER

The first programmable logic controllers had many important developments incorporated into them:

1. The PLC is a hardened industrial computer designed to withstand the harsh factory environment.
2. PLCs are reusable. They contain a changeable program that eliminates extensive rewiring and component changes.
3. PLCs offer easy troubleshooting.
4. PLCs feature easy installation and small size.

Let us look at each of these innovations.

The PLC Is a Hardened Industrial Computer

A PLC is made so that it can survive in the manufacturing environment. We need to differentiate between a PLC as an industrial computer and an industrial computer-programming terminal. Even though a PLC is an industrially hardened computer, an industrial computer-programming terminal is similar to an industrially hardened personal computer.

PLCs are small, easy-to-install units. A PLC is easy to install because all input and output connections are connected to terminal strips in a central location. The relays have been replaced by instructions programmed inside the PLC unit. Arranging and, later, rearranging relay contact sequences to program a replacement **instruction** is much easier than rewiring together relays and mechanical timers, counters, or sequencers.

A PLC's Control Sequence Is Easily Changeable through Programming

The PLC is easily programmed and reprogrammed. This new device's programs are developed directly from the standard ladder-diagram format, with which electrical maintenance personnel are already familiar. As an industrial computer, the PLC is able to replace such mechanical functions as relay control, timers, counters, and sequencers. As time progressed and electronic and microprocessor technology advanced, new functionality was added to the PLC. Advanced functions were added, such as arithmetic, data manipulation, shift registers, data storage, ladder diagram programming using the personal computer, and communication links and networking to other PLCs and personal computers.

Easier Troubleshooting

Troubleshooting became easier, as each input is wired separately to its own input screw terminal. Notice in Figure 1-11 that SW 1 and SW 2, although physically connected in a schematic diagram, and the PLC user ladder program are separate inputs when wired to the PLC. The separately wired input signals are recombined to represent the original schematic or user ladder program using the programming software's instructions. Traditional hardware relays are now programmed inside the unit. Changing an instruction or sequence is as simple as changing the program (much easier than rewiring hardware relays).

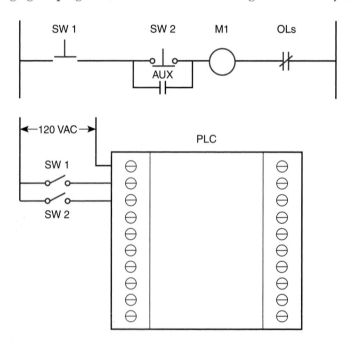

Figure 1-11   Correlating ladder program rung to actual PLC wiring.

The six major sections of a PLC are:

1. sensing inputs or controlling hardware
2. PLC input hardware
3. the controller or CPU
4. handheld programming device or personal computer
5. output PLC hardware
6. hardware output devices

To understand how a PLC works with these signals, we will look at the major sections that make up a typical PLC (see Figure 1-12).

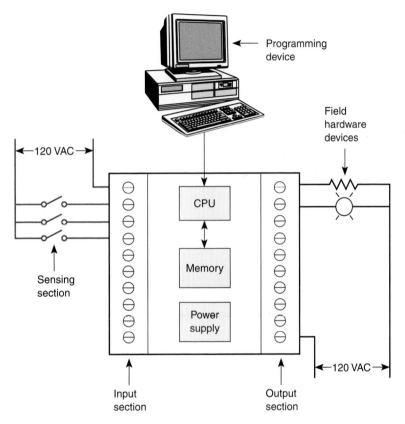

Figure 1-12   Programmable controller block diagram.

A programmable controller is made up of the following sections, each of which has a unique job in its operation.

1. *The sensing section:* The sensing section is made up of limit switches, pressure switches, photoelectric sensors, push buttons, and so forth. These incoming hardware devices provide input signals. Devices such as the push button, limit switch, or photoelectric sensor are field input devices. The term "field input" refers to hardware items providing incoming signals that are tangible items that you physically connect to the PLC.

2. *Input section:* The input section of the PLC contains two major areas. First, the physical screw terminals where incoming signals (inputs) from field input devices like a limit switch, for example, are attached to the PLC. Figure 1-13 illustrates a product sitting on a conveyor. When the conveyor moves the product into position, the sensor will send an input signal into input screw terminal number zero on the PLC input section.

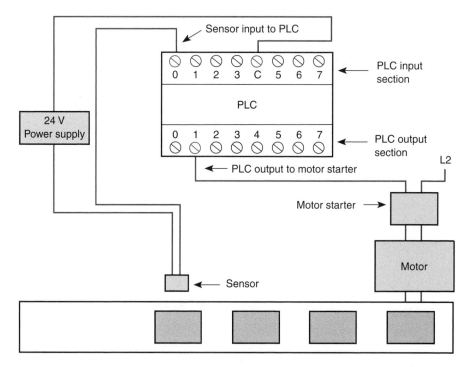

Figure 1-13   Product sensed as in position on the conveyor will send an input signal to the PLC.

The second portion of the input section is the PLC's internal conversion electronics. The function of the input section's electronic components is to convert and isolate the high-voltage input level from field devices. High-voltage signals from field devices are converted to +5 volts direct current (VDC) for a valid ON input signal, and 0 VDC for a valid OFF input signal. Incoming signal conversion and isolation are necessary, as the solid-state microprocessor components operate on +5 VDC, whereas an input signal may be 24 VDC, 120 volts alternating current (VAC), or 220 VDC. Inputting a 120 VAC signal, for example, into a 5 VDC circuit will quickly destroy your PLC.

3. *Controller:* The controller is commonly called the central processing unit (CPU) or, simply, the processor. This is the brain, or microprocessor, that controls or supervises the entire process. The CPU solves the user program and updates the status of the outputs.

4. *Programmer:* The programmer is the device whereby the programmer or operator can enter or edit program instructions or data. The programmer can be a handheld unit, as in Figure 1-1, a personal computer, or an industrial computer programming terminal.

5. *Output section:* The result of looking at, or *reading*, the ON or OFF status of the inputs and using this information to solve the user ladder program is to send updated signals to the output section. The output section is simply a series of switches, one for each output point, that are controlled by the CPU and are used to turn output field devices

ON or OFF. In the example in Figure 1-13, output screw terminal one controls the field output device, the conveyor motor starter.

6. *Field hardware devices:* The PLC will stop the conveyor motor by deenergizing the motor starter coil when the product is sensed. The motor starter is the device controlled by the PLC. The devices that are controlled by the PLC's output section screw terminals are the field hardware devices.

Now that we have a basic understanding of how a PLC operates, let us look at how PLCs are classified into groups. First we will look at grouping by size, then by physical configuration.

PLC Size Classifications

Dividing PLCs into groups depending on size is a difficult task, as everyone has a different definition on how the physical PLCs should be grouped. Traditionally, there has been agreement on the following size categories: micro, small, medium, large, and, in some cases, very large. In many cases there is overlap between different PLC families within the same manufacturer. Overlap can be a result of adding options to a smaller PLC. In some cases adding a different modular processor can increase I/O count, memory size, additional, and sometimes advanced programming language instructions, and network connectivity.

Figure 1-14 is an example of a very large PLC, the Rockwell Automation ControlLogix. ControlLogix can control up to 128,000 inputs and outputs or up to 4,000 analog I/O points. Memory size can be as much as 8M bytes.

Today, there are many considerations in addition to physical I/O count when determining PLC capabilities. What once might have been considered a medium-sized PLC supports I/O counts in excess of 2048. As an example, the SLC 500 PLC with maximum I/O will support 4096 inputs and outputs while the Rockwell Automation/Allen-Bradley ControlLogix PLC will support up to 128,000 I/O points. As a result of advancements in microprocessor and PLC technology, the traditional PLC grouping by size is a bit outmoded.

Figure 1-14   Thirty-two bit Rockwell Automation ControlLogix PLC. (Used with permission of Rockwell Automation, Inc.)

Classifying modern PLCs must take into consideration (in addition to I/O count) memory size, instruction set size, and communication and networking capabilities. Figure 1-15 provides an overview of selected PLCs. Keep in mind, as you review the chart, that all manufacturers have PLCs in each size category.

Manufacturer	PLC	Memory	I/O Count	Communications Options
General Electric	VersaMax Nano	2 K	6 Input 4 Output	Serial
Siemens	Simatic S7-200, CPU 221	4 K Program 2 K Data	6 Input 4 Output	Optional RS 485
Mitsubishi	FX0	1.6 K	Up to 16 Inputs Up to 14 Outputs	Serial, Profibus-DP, CC-Link Network
Rockwell	MicroLogix 1000	1 K	Up to 20 Inputs Up to 12 Outputs	Serial, Data Highway 485, DeviceNet
General Electric	VersaMax Micro	9 K	Up to 84	Serial, RS-485
Omron	CPM1A	3 K	100	Host Link, NT Link, 1:1 Link
Mitsubishi	FX2N Super Micro	Up to 16 K	Up to 256	Serial and RS-422/485, Profibus-DP and CC-Link
Omron	CQM1	13 K	Up to 256	Device Net, CompoBus/S, Host Link, NT Link, 1:1 Link
Siemens	Simatic S7-300	6 to 512 K	Up to 512 discrete or 128 analog	Point-to-point links, AS-Interface, Profibus-DP, Profibus-FMS, Industrial Ethernet
Rockwell	SLC 500	Up to 64 K	Up to 4096 Inputs and Outputs	Serial, DH-485, DH+, Device Net, Control Net, Ethernet
Omron	CVM1	Up to 62 K	Up to 2048	Ethernet, Sysmac-Net, Sysmac-Link, Controller Link, Device Net, Host Link, NT Link
Mitsubishi	AnN Series	Up to 320 K	Up to 2048	Modbus, Profibus-DP, Ethernet
Rockwell	ControlLogix	750 K thru 8 M bytes	Up to 128,000	Serial, DH-485, DH+, Device Net, Control Net, Ethernet

Figure 1-15   An overview of selected PLCs.

Figure 1-16 illustrates the Siemens Simatic S7 Family of PLCs. Starting at the top is the high-end S7-400, the midrange S7-300, and, at the bottom, the S7-200 Micro PLC.

Now that we have a basic understanding of how a PLC operates, let us take a look at fixed versus modular styles of PLCs.

Figure 1-16   Siemens S7 Family of Programmable Controllers. (Courtesy of Siemens Energy & Automation, Inc.)

INTRODUCTION TO FIXED AND MODULAR PLC HARDWARE

PLC hardware falls into two physical configurations, *fixed* and *modular*. A fixed PLC has all of its components—the input section, CPU and associated memory, power supply, and output section—built into one self-contained unit. All input and output screw terminals are built into the PLC package and are fixed, not removable. This style of PLC is also called a *packaged controller*.

The modular PLC comes as separate pieces. A modular PLC is purchased piece by piece. There may be two or three power supplies to choose from, a handful of different processors (CPUs), many separate input and output cards or modules, and a selection of assemblies, called racks, chassis, or baseplates to hold the pieces together. When purchasing a modular PLC you select and purchase the specific pieces you need to build the PLC specifically for the needs of your control situation.

Fixed PLCs

A fixed PLC consists of a fixed, or built-in, input and output section. The microprocessor section and power supply section are also included within this self-contained package. There is one fixed nonremovable screw terminal strip containing all input signal screw connections, and another terminal strip containing all output control signal screw terminals. The original packaged PLCs were referred to simply as fixed or packaged controllers. With miniaturization of modern electronics components, today's more compact fixed PLCs are referred to as **micro PLCs.**

Figure 1-17 illustrates an Omron Sysmac CPM1 Series fixed micro PLC installed in an electrical panel. Notice the input field device wires going into the top of the PLC. The wires coming out the bottom of the PLC will send outgoing, or output, signals to field devices.

Figure 1-17   Omron CPM1 fixed micro PLC. (Courtesy of Omron Electronics, Inc.)

Figure 1-18 identifies the major parts of the CPM1 micro PLC. Notice that the input and output terminals are built into the assembly and there is a peripheral port for connecting a *communication* cable. Also notice that input, output, and PLC status lights

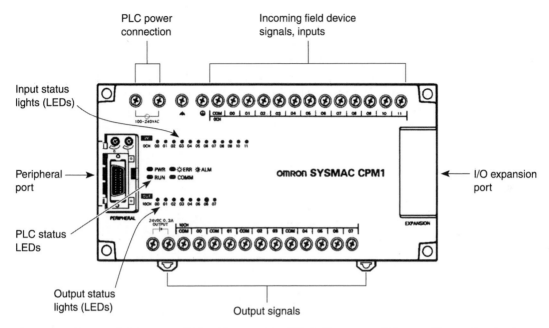

Figure 1-18   Omron CPM1 micro PLC with parts identified. (Courtesy of Omron Electronics, Inc.)

(LEDs) are built into the unit to provide operation information to electrical and maintenance personnel.

Figure 1-19 illustrates an Allen-Bradley SLC 500 fixed PLC. The area labeled "output terminals" is a door hinged on the bottom. Opening the door reveals the built-in output screw terminals. The area labeled "input terminals" is a door hinged on the top. Opening this door reveals all of the built-in input screw terminals. Notice that the power supply and processor (CPU) are contained in the assembly.

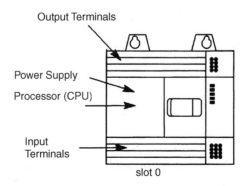

Figure 1-19   Allen-Bradley SLC 500 fixed PLC with the major parts labeled. (Used with permission of Rockwell Automation, Inc.)

Modular I/O PLCs

Modular PLCs have their I/O points on plug-in type, removable units called I/O modules. PLCs with modular inputs and outputs consist of a chassis, rack, or baseplate where the power supply, CPU, and all input and output modules are present as separate hardware items. Figure 1-20 shows a modular Allen-Bradley SLC 500 PLC, with parts labeled, assembled to make a working PLC.

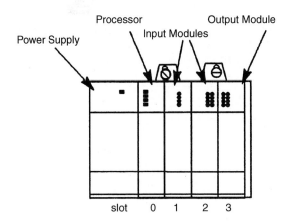

Figure 1-20   Allen-Bradley SLC 500 modular PLC. (Used with permission of Rockwell Automation, Inc.)

When I/O is modular, the user can mix input and output types in the rack or baseplate to meet specific needs. There are usually few limitations on the mix or positioning of I/O modules. Common racks, chassis, or baseplates hold 4, 5, 7, 8, 10, 13, or 16 I/O modules. Typical modules will contain 4, 8, 12, 16, or 32 I/O points.

Figure 1-21 is a Siemens Simatic S7-300 modular PLC mounted on a DIN rail. The power supply and CPU are on the far left of the assembly. The right-most eight sections are the input or output modules.

Electrical connections between each module and the CPU are made by two mating plugs. One plug is on a printed circuit board sticking out the back of each module. A printed circuit board that the modules plug into runs along the back of the rack or is built into a baseplate; this is called the **backplane.** The backplane conveys signals from the CPU to each module and from the module to the CPU. Other PLCs that do not slide into a rack or chassis or clip onto a baseplate simply clip together on a DIN rail.

The DIN rail is the metal track or rail attached to the back of an electrical panel, where devices can be easily clipped to or removed from the rail. The advantage of the DIN rail is that it permits the easy snap-on and removal of hardware devices either at installation or for maintenance or replacement. Figure 1-22 illustrates an Omron CQM1 PLC being installed on a DIN rail.

Figure 1-21   Siemens Simatic S7-300 modular PLC. (Courtesy of Siemens Energy &
Automation, Inc.)

Figure 1-22   Omron CQM1 PLC being installed on a DIN rail. (Courtesy of Omron
Electronics, Inc.)

Pictured in Figure 1-23 is a rear view of an Omron CQM1 modular PLC being installed on a DIN rail.

The power supply, which is also modular, is typically hooked up to line voltage. The
function of the PLC power supply is to convert line voltage to low-voltage DC and then
isolate it for use to operate the CPU and any associated I/O modules electronics in the
rack or chassis. In most cases, power must be supplied to all field input and output devices
from a source other than the PLC power supply.

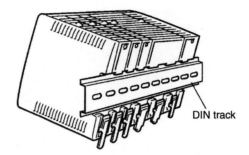

Figure 1-23 Rear view of an Omron CQM1 modular PLC being installed on a DIN rail. (Courtesy of Omron Electronics, Inc.)

Power supply placement is dictated by the PLC manufacturer. The power supply is installed either on the right side, left side, or in any position in the modular rack or chassis. Some PLCs allow the power supply to be either installed in the rack or chassis or mounted as a standalone device outside the rack or chassis. This configuration frees up a slot in the rack or chassis for an additional I/O module. Figure 1-24 illustrates an SLC 500

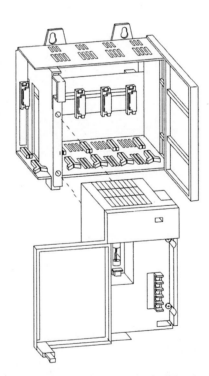

Figure 1-24 SLC 500 power supply and a four-slot rack. (Used with permission of Rockwell Automation, Inc.)

power supply about to be installed on a four-slot chassis. The power supply slides onto the left side of the four-slot chassis. Two screws secure the power supply to the chassis. Notice the power connection terminal strip in the lower right-hand corner of the power supply. A fuse is in the upper left-hand corner.

The CPU is typically installed in either the right or left position, next to the power supply. Again, the CPU position is dictated by the PLC manufacturer. The remaining slots in the rack, chassis, or baseplate can hold I/O modules, usually in any mix of input or output. Most modules can be placed in any slot; however, there are some specialized modules that have rules concerning their placement.

Figure 1-25 illustrates an Allen-Bradley four-slot modular chassis with the power supply installed on the far left. The CPU is installed next to the power supply. An I/O module is shown being installed into the rack next to the CPU. Notice the two connectors on the backplane board in the rear of the rack.

Now that we have looked at PLC packaging formats, we need to look at the processor or central processing unit (CPU) and its function and makeup.

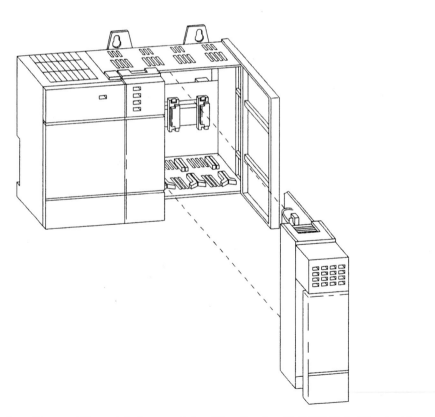

Figure 1-25   Installing an I/O module in a modular Allen-Bradley SLC 500 PLC. (Used with permission of Rockwell Automation, Inc.)

THE CENTRAL PROCESSING UNIT

The central processing unit is built into the single-piece fixed PLC. When working with modular PLCs, the CPU is typically a plug-in module just like with an I/O modular unit.

The CPU works like the human brain works to direct the rest of the body. In actuality, the CPU is a solid-state, microprocessor integrated circuit chip. The microprocessor integrated chip is placed on a printed circuit board with other supporting and interface chips to build the PLC's CPU, or processor module. The CPU is built into a fixed PLC, whereas the CPU is a separate module inserted into a PLC chassis. Figure 1-26 is an illustration of an Allen-Bradley SLC 500 modular CPU. There are six LEDs near the top of the module. These LEDs provide information to the electrical or maintenance worker regarding the following codes: whether the processor is running (RUN), faulted (FLT), battery condition (BATT), forces in effect (FORCE), Ethernet channel status (ENET), and RS-232 channel status (RS232). The CPU comprises two components, the controller and the memory system.

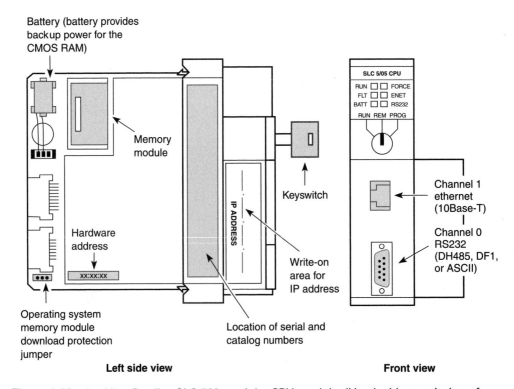

Figure 1-26   An Allen-Bradley SLC 500 modular CPU module. (Used with permission of Rockwell Automation, Inc.)

The controller is the microprocessor, or brain, that supervises all operations in the system. The CPU reads, or gathers, information from external sources such as input devices and stores this information in memory for later use by the CPU. When done solving the user program, the CPU will **write,** or send, data out to external devices such as output modules and field hardware devices.

The memory system has to provide the following:

1. storage for the user program
2. storage of the input status file data: The input status file consists of memory locations that store the ON or OFF status of each field input device.
3. storage of the output status file data: The output status file consists of memory locations that store the ON or OFF status of field hardware devices as the result of solving the user program. Data in the output status file is waiting to be transferred to the output module's switching device. The output module's switching device for each output point will turn power on or off to each field output device.
4. data storage: The data storage area of memory is used to store numerical data that may be used in math calculations, recipe ingredient weights, bar code data being input to the PLC, and similar functions.

Input Modules

Input modules serve as the link between field devices and the PLC's CPU. Each input module has a terminal block for attaching input wiring from each individual field input device. Typical input modules have either 8, 16, or 32 input terminals.

The main function of an input module is to take the field device input signal, convert it to a signal level that the CPU can work with, electrically isolate it, and send the signal, by way of the backplane board, to the CPU. The input signal, once received on the CPU module, must be stored somewhere to await processing by the CPU. The memory storage area for input signals is called the input status file (sometimes called the input status table). Each input screw terminal on each input module has one memory location in the input status file. Each memory location stores the binary equivalent of the ON or OFF status of its associated input signal.

Figure 1-27 illustrates a mechanical limit switch as an input to a PLC input module. Notice the 24-VDC power supply connections between the input module and the limit switch. The limit switch is simply an on-off switch providing a signal to the input module. Each input module screw terminal has a unique identifier called an **address.** The limit switch in Figure 1-27 is connected to the input module screw terminal addressed as zero. Figure 1-27 illustrates how an input module provides a connection for the limit switch's signal to be made available to the CPU for interaction with the user program.

Output Modules

Output modules serve as the link between the PLC's microprocessor and hardware field devices. Each output module has a terminal block for attaching output wiring to go to

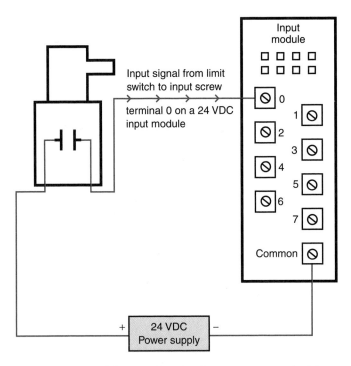

Figure 1-27   Limit switches 24-VDC input signal to input screw terminal 0 on an input module from a modular PLC.

each individual field output device. Typical output modules have either 8, 16, or 32 output terminals.

The output signal, once received from the CPU, must be stored before being sent to each output module's output screw terminals. The storage area for output signals is called the output status file. Each output screw terminal on each output module has one memory location in the output status file.

The main function of an output module is to take the CPU's control signal (sent by way of the backplane), electrically isolate it, and energize (or deenergize) the module's switching device to turn on (or turn off) the output field device. The output module provides a connection to a motor starter for CPU control as a result of interaction with the user program. Figure 1-28 is an illustration of an eight-point output module. Keep in mind that each point on an output module is basically a switching device turned on or off by the CPU as a result of the executed user ladder program. Notice that line 1 and line 2 power is supplied by the user and switched by the module to turn the motor starter on or off.

Before we explore the workings of status files and other data files, we will explore binary numbers and other numbering methods in Chapter 4. With a good foundation in these numbering systems we will go on to binary data representation in different register formats.

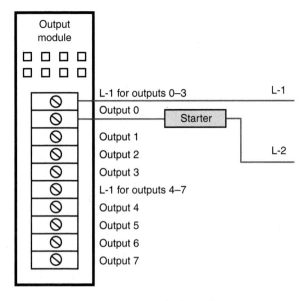

Figure 1-28   Output module wiring to a motor starter coil.

SUMMARY

The programmable logic controller has made it possible to control large processes and machines with less physical wiring and lower installation and maintenance costs as compared to standard electromechanical control devices. The ability to quickly modify a process by changing the PLC program is one of the major advantages of PLCs versus conventional, relay-based control systems. This feature was clearly illustrated in the Caterpillar story. In summary:

1. The PLC is a computer, and someone had to tell it what to do through a program that was developed and then entered into its memory.
2. A PLC, like other computer devices, does exactly what the human programmer tells it do, and nothing else.
3. Programmable controllers are reusable; they can be easily reprogrammed as production changes dictate.
4. Programmable controllers come in two configurations, fixed and modular.
5. A fixed programmable controller has the I/O, processor, and power supply all built into a single, factory-assembled unit.
6. A modular programmable logic controller consists of a rack, power supply, CPU, and input and output modules.
7. After either assembling the separate parts to make a modular PLC or choosing a fixed PLC, you will have a complete unit that, when connected to field input devices through a program that you develop, will control output devices as specified by your control objective.

REVIEW QUESTIONS

1. Why was the programmable logic controller developed?
2. List four important developments incorporated into the first programmable controllers.
3. Define programmable controller.
4. List the major differences between a modular PLC and a fixed PLC.
5. Develop a block diagram of a typical PLC.
6. Explain the function of each block from question 5.
7. Incoming control signals are called _____.
8. The user program directs the PLC as to how it is to control field devices such as _____.
9. Signals going out of the PLC to control field devices are called _____.
10. Typical racks or bases hold how many I/O modules?
11. Do the CPU and power supply reside in a slot in the typical rack or base?
12. The _____ is nothing more than a list of instructions telling the PLC what to do.
13. When the PLC is running and following the programmer's instructions, this is called _____.
14. The instructions you wish your PLC to carry out are transferred to the memory of the controller from either _____ or _____.
15. Transferring the PLC program from a personal computer's memory to PLC memory is called _____.
16. Before downloading a user program, the processor must be in _____.
17. If all input and output signals are wired to the correct screw terminals, the processor can be put in _____.
18. A PLC is easy to install because all input and output connections are connected to _____.
19. As time progressed and electronic and microprocessor technology advanced, new functionality was added to the PLC. List several advanced functions added to the PLC along with communication links with other PLCs and personal computers.
20. The function of the _____ electronic components is to convert and isolate the high-voltage input level from field devices.
21. Incoming signal _____ and _____ are necessary as the solid state microprocessor components operate on +5 volts DC.
22. The CPU solves the user program and updates the status of the _____.
23. The _____ is the device whereby the programmer or operator can enter or edit program instructions or data.
24. The PLC programmer can be a(n) _____ or _____ with the proper application _____ connected to the PLC CPU via a(n) _____.
25. Input modules serve as the link between _____ and the PLC's CPU.

26. Each input module has a(n) _____ for attaching input wiring from each individual field input device.

27. Typical input modules have either _____, _____, or _____ input terminals.

28. The main function of a(n) _____ is to take the field device input signal, convert it to a signal level that the CPU can work with, _____ it, and send the signal by way of the _____ to the CPU.

29. The _____, once received by the CPU module, must be stored somewhere awaiting processing by the CPU. The storage area for these signals is called the _____.

30. Each input screw terminal on each input module has _____ in the input status file. Each memory location stores the _____ of the ON or OFF status of its associated input signal.

31. _____ modules serve as the link between the PLC's microprocessor and the hardware field devices.

32. Typical output modules have either _____, _____, or _____ output terminals.

33. The _____ provides a connection to a motor starter for CPU control as a result of interaction with the user program.

34. The output signal, once received from the CPU, must be stored somewhere before being sent to each output module's output screw terminals. The storage area for output signals is called the _____.

IF PASSWORD IS CORRECT GO TO DRIVE SET UP SCREEN

0022

— EQU —
Equal
Source A   N12:21
             7777<
Source B   32767
             32767<

B3:10/0
[OSR]

B3:1/15
( )

RESET PASSWORD
ATTEMPTS COUNTER
C5:5
(RES)

0023

al
A   N12:21
      7777<
B   32767
      32767<

B3:10/1
[OSR]

— CTU —
Count Up
Counter   C5:5

(CU)

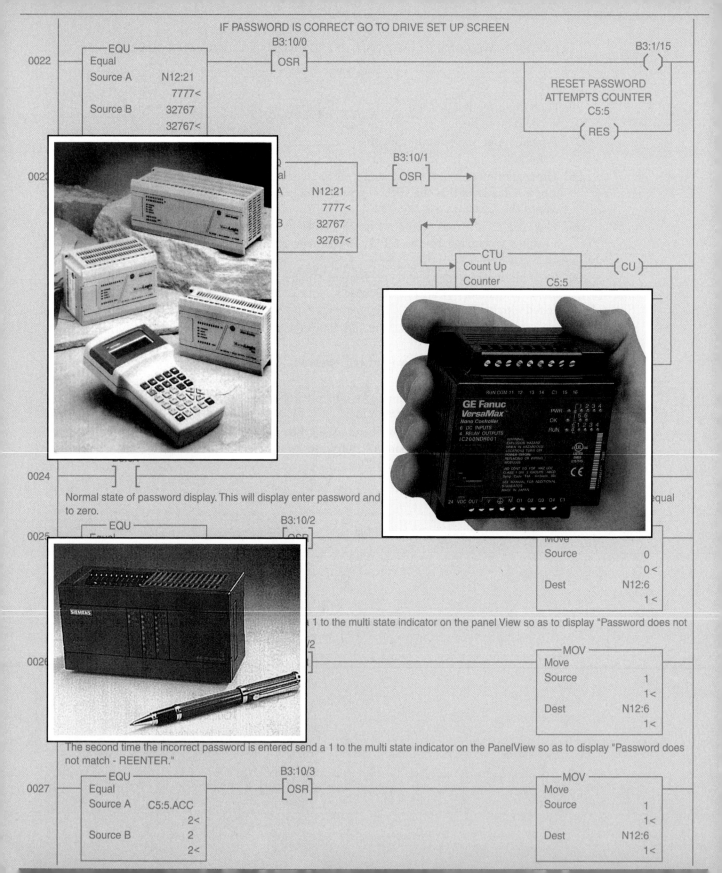

0024

] [

Normal state of password display. This will display enter password and   equal
to zero.

0025

— EQU —
Equal

B3:10/2
[OSR]

Move
Source   0
                    0<
Dest   N12:6
                    1<

a 1 to the multi state indicator on the panel View so as to display "Password does not

0026

/2

— MOV —
Move
Source   1
                    1<
Dest   N12:6
                    1<

The second time the incorrect password is entered send a 1 to the multi state indicator on the PanelView so as to display "Password does not match - REENTER."

0027

— EQU —
Equal
Source A   C5:5.ACC
              2<
Source B   2
              2<

B3:10/3
[OSR]

— MOV —
Move
Source   1
                    1<
Dest   N12:6
                    1<

CHAPTER

Micro Programmable
Logic Controllers

OBJECTIVES

After completing this chapter, you should be able to:

- define what a micro PLC is
- explain the advantages of using a micro PLC in new machine development
- explain the difference between a micro PLC and a modular PLC
- look at an overview of selected manufacturers' micro PLCs

INTRODUCTION

As a result of advancements in microprocessor technology, the size and power of the newer PLCs have resulted in a smaller PLC package size and more computer power and features than ever before. These new, smaller, and more powerful PLCs are called micro PLCs, or simply micros. This chapter will introduce you to the world of the micro PLC.

Even though original equipment manufacturers, called **OEMs,** have been incorporating PLCs into control process lines, up to and including entire plants, for many years, the cost and size of the available PLCs in most cases prohibited incorporating them into small machines or small applications. Equipment manufacturers found that developing a small machine where the cost of incorporating a PLC could easily make up 20 to 25% of the machine cost was simply not justifiable. They needed a small PLC with basic functionality that would not only be low-cost, say a few hundred dollars, but small in size, too. They needed a small controller to replace relays, dedicated timers, and counters.

With the introduction of the small, basic-featured PLC in the late 1980s, OEMs had the capability to add or increase automation to small or simple machines where hardwired relay control had been the norm. These small PLCs were called *micro PLCs* or simply *micros*.

Even though the early micro PLCs were simply relay, timer, and counter replacers, modern micro PLCs have evolved parallel to the microprocessor and personal computer advancements. Today's micro PLC has not only grown in computer power but has shrunk in size and cost. Modern micro PLCs have advanced features such as data manipulation instructions, math, high-speed counters, sequencers, subroutines, interrupts, personal computer programmability, and network connectivity. These advanced features were only found on their larger cousins just a few years earlier.

The micro PLC is typically 32 input and output points or less. Micro PLCs come as self-contained units with the processor, power supply, and I/O built into one package. Since all the pieces are contained in a single, nonchangeable package, these PLCs are called fixed, or packaged controllers. Figure 2-1 shows four different manufacturers' micro PLCs.

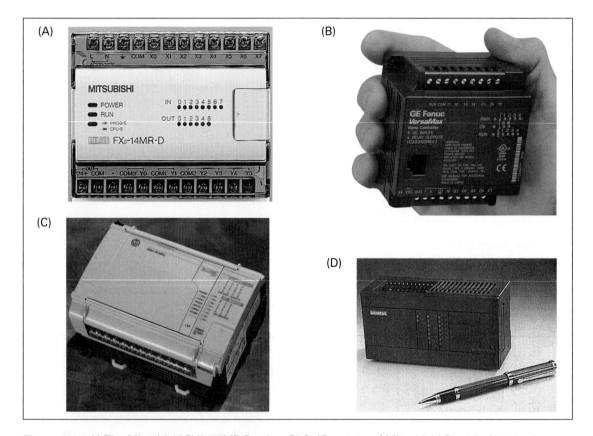

Figure 2-1   A) The Mitsubishi FX0-14MR-D micro PLC. (Courtesy of Mitsubishi Electric Automation, Inc.); B) General Electric's VersaMax Nano PLC. (Courtesy of GE Fanuc Automation); C) The Allen-Bradley MicroLogix 1500. (Used with permission of Rockwell Automation, Inc.); D) The Siemens S7-200 micro PLC. (Courtesy of Siemens Energy & Automation, Inc.)

The advantage of the micro controller is that the package is smaller, less costly, and easier to install than its larger, modular cousin.

Control Engineering magazine categorized micro PLCs as less than 128 I/O points, while medium PLCs contained 128 to 512 I/O and large PLCs over 512 I/O points. The magazine reported a survey on PLC usage in its February 1998 issue where 61% of the respondents reported they purchased PLCs with up to 128 I/O points, 29% purchased PLCs in the 129 to 512 I/O point range, and 10% responded that they purchased PLCs in the 512 I/O and greater range. In the March 2000 issue, *Control Engineering's* 2000 study revealed that 77% of the respondents used micro PLCs more than other sizes. Even though this information is somewhat dated, it shows a definite interest in the smaller PLCs.

OVERVIEW OF SELECTED MICRO PLCs

The following text is intended to provide an introductory overview of different manufacturers' micro PLCs. It is a basic introduction to some of the programming and configuration options available. We will look at, in alphabetical order, the GE Fanuc Automation VersaMax Nano and Micro, along with the GE Series 90 Micro, Mitsubishi FX2N Super Micro, Omron's CPM1A, Rockwell Automation's MicroLogix 1000 and 1500, and Siemens Simatic S7-200.

GE FANUC AUTOMATION VERSAMAX CONTROLLERS

General Electric VersaMax Nano and Micro controllers are relatively new members of the GE family of small programmable logic controllers. The VersaMax Nano PLC is a palm-sized compact unit measuring only 3 inches wide, 3.20 inches high, and 1.88 inches deep. This unit will either snap on to a DIN rail or screw directly into a panel. The VersaMax Nano has six 24-volt DC inputs and either four 24-volt DC or relay outputs. The Nano is powered by a 24-volt DC power supply. This member of the GE family is not expandable beyond the ten fixed I/O points. Photo (B) in Figure 2-1 shows the General Electric Nano Controller with six inputs and four outputs. Note the small size of this compact PLC. The VersaMax Nano has 1 K words of memory and executes logic at a rate of 1.2μ per Boolean operation. Like many of the newer, smaller PLCs, the VersaMax Nano has many advanced PLC functionality features including PID, floating-point math, subroutines, serial communication read and write commands, two 10-KHz high-speed counters, and three 5-kHz pulse wave modulated (PWM) pulse train outputs for variable frequency drive control.

Another member of the VersaMax family is the VersaMax Micro. The Micro is available in the configurations shown in Figure 2-2.

Measuring 5.5 inches wide, 3 inches high, and just over 3 inches deep, the 28-point unit contains all the features and capabilities of the Nano but with 9 K words of memory, 1.0μ per Boolean operation, a real-time clock, and motion control capabilities. Figure 2-3 shows the GE Fanuc Automation VersaMax Micro Controller with 16 DC inputs, 11 relay outputs, and 1 DC output.

VersaMax Micro PLC			
I/O Points	Power Supply	Inputs	Outputs
14	120/240 VAC	8 at 120 VAC	6 at 120 VAC
	24 VDC	8 at 24 VDC	6 at 24 VDC
	120/240 VAC	9 at 24 VDC	6 Relay
	24 VDC	8 at 24 VDC	6 Relay
28	120/240 VAC	16 at 24 VDC	1 at 24 VDC 11 Relay
	24 VDC	16 at 24 VDC	1 at 24 VDC 11 Relay
	120/240 VAC	16 at 120 VAC	12 at 24 VAC
	24 VDC	16 at 24 VDC	12 at 24 VDC
	120/240 VAC	13 at 24 VDC 2 Analog	9 Relay Out 1 at 24 VDC 1 Analog

Figure 2-2   VersaMax Micro PLCs. (Data compiled from GE Fanuc Automation data)

Figure 2-3   General Electric VersaMax Micro Controller. (Courtesy of GE Fanuc Automation)

VersaMax Micro I/O Expansion

The VersaMax Micro 14-point units can be expanded using up to four expansion units for a total of 70 I/O points. The 28-point units can also be expanded with up to four expansion units to provide a total of 84 I/O points. VersaMax expansion units are listed in Figure 2-4.

VersaMax Expansion Units			
I/O Points	Power Supply	Inputs	Outputs
14	120/240 VAC	8 at 24 VDC	6 Relay
	24 VDC	8 at 24 VDC	6 Relay
	24 VDC	8 at 24 VDC	6 Relay

Figure 2-4   VersaMax expansion units. (Data compiled from GE Fanuc Automation data)

Programming the VersaMax PLCs

VersaPro is GE's newest windows PLC programming software for Series 90-30 and VersaMax PLCs. The software will run on Windows 95 using a 486 processor with a minimum of 16 MB RAM. Logicmaster software can easily be imported into VersaPro. The programming instructions are divided into two groups, relay or Boolean and function block. Relay instructions are bit instructions. Function blocks work on bit, word, and double word data types.

Figure 2-5 shows a VersaPro ladder windows logic screen. Notice the ladder window, left, and an address window, right.

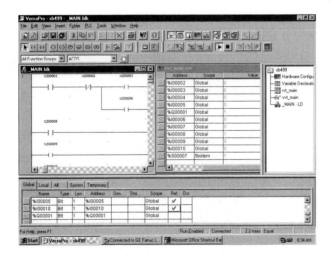

Figure 2-5   VersaPro ladder window. (Courtesy of GE Fanuc Automation)

Figure 2-6 illustrates a ladder program with a function block instruction and the function block instruction properties box.

Operator Interface to VersaMax Nano or Micro

The VersaMax data panel operator interface devices allow an operator to change timer and counter values, and to register values without the need for a programming device.

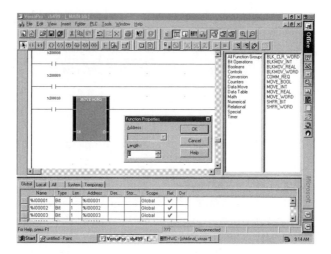

Figure 2-6 VersaPro function block programming. (Courtesy of GE Fanuc Automation)

Alarm and informational messages can be sent from the PLC controller and displayed on a 2-line by 16-character, 4-line by 16-character, or 4-line by 20-character display, depending on the data panel selected (See Figure 2-7).

Model	Function Keys	Display Lines	Stored Messages	Data Entry Key Pad
DP 20	6 Operation keys	2 × 16 character	Messages stored in PLC	No
DP 45	6	2 × 16 character	Up to 200	No
DP 65	8	4 × 16 character	Up to 200	No
DP 85	8	4 × 20 character	Up to 200	Numerical, 0–9

Figure 2-7 VersaMax data panels. (Data compiled from GE Fanuc Automation literature)

Figure 2-8 shows a VersaMax DP20 data panel interface terminal. This operator unit is used to change timer, counter, or register values without the need for a programming device. The DP 20 comprises an LCD backlit display for displaying up to 2 lines of 16 characters. The DP 20 cannot store messages. Messages are stored and sent from the PLC via a serial communication link.

GE FANUC AUTOMATION SERIES 90 MICRO PLC

The GE Fanuc Automation Series 90 Micro PLC is an older member of the GE PLC Series 90 family. The Series 90 Micro is a compact, fixed-I/O style micro PLC designed for either 35mm DIN rail or panel mounting. The Series 90 Micro is available with either

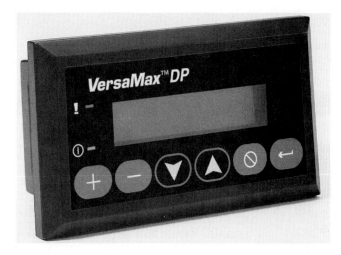

Figure 2-8 General Electric's VersaMax DP 20 interface terminal. (Courtesy of GE Fanuc Automation)

14, 23, or 28 I/O points. Figure 2-9 illustrates a typical 14-I/O-point Series 90 Micro PLC. Input connections are on the top of the PLC, whereas power supply connections and output connections are on the bottom of the PLC. The basic unit has nonremovable terminal strips. The Micro PLC illustrated in Figure 2-9 has optional removable terminal strips.

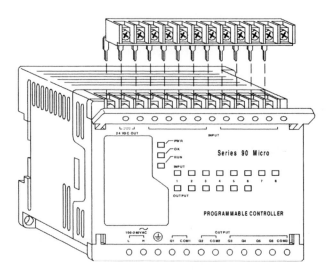

Figure 2-9 General Electric Series 90 Micro PLC. (Courtesy of GE Fanuc Automation)

Status Indicators

Referring to Figure 2-9, notice the status indicators near the top left side of the unit. Figure 2-10 describes the function of each.

Status Indicator	State	Status Indicator Function
PWR	Lighted	Power supply is receiving power and operating correctly.
	Dark	Power supply is not powered or has faulted.
OK	Blinking	PLC is going through self-diagnostics.
	Steady	PLC passed self-diagnostic tests.
RUN	Lighted	PLC is in RUN mode and executing user program.
	Blinking	OK and RUN indicator blinking means fault detected during self-diagnostic tests.
Individual input	Lighted	Each input point associates with an input status indictor. When lighted, PLC detects an input signal.
	Dark	If indicator is not illuminated, PLC does not detect a valid input signal.
Individual output	Lighted	Each output point associates with an output status indicator. When lighted, PLC is commanding the output point associated with the lighted status indicator to turn ON.
	Dark	If indicator is not illuminated, PLC has commanded the associated output point to turn OFF. All outputs also turn OFF in Stop or I/O Disable Mode.

Figure 2-10   Series 90 Micro PLC status indicators. (Data courtesy of GE Fanuc Automation)

Configuring and Programming the Series 90 Micro

The Series 90 Micro can be configured and programmed using a handheld programmer, an IBM personal computer, a Workmaster II, or CIMSTAR I industrial computer.

Communicating with the Series 90-20

Figure 2-11 shows the 15-pin, D-shell-type connector on the front of the Series 90 micro that is used to connect to an RS-422 compatible serial port to communicate with either the handheld programmer or LogicMaster 90-30/20/Micro software. The 14-point micro PLC has one serial port as illustrated in Figure 2-11, and the 23- and 28-point micro PLCs have two serial ports.

Field Wiring Interface

Figure 2-12 illustrates a 14 I/O point Series 90 Micro. Notice that inputs I1 through I4 share a common connection, COM 1, while I5 through I8 have their own common

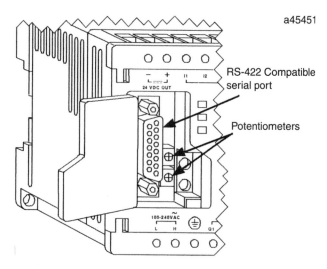

a45451

RS-422 Compatible
serial port

Potentiometers

Figure 2-11 Series 90 Micro PLC serial port. (Courtesy of GE Fanuc Automation)

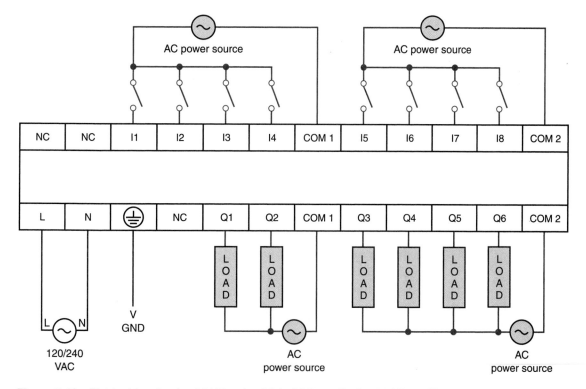

Figure 2-12 Field wiring for the 14 I/O point AC in / AC out Series 90 Micro. (Courtesy of GE Fanuc Automation)

connection. The micro PLC is connected power-to-power to the L and N terminals at the lower left corner. Outputs Q1 and Q2 share output common COM 1, while outputs Q3 through Q6 share output common COM 2. The separate input and output commons provide the option to connect to different AC sources if needed.

THE MITSUBISHI FX2N SUPER MICRO

The Mitsubishi FX2N family of high-performance Super Micro controllers is available with 16, 32, 48, 64, 80, and 128 I/O points. The FX2N has a full range of 152 programming instructions. There are 27 basic relay type, 18 comparison, and 107 applied instructions.

The FX2N-16MR, AC-powered, 8-input and 8-output unit measures only 5.2 inches wide by 3.6 inches high and 3.4 inches deep. Figure 2-13 is an FX2N-80MR AC-powered PLC base unit with 40 inputs and 40 outputs. The 40MR unit measures only 11.4 inches wide by 3.6 inches high and 3.4 inches deep.

Figure 2-13 The Mitsubishi FX2N-80MR Super Micro PLC supporting 80 I/O points. (Courtesy of Mitsubishi Electric Automation, Inc.)

Figure 2-14 points out the features of the Mitsubshi FX2N Super Micro PLC. See the legend below for feature identification. This data was collected from the FX2N Series Programmable Controller Installation Notes manual.

- A. 35mm DIN rail for mounting PLC
- B. PLC mounting holes for direct mounting
- C. Input screw terminals
- D. Input screw terminals cover
- E. Input point status indicators
- F. I/O expansion bus cover
- G. PLC status indicators
- H. Programming port cover
- J. Top panel
- K. Output screw terminals
- L. Output screw terminals cover
- M. Clips to hold PLC to DIN rail
- N. Output point status indicators
- P. Memory backup battery

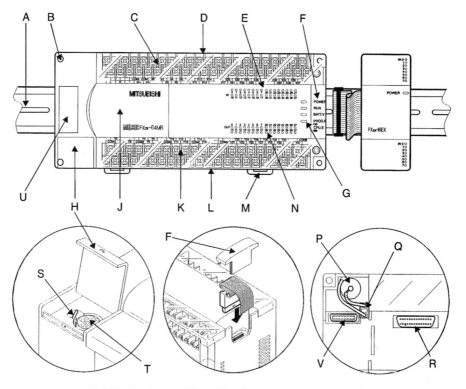

Figure 2-14 Mitsubishi FX2N Super Micro PLC features. (Courtesy of Mitsubishi Electric Automation, Inc.)

Q. Connector for battery or super capacitor for memory backup
R. Memory cassette port
S. Run/Stop switch
T. Programming port
U. Cutout for extension board
V. Extension board connector

FX2N Adapter Cards

There are a number of optional adapter cards available for expanding communication options, or fine-tuning system parameters. The adapter boards attach to the cutout for the extension board area on the left side of the base unit. Adapter boards are powered by the base unit's power supply. Figure 2-15 provides an overview of selected FX2N adapter cards and their uses.

Figure 2-16 shows one of the available adapter cards, the FX2N-8AV-BD. This adapter board is the manual adjustable potentiometer input board. Using this adapter board provides the ability to fine-tune system parameters or timers and counter settings without a programming device or operator interface.

Description	Adapter Board Use
RS-232C Serial Port	Communicate with other RS-232 devices such as a printer or bar code reader
RS-422 Serial Port	Alternate programming port
RS-485 Serial Port	Multidrop network, parallel link, PLC-to-PLC network
Potentiometer Input	Fine-tune system parameters or timers and counters

Figure 2-15 Selected FX2N adapter boards and their uses. (Data compiled from Mitsubishi literature)

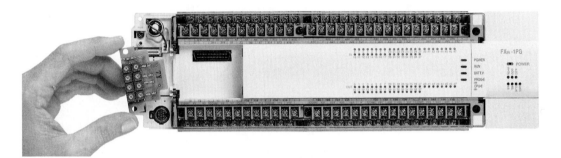

Figure 2-16 FX2N manual adjustable potentiometer input board. (Courtesy of Mitsubishi Electric Automation, Inc.)

FX2N Programming

Mitsubishi offers handheld, pendant-style programming along with two Windows-based software packages. FX-WIN is an easy-to-use programming tool for use with FX-family PLCs, while MM+ is a powerful Windows-based software package used to program all FX and the larger modular A series PLCs. Figure 2-17 illustrates a Windows screen.

I/O Expansion

An FX2N unit's I/O count can be expanded up to 256 I/O points using a base unit and either AC-powered expansion units or input and output expansion blocks. Powered expansion units are used when the I/O expansion capacity for the base unit is exhausted. The powered expansion unit provides the additional power required to add additional I/O up to a maximum of 256 I/O.

Unpowered extension blocks are used to add additional I/O points to the base CPU unit. Unpowered extension blocks can be added if the base unit's power supply loading, 256 I/O count, is not exceeded. These extension blocks come in 8 or 16 inputs and outputs or 4 input and 4 output versions. Figure 2-18 illustrates a FX2N-32ER 16-input and 16-output-powered extension unit with one FX2N-16EX (16-input) and one FX2N-16EY

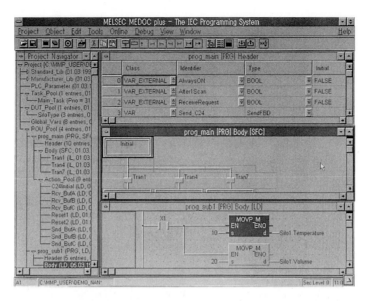

Figure 2-17   Mitsubishi Windows software screen. (Courtesy of Mitsubishi Electric Automation, Inc.)

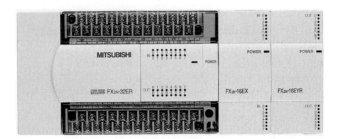

Figure 2-18   Mitsubishi FX2N-32ER 16-input and 16-output-powered extension unit with FX2N-16EX and FX2N-16EY unpowered extension blocks. (Courtesy of Mitsubishi Electric Automation, Inc.)

(16-output) unpowered extension block. This expanded PLC supports 32 inputs and 32 outputs.

Special Function I/O

Special function blocks are available to increase the capabilities of the Mitsubushi FX2N Super Micro PLC. Up to eight special function blocks are allowed per system if power supply loading is not exceeded. A sampling of the available special function blocks for the FX2N PLC is shown in Figure 2-19. Figure 2-20 illustrates the FX2N-1HC one-channel, high-speed counter block installed on the right-hand side of an FX2N-32MR.

Special Function Blocks for the FX2N PLC
Four channel analog input
Four channel analog output
Thermocouple input block
High-speed counter block
Pulse train output block
RS-232 communication block

Figure 2-19 Special function blocks for the FX2N PLC.

Figure 2-20 FX2N-32MR with 1HC high-speed counter block. (Courtesy of Mitsubishi Electric Automation, Inc.)

OMRON'S CPM1A SERIES MICRO PLCs

The CPM1A is Omron's family of micro PLCs. These micro PLCs are available in four different CPU models with either 10, 20, 30, or 40 I/O available. The 30- and 40-I/O point models are expandable. The CPM1As are one-piece, fixed-I/O units with either relay outputs or sinking or sourcing transistor output models.

Figure 2-21 contains a 20-I/O-point CPM1, top center, a 10-I/O-point CPM1A, lower left, and a 30-I/O-point CPM1, lower right. Up to 3 analog I/O units can be connected to provide analog inputs and outputs. Each unit provides 2 analog inputs and 1 output. Up to 6 analog inputs and 3 outputs are available by using 3 analog I/O units. Figure 2-22 illustrates a CPM1A-10CDR 6-input and 4-output micro PLC with features identified.

Programming the CPM1A

Two options are available for programming the CPM1A, a programming console (handheld programmer) or SYSMAC Windows-based programming software.

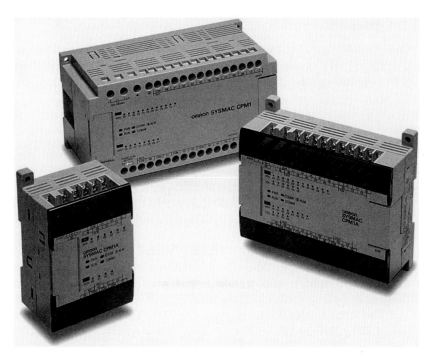

Figure 2-21   Omron CPM1 and CPM1A micro PLCs. (Courtesy of Omron Electronics Inc.)

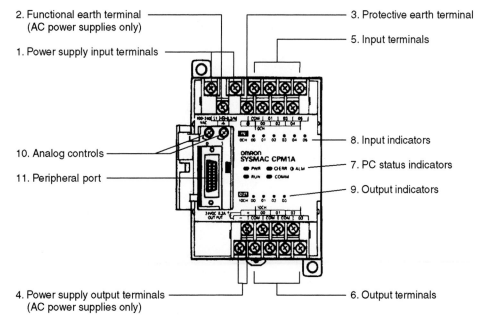

Figure 2-22   CPM1 ten I/O micro PLC features. (Courtesy of Omron Electronics Inc.)

CPM1A I/O Expansion

Only the 30- and 40-I/O-point models of the CPM1A Micro PLC are I/O-expandable. The CPM1A I/O expansion module contains 12 input points and 8 output points. Up to 3 I/O expansion modules can be connected to the right-hand side of the 30- or 40-I/O CPM1A.

ROCKWELL AUTOMATION'S MICROLOGIX 1000 MICROCONTROLLER

The Rockwell Automation/Allen-Bradley MicroLogix 1000 micro PLC is available in 14 models ranging from 10-, 16-, and 32-point discrete I/O configurations to 3 analog models with up to 5 analog and 20 discrete I/O. Input options include AC or DC along with current or voltage analog. Output options include triac, relay, or MOSFET along with current or voltage analog.

Figure 2-23 shows the Allen-Bradley MicroLogix 1000 family of PLCs. The top unit is a 20-input, 12-output, MicroLogix 1000 microcontroller. The center two MicroLogix 1000s are 10-input, 6-output units. A handheld programming terminal is pictured at the bottom.

The MicroLogix 1000 is truly a micro PLC. The 16-I/O, DC-powered PLC measures only 4.72 by 3.15 by 1.57 inches.

Figure 2-23 Three MicroLogix 100 micro controllers and a handheld programmer. (Used with permission of Rockwell Automation, Inc.)

Programming the MicroLogix 1000 can be accomplished using either the handheld programming terminal, as illustrated in Figure 2-23; Rockwell Software's RSLogix 500; Windows 95, 98, 2000, or NT software; or A.I. 500 DOS programming software. The MicroLogix 1000 can execute a 500-instruction program using its 1 K user memory in only 1.56 μs with a throughput of 1.85 μs. When you need additional I/O capabilities, the MicroLogix 1500 is the next step up from a MicroLogix 1000.

ROCKWELL AUTOMATION'S MICROLOGIX 1500 MICROCONTROLLER

The MicroLogix 1500 is the most powerful of the MicroLogix family. The MicroLogix 1500 consists of three base units that contain the power supply and include up to 28 embedded I/O points. Two processor units are available, which slide into the selected base unit. The 1764-LSP processor has 7 K user memory; whereas the 1764-LRP processor supports 14 K user memory. Since the MicroLogix 1500 is a rackless PLC, expansion is as simple as snapping together Bulletin 1769 Compact I/O modules.

The processor can be easily updated with the latest features by simply downloading the latest operating system upgrade from the Internet. The processor is field-upgraded using a flash upgrade utility.

All MicroLogix PLCs and the SLC 500 modular processors share the Rockwell Software RSLogix 500 programming software. Figure 2-24, left, illustrates a MicroLogix base unit with three I/O expansion modules on its right side. Two loose I/O expansion modules are illustrated, center, while the top right illustrates an expanded MicroLogix 1500 base unit further expanded into an additional bank of I/O using an expansion cable. Notice the left-most module in the expansion chassis. This is a communication module that will allow connecting to a DeviceNet Network. Figure 2-25 provides an overview of MicroLogix 1500 PLC base units.

Figure 2-24   Allen-Bradley MicroLogix 1500 PLCs. (Courtesy of Allen-Bradley, a Rockwell Automation business)

MicroLogix 1500 PLC Base Unit Specifications			
PLC Part Number	1764-24BWA	1764-24AWA	1764-28BXB
PLC Operating Power	85 V/265 VAC	85 V/265 VAC	20.4 V to 30 VDC
Inputs	12	12	16
Input Type	24-VDC sink/source	120 VAC	24-VDC sink/source
Outputs	12	12	12
Output Type	Relay	Relay	6 Relay 6 FET Transistor

Figure 2-25   Base unit specifications for Rockwell Automation's MicroLogix 1500. (Used with permission of Rockwell Automation, Inc.)

Figure 2-26 shows the installation of the processor unit onto the base unit.
Figure 2-27 illustrates features of the 1500 MicroLogix controller.
Figure 2-28 identifies each called-out feature of the PLC shown in Figure 2-27.

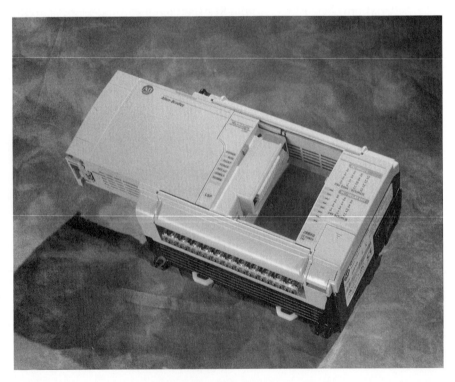

Figure 2-26   Installing the processor unit into the MicroLogix 1500 PLC Base unit. (Used with permission of Rockwell Automation, Inc.)

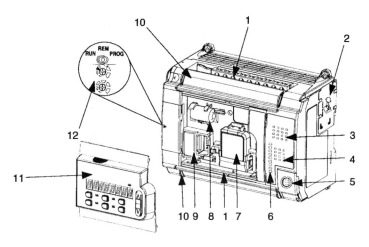

Figure 2-27   MicroLogix 1500 hardware features. (Used with permission of Rockwell Automation, Inc.)

Feature Number	Feature Description	Feature Number	Feature Description
1	Removable I/O Terminal Blocks	7	Optional Memory Module and Time Clock
2	I/O Expansion Interface	8	Optional Extra Battery
3	Input Status Indicators (LEDs)	9	Battery
4	Output Status Indicators (LEDs)	10	I/O Terminal Doors
5	Communication Cable Connection	11	Optional Data Access Terminal
6	PLC Status Indicators	12	PLC Mode Switch and Adjustment Trim Pots

Figure 2-28   Features of the PLC shown in Figure 2-27.

Expanding the MicroLogix 1500 I/O Count

I/O expansion is accomplished by connecting I/O modules to the right-hand side of the MicroLogix 1500 base unit. Compact I/O enables expansion of the MicroLogix 1500 up to 256 I/O points. Up to 16 modules can be directly addressed by the PLC. Compact I/O modules require no rack or chassis. The modules simply slide into position from the front on upper and lower tongue-and-groove slots as illustrated in Figure 2-29. Once in position, the module locks into place to provide a one-piece unit that can be handled until it is installed into its cabinet, either on a DIN rail or mounted directly to the panel back plate.

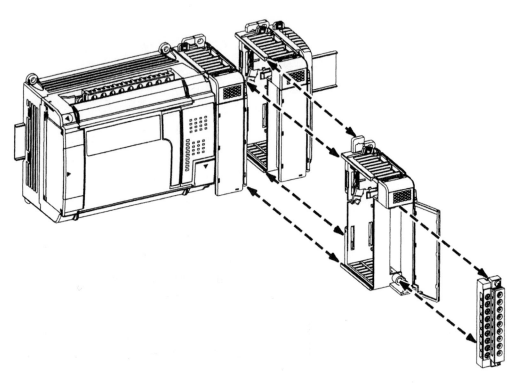

Figure 2-29   MicroLogix 1500 expansion module installation. (Used with permission of Rockwell Automation, Inc.)

MicroLogix 1200

The newest addition to Rockwell Automation's micro PLC family is the MicroLogix 1200. The MicroLogix 1200 is an integrated package with the processor, power supply, and I/O embedded into a small footprint measuring as little as 3.54 inches by 4.33 inches. Two I/O counts are available, 24-point and 40-point. The 1200's rackless design provides for easy expansion, with up to six expansion I/O modules per PLC system for a maximum I/O count of 88. The MicroLogix 1200, just like the 1000 and 1500, uses the same programming software as the SLC 500 modular PLC—RSLogix 500 software.

Figure 2-30 illustrates a MicroLogix 1200 base unit with 14 inputs and 10 outputs. Two I/O expansion modules are clipped to the right side. The left expansion module is a 16-point DC input module; the right expansion module is a four-channel analog input module.

Figure 2-31 contains selected specifications for Rockwell Automation's MicroLogix 1000, 1200, and 1500 PLCs.

Rockwell Automation's Pico Controller

The Allen-Bradley Pico Controller is a small DIN-rail or panel-mounted controller with basic PLC functionality. The pico controller performs simple logic, timing, counting, and

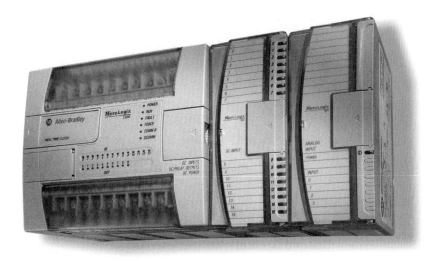

Figure 2-30   Allen-Bradley MicroLogix 1200 PLC with two slots of I/O expansion. (Used with permission of Rockwell Automation, Inc.)

	MicroLogix 1000	MicroLogix 1200	MicroLogix 1500	
			1764-LSP	1764-LRP
User Program and Data Memory	Up to 1 K	Up to 6 K	Up to 7 K	Up to 12 K
Memory Backup	EEPROM	EEPROM	Battery	Battery
		Memory Module		
I/O Count	Up to 32	Up to 88	Up to 256	
Analog I/O	Embedded	Expansion		
Programming Software	RSLogix 500 Windows 98, NT, 2000, XP			
	A.I. 500 (DOS)	DOS software programming not available		

Figure 2-31   Selected MicroLogix PLC specifications. (Used with permission of Rockwell Automation, Inc.)

shift register functions. The pico also has a built-in real-time clock. Programming is as simple as pressing the correct series of buttons on the front of the unit.

Figure 2-32 illustrates the family of Allen-Bradley pico controllers. Four of the units in Figure 2-32 have a small status display on the front left of the unit. Directly to the right of the LCD status display are four buttons around a center round button. These five buttons are used to develop the PLC program or "draw the circuit" as described by Allen-Bradley literature. This keypad is used to access the main menu. From the main menu, the user can perform operations such as setting a password or internal clock, putting the unit into run mode, or deleting the program.

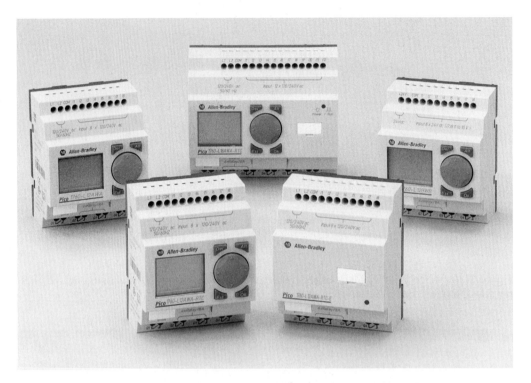

Figure 2-32 Allen-Bradley Pico Controller family. (Used with permission of Rockwell Automation, Inc.)

Even though the front push buttons can be used to program the unit, Windows-based PicoSoft programming software is available for personal computer program development. Referring to Figure 2-32, the front right-hand unit has no keypad or LCD display; this unit must be programmed using software.

The pico controller is available with the specifications shown in Figure 2-33.

SIEMENS SIMATIC S7-200 MICRO PLC

The S7-200 is part of the Siemens series of micro PLCs. The S7-200 series is a fixed-style PLC with the CPU, power supply, and discrete I/O built into one compact unit. Figure 2-34 illustrates the features of the S7-200 CPU.

Features identified in Figure 2-34 are:

Status LEDs	Status lights (LEDs) provide information as to whether the CPU is running, stopped, or faulted. There is also an LED for the ON or OFF status of each local I/O point.
Cartridge	Insertion of memory cartridge or battery cartridge for CPU program data retention. See following section on memory retention using cartridges.

Pico Controller Specifications						
Part Numbers	1760-L12BWB-NC	1760-L12BWB	1760-L12AWA-NC	1760-L12AWA	1760-L12AWA-ND	1760-L18AWA
Power Supply	24 VDC	24 VDC	120/240 VAC	120/240 VAC	120/240 VAC	120/240 VAC
Digital Inputs	Up to 8 at 24 VDC		8 120/240 VAC			12 at 120/240 VAC
Analog Inputs	Up to 2 at 0-10 volts		No	No	No	No
Relay Outputs (8 amp)	4	4	4	4	4	6
Keypad	Yes	Yes	Yes	Yes	No	Yes
LCD Display	Yes	Yes	Yes	Yes	No	Yes
Real-Time Clock	No	Yes	No	Yes	Yes	Yes
Programming Software	PicoSoft for Windows 95, 98, or NT					
Size	2.81 inches wide x 4.33 inches high x 2.28 inches deep					4.23 inches wide

Figure 2-33   Data compiled from Allen-Bradley Pico Controller Data.

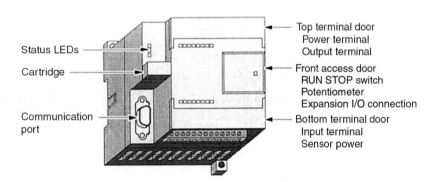

Figure 2-34   S7-200 CPU features. (Courtesy of Siemens Energy & Automation, Inc.)

Communications Port	Allows you to connect the CPU to a programming device or other devices. See Figure 2-35.
Bottom Terminal Door	Input screw terminals and power for sensors are under the door.
Front Access Door	Run/Stop switch, potentiometer, and I/O expansion connector.
Top Terminal Door	Incoming PLC power and output screw terminals.

Memory Retention Using Cartridges

The S7-200 CPU provides several methods to permanently save the ladder program, program data, and CPU configuration data:

1. The CPU has a built-in memory chip (EEPROM) to provide permanent memory storage of the user program, CPU configuration, and user-selected data.
2. There is also a super capacitor built into the CPU unit that will maintain the contents of temporary memory (RAM) any time power is removed from the unit.
3. One of two optional battery cartridges can be installed into the cartridge slot on the CPU unit to maintain temporary memory after power has been lost or disconnected. One battery cartridge contains only a battery, while the second also provides a clock and calendar.
4. An optional EEPROM memory cartridge can be installed in the CPU cartridge slot to provide portable memory storage. The user program and CPU configuration, along with data stored in the internal EEPROM memory, can be loaded to or from the optional memory cartridge. The portable memory cartridge can then be removed for safekeeping or used to load the user program and associated saved data into another S7-200 CPU unit.

Figure 2-35 shows a configuration with a personal computer connected to several S7 200 CPUs on a network. An operator at the personal computer can communicate with any CPU on the network, even though the STEP7-Micro/WIN 32 software can only communicate with one CPU at a time.

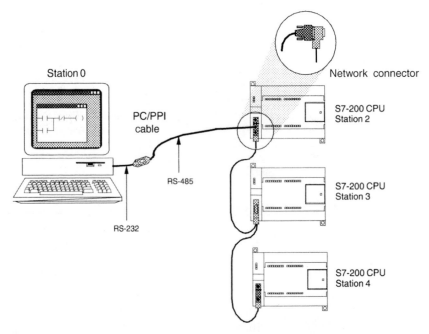

Figure 2-35   Communicating between a personal computer and multiple S7-200 CPUs on a network.

S7-200 Specifications

The low cost, small size, ability to expand, and extensive instruction set make the S7-200s well-suited for small control applications. Figure 2-36 shows one member of the Siemens S7-200 family. Notice the PLC size compared to the pen shown in the foreground.

There are a number of different CPU sizes and voltages. This flexibility makes it easy to incorporate the S7-200 into almost any control situation for a micro PLC. CPUs are available in two families, the 21x series of CPUs and the 22x series of CPUs. The 21x family includes the CPU 210, 212, 214, 215, and 216. The 22x series are the newer S7-200 CPU family members. The 22x family includes the CPU 221, 222, and 224. Figure 2-37 presents selected features of the CPU 22x family.

The 22x series of CPUs has used advances in microprocessor technology to pack more features and performance into a CPU footprint that is 44% smaller than that of earlier CPUs. As an example, the 221 CPU is only 3.6 inches wide by 3.2 inches high and 2.4 inches deep.

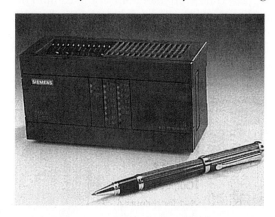

Figure 2-36   Simatic S7-200 Micro PLC. (Courtesy of Siemens Energy & Automation, Inc.)

Features	CPU 221	CPU 222	CPU 224
Program Memory	2048 words		
User Data Memory	1024 words		2560 words
Memory Type	EEPROM		
Memory Cartridge	EEPROM		
Local I/O	6 In/4 Out	8 In/6 Out	14 In/10 Out
Expansion Modules	None	2	7
Analog I/O	None	12 In/10 Out	12 In/10 Out
Communication Ports	1- RS-485		
Boolean Execution Speed	.37µs/instruction		

Figure 2-37   Siemens S7-200 CPU selected specifications. See PLC technical manuals for complete specifications. (Data compiled from Siemens Technical data sheets)

Programming the S7-200

Programming S7-200 micro PLCs is as simple as having a personal computer with STEP7-Micro/WIN programming software, a programming cable, and the S7-200 CPU. There are two varieties of programming software depending on whether you are using Windows or DOS. STEP7-Micro/WIN programming software is a 32-bit programming package for Windows 95, 98, and NT operating systems. STEP7-Micro/DOS programming software is used in the DOS environment.

S7-200 I/O Expansion

The power supply contained in the CPU provides power not only to the base unit but also to any connected expansion modules. The S7-22x CPUs, with the exception of the CPU 221, can be expanded. The CPU 222 can be expanded up to 46 I/O (24 in and 22 out), while the CPU 224 can be expanded up to a maximum of 120 I/O (62 in and 58 out).

There are two methods of expanding the S7-200 CPU. The first is to build a flexible ribbon cable with a mating connector into the expansion module for easy connection to the CPU or other expansion module. This expansion method creates a compact single unit as illustrated in Figure 2-38.

Figure 2-38   The darker unit, center left, is a S7-200 CPU with two expansion modules. (Courtesy of Siemens Energy & Automation, Inc.)

The second option is to use the 31-inch I/O expansion cable. Using this cable provides the flexibility of adding additional expansion modules to the already expanded S7-200, for example, as illustrated in Figure 2-38, but allows the expansion modules to be mounted up to 31 inches away on the panel backplate or the DIN rail. Figure 2-39 lists the currently available expansion modules.

Siemens S7-200 Expansion Modules
24-VDC Digital, 8 Inputs
24-VDC Digital, 8 Outputs
8 Relay Outputs
24 VDC Digital Combination, 8 Inputs and 8 Outputs
24 VDC Digital Combination, 8 Inputs and 8 Relay Outputs

Figure 2-39   Siemens S7-200 Expansion Modules.

SUMMARY

Due to advancements in microprocessor technology, the size and power of the newer PLCs have resulted in a smaller PLC package size and more computer power and features than ever before. These new, smaller, and more powerful PLCs are called micro PLCs.

Builders of small machines also wanted a small, inexpensive, basic-function controller with basic relay, timer, and counter capabilities to replace hardwired relays, timers, and counters. These newer, smaller PLCs were given names such as nano and pico controllers. Nano and pico controllers typically have only a handful of I/Os. Six inputs and 4 relay outputs, 8 inputs and 4 relay outputs, or 12 inputs and 6 relay outputs are typical. Now almost any machine can have PLC intelligence built in.

REVIEW QUESTIONS

1. Why did OEMs want a smaller, basic-functionality PLC?
2. List four features OEMs wanted in a smaller PLC controller.
3. Today's micro PLCs have advanced features such as _____.
4. The micro PLC typically has _____ I/O points or less.
5. Micro PLCs come as self-contained units. The _____, _____, and I/O are built into one package.
6. Since all pieces are built into a single non-changeable unit, these PLCs are called _____, or _____ controllers.
7. The smaller, basic-function controllers, which have less functionality than a micro controller, have names such as _____ or _____ controllers.
8. These smaller, limited-functionality controllers are simply relay replacers with basic relay instructions plus _____ and _____ instructions.
9. Pico or nano controllers support I/O counts such as _____.
10. Most smaller controllers only have relay outputs. Explain why these units typically come with relay outputs.

IF PASSWORD IS CORRECT GO TO DRIVE SET UP SCREEN

0022
```
    ┌─ EQU ──────────────┐              B3:10/0                                    B3:1/15
    │ Equal              │              ─[OSR]─                                    ─( )─
    │ Source A    N12:21 │
    │              7777< │                                           RESET PASSWORD
    │ Source B    32767  │                                           ATTEMPTS COUNTER
    │             32767< │                                                  C5:5
    └────────────────────┘                                             ─( RES )─
```

0023
```
    ┌─ GRT ──────────────┐    ┌─ NEQ ──────────────┐    B3:10/1
    │ Greater Than (A>B) │    │ Not Equal          │    ─[OSR]─
    │ Source A    N12:21 │    │ Source A    N12:21 │
    │              7777< │    │              7777< │
    │ Source B        0  │    │ Source B    32767  │
    │                 0< │    │             32767< │
    └────────────────────┘    └────────────────────┘
                                                          ┌─ CTU ──────────────┐
                                                          │ Count Up           │    ─( CU )─
                                                          └────────────────────┘
```

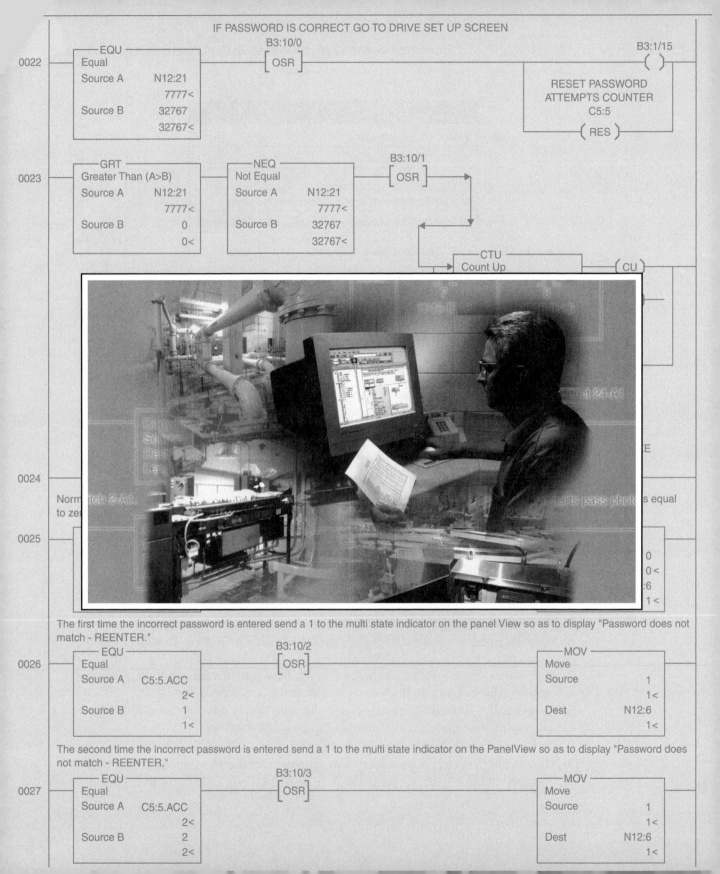

0024

Norm...tch 2-A1.   ...parts pass photo...s equal
to zer...

0025
```
                                                                                        0
                                                                                        0<
                                                                                       :6
                                                                                        1<
```

The first time the incorrect password is entered send a 1 to the multi state indicator on the panel View so as to display "Password does not match - REENTER."

0026
```
    ┌─ EQU ──────────────┐              B3:10/2                    ┌─ MOV ──────────────┐
    │ Equal              │              ─[OSR]─                    │ Move               │
    │ Source A    C5:5.ACC│                                        │ Source          1  │
    │                 2< │                                         │                 1< │
    │ Source B        1  │                                         │ Dest       N12:6   │
    │                 1< │                                         │                 1< │
    └────────────────────┘                                        └────────────────────┘
```

The second time the incorrect password is entered send a 1 to the multi state indicator on the PanelView so as to display "Password does not match - REENTER."

0027
```
    ┌─ EQU ──────────────┐              B3:10/3                    ┌─ MOV ──────────────┐
    │ Equal              │              ─[OSR]─                    │ Move               │
    │ Source A    C5:5.ACC│                                        │ Source          1  │
    │                 2< │                                         │                 1< │
    │ Source B        2  │                                         │ Dest       N12:6   │
    │                 2< │                                         │                 1< │
    └────────────────────┘                                        └────────────────────┘
```

CHAPTER

3

Programming a Programmable Controller

OBJECTIVES

After completing this chapter, you should be able to:

- describe the available options for programming a PLC
- explain what on-line is in comparison to off-line
- list the advantages of software programming with a personal computer over handheld programming terminal
- explain the differences when interfacing a notebook personal computer to a PLC
- discuss open or soft PLC programming. Explain the advantages and disadvantages of programming using this method
- provide an overview of the IEC 1131-3 standard

INTRODUCTION

The PLC can do nothing without someone developing a program and loading this user program into the CPU's memory. Once the CPU has the program in memory and has been put into run mode, it can look at inputs and, as a result of solving the user program ladder logic instructions, it can control the outputs and their associated field devices.

There are multiple ways to program a PLC.

1. One of the oldest methods of programming a PLC is by pressing buttons on a handheld programming terminal to enter a user program.
2. The most popular method of PLC programming is using a personal desktop computer and either a **DOS** or Windows operating system to run the manufacturer's

software for that specific PLC. Figure 3-1 illustrates the choice of an SLC 500 handheld terminal or a personal computer for programming.

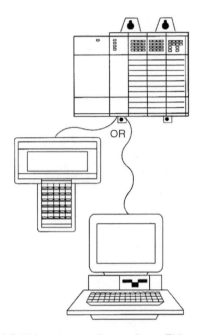

Figure 3-1   Allen-Bradley SLC 500 programming options. Either a handheld programmer or a personal computer can be used to program the PLC. (Used with permission of Rockwell Automation, Inc.)

3. PLC programming can be accomplished using a notebook personal computer running PLC programming software and a Personal Computer Memory Card International Association (PCMCIA) interface card, or, in some cases, a direct connection between the personal computer serial port and the PLC CPU.
4. Using an industrial computer and the PLC manufacturer's program **software.**
5. Using third-party "open software" and running a personal computer as the PLC's CPU.

This chapter will explain the various methods available for programming a PLC and the advantages and disadvantages of each. We will also introduce open control and show how it correlates with IEC 1131-3 standards and with soft PLC programming solutions.

HANDHELD PROGRAMMING TERMINALS

One of the oldest methods of entering programming instructions into a PLC's CPU is to use the manufacturer's handheld programming terminal. A handheld programmer (HHP) or handheld terminal (HHT) is a compact, portable, easy-to-use unit containing multicolored, multifunction keys and a display screen.

Figure 3-2 shows an Allen-Bradley SLC 500 handheld programming terminal (HHT) connected to a modular SLC 500 PLC through a 1747 C10 cable.

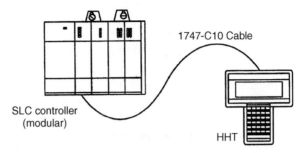

Figure 3-2 Allen-Bradley SLC 500 handheld programming terminal (HHT) connected to a modular SLC 500 PLC through a 1747 C10 cable. (Used with permission of Rockwell Automation, Inc.)

Instruction entry using a handheld programmer is as simple as pressing a series of buttons on the programmer. The typical handheld programmer has keys that are grouped and color coded by function **(function keys).** There are usually keys for entering and editing instructions, navigation keys for moving around the program, and troubleshooting keys. Figure 3-3 illustrates an Allen-Bradley SLC 500 handheld programming terminal. Notice the display area and multifunction keys.

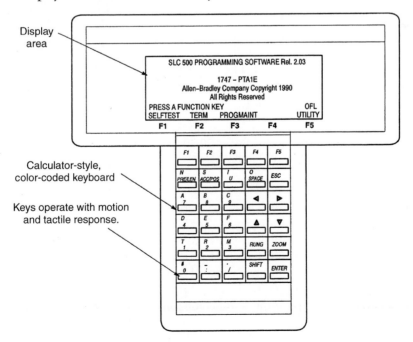

Figure 3-3 Allen-Bradley SLC 500 handheld terminal (HHT). (Used with permission of Rockwell Automation, Inc.)

The programmer is connected to the CPU either via a cable or, in some cases, by connecting the programmer to the CPU. As an example, the Modicon Micro PLC handheld programmer is attached directly to the front of the PLC and thus to the CPU.

Each handheld programming terminal will work only with the manufacturer's PLC for which it was designed. The General Electric handheld programmer from Figure 3-4 will not work in place of the SLC 500 handheld programmer in Figure 3-2, nor will the Allen-Bradley handheld programming terminal work in place of the General Electric programmer to program the GE Series 90-30 PLC. Likewise, the Allen-Bradley MicroLogix 1000 PLC's HHP, as illustrated in Figure 3-5, will program only the MicroLogix 1000.

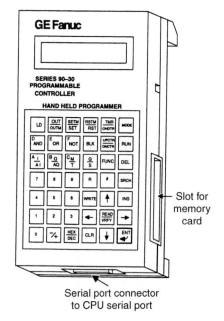

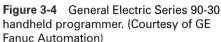

Figure 3-4   General Electric Series 90-30 handheld programmer. (Courtesy of GE Fanuc Automation)

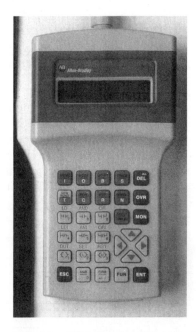

Figure 3-5   The Allen-Bradley MicroLogix 1000 handheld programmer. (Used with permission of Rockwell Automation, Inc.)

The HHP display will also vary depending on the particular HHP you may be using. Some displays will only display the last instruction that has been programmed, while other units' displays will display from two to four rungs of ladder logic. Figure 3-6 is a sample display from an SLC 500 handheld programmer showing a typical rung of logic.

Figure 3-7 illustrates two rungs of logic on the handheld programmer display. Notice the cursor positioning (with the arrow keys).

Figure 3-8 displays the zoom feature available with the Allen-Bradley SLC 500 handheld programmer. The zoom feature is activated after placing the cursor on the instruction for which you want additional information. You can zoom in and view information associated with the instruction.

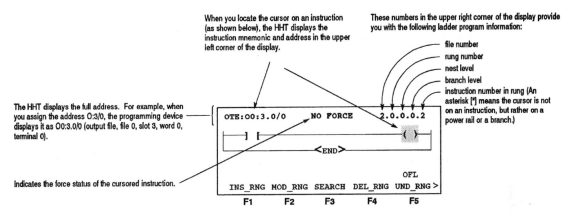

When you locate the cursor on an instruction (as shown below), the HHT displays the instruction mnemonic and address in the upper left corner of the display.

These numbers in the upper right corner of the display provide you with the following ladder program information:

file number
rung number
nest level
branch level
instruction number in rung (An asterisk [*] means the cursor is not on an instruction, but rather on a power rail or a branch.)

The HHT displays the full address. For example, when you assign the address O:3/0, the programming device displays it as O0:3.0/0 (output file, file 0, slot 3, word 0, terminal 0).

Indicates the force status of the cursored instruction.

Figure 3-6 SLC 500 HHP sample display. (Used with permission of Rockwell Automation, Inc.)

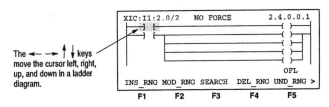

The ← → ↑ ↓ keys move the cursor left, right, up, and down in a ladder diagram.

Figure 3-7 Two rungs shown on a handheld programmer display.

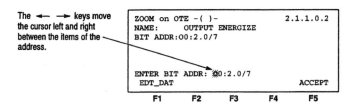

The ← → keys move the cursor left and right between the items of the address.

Figure 3-8 Zoom in to view additional information about the instruction selected by the cursor.

One advantage of a handheld programmer is its compact size. The small size of many HHPs makes them extremely portable; many will fit into your pocket. The Allen-Bradley MicroLogix 1000 HHP measures less than 4 inches by 7 inches. Maintenance personnel like HHPs for their compactness and ruggedness. HHPs are popular tools for troubleshooting and for editing or modifying programs either on the factory floor or at remote locations. If a program change is needed in a remote location, simply disconnect the HHP and carry it there. Then plug the HHT into the CPU and either upload the current program for editing or download a new or modified program.

There are two types of HHPs, the **dumb terminal** and the **smart terminal.** Only a smart terminal will allow you to physically disconnect an HHP and download a program at a remote location. Next we will look at the differences between dumb and smart terminals.

Dumb Programming Terminals

Some handheld programming terminals, especially older models, are no more than a plastic shell with a small keyboard and display. These devices are only used to transfer the data represented by key closures directly to the CPU. Programs are developed by pressing the correct sequence of keys, which the CPU accepts and stores directly into memory. A programming terminal such as this has no intelligence or memory of its own. As a result, it is called a dumb terminal. A dumb terminal cannot be used to program, edit, or monitor a program if it is not physically hooked up to the CPU.

Smart, or Stand-Alone, Programming Terminals

Most newer, handheld programming terminals have their own built-in, or "on board" microprocessor and memory. This type of programming terminal is called a smart programming terminal. A smart programming terminal can be used to develop programs without being attached to a PLC. Sometimes the term "stand-alone terminal" is used to signify that the programmer can be used even if it is not attached to a PLC.

The advantage of a smart HHP is memory retention. A smart programming device will remember the program that is stored in its memory (assuming the battery is good or power is being supplied).

When connected to a PLC, a communication link is created so that the HHP (a smart device) can communicate to the CPU. The process of transferring a copy of a user program from an HHP to a CPU is called downloading. If you download a program from an HHP to a CPU, only a copy of the program will be transferred. The HHP does not lose its copy of the program. By sending a copy of the program from the HHT to the CPU you can download a copy to many machines. This is a nice feature if you are a machine builder and need to program a number of machines.

If you need to look at a copy of a program that is currently running in the PLC, you can transfer a copy of the CPU's program to the HHP by uploading the CPU's program into the handheld programmer's memory. Note that most CPUs and HHPs can only store one program at a time. If you transfer a program from one to the other, the current program in the receiving device will be replaced by the new program. Replacing one program with another from either an upload or a download will cause the old program to be written over with the new program. A program that is written over is lost forever.

Advantages of Using a Smart, Handheld Programming Terminal

Maintenance and electrical support personnel like to use smart HHPs in the factory environment because of the following features:

1. easy transfer of the PLC program to the HHT for editing or troubleshooting
2. easy transport of a program to the field to update a current machine's program

3. rugged and industrially hardened for the factory environment
4. low cost; cheaper than a notebook computer
5. easy to use and easy to learn; no software required
6. compact size: some will fit in your pocket
7. program storage
8. monitor resident PLC program for troubleshooting

Disadvantages of Using a Smart, Handheld Programming Terminal

The following list contains the disadvantages of using an HHP versus a portable notebook personal computer:

1. Not all CPUs in a manufacturer's PLC will support handheld programming.
2. An HHP will hold only one program at a time whereas a notebook or desktop personal or industrial computer can hold many programs on a hard drive.
3. Handheld programmers usually require more keystrokes to enter and get the same information when compared to a notebook, desktop, or industrial computer.
4. HHPs have limited capability to display ladder rungs due to screen size. If the HHP displays rungs of logic, in most cases only about half the number of rungs will display, in contrast to a notebook computer.
5. Documentation is not displayed.
6. If there are different manufacturer's PLCs in the field, more than one HHP will be needed.
7. If the battery dies, a program stored in memory will be lost.

DESKTOP PERSONAL COMPUTERS (PCs) AND PLC PROGRAMMING

Even though the handheld programmer can be used to program a PLC, the most common use of the handheld programmer is as a troubleshooting tool. Today, a personal computer is used to develop and edit most PLC programs. A standard IBM-compatible personal computer is used to run PLC programming software purchased as part of your PLC hardware and software package. PLC software is specifically designed to program a certain family of PLCs from a specific manufacturer. Most PLC program development software packages will not allow you to develop programs on another manufacturer's PLCs. In some cases, a single manufacturer will have multiple PLC families, each of which will require its own software to program.

PLC program development software is used to create, edit, document, store, and monitor PLC user programs. Printed hard-copy reports containing ladder diagrams, system configurations, documentation, and the contents of each data file are available from most software packages. Printouts are generated in the software and printed on your personal computer's printer. In addition, PLC software packages provide troubleshooting assistance through monitoring a PLC program while it is running. Built-in search functions and forcing I/O features can streamline troubleshooting.

Editing a PLC ladder program with a software package is as easy as the cut-and-paste function used when you need to duplicate, cut, or move a ladder rung.

Most software instructions are based on graphical symbols derived from the old, familiar ladder diagrams. For the most part, maintenance or electrical personnel with a good understanding of standard relay ladder diagrams will have few problems developing or troubleshooting basic PLC ladder diagrams using software developing ladder logic. PLC program development software packages have many advantages over handheld programmers.

Advantages of Software Programming

1. Personal computer monitors can display multiple rungs of logic.
2. Personal computer monitors allow maintenance personnel to monitor multiple program rungs for troubleshooting purposes. Scrolling through a program rung by rung is as simple as pressing either the up or down arrow key.
3. Programs can be stored on the computer's hard drive.
4. Programs can be transferred from the hard drive to a floppy drive, CD-ROM, DVD, or memory stick.
5. Programs transferred to a floppy drive can be easily transported.
6. Rung comments, instruction comments, or symbols can easily be added to any rung.
7. Personal computer software offers cut-and-paste features for easy editing.
8. Rung comments, symbols, and other related text can be displayed with the associated rung.
9. Data tables can be easily monitored.

Connecting from a Personal Computer Serially to a PLC

The simplest way to connect a personal computer to a PLC processor with a serial port for software upload, download, or monitoring, is through a serial cable connection. The PLC 5, ControlLogix, SLC 500 5/03, 5/04, and 5/05 modular processors, along with the MicroLogix family of PLCs, can be programmed through their serial ports. The serial cable is connected to your personal computer's COM 1 or COM 2 port and the PLC CPU's serial communication port. The proper serial cable must be used for connection between the personal computer and the PLC. Figure 3-9 illustrates an Allen-Bradley SLC 5/04 process connected serially to a desktop personal computer. A 1747-CP3 cable is available from Allen-Bradley for connection between the desktop computer and the processor's serial communication port, channel zero. A null-modem cable, available at any computer store, can also be used for this type of connection.

RS Linx is Allen-Bradley's communication driver for RS Logix software. The RS Linx serial driver auto-configuration feature is used to set up the communication parameters. When communication is established, the personal computer and PLC will exchange signals with each other. These signals will let the other know: when it is ready to accept data, that it is OK to send data now, and the data you sent was OK, thank you. This exchange of signals is called *handshaking*. When the personal computer and PLC can successfully communicate, data can be uploaded or downloaded.

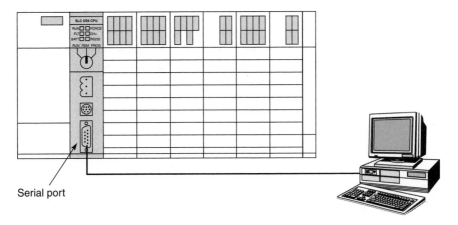

Serial port

Figure 3-9   Serial connection between a desktop personal computer and an Allen-Bradley SLC 5/04 processor. (Used with permission of Rockwell Automation, Inc.)

Connecting from a Personal Computer to a PLC through a Hardware Interface

To communicate with some PLC processors, a special interface converter box needs to be connected between your personal computer's COM port and the PLC controller.

The interface converter changes the serial RS-232C signal from the personal computer's serial port to a signal understood by the PLC. Figure 3-10 illustrates a personal computer interface to an SLC 500 PLC through a 1747-PIC interface converter. The 1747-PIC changes the RS-232 signal from the personal computer's COM port to Data Highway 485 protocol.

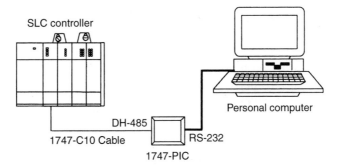

Figure 3-10   Allen-Bradley's 1747-PIC interface converter connecting a personal computer and the SLC 500 PLC. (Used with permission of Rockwell Automation, Inc.)

The IC690ACC900, RS-422/RS-485 to RS-232 converter is the interface converter used to interface a personal computer to a General Electric Series 90-30 or Series 90-70 PLC. The converter converts the RS-232 interface from a standard personal computer serial port to an RS-422/RS-485 (the signals needed to communicate to the PLC). The cable between the converter and the PLC can be up to 10 feet without an external +5 VDC power supply. Using an external power supply, the cable between the converter and the PLC can be up to 1,000 feet. A cable of up to 50 feet can connect the converter to the programming computer's serial port. Figure 3-11 illustrates the connection between the personal computer and the Series 90-30 PLC.

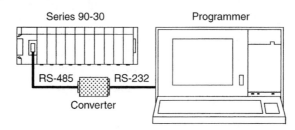

Figure 3-11   General Electric's RS-422/RS-485 to RS-232 interface converter connecting a personal computer and a series 90-30 PLC. (Courtesy of GE Fanuc Automation)

Connecting to a PLC through Internal Hardware Interface Cards in Your Personal Computer

Desktop personal computers have internal slots for installing **I/O interface** cards. Interface cards convert the RS-232C serial signals from the personal computer to signals the PLC can understand. In most cases, you will need to purchase the proper interface card from your PLC manufacturer for interface to your particular PLC.

The Allen-Bradley 1784-KTX interface card is one example of the many available interface cards. The KTX card is a half-size interface card that must be installed into a 16-bit ISA or EISA expansion slot inside your personal computer. Figure 3-12 illustrates the connection end of a KTX interface card. There are three connections, called channels. Each channel has a specific interface that it will provide.

Channel 1A is typically used to interface to a SLC 500 5/04 processor, a PLC 5 processor or a ControlLogix DH+/RIO communications module. This communication module allows a newer ControlLogix PLC to be connected to a DH+ (Data Highway Plus) or RIO (remote I/O) network and communicate with SLC 500 or PLC 5s. Channel 1B is used to interface older Allen-Bradley PLC-2s or PLC-3s.

Figure 3-13 shows a personal computer with a KTX interface card connecting to the SLC 500 Data Highway 485 network through channel 1C. Figure 3-14 illustrates the communication ports on a modular SLC 500 5/01 or 5/02 CPU and an SLC 500 fixed controller. Notice that three SLC 500 PLCs make up this network.

The communications cable comes from KTX channel 1C, the bottom connection or channel (illustrated in Figure 3-13) to one of the three 1747-AIC isolated-link couplers. The communication cable (a Belden #9842 cable) is daisy-chained from one link coupler to the next. There is one link coupler for each SLC on the network. The primary function of the link coupler is to amplify the signal and to isolate each PLC from the network. Each PLC and the personal computer programming terminal on a network is called a node. The node address uniquely identifies each connection to the network. The node address is set up in the programming software.

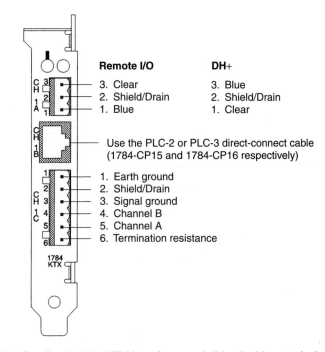

Remote I/O

3. Clear
2. Shield/Drain
1. Blue

DH+

3. Blue
2. Shield/Drain
1. Clear

Use the PLC-2 or PLC-3 direct-connect cable
(1784-CP15 and 1784-CP16 respectively)

1. Earth ground
2. Shield/Drain
3. Signal ground
4. Channel B
5. Channel A
6. Termination resistance

Figure 3-12   Allen-Bradley's 1784-KTX interface card. (Used with permission of Rockwell Automation, Inc.)

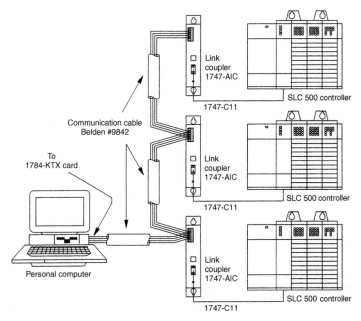

Figure 3-13   Communicating to multiple SLC 500s through a KTX card and Allen-Bradley's Data-Highway 485 network. (Used with permission of Rockwell Automation, Inc.)

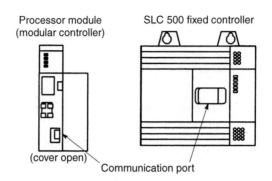

Figure 3-14   SLC 500 fixed and modular processor communications ports. (Used with permission of Rockwell Automation, Inc.)

The 1784-KTX is only one of three variations of the KTX family of interface cards. Personal computer expansion cards also come in Ethernet, ControlNet, and DeviceNet for installation into desktop personal computers and industrial computer expansion slots.

NOTEBOOK PERSONAL COMPUTERS AND PLC PROGRAMMING

Interface between a notebook personal computer and most PLCs can be achieved through one of three methods: direct serial connection, hardware interface devices, and a PCMCIA card. First we will look at the serial connection.

PLC Direct Serial Connection

Notebook computers usually have one serial port, COM 1. The notebook computer's serial port is just like those on a desktop computer. A direct connection between the notebook computer and the PLC is simply made by using a null-modem cable.

Hardware Interface

Because notebook computers have a COM port, a hardware interface, like the Allen-Bradley 1747 PIC or the General Electric interface converter, can be easily interfaced between the notebook computer and the PLC.

PCMCIA Card

Notebook personal computers are too small to have interface cards installed internally. Another method had to be developed. A small (credit-card sized), plug-in interface card was developed, called the PCMCIA card.

What Is a PCMCIA Card?

The Personal Computer Memory Card International Association (PCMCIA) is a standards committee formed in 1989 by leading manufacturers of personal computers and components. The standards committee was to govern the size, power, signal specifications, and software standard for removable memory cards. The standards were to ensure that users could feel confident that personal computers and peripherals would work together. Figure 3-15 shows Allen-Bradley's PCMCIA interface card for notebook computers.

Figure 3-15   Allen-Bradley's PCMCIA interface card, called a 1784 PCMK card. (Used with permission of Rockwell Automation, Inc.)

Initially these memory cards provided an interface between the notebook personal computer and fax modems, local area networks (LANs), and cellular uplinks. As time went on, PCMCIA cards were developed that allowed interfacing to other data-generating devices such as the PLC. Today, the PCMCIA card can be used to add memory storage and I/O capabilities to notebook and smaller computers.

As for the size of a PCMCIA card, they are about the size of a credit card (54 mm by 85.6 mm). The card thickness varies with the type of card. PCMCIA cards come in three types (type I, type II, and type III).

Type I cards are 3.3 mm thick and are typically used for data storage.

Type II cards are 5.0 mm thick and are used for modem and LAN I/O interface.

Type III cards are 10.5 mm thick. The 10.5 mm cards are used for larger data storage or I/O requirements.

All three types of PCMCIA cards will fit in the same slot of your personal computer if the slot is thick enough. If your computer will accept a type II card, then the thinner type I card is also usable. All three cards use the same 68-pin connector. Each connector has two rows of 34 pins.

Card Insertion and Communication   When you insert a PCMCIA card into the slot, the card's connector mates with a socket. The socket provides the connection between the card and the personal computer.

Communication between the card and the personal computer is provided through software interfaces defined by the PCMCIA standards. These software interfaces are called "card and socket services." Card services provides the access to your systems resources, such as memory. Socket services identifies how many sockets are present in your computer and detects whether a card is inserted into one of the available sockets. Socket services also controls the communication between the interface card and your personal computer.

Interfacing the PCMCIA Card to Your PLC   Allen-Bradley has a PCMCIA card called the 1784-PCMK card. A type II card, it serves as an interface between a programmable controller and a personal computer used as a programming terminal. The 1784-PCMK cards are available to interface either to the Data Highway Plus network (the PLC 5 connection) or Data Highway 485 (the SLC 500 network). The PCMK card is inserted into your personal computer's PCMCIA slot, as illustrated in Figure 3-16.

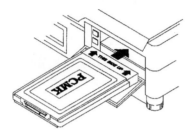

Figure 3-16   PCMK card insertion into a personal computer's PCMCIA slot. (Used with permission of Rockwell Automation, Inc.)

One end of the interface cable is attached to the PCMK card, as illustrated in Figure 3-17. The other end of the interface cable is attached to the PLC's processor, as illustrated in Figure 3-18.

After setting up your computer and the communications drivers so that communication can take place between the personal computer, the PCMK card, and the PLC's CPU, you are ready to upload, download, or monitor the PLC's program on your personal computer.

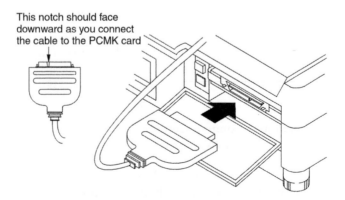

Figure 3-17   Interface cable attachment to a Series A PCMK card. (Used with permission of Rockwell Automation, Inc.)

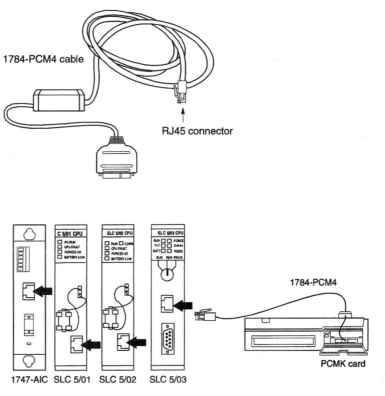

Figure 3-18   Interface cable from PCMK card attached to different Allen-Bradley SLC 500 CPUs. (Used with permission of Rockwell Automation, Inc.)

Computers used in the factory environment fall into two groups, the standard desktop or office personal computer, sometimes called a "white box," and the industrial computer. Industrial computers, like the PLC, have been industrially hardened to survive the factory environment.

PCMCIA cards also come in other network variations such as Ethernet, ControlNet, and DeviceNet. The cards look very similar except for the markings on the front of the card, which will include the card's part number, and the physical connection for the cable to interface between your personal computer and the network. While older notebook computers required a PCMCIA for Ethernet, most newer notebook computers come with an integrated Ethernet port. If using a ControlNet network, a 1784-PCC card is used to interface. To interface to a DeviceNet network using a notebook computer, a 1784-PCD card is one of the available interface choices.

WINDOWS CE-BASED HANDHELD PERSONAL COMPUTERS

The latest PLC maintenance tool is the handheld, sometimes called a palm-top computer. Since most newer modular PLC processors do not support handheld programmers, there was a need for a compact device that could be easily carried around by maintenance personnel. The desktop personal computer was not designed to be moved around from place to place on the factory floor and, although more convenient, the notebook computer has a tendency to disappear from its workstation if not guarded closely. The palm-top personal computer is small enough to be carried around in your pocket. A handheld computer such as the Hewlett-Packard Jornada can measure as small as $7.4 \times 3.7 \times 1.3$ inches and weigh just over one pound. Handheld computers use the Windows CE operating system.

What is Windows CE?

The Microsoft Windows CE operating system is a new compact operating system, introduced in 1996, making it possible for communication, entertainment, and mobile computing devices that talk to each other to share and exchange information with Windows-based personal computers. Windows CE is a 32-bit multitasking, multithreaded operating system. As a compact, portable operating system, Windows CE offers high performance in limited memory devices such as handheld computers. The first handheld computers with the CE operating system began shipping in late 1996. Windows CE-based handheld personal computers do not have disk drives. The built-in serial ports provide easy interfacing to many types of industrial equipment and to PLCs. Significantly, the handheld computer can be docked to a Windows-based desktop computer for easy sharing of information.

INDUSTRIAL COMPUTERS

Notebook and desktop personal computers are not designed for continuous use in the manufacturing environment. When a computer needs to reside in the harsh manufacturing environment, an industrially hardened computer that is designed to withstand plant floor stress is the correct choice. Industrial computers are designed to withstand the dirt, shock, vibration, high temperatures, and washdowns found in the factory environment by incorporating the following features:

- air filters on intake fans
- fans with ball bearings
- National Electrical Manufacturers Association (NEMA) 12, 4, and 4X enclosure ratings
- shock-mounted hard drives
- 0 to 55 degrees centigrade compared to 0 to 40 degrees centigrade operating environment
- special hard drives rated for longer mean time between failures (MTBF)
- integrated mouse on front panel
- hazardous environment rating Class I, Division 2
- industrial computers are modular, which results in faster and easier repair
- industrial computers do not need to be placed in a fan-cooled or air-conditioned enclosure with a viewing window
- industrial computers are available in two versions

Panel-Mount Industrial Computer

One version of the industrial computer has the computer and monitor built into one integrated unit, which is panel mountable. These computers are rated NEMA 4/12/13. Figure 3-19 is an Allen-Bradley panel-mountable industrial computer.

Most industrial computers offer complete flexibility when ordering the computer for your application. When specifying an industrial computer from a manufacturer the following features are usually selected with multiple options for each:

- Pentium processor, clock speed selectable
- monitor options include CRT display or touch screen
- select the type of video card and size of video memory
- select the size of RAM memory
- is a floppy drive needed?
- select hard drive size
- interface card options, such as Ethernet network interface cards

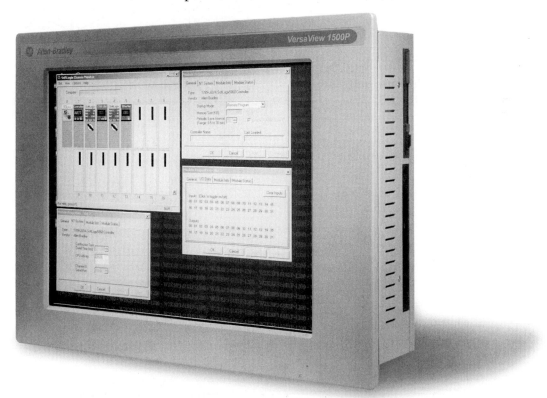

Figure 3-19   Rockwell Automation VersaView 1500 P industrial computer. See Figure 3-20 for features identification. (Used with permission of Rockwell Automation, Inc.)

- select operating system such as Windows 95 or Windows NT
- select operating software and drivers for PLCs or man-machine interface devices (MMIs)

The top portion of Figure 3-20 illustrates the 1200P, 1500P, and 1700P Rockwell Automation VersaView 6181P integrated display computers. These integrated computers

Top and Side View

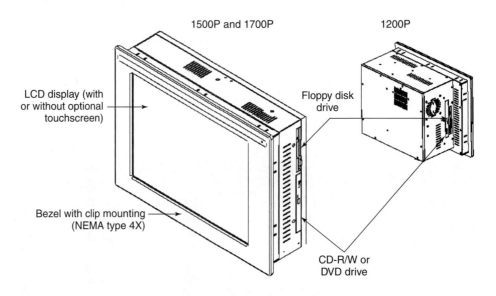

Bottom View

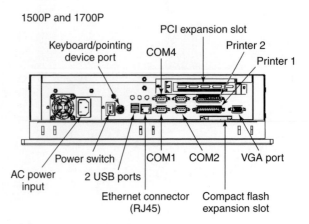

Figure 3-20   Rockwell Automation VersaView 6181P integrated display computers. (Used with permission of Rockwell Automation, Inc.)

come in 10-, 12.1-, 15-, and 17-inch diagonal LCD displays with or without touch screen. Computer operating systems include Windows 2000 or Windows XP. The standard processor is a Pentium III Celeron 850MHz or a Performance Pentium III at 1.2 GHz. The bottom portion of the figure illustrates the 1500P and 1700P connection points on the bottom of the unit. As an example, the 1700P, with its 13.0×10.7-inch display dimensions, is only 14.02 inches high by 17.8 inches wide and 4.29 inches deep. The 1700P weighs 28 pounds.

The Rockwell Automation/Allen-Bradley RAC 6181 panel-mount industrial computer, as shown in Figure 3-21, is a rugged industrial computer in a small package. The RAC 6181 is powered by a number of available Pentium processors and is designed to run on either Windows 98 or Windows NT. Measuring 12.75 inches wide by 10.25 inches high and only 5.5 inches deep, this 10.4-inch color thin film transistor (TFT) active matrix flat-panel display is much smaller and lighter weight than other industrial workstations. Being only 5.5 inches deep, the 6181 will fit into almost any panel.

Figure 3-21 Rockwell Automation's RAC 6181 panel-mount family of industrial computers. (Used with permission of Rockwell Automation, Inc.)

The center object in Figure 3-21 shows a side view of the computer and its connections. Figure 3-22 provides a better view of, and identifies, each connection point or port.

Industrial Rack-Mount Computer

A rack-mounted computer can be a separate component either mounted in a standard 19-inch rack or embedded inside an industrial enclosure. A separate monitor (either a

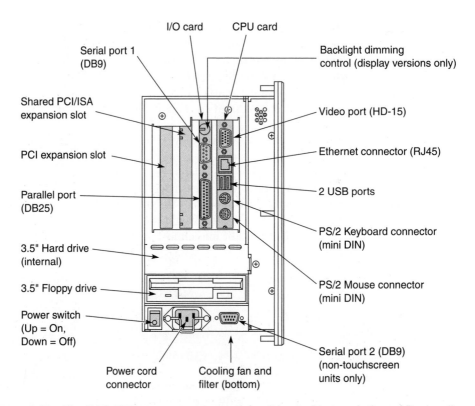

Figure 3-22   The RAC 6181 computer connections. (Used with permission of Rockwell Automation, Inc.)

display or touch screen) can be panel mounted. Figure 3-23 is an industrial rack-mounted industrial computer (lower left). An accompanying 20-inch color monitor is displayed in the upper right. The major difference between the two versions of industrial computers is the way in which they are mounted, and whether the monitor is separate or integrated into a single package. When ordering the rack-mount industrial computer, the same options listed for the integrated computer and monitor are available.

The standard home or office mouse and keyboard were never designed to stand up to the dirt, liquids, gases, and other contaminants found in the industrial environment. The industrial mouse is constructed of stainless steel and heavy rubber. Being completely sealed, many industrial mice are completely submersible. The keyboard must also be designed to stand up to the harsh industrial environment. Industrial keyboards are spill-resistant and shielded against EMI and EFI electrical noise. Both the keyboard and mouse can be purchased as tabletop or panel-mount devices. Figure 3-24 illustrates an industrial mouse and keyboard interfaced to the RAC 6181 industrial computer.

Figure 3-23   Allen-Bradley's integrated industrial computer with a 14-inch CRT screen. (Used with permission of Rockwell Automation, Inc.)

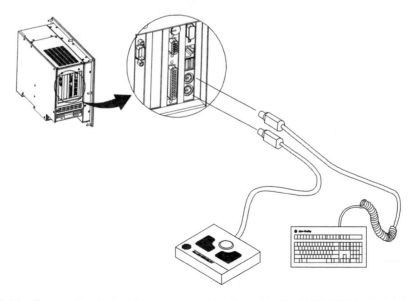

Figure 3-24   Connecting industrial mouse and keyboard to the RAC 6181 industrial computer. (Used with permission of Rockwell Automation, Inc.)

INDUSTRIAL MONITORS

Industrial monitors are panel mounted and used in conjunction with an industrial rack-mount computer such as the one illustrated in Figure 3-23. Figure 3-25 illustrates the Rockwell Automation/Allen-Bradley family of 6185 Industrial TFT flat-panel monitors. These industrial monitors are available as either 10.4-, 15-, 18.1-, or 20.1-inch diagonal displays with or without touch screen. With profiles as slim as 3 inches, and weighing as little as 6.5 pounds, they easily fit into just about any panel. Industrial monitors are typically rated NEMA 4/12 with a stainless steel NEMA 4X version available.

Figure 3-25 The Allen-Bradley family of 6185 TFT flat panel monitors. (Used with permission of Rockwell Automation, Inc.)

OPEN PLC SYSTEMS

Not long ago, selecting a PLC control system was as simple as choosing the PLC manufacturer. Selecting a single PLC manufacturer locked the user into that vendor's hardware, software, and programming methods, and in some cases dictated what peripheral human-machine interface devices could be added and over which networks they could be interfaced. When adding to an existing PLC system, a single brand is typically used because:

- Hardware and software proficiency usually extends to the last brand programmed, installed, or maintained.
- Other PLC hardware in the plant may be all the same brand.
- Spare parts on hand will not fit other brands.

- Terminology is not always the same.
- Maintenance personnel are familiar with current hardware, installation, and wiring. Retraining on a new PLC brand can be expensive.
- Programmers are familiar with programming software for program development and editing.
- There may be finger-pointing when mixing manufacturers and looking for support when there are problems, whereby brand A will point to brand B as being the source of the problem.

Even though PLCs from different manufacturers function similarly, supporting multiple brands can be a real maintenance pain in the neck and a programming nightmare. Each PLC has its own proprietary software packages and programming techniques. In some cases, each PLC brand has its own proprietary network. In the past many PLCs have been called closed systems because you must use the software that was manufactured specifically for that manufacturer's PLC.

PLC software has evolved at a rapid pace over the last 10 to 15 years. Today the costs for PLC hardware are declining with the emergence of smaller, smarter, and faster "micro" PLCs for less money, while software is becoming more sophisticated as PLC functionality increases. As a result, software is becoming a proportionally larger cost item in a PLC system. Few would disagree that software development, debugging, installation, control system commissioning, troubleshooting, and future system modification would be greatly simplified if there were some standards governing the development, usability, and reusability of software.

With this in mind, a group of technical experts was commissioned in 1979 to develop the first draft of a comprehensive programmable controller standard. This standard would attempt to standardize PLC programming and consequently make a program developed on one system usable on other PLC platforms with minimum modification. The first draft from this committee was issued in 1982. This standardization is known as the IEC 1131 standard.

The IEC 1131 standard for programmable controllers comprises five parts:

Part 1: General Information

Part 2: Hardware Requirements

Part 3: Programming Languages

Part 4: User Guidelines

Part 5: Communication

Part three of the standard, IEC 1131-3 (Programming Languages) has attracted the most attention from the international community.

THE IEC 1131-3 PROGRAMMING STANDARD

Standard IEC 1131-3 defines a consistent set of programming languages for programmable controllers. The specification consists of four traditional languages and one higher-level programming language. The languages are broken down into two graphical languages, Ladder Diagram and Function Block Diagram, and two text-based languages, Instruction List and Structured Text, along with flowchart-type programming called Sequential Function Chart programming. Rockwell Automation's SLC 500 PLC is programmable only with ladder logic, whereas the PLC 5 programs in three languages: ladder, structured text, and sequential function chart. The newer Rockwell Automation ControlLogix PLC programs in ladder, structured text, sequential function chart, and function block. The five programming languages are defined as follows.

Ladder Diagram (LD)

Ladder diagram programming is similar to relay ladder logic. When most people in North America think of PLC programming, they associate ladder rungs and ladder programming with PLCs under the IEC 1131-3 standard. Individuals familiar with relay ladder diagrams or PLC ladder programming can continue to program with relay ladder logic.

Ladder logic comprises rungs, instructions, and branches. Figure 3-26 illustrates the main components from Rockwell Automation's RSLogix 5000 software. Logix 5000 software is used to program the ControlLogix PLC.

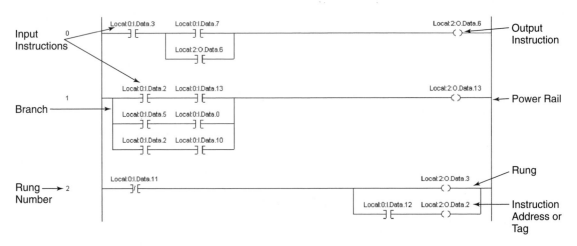

Figure 3-26   RSLogix 5000 ladder logic components.

The MEDOC Plus programming software has been developed to improve programming productivity by incorporating the IEC1131-3 standard. IEC1131-3 programming languages such as ladder, instruction list, function block diagram, and sequential function chart, in addition to user-defined function, are incorporated into this software package.

Advanced features supported by ladder diagram programming include function blocks, which simplify process-orientated applications. ASAP Inc. of Chagrin Falls, Ohio, is one of many software developers. ASAP has an open-control software solution called ASIC-100 software. ASIC Relay Ladder Logic programming software offers a comprehensive set of control elements and function blocks. These elements are organized into a palette for easy

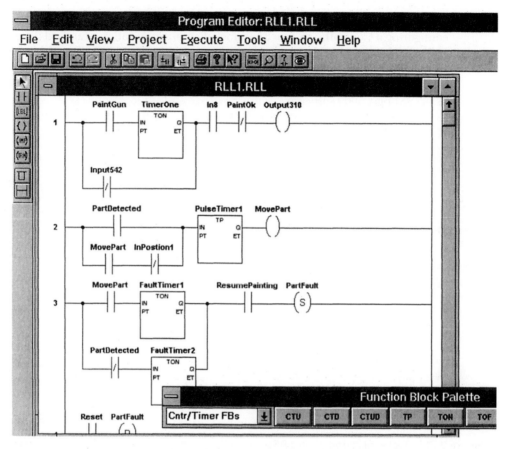

Figure 3-27 ASIC-100 Relay Ladder Logic uses an extensive palette of IEC-1131 standard function-blocks in a user-friendly editor. (Courtesy of ASAP Inc., Chagrin Falls, Ohio)

access. Power flow, I/O forcing, and on-line changes are provided during run time. See Figure 3-27 for an example of ASIC-100 relay ladder logic. Notice the function-block palette in the lower right-hand corner. This function-block palette illustrates counter and timer function blocks. As a programmer, you simply click on the desired instruction in the palette with your mouse to insert the instruction into your ladder logic.

Function-Block Diagram (FBD)

Function-block diagram programming is based on a graphical language widely used in Europe. New applications using a newer PLC like ControlLogix for process control and variable frequency drive systems will use function-block programming, as it contains new, more powerful instructions that are not available in ladder logic programming. Function-block programming does not contain ladder rungs (normally open or normally closed contact symbols that are used in ladder programming). All function-block instructions are blocks or boxes displayed on sheets, similar to sheets of paper. A function-block diagram application comprises a series of sheets. A sheet is equivalent to a piece of paper. When configuring a ControlLogix function-block diagram application, standard paper sizes such as $8\frac{1}{2} \times 11$, $8\frac{1}{2} \times 14$, 11×17, and standard metric paper sizes are available for selection. In many applications, many sheets make up the function-block application. There is no limit to the number of sheets a ControlLogix function-block application can have when using RSLogix 5000 software.

Figure 3-28 shows an RSLogix 5000 function-block sheet.

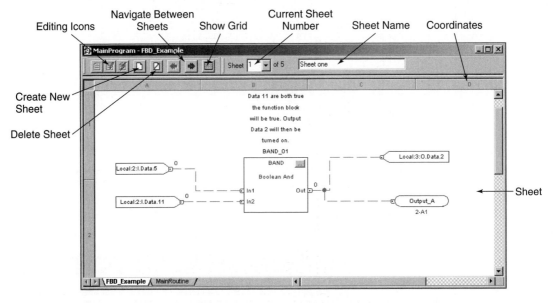

Figure 3-28   Rockwell Automation's RSLogix 5000 function-block diagram sheet components.

The function-block diagram is a graphical representation of executable code. A function-block diagram comprises a series of blocks representing instructions connected by lines called wires. Many function-block instructions are based on familiar relay ladder logic instructions; however, many function-block instructions have additional parameters that are selectable by the user. A function-block diagram is made up of the following elements:

Input reference	The input reference contains the address or tag where the input information is coming from. This can be an I/O address, other data table type address, or constant.
Function blocks	All instructions are represented as a function block or "box."
Output references	The output reference contains the address where the output information is output to. This can be an I/O address, or other data table type address.
Solid lines (wires)	Whole number data is represented as a solid wire.
Dotted lines (wires)	Bit data is represented by a dotted wire.

Output wire connector   An output wire connector is used to send data to another sheet on the function-block diagram without having to draw wires off the end of the sheet.

Input wire connector   The input wire connector is the target for the output wire connector typically on another sheet, but can also be used on the same sheet to eliminate the clutter of many wires. This is referred to as a hot link.

Instruction properties   Click on the icon in the upper right corner to view or set up instruction properties.

Figure 3-29 illustrates an example of the main elements of a simple function-block diagram for Rockwell Automation's RSLogix 5000 software.

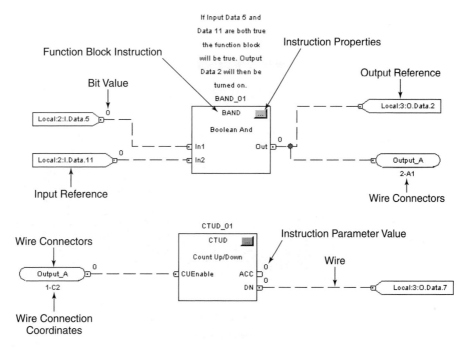

Figure 3-29   Function-block diagram main elements.

Figure 3-30 is a printout of a Rockwell Automation ControlLogix PLC function-block diagram. The execution order is to start at the left and flow to the right. The input reference data 12 represents input 12. This input bit information flows into the TONR (timer on-delay with reset) timer function block via the dotted line. The dotted line identifies this as bit or boolean (BOOL) data. When done, the timer instruction feeds into the CTUD (count up and/or count down counter) count up enable (CUEnable) pin from the timers done (DN) pin via the dotted line. As the counter increments, the output reference is updated through a solid line or wire. A solid wire represents a value.

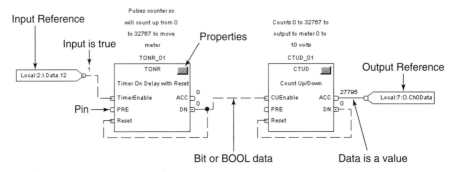

Figure 3-30   Basic RSLogix 5000 function-block logic.

Notice the box with the dots in the upper right hand corner of the TONR function-block. This instruction is a timer on delay with a built-in reset. By clicking on this box, the instruction properties can be monitored or modified. Refer to Figure 3-31 for the Properties view for the TONR function-block instruction.

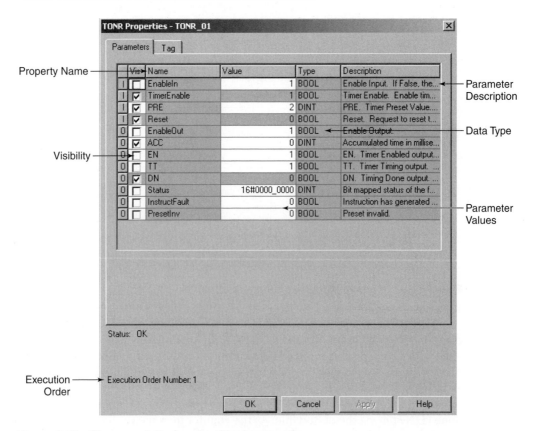

Figure 3-31   Timer on delay function-block properties.

The column entitled Name is the name of the parameter. The value column displays the current value for each parameter, while the Type column identifies the type of data represented by the parameter. The ControlLogix PLC uses the term BOOL (boolean) to identify bit data. A DINT is a double integer, 32 bits.

One nice feature of the function-block diagram is the option to customize which parameters are displayed on the function block. The Vis or visibility column will define which parameters and their associated pins are displayed on the function block. Selecting only the desired parameters with a check in the Visibility box reduces the number of connection points, called pins, on the function block, thus giving a cleaner function-block diagram. The description information describes the function of each parameter. The ControlLogix programming lab manual that accompanies this text contains additional information, along with introductory lab exercises for programming ControlLogix function-block diagrams.

Instruction List (IL)

Instruction list is a low-level, assembly-type language. Instructions are organized into a list-like format. Instruction list programming allows only one operation to be performed per line. As an example, storing a value in a memory location would be a single operation. Instruction list programming is usually used in smaller applications.

Structured Text (ST)

Structured text is an English-like programming language that resembles BASIC programming. Structured text programming can be used to perform most of the tasks currently done with ladder logic. Figure 3-32 illustrates a timer ladder rung. The structured text program equivalent follows.

IF (I:1/0) AND (!T4:0.DN) THEN

TON (T4:0, 1.0, 4, 0) ;

The first line includes our two input instructions. The exclamation mark before the T4:0.DN identifies the input instruction as an XIO bit instruction. Our timer instruction is on the second line. The TON instruction is identified before the parentheses, while the timer address, time base, and preset and accumulated values are listed inside the parentheses.

Figure 3-32   Ladder logic to be represented as a structured text program.

Sequential Function Chart (SFC)

Sequential function chart (SFC) programming is similar to flowchart programming. SFC programs consist of steps and transitions. Each step is represented by a box that contains one or more major actions. When all actions in the box are satisfied, the box is exited. A transition step must be true before moving on to the next step. Once leaving a particular step, the processor executes the next step. Previous steps are no longer executed. ASAP Inc.'s ASIC 100 is a sophisticated flowchart-like programming language that offers conditional branching, parallel branching, control loops, and jumps. ASIC 100's patented extensions to SFC add motion control capability, icon-based programming, new control functionality, and enhanced diagnostics. A sample of ASIC 100's Sequential Function Chart programming is illustrated in Figure 3-33.

Programming with a sequential function chart lets you develop segmented programming. Rather than developing one long ladder program, the program can be divided into

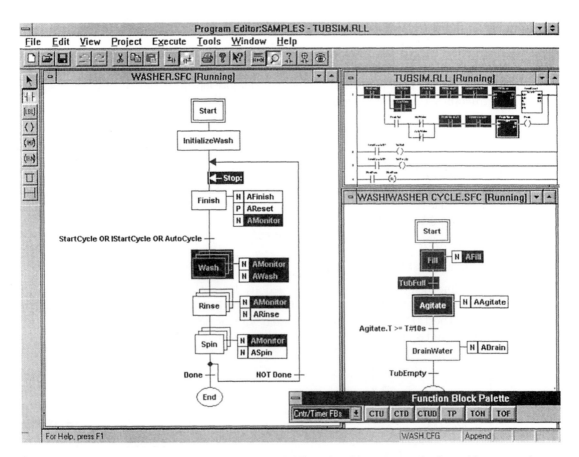

Figure 3-33 ASIC-100 provides IEC1131 Sequential Function Chart programming with patented extensions. (Courtesy of ASAP Inc., Chagrin Falls, Ohio)

several sequences or steps (see Figure 3-34). There is a transition between each step. The advantage to sequential function charts is that only the logic in the active step is scanned until it is time to transition to the next step. Compare this to scanning the entire ladder program when only a few rungs are active.

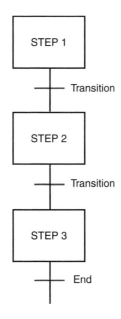

Figure 3-34 Three-step sequential function chart with transition points identified.

Each step corresponds to a control task, for example, turning on a mixer for a specified time. As a result, sequential function chart programs are good for batch sequencing. Each step comprises rungs of ladder logic. Figure 3-35 illustrates the logic contained in step one of Figure 3-34.

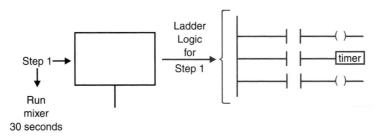

Figure 3-35 Viewing logic inside step one.

The logic in the current step runs continuously from rung one to the last rung. When the last rung of a step is scanned, the processor will look at the transition to see if the processor is to go on to the next step. The transition is the logic condition that directs the processor to progress to the next step. The transition can be as simple as one rung of logic. When the transition logic is true, the processor will scan step 2 repeatedly until its transition is true.

The PLC processor checks the current step's transition at the end of scanning each rung of the step. If the transition is true, the processor goes on to scan the logic in the next step. If the transition is false, the current step's logic is rescanned (see Figure 3-36).

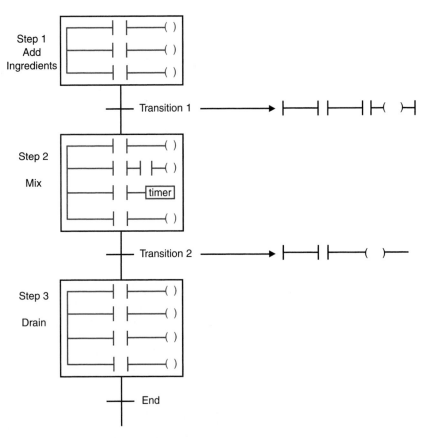

Figure 3-36   Entire sequential function chart program with logic rungs visible.

Analysis of Figure 3-36

Step 1, which comprises rungs one, two, and three, is continuously scanned until all ingredients are added. The transition rung will be true when all ingredients have been added. With transition 1 true, the processor leaves step 1 and begins scanning step 2. Step 2 is the mixing sequence. This step will be continuously scanned until the mixing time has elapsed. The timer will make the rung in transition 2 true when the mixing time expires. With transition 2 true, the processor will stop scanning step 2 and move on to step 3. Step 3 will execute until the transition is true, at which time the sequence will stop. The PLC processor will now look at step 1 for a start signal to begin the sequence over again.

Figure 3-37 illustrates a sequential function chart window from Mitsubishi's FX2N Series of Super Micro Programmable Controllers FX-WIN Windows-based programming package.

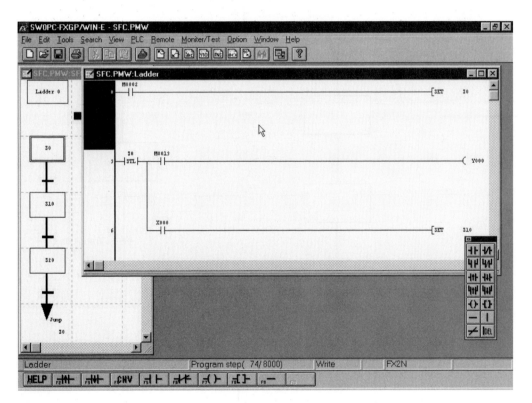

Figure 3-37   Sequential function chart window from Mitsubishi's FX-WIN Windows-based programming package. (Courtesy of Mitsubishi Electric Automation, Inc.)

Notice the boxes representing the steps and the transition points on the left window. The topmost window identifies an SFC PMW:Ladder, which shows the ladder rungs contained within one of the step boxes. The smaller window near the bottom right contains the ladder instructions that currently can be dragged and dropped onto the ladder rungs.

Selection Branch

A selection branch is equivalent to OR logic. When step 1 in Figure 3-38 is completed, the processor will look at the transition at the beginning of each of the branches. The transition that becomes true first will determine the direction of program flow. If both transitions A and B become true at the same time, priority will be given to the left-most one, transition A.

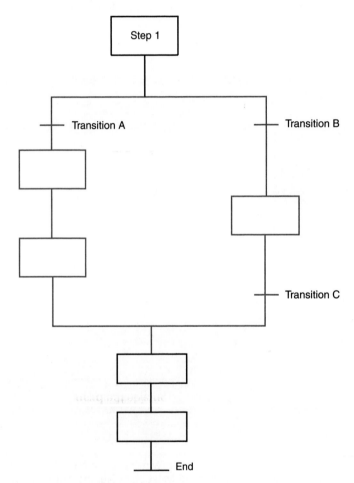

Figure 3-38   Sequential function chart OR logic.

Simultaneous Branch

A simultaneous branch executes parallel paths simultaneously. In Figure 3-39, when step A is scanned and transition B is true, the parallel path of steps C and F will be run at the same time as the path with step D.

The transitions B and G are both outside the **branch.** Each time the processor finishes executing logic in the simultaneous branch, the common transition, G, in Figure 3-39 will be tested to see whether it is true. If it is true and both branches have been executed at least once, the control will pass to step H.

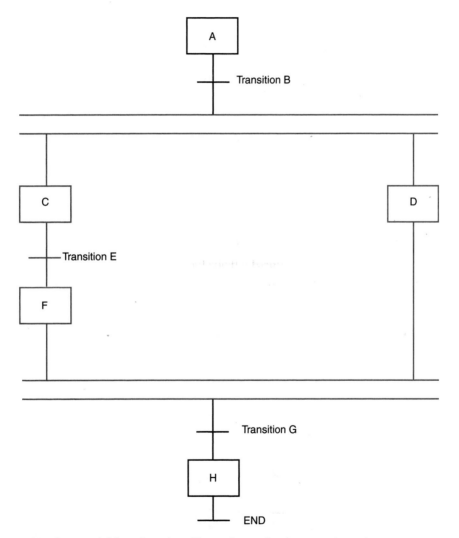

Figure 3-39 Sequential function chart illustrating a simultaneous branch.

There are many answers to the question of what hardware to use to program a PLC. We have looked at the traditional methods of PLC programming in this chapter. However, there is a relatively new solution to PLC programming and control. What if we combine the separate personal computer as a programming terminal and the PLC's CPU? In this way, we can eliminate the PLC's CPU and directly control our system from our personal computer. In the past, the standard personal computer was a fragile device with questions regarding reliability and high failure rates. The unpredictability of the personal computer made it a risky choice as a factory floor controller. As a result, the electricians and maintenance personnel who installed and maintained PLC control systems—and the engineers who designed them—had little confidence that a personal computer could do the job of an industrially hardened, easily maintainable, expandable, reliable PLC and its CPU.

Today's personal computer hardware has been standardized and made more reliable, and computational ability has been increased. With computer processing speed and memory size increased, along with the increased ability to connect to networks like Ethernet, why not consider using a personal computer as the control processor? The next section will explore the issues of replacing your PLC's CPU with a personal computer.

SOFT PLC OR OPEN-ARCHITECTURE CONTROL

Today the procedure for selecting a PLC and its associated control system hardware and software is much different than selecting a single manufacturer and its proprietary hardware and software offerings. The first selection is made between the traditional real-time PLC solution and the newer, personal computer (called a white box) solution. Using a personal computer in place of the traditional PLC and its CPU to control I/O is called a "soft PLC" or "soft logic PLC."

When selecting the traditional PLC for the control solution, ladder logic, or any combination of the five languages specified in the IEC 1131-3, standards is used. On the other hand, if the white box, desktop-type personal computer is selected, programming languages such as ladder logic, structured text, or sequential flow chart languages are used. The white box (personal computer) or industrial computer platform will typically run Windows NT with a Pentium-class processor. With the personal computer as the processor, there is no typical rack or base containing a power supply, the CPU, and the I/O modules. The personal computer, through internally mounted interface cards, such as the Allen-Bradley 1784-KTX or KTS, can communicate directly to I/O modules residing in a remote I/O network. A remote I/O network may consist of operator interface devices and variable-frequency drives along with a remote, I/O-configured PLC chassis and I/O modules. A remote I/O PLC chassis is, in some cases, the same chassis used for a local PLC. The chassis will need a power supply, the typical I/O modules, and, in some cases, a communication adapter module. The communication module will be the communication link between the remote chassis and the remote I/O (RIO) link. Figure 3-40 is an overview of a soft-PLC, white box system. The personal computer may have a component such as a 1784-KTX interface card installed to provide interface to the remote I/O network.

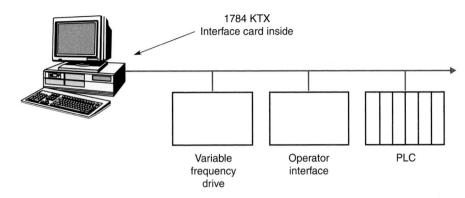

Figure 3-40   Overview of a soft logic remote I/O PLC system.

ONE STEP FURTHER: PRODUCTION DATA FROM PLC TO OFFICE DATABASES

In the typical manufacturing environment, much information is needed by production, planning, inventory tracking, scheduling, and management personnel. The production supervisor needs to know how many parts were produced on each conveyor line during each shift. How can this information be provided? As the production supervisor you could:

1. Have an employee physically count parts as they are produced and supply you with a report.
2. Have mechanical counters count parts and have someone record counter data, produce a report, and provide it to you.
3. Have a PLC count the parts and have someone monitor the PLC's data table, generate a report, and provide it to you.
4. Have a PLC count the parts and automatically send the data to a spreadsheet on your desktop computer.

In the typical manufacturing environment, the processor in your PLC contains a large amount of production data. Data stored in the processor could include production data for each shift of each workday. Production data could include the number of each part made, number of parts rejected, types of parts made, logging equipment uptime for maintenance scheduling, production reports by station or by operator, and inventory or material usage tracking. The PLC can be used to track production and store the resulting data in its internal registers or data tables.

In today's fast-paced world, the most efficient solution is to automate as much as possible so as to provide maximum production information in the fastest and most accurate manner and at the lowest cost. The key to exchanging the data between your PLC and a personal computer spreadsheet is dynamic data exchange (DDE) using object linking and embedding (OLE).

Dynamic Data Exchange

Dynamic data exchange (DDE) and object linking and embedding (OLE) are part of the Microsoft Windows operating system. The software package that links the PLC to the personal computer is a dynamic data exchange server. This allows Microsoft Windows DDE-compliant applications to exchange data with your PLC's processor (provided your PLC programming software is also DDE compliant). Exchanging data from the plant floor to a supervisory computer allows data logging, data display, trending, downloading of recipes, setting of selected parameters, and availability of general production data.

OLE comprises two things: object linking and object embedding. Using the Microsoft Office Suite, you can embed data from an integer file representing timer or counter production data for each work shift into an Excel spreadsheet. When data is embedded into an Excel worksheet, for example, a snapshot of the data contained in the original data table, perhaps an integer file, is copied into the spreadsheet. A snapshot of data would represent a copy of the data contained in the integer file at the moment the data was pasted into the spreadsheet. Being a snapshot, the data will not be updated as new information is loaded into the integer file.

When a data table is linked to a spreadsheet, at any time data is updated in either the spreadsheet or the PLC data table, the updated data is reflected in the other location as well. If the linked Microsoft Office application is not running as PLC data tables are updated, the MS Office application, such as an Excel spreadsheet, will have its "links updated" as the spreadsheet program is opened.

SUMMARY

Personal computers advanced in functionality with the advent of the Windows operating system in the late 1980s, which allows PCs to run more applications. As a result of the Windows operating system and advances in the personal computer, the early 1990s saw the mature personal computer introduced on the factory floor. As the personal computer moved onto the factory floor, automation software development for the PLC, operator interface, and drive control mushroomed. As operating systems advanced and Windows was introduced, the laptop computer became powerful enough for automation software development and troubleshooting and thus became a major player in automation design, installation, start-up, and troubleshooting.

Between IEC 1131-3 programming language standardization and the development of Windows NT, DDE, and OLE, the factory floor became better controlled and data was more easily shared with other personal computers.

Today there are many ways to program a PLC. The possibilities range from handheld terminals to personal computers. Not all of the more advanced PLC processors support programming with a handheld programming device.

Laptop personal computers with PCMCIA cards and desktop personal computers with internal interface cards like the 1784-KTX cards are popular programming choices today.

An integrated industrial computer with built-in CRT or a rack-mount industrial computer with a separate monitor, factory-hardened keyboard, and pointing device can be used for typical factory floor installation.

REVIEW QUESTIONS

1. What is the difference between a smart and a dumb programming terminal?
   A. A dumb terminal does not have communication capabilities.
   B. A dumb terminal is a smart terminal that is not working properly.
   C. A smart terminal has its own microprocessor.
   D. A smart terminal is a dumb terminal with communication capabilities.
   E. A smart terminal has its own built-in memory for program storage.
   F. C and D are correct.
   G. C and E are correct.

2. When downloading a program from an HHP programming terminal to a CPU:
   A. The program in the HHP is physically transferred to the PLC's CPU and is no longer in the HHP's memory.
   B. A copy of the program is written to the PLC's CPU. The program is still in the HHP's memory.
   C. A copy of the program is read from the PLC's CPU and stored in the HHP's memory.
   D. A copy of the program is read from the PLC's CPU and stored in the HHP's memory. The CPU program is no longer in the CPU's memory.

3. List three reasons why maintenance personnel would prefer an HHP when working on the factory floor or in the field.

4. Downloading a program:
   A. loads a copy of the program from the PLC to the programming terminal
   B. loads a copy of the program from the programming terminal to the PLC's CPU
   C. Both A and B are correct: the program can be downloaded from either

5. When you purchase a new PLC, take it out of the box, and plug it in, the PLC will:
   A. go into run mode and execute a default program loaded at the factory
   B. go into program mode and signal you to enter a program
   C. do absolutely nothing except wait for you to give it some direction
   D. some PLCs will fault, as there is no program in memory
   E. A or D is correct, depending on the PLC manufacturer
   F. C and D are correct

6. List four disadvantages maintenance or electrical personnel would consider before purchasing an HHP.

7. If you have a handheld programming terminal from one manufacturer:
   A. You can program only that manufacturer's PLCs.
   B. You can program any new-generation PLC from any manufacturer.
   C. You can program any PLC.
   D. You can program only the specific PLC for which the handheld terminal was designed.
   E. You may be able to program other manufacturer's PLCs; check your PLC specification sheet.
   F. You can program any PLC currently on the market with a special modification kit available from the HHP manufacturer.

8. Uploading a program:
   A. loads a copy of the program from the PLC to the programming terminal
   B. loads a copy of the program from the programming terminal to the PLC's CPU
   C. uploads a copy of the program from the programming terminal to the CPU
   D. Both A and B are correct: the program can be uploaded from either

9. PLC software packages provide troubleshooting assistance through:
   A. monitoring a PLC program while it is running
   B. built-in search functions
   C. automatic error checking and correction
   D. forcing I/O to test logic and outputs
   E. A, B, and C
   F. A, B, and D

10. Moving or deleting rungs of a PLC ladder program with a software package is as easy as _____ and _____.

11. Most software instructions are based on graphical symbols that are derived from the old, familiar _____.

12. Most maintenance or electrical personnel with a good understanding of standard _____ diagrams will have few problems developing or troubleshooting basic PLC ladder diagrams using software-developing ladder logic.

13. There are multiple ways to program a PLC using a personal computer. Which of the methods listed here are acceptable?
    A. a personal computer using the COM port and a straight-through serial cable
    B. a notebook personal computer with a PCMCIA card
    C. a personal computer using the COM port and a null-modem cable
    D. a personal computer and a conversion accessory made by the manufacturer
    E. A, B, and C
    F. B, C, and D

14. At the remote location, plug the HHT into the CPU and either _____ the current program for editing or _____ a new or modified program.

15. There are two types of HHPs, _____ terminals and smart terminals.

16. Only a _____ terminal will allow you to physically disconnect an HHP and download a program at a remote location.

17. Some _____ programming terminals, especially if they are older models, are no more than a plastic shell with a small keyboard and display. These devices are only used to transfer the data represented by key closures directly to the CPU.

18. Programs entered using a _____ are developed by pressing the correct sequence of keys, which the CPU accepts and stores directly into memory.

19. A dumb terminal cannot be used to program, edit, or monitor a program if it is not _____ _____ to the CPU.

20. Most newer handheld programming terminals have their own built-in or on-board microprocessor and memory. This type of programming terminal is called a _____ programming terminal.

21. A smart programming terminal can be used to develop programs without being physically _____ to a PLC.

22. When not attached to a PLC, a smart, handheld programming terminal must have a battery so its _____ _____ _____ _____ _____.

23. For remote programming, where the HHP is not attached to a PLC's CPU, the HHP is plugged into a _____ _____, which plugs into the 110 VAC standard wall plug.

24. The advantage of a smart HHP is memory retention. A smart programming HHP will remember the program that is stored in its memory, assuming the _____ is good or power is applied.

25. The process of transferring a copy of a user program from an HHP to a CPU is called _____.

IF PASSWORD IS CORRECT GO TO DRIVE SET UP SCREEN

0022
—EQU—
Equal
Source A   N12:21
                7777<
Source B   32767
            32767<

B3:10/0
[OSR]

B3:1/15
( )
RESET PASSWORD
ATTEMPTS COUNTER
C5:5
(RES)

0023
—GRT—
Greater Than (A>B)
Source A   N12:21
                7777<
Source B   0
            0<

—NEQ—
Not Equal
Source A   N12:21
                7777<
Source B   32767
            32767<

B3:10/1
[OSR]

C5:5
3 <
2 <
(CU)
(DN)

−1
−1<
N12:21
7777<

NOTIFICATION
B3:0/7
0024
] [

HANDSHAKE
B3:0/8
( )

Normal state of passwo...   ...try attempts counter is equal
to zero.

0025
—EQU—
Equal
Source A   C5:5

Source B

—MOV—
Move
Source   0
            0<
Dest   N12:6
            1<

The first time the incorr...   ...s to display "Password does not
match - REENTER."

0026
—EQU—
Equal
Source A   C5:5.ACC
                2<
Source B   1
            1<

B3:10/2
[OSR]

—MOV—
Move
Source   1
            1<
Dest   N12:6
            1<

The second time the incorrect password is entered send a 1 to the multi state indicator on the PanelView so as to display "Password does
not match - REENTER."

0027
—EQU—
Equal
Source A   C5:5.ACC
                2<
Source B   2
            2<

B3:10/3
[OSR]

—MOV—
Move
Source   1
            1<
Dest   N12:6
            1<

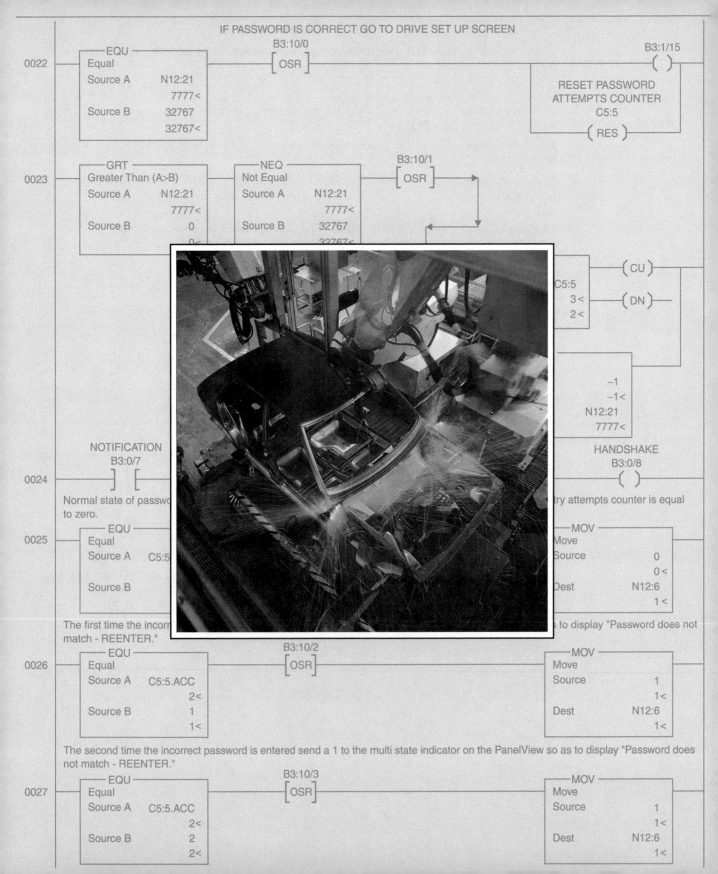

CHAPTER

4

Number Systems

OBJECTIVES

After completing this chapter, you should be able to:

- understand decimal, binary, octal, hexadecimal, and binary-coded decimal numbers
- convert from one number system to another
- understand how BCD thumbwheels interface to a PLC
- explain how a BCD digital output display interfaces to a PLC

INTRODUCTION

To effectively choose, install, maintain, and troubleshoot today's PLCs, you must possess basic PLC programming skills. To **program** the modern PLC, you must understand the different methods by which internal data is represented. Typical data representation inside a PLC data table can include binary, binary-coded decimal, decimal, octal, hexadecimal, or some combination of these.

Programmable controllers do not understand our everyday human languages such as spoken words or numbering conventions. Thus, programmable controllers cannot add, subtract, multiply, divide, or otherwise manipulate the decimal numbers to which we are accustomed. The PLC, like any microprocessor-based device, works in the world of zeros and ones. PLCs process ones and zeros using the **binary number system** and binary arithmetic.

Alphanumeric characters that are entered into a computer from a keyboard are converted into unique binary patterns representing each keyboard character. The computer manipulates this binary data as directed by the **user program** and converts it back to decimal data for output to a monitor or printer. The PLC works the same way when a human inputs decimal data through an operator interface device or keyboard to change the preset values. The PLC reconverts this internal binary data to decimal data before outputing the modified values to the operator interface device for human viewing.

The decimal, binary, octal, and hexadecimal systems are common number systems used with PLCs. Binary coding of decimal numbers (BCD) and binary coding of seven or eight bits to represent alphanumeric characters, as in the American Standard Code for Information Interchanges **(ASCII),** are common coding techniques. This chapter will investigate most of these different data representations.

Upon beginning to read this chapter, you are probably asking why you should study and understand different number systems. Some important reasons follow:

1. Internal PLC data tables represent data in binary format.
2. Some PLCs use decimal numbering to represent addresses.
3. Other PLCs use octal numbering to represent addresses.
4. Operator interface devices, such as thumbwheels and seven-segment output displays, have their own binary coding, which allows for an easy human-machine electrical interface.
5. Analog input data is converted to binary data as it is entered into the PLC.
6. Analog output data must be converted from binary data as the signal is prepared for output from the PLC to the analog output hardware device.
7. Many PLCs display error codes in hexadecimal.
8. Some PLCs have multiple instructions that require hexadecimal masks for filtering data.
9. Hexadecimal error codes may need to be converted to decimal as part of a fault routine.
10. When configuring, some PLC analog I/O cards require a 16-bit word to be constructed from a table of operating options. This 16-bit word will be converted to decimal before being programmed on the ladder rung that sends the configuration information to the analog module.

NUMBER SYSTEM CHARACTERISTICS

There are many ways to use numbers to count. In our everyday lives we depend on the decimal number system to count everything from quantity, weight, and speed to money. Our method of counting is based on ten, which is why it is referred to as base ten. The base, or radix, of a number simply identifies how many unique symbols are used in that particular number system. Decimal has ten unique symbols used for counting, starting with zero and ending with nine. Notice that the largest-value symbol is one value less than the base. The number systems typically encountered when working with PLCs include binary (base two), decimal (base ten), octal (base eight), and hexadecimal (base 16). Figure 4-1 lists each number system, its base, how many digits (or valid symbols) are available, and the range of the available symbols.

Ask yourself, where are these different number systems used and how will they affect me?

As a PLC programmer, or electrical or maintenance individual working with PLCs, you will someday find yourself attempting to understand information stored inside the PLC in

NUMBER SYSTEM CHARACTERISTICS			
System	Base	Digits	Range
Binary	2	2	0 and 1
Octal	8	8	0 through 7
Decimal	10	10	0 through 9
Hexadecimal	16	16	0 through 9 and A through F

Figure 4-1 Number systems typically used with PLCs.

different number systems, program or configure PLC hardware, understand or program instructions that use masks, or set up or troubleshoot a network. Below are a few other reasons specific number system understanding is important:

1. Computers and PLCs only understand binary information. In many cases humans have to understand and convert decimal information to and from binary to understand what is happening inside the PLC.
2. If using an octal-based PLC such as a PLC 5, I/O addresses are octal-formatted.
3. When using a PLC 5 or SLC 5/04 processor and Data Highway Plus as your communication network, station addresses are octal.
4. Even though decimal numbers are familiar to humans, PLCs for the most part only understand binary. In many cases information must be converted back and forth.
5. Hexadecimal information is found in processor or I/O module error codes, instruction masking, and DeviceNet electronic data sheet file identification, to mention a few.
6. You might use a ControlLogix PLC to communicate with an existing Data Highway Plus or Remote I/O network, both of which are addressed in octal.

In this chapter we will explore the world of numbers as used by different PLCs. We will introduce number systems typically used by PLCs, examine methods of conversion from one to the other, and look at examples of how selected number systems are used in PLC applications. Let us start by looking closely at the decimal number system, with which we all are familiar.

THE DECIMAL NUMBER SYSTEM

In our everyday life we are accustomed to counting in the decimal system; that is, 0, 1, 2, 3, 4, 5, 6, 7, 8, 9, 10, 11, and so forth. We use ten symbols, zero through nine. After the number nine, we use various combinations of these symbols to express numerical information. Our decimal system is configured in what is called base ten. The characteristic that distinguishes different number systems is the base, or radix, which simply identifies the number of symbols (numbers) used to represent a given quantity. The base, or radix, also tells us what weight each digit position represents in relation to the other digits.

Let us look at an example. In base ten, the number 123 in reality represents three values:

Here we have a 1 in the 100s place. This equals:	100
We have a 2 in the 10s place. This equals:	20
We have a 3 in the 1s place. This equals:	3
Adding these values, we get the following:	123

We call this number "one hundred twenty-three."

To signify that the number 123 is in base ten, it can also be written as 123_{10}. The subscript, 10, indicates that the numerical value is represented in base ten. In our daily life it is accepted that the numbers we encounter are decimal, so the base, or radix, is usually not included.

DECIMAL PLACE VALUES

Each physical number position has a weighted value. Each successive decimal number position, from right to left, is ten times greater than the previous position, as we will illustrate. We calculate the powers of ten to arrive at each decimal place value:

Powers of ten:	$\rightarrow$	10^4	10^3	10^2	10^1	10^0
Place value:	$\rightarrow$	10,000	1,000	100	10	1

The first placeholder, or least significant digit, is 10^0, which equals 1. This is our 1s place. When counting, we count from 0 through 9. When we reach 9, we run out of 1s. The next number after 9 is indicated by placing a 1 in the second number position, the 10s place, while our 9 advances one count, to 0. The second place value equals 10^1 or $10 \times 1 = 10$. This is the 10s place. This leaves us with a number that looks like 10. We call this "ten" because we have one 10 and no 1s.

As we count, we have one 10 and one 1, which we call "eleven"; one 10 and two 1s which we call "twelve"; and so on, until we reach one 10 and nine 1s. We have run out of numbers in the 1s place, and so, once again, we carry our count into the 10s place. The next number, then, is two 10s and zero 1s. We call this number "twenty." Eventually, when we reach nine 10s and nine 1s (99), we have to go into the third (100s) place to continue counting.

The third place value, the 100s place, is written 10^2, or $10 \times 10 = 100$. The number in this place tells us how many 100s we have. The third place is 10^3 or $10 \times 10 \times 10 = 1,000$. This is the 1,000s place. The fourth place is 10^4, or $10 \times 10 \times 10 \times 10 = 10,000$. This is the 10,000s place. Thus, the number 12,345 should be easily identified, as shown in Figure 4-2.

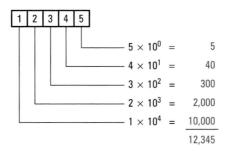

Figure 4-2   Derivation of the decimal number 12,345.

BINARY AND BINARY-CODED DECIMAL NUMBERS

In the next few pages we are going to introduce the world of computer numbers: the binary and binary-coded decimal number systems.

Binary Numbers

Binary numbering uses only two digits, 0 and 1. Although humans understand and use the decimal number system, microprocessor-controlled devices such as digital computers and programmable controllers understand binary signals instead. Computer-based devices are designed to understand and operate with binary information, as it is easier to design a machine to distinguish between two conditions rather than ten, which would be needed if the decimal system were used. Also, most basic industrial control devices operate on two conditions, such as open or closed, on or off. As a result, to design a machine to effectively control two-state devices using a two-state signal such as binary was an easy decision.

Computers do accept decimal numbers as inputs and also produce decimal outputs to satisfy the human operator; however, values are converted by the input circuitry to a binary format acceptable to the central processing unit (CPU). The output circuitry then converts binary signals from the CPU to decimal values recognized by humans.

Digital computers and other microprocessor-controlled equipment, such as programmable logic controllers, only understand and operate on binary values (1s and 0s). These binary digits are called bits. The term bit is derived from BInary digiT. A single bit will be a single one or zero.

Bit Position and Weighting

Computers work with groups of binary digits (bits) organized into words to either store or manipulate information. For many years the most modern PLCs were 16-bit machines. The SLC 500, MicroLogix, and PLC 5 are examples of a 16-bit PLC. A 16-bit PLC works with information in a group of 16 bits called a word. Newer PLCs like Rockwell Automation's

ControlLogix are 32-bit machines; they handle information in a group of 32 bits, known as a double word.

The physical position of the bits in a 16-bit word is important. Each physical position has a weighted value. Since computers only understand binary, or base two, each successive position from right to left is two times greater than the previous position. Figure 4-3 illustrates the bit positions for a 16-bit word.

2^{15}	2^{14}	2^{13}	2^{12}	2^{11}	2^{10}	2^9	2^8	2^7	2^6	2^5	2^4	2^3	2^2	2^1	2^0

Figure 4-3   A 16-bit binary word bit weighting.

The first position, the LSB, is the 1s place, or 2^0. The next place is the 2s place, or 2^1 ($2 \times 1 = 2$). The third place is the 4s place, or 2^2 ($2 \times 2 = 4$). The fourth place is the 8s place, or 2^3 ($2 \times 2 \times 2 = 8$), and so on. Figure 4-4 illustrates the decimal-weighting value of each bit position.

Value	Decimal	Value	Decimal
2^0	1	2^8	256
2^1	2	2^9	512
2^2	4	2^{10}	1,024
2^3	8	2^{11}	2,048
2^4	16	2^{12}	4,096
2^5	32	2^{13}	8,192
2^6	64	2^{14}	16,384
2^7	128	2^{15}	32,768

Figure 4-4   Binary place values converted to decimal for each binary place holder.

The rightmost bit, which represents the lowest value, is called the least significant bit (LSB). The least significant bit is the 1s position (refer to Figure 4-4). The most significant bit (MSB) position is the left-most bit, or bit 15. In a 16-bit word, the most significant bit holds the value of 32,768.

Remember how decimal numbers such as 12,345 from Figure 4-4 went together? It should not be too difficult to convert binary to decimal. As an example, let's look at the binary value 0000000010101010. If we determine each bit's value similar to the way we did for the decimal number in Figure 4-2, it should be easy to determine the decimal value of this binary value. See Figure 4-5.

The value of the binary word 0000 0000 0000 0000 1010 1010 is equal to 170 in decimal. If there were a 1 in all 16 positions, and you added all the values, the decimal value

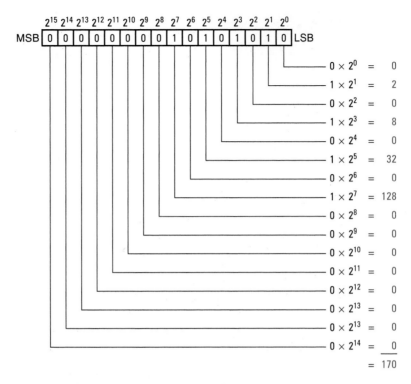

Figure 4-5 Decimal place value equivalent.

for that binary value would be 65,535. The decimal values that could be represented in a 16-bit word range from 0 to +65,535.

Figure 4-6 gives a sample of selected decimal numbers (down the left column) and shows how they correlate to their respective binary bit patterns.

Considering that 16 bits can represent a decimal value from 0 to +65,535, how do we represent negative numbers in a PLC? This particular method of data representation does not allow for negative numbers. When all 16 bits are used for the value in this manner, all values are positive: this is a 16-bit unsigned integer. Using unsigned integers for a process such as flash-freezing food products as they were manufactured, there would be no easy way for the PLC to monitor or control freezer temperature, which ranges between −20 and −40 degrees Fahrenheit. We need a way to incorporate negative values into the PLC. A value must be identified as either positive or negative. When working with decimal numbers, we place either a plus or minus in front of a decimal value to identify a positive or negative value. Unfortunately, the PLC does not have the ability to use a plus or minus sign, as it only understands 1s and 0s. What if we take the left-most bit position, and rather than it being a place value in the number represented, we use this bit to represent the sign of the number? Let's use a 0 in bit position 15 to represent a positive number and a 1 to represent a negative number. Data represented in this fashion is called a 16-bit

Powers of two	2^4	2^3	2^2	2^1	2^0
Place value	16	8	4	2	1
A: DECIMAL 1 =					1
B: DECIMAL 2 =				1	0
C: DECIMAL 3 =				1	1
D: DECIMAL 4 =			1	0	0
E: DECIMAL 5 =			1	0	1
F: DECIMAL 6 =			1	1	0
G: DECIMAL 7 =			1	1	1
H: DECIMAL 8 =		1	0	0	0
I: DECIMAL 21 =	1	0	1	0	1

Figure 4-6   Comparison of decimal to binary numbers.

signed integer. Figure 4-7 is similar to Figure 4-5, except now we have a way to identify if the value is a positive or negative value.

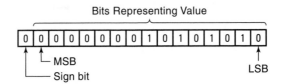

Bits Representing Value

MSB

Sign bit

LSB

Figure 4-7   A 16-bit signed integer where bit 15 represents the sign of the binary value.

Now the lower fifteen bits (0–15) represent the value. The range of the positive data values will be less than before, as we have one less bit. Figure 4-8 illustrates a 16-bit signed integer along with the place values for each bit.

Notice that the maximum value that can be stored in a 16-bit word is 32,767. The range of data table information stored in a 16-bit signed integer is −32,768 up to +32,767. It is important that you be familiar with this concept, as this is the basic format information is stored in an SLC 500, MicroLogix, PLC 5, and other popular 16-bit PLCs. Binary files, timers and counters, and integer file information are represented as 16-bit signed integer data.

The ControlLogix, being a 32-bit PLC, uses data in the same general format, except that groups of 32 bits are used to store or manipulate information. Since 32 bits is 2 times 16, 32-bit information is known as a double word, or double integer. Figure 4-9 shows a 32-bit signed double integer for a ControlLogix PLC. The data range for a 32-bit signed integer is −2,147,483,640 to +2,147,483,647.

Information is still represented as a signed integer, now a 32-bit signed integer. The MicroLogix 1200 and 1500 have the capabilities in some situations to use 32-bit data. For these two PLCs, 32-bit information is called a long word.

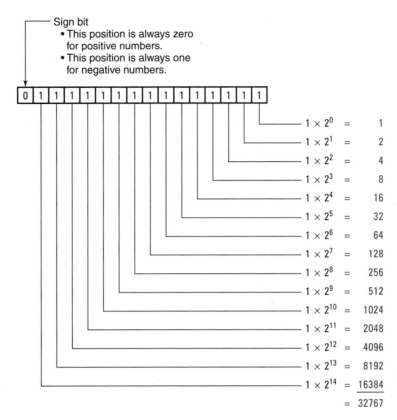

Figure 4-8   A 16-bit signed integer format and bit values.

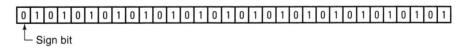

Figure 4-9   A 32-bit signed integer with a data range of $-2{,}147{,}483{,}648$ to $+2{,}147{,}483{,}647$.

Converting 16-bit words, either signed or unsigned integers, is the same as long as you are converting positive numbers. When using 16-bit signed integers, negative numbers are represented as 2's complement binary.

Binary Coding

For a PLC to control any process or application, it must be able to receive signals in the form of inputs from outside devices. After the CPU processes the incoming data, the resulting information is sent to field hardware devices in the form of outputs. As humans, we prefer to work with decimal numbers, while the PLC only understands binary information. Although the CPU has no difficulty working with the binary data, problems occur when the machine has to interface with humans. Binary-to-decimal conversion is difficult for the CPU to accomplish, especially when working with large decimal numbers. To solve this interface problem, unique combinations, or groups, of 1s and 0s have been developed to represent decimal numerical data. The process of assigning these unique codes results in binary-coded decimal (BCD) values.

Thumbwheels are used to allow the human operator to input numerical data into the programmable controller. A typical form of thumbwheel interface to a programmable controller is the BCD thumbwheel. Figure 4-10 illustrates a four-digit thumbwheel that will input the number 1,234 into the PLC as a binary-coded decimal code. Notice that each thumbwheel digit inputs four pieces of information. The input information is a binary representation of the decimal digit dialed on the thumbwheel. Notice each digit inputs the 1s, 2s, 4s, and 8s places.

There are many different types of thumbwheels available in many numbering formats; decimal, hexadecimal, and octal are a few, in addition to **BCD.** BCD thumbwheels

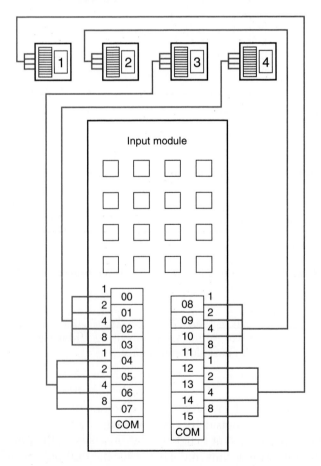

Figure 4-10   Four-digit BCD thumbwheel.

are selected to interface to PLCs, many of which include an input conversion instruction in their instruction set that will convert BCD thumbwheel input data into decimal data. Moreover, there also usually exists a conversion instruction to convert decimal data table values to BCD outputs.

Binary-Coded Decimal (BCD) Numbers

Binary-coded decimal numbers were designed to help assist humans in their interface with the computer. BCD is a four-bit code assigned to the decimal numbers 1 through 9. In Figure 4-11, the decimal numbers 0 through 9 and their corresponding BCD values are listed.

BCD NUMERICAL REPRESENTATIONS		
Decimal	Binary	BCD
0	0	0000
1	1	0001
2	10	0010
3	11	0011
4	100	0100
5	101	0101
6	110	0110
7	111	0111
8	1000	1000
9	1001	1001

Figure 4-11   Comparison of BCD to decimal and binary numbers.

Each single-digit thumbwheel has an internal wheel with the numbers 0 through 9. The wheel is moved to the correct numerical position with the thumb. Internal to the thumbwheel switch is a series of four switches, one each for the 8s, 4s, 2s, and 1s binary positions. Each switch is connected as an input to a PLC **input module,** as illustrated in Figure 4-12.

The four mechanical switches automatically close as the decimal number is selected. These internal switches provide the proper electrical continuity for the 1s, 2s, 4s, and 8s binary bit patterns for the selected decimal number. The appropriate binary bit pattern is then electrically input into the PLC input module.

Figure 4-12 illustrates a single-digit thumbwheel dialed to the decimal number 4. Notice the small circuit board on the back of the thumbwheel. This circuit board has five connections, one for each bit's weight plus a common. Each of the four connections—the

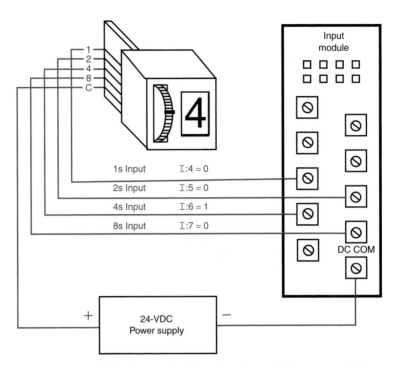

Figure 4-12   A single-digit BCD thumbwheel interfaced to a PLC input module.

1s, 2s, 4s, and 8s circuits—are individually wired into input 4 (I:4), input 5 (I:5), input 6 (I:6), and input 7 (I:7).

In the figure, current from a 24-VDC power supply will flow into the common connection on the back of the thumbwheel. Each circuit carries its corresponding bit's ON or OFF signal directly into the input module. For the decimal digit 4, as dialed on the thumbwheel in Figure 4-12, the input bit pattern is 0100 (the only value is 1 in the 4s place). The input bit pattern will be seen by the input module as zero volts on inputs I:4, I:5, and I:7. However, input I:6 will input a 24-VDC signal from the 24-VDC power supply through the closed thumbwheel switch representing the 4s value and into the input module. The PLC will place this bit pattern into the input status file.

Figure 4-13 illustrates the bit patterns for the BCD numbers 1 through 9.

After selecting the 9s digit, the next selection is 0. To dial in the decimal number 12, for example, the operator sets the 1s place thumbwheel digit to the 2 position and the 10s place thumbwheel digit to the 1 position. The 10s digit is set for one 10, while the 1s digit is set for two 1s.

Figure 4-14 illustrates two thumbwheels interfaced to a generic eight-point 24-VDC input module. The thumbwheels are dialed to decimal 57. This two-digit thumbwheel will consume all eight points on the input module.

Any digit greater than 9 cannot be dialed into a single-digit BCD thumbwheel. As a result, any digital code representing a numerical value greater than 9 would be read as

	Binary-Coded Decimal Number Representations			
Decimal	8	4	2	1
1	0	0	0	1
2	0	0	1	0
3	0	0	1	1
4	0	1	0	0
5	0	1	0	1
6	0	1	1	0
7	0	1	1	1
8	1	0	0	0
9	1	0	0	1

Figure 4-13   Binary-coded decimal number bit patterns compared to their decimal equivalents.

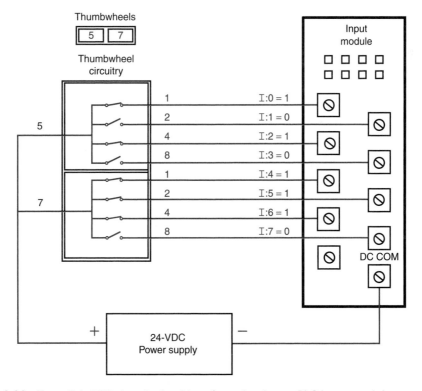

Figure 4-14   Two-digit BCD thumbwheel interface circuitry to PLC input module.

invalid. Even though single-digit BCD codes above 9 are invalid, however, there are valid binary numerical values for the quantities 10 through 15. Figure 4-15 illustrates BCD invalid codes and valid binary values for the numerical representations 10 through 15. If you attempt to convert an invalid BCD code to an integer with an SLC, PLC 5, or Control-Logix PLC the processor will fault.

INVALID BCD CODES		
Decimal	Binary	BCD
10	1010	1010
11	1011	1011
12	1100	1100
13	1101	1101
14	1110	1110
15	1111	1111

Figure 4-15   BCD invalid codes compared to valid decimal and binary numbers.

BCD Number Representation of Numbers Between 10 and 99

Just as the number 10 must be represented by two decimal digits, the proper BCD representation of the quantities 10 through 99 also requires two sets of four-bit BCD codes. The correct BCD representation of the quantity 10 is:

$$1 \qquad\qquad 0$$
$$0001 \qquad\qquad 0000$$

The correct BCD code for the quantity 11 is:

$$1 \qquad\qquad 1$$
$$0001 \qquad\qquad 0001$$

Figure 4-16 illustrates the number 11 dialed on a four-digit thumbwheel.

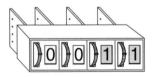

Figure 4-16   A thumbwheel dialed to decimal 11.

Decimal and BCD numerical representation for quantities of 10 and greater are listed in Figure 4-17.

BINARY-CODED DECIMAL NUMBERS							
Decimal Number	BCD 10s Digit Representation				BCD 1s Digit Representation		
	8	4	2	1	8	4	2	1
9	0	0	0	0	1	0	0	1
10	0	0	0	1	0	0	0	0
11	0	0	0	1	0	0	0	1
13	0	0	0	1	0	0	1	1
20	0	0	1	0	0	0	0	0
25	0	0	1	0	0	1	0	1
33	0	0	1	1	0	0	1	1
52	0	1	0	1	0	0	1	0
68	0	1	1	0	1	0	0	0
76	0	1	1	1	0	1	1	0
80	1	0	0	0	0	0	0	0
99	1	0	0	1	1	0	0	1

Figure 4-17   Two-decimal digit BCD representation.

BCD Representation of the Decimal Numbers 100 Through 999

Numerical values from 100 to 999 require three sets of BCD thumbwheel digits to be dialed up to the desired thumbwheel code for proper representation. The following shows a three-digit decimal representation converted to BCD:

$$100_{10} = \quad\quad 1 \quad\quad\quad 0 \quad\quad\quad 0$$
$$0001 \quad\quad 0000 \quad\quad 0000$$

$$743_{10} = \quad\quad 7 \quad\quad\quad 4 \quad\quad\quad 3$$
$$0111 \quad\quad 0100 \quad\quad 0011$$

Figure 4-18 illustrates the decimal number 743 dialed on a four-digit thumbwheel.

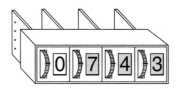

Figure 4-18   A thumbwheel dialed to decimal 743.

BCD Representation of Four-Digit Decimal Values

Numerical values from 1,000 to 9,999 require four BCD thumbwheel digit sets to be dialed up to the desired thumbwheel BCD code for proper representation. Figure 4-19 is an example of four-digit decimal representation in BCD:

$$9{,}652_{10} = \quad \begin{array}{cccc} 9 & 6 & 5 & 2 \\ 1001 & 0110 & 0101 & 0010 \end{array}$$

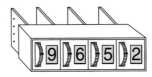

Figure 4-19   A thumbwheel dialed to decimal 9652.

Applications of BCD

Typical input applications for PLCs include the entry of decimal data such as count, volume, and weight. These values have traditionally been accomplished by using BCD thumbwheel switches.

A traditional output device used to communicate to the outside world is the seven-segment display. Here BCD data output from the CPU is converted to signals that will select and drive the proper segments to provide humans with a decimal readout from a machine or process. Figure 4-20 illustrates a sixteen-point output module connected to a four-digit seven-segment LED display. The BCD bit pattern to illuminate the digits is shown below the LED digits.

Although thumbwheels and seven-segment displays are still used today, many applications have moved toward more modern operator interface devices. Today's operator interface devices are smart devices; that is, they typically have a microprocessor built into each unit. Operator interface devices are linked back to the PLC through a communication link. Also known as human-machine interface (HMI) devices, they range from hand-held keypad units with up to four lines of display text to 5-, 9-, 12-, or 14-inch-diagonal color or amber graphic display screens with keypads or touch-screen displays.

Now that we have a good understanding of the binary and BCD systems, we will investigate the octal number system.

THE OCTAL NUMBER SYSTEM

Computers cannot work directly with decimal numbers and binary numerical representations, due to the long, random 1/0 sequences, which become difficult for humans to manage. To help humans work with computers, shorthand methods representing these binary values have been developed, one of which is the **octal number system** (base eight). Being base eight, valid octal numbers are zero through seven. Remember, the span of valid numbers for any specific base is always one less than the base.

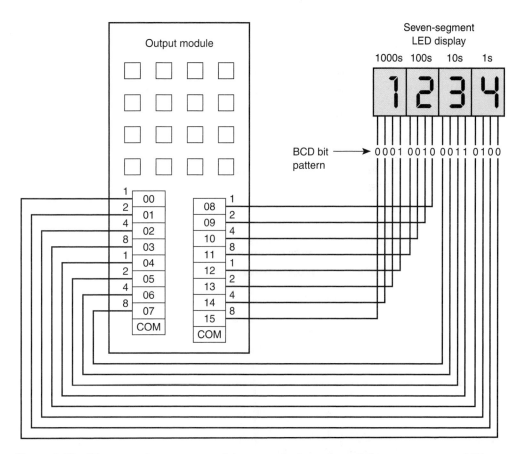

Figure 4-20 Sixteen-point output module connected to a four-digit seven-segment LED display.

The octal number system is related to the binary system in that base two numbers and base eight numbers are both based on powers of two. As a result, binary bits can also be used to represent octal numbers. Three binary bits can be used to represent the values 0 through 7, as illustrated in Figure 4-21.

Decimal	Binary	Decimal	Binary
0	000	4	100
1	001	5	101
2	010	6	110
3	011	7	111

Figure 4-21 Decimal numbers 0 through 7 represented with three binary bits.

In octal, the place values are $8^0 = 1$, $8^1 = 8$, $8^2 = 64$, $8^3 = 512$, $8^4 = 4,096$, etc. Figure 4-22 illustrates the respective weights of each octal digit position.

OCTAL PLACE VALUES				
Powers of eight	8^3	8^2	8^1	8^0
Place value	512	64	8	1

Figure 4-22   Octal number system digit place values.

Decimal and octal numbers are equal for the values 0 through 7. When we reach 7 in octal, we run out of digits in the 1s place, much like we encountered when we reached 9 in the decimal number system. In both cases we need to advance our count into the next place value. In octal, we advance from 7 to 10. A 10 in octal equals one 8 and zero 1s.

VALID OCTAL NUMBERS

The following is a list of valid octal numbers:

0–7	100–107	200–207
10–17	110–117	210–217
20–27	120–127	220–227
30–37	130–137	230–237
40–47	140–147	240–247
50–57	150–157	250–257
60–67	160–167	260–267
70–77	170–177	270–277

Notice that there are no 8s or 9s anywhere in the octal number system. Figure 4-23 lists octal numbers through 30_8, along with their decimal equivalents.

OCTAL-TO-DECIMAL CONVERSION

Let us convert the octal numerical value 157 to decimal (see Figure 4-24).

The octal number 157 is broken down into: seven 1s; 5 in the 8s place, which is equal to 5 times 8, or 40; 1 in the 64s place, which is equal to 64. Adding the values from Figure 4-24, the decimal equivalent is 111.

BINARY-TO-OCTAL CONVERSION

Octal numbers can be used as a convenient replacement for binary numbers. This is possible because binary numerical representation using three bits can represent a total of eight numbers, as shown here:

000 = 0	100 = 4
001 = 1	101 = 5
010 = 2	110 = 6
011 = 3	111 = 7

DECIMAL AND OCTAL COMPARISON TABLE		
Decimal Number	Octal Number	Octal Explanation and Decimal Equivalent
1	1	
2	2	
3	3	
4	4	
5	5	
6	6	
7	7	
8	10	One eight and no ones = 8 in decimal.
9	11	One eight and one ones = 9 in decimal.
10	12	One eight and two ones = 10 in decimal.
11	13	One eight and three ones = 11 in decimal.
12	14	One eight and four ones = 12 in decimal.
13	15	One eight and five ones = 13 in decimal.
14	16	One eight and six ones = 14 in decimal.
15	17	One eight and seven ones = 15 in decimal.
16	20	Two eights and no ones = 16 in decimal.
17	21	Two eights and one one = 17 in decimal.
18	22	Two eights and two ones = 18 in decimal.
19	23	Two eights and three ones = 19 in decimal.
20	24	Two eights and four ones = 20 in decimal.
21	25	Two eights and five ones = 21 in decimal.
22	26	Two eights and six ones = 22 in decimal.
23	27	Two eights and seven ones = 23 in decimal.
24	30	Three eights and no ones = 24 in decimal.

Figure 4-23   Decimal and octal comparison and explanation.

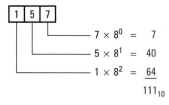

Figure 4-24   An octal number converted to decimal.

Accomplishing binary-to-octal conversion is really quite simple. As an example, let's convert the binary number 101010101 to octal. The easiest way is to start with the LSB and break the binary number into groups of three bits, as shown in Figure 4-25.

101010101		
101	010	101

Figure 4-25   A binary number broken into groups of three bits, starting with the LSB.

Now replace the three-digit binary numerical representations with their octal equivalents, as in Figure 4-26.

Binary	101	010	101
Octal equivalent	5	2	5
Result	$101010101_2 = 525_8$		

Figure 4-26   Replacing the binary numerical representations with their octal equivalents.

In octal, 101 binary becomes 5, 010 becomes 2, and 101 becomes 5. The conversion is complete: 101010101_2 equals 525_8.

OCTAL-TO-BINARY CONVERSION

To convert an octal number to binary, simply reverse the binary-to-octal conversion process you just learned. For example, let us convert 271_8 to binary. Just as you converted binary numbers to octal numbers, replace the octal numerical representations with their binary equivalents (see Figure 4-27).

Octal number	2	7	1
Binary equivalent	010	111	001
Result	$271_8 = 010111001_2$		

Figure 4-27   Replacing the octal numerical representations with their binary equivalents.

The octal number 2 becomes 010 in binary, 7 becomes 111, and 1 becomes 001. The conversion is complete: 271 octal equals 010111001 binary.

THE HEXADECIMAL NUMBER SYSTEM

Each decimal digit in the BCD number system is represented by a group of four binary bits. Even though BCD only uses ten 4-bit combinations to represent the decimal numbers 0 through 9, four bits can provide up to 16 different bit combinations. When using BCD numbering, the last six 4-bit combinations, 1010 through 1111, are invalid codes.

The **hexadecimal number system** was developed to use the remaining six binary bit patterns. In reality, the hexadecimal number system is an extension of BCD. Unlike in BCD, where the last six codes (1010 through 1111) are invalid, the hexadecimal system uses these bit combinations. Hexadecimal uses the first ten symbols from the decimal system (0 through 9), and the first six characters from the English alphabet (A through F), which were selected as the single characters to represent the values 10 through 15. These characters were selected because they are common characters on all keyboards and printers; this was the advantage of using letters rather than designing new characters.

Hexadecimal, or hex as it is sometimes called, is based on 16 digits. The hexadecimal and binary number systems are also related as they are both based on a power of two ($2 \times 2 \times 2 \times 2 = 16$). Like other number systems, hexadecimal is based on the progression of powers of its base, which is 16. Figure 4-28 compares decimal, BCD, and

DECIMAL, HEXADECIMAL, AND BCD COMPARISON TABLE		
Decimal value	Hexadecimal value	BCD value
1	1	0001
2	2	0010
3	3	0011
4	4	0100
5	5	0101
6	6	0110
7	7	0111
8	8	1000
9	9	1001
10	A	0001 0000
11	B	0001 0001
12	C	0001 0010
13	D	0001 0011
14	E	0001 0100
15	F	0001 0101

Figure 4-28   Decimal, hexadecimal, and BCD comparisons.

hexadecimal numbers. Notice that BCD and decimal values greater than nine comprise two digits, whereas hexadecimal values are one digit.

Hexadecimal place value weights are illustrated in Figure 4-29.

HEXADECIMAL PLACE VALUES				
Powers of 16	16^3	16^2	16^1	16^0
Place value	4,096	256	16	1

Figure 4-29   Place values of hexadecimal as a result of progression of powers of base 16.

Hexadecimal place values are base 16. The place value weights are determined as a result of the progression of powers of base 16: 16^0, 16^1, 16^2, 16^3, 16^4, and so on. This was illustrated in Figure 4-28. The highest single-digit value allowed in any position is 15; then, the numbering rolls over to 0 and starts again.

Example 1:

Convert 123_{16} into its decimal equivalent, shown in Figure 4-30.

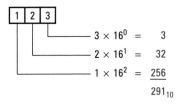

Figure 4-30   Hexadecimal 123 converted to decimal.

Binary-to-Hexadecimal Conversion

With four bits, we can make 16 different bit combinations. One method of converting binary to hexadecimal is to start with the LSB and break the binary number into groups of four bits.

$$1001000110110101_2 = \underline{\qquad ? \qquad} \text{ base 16}$$

Start with the least significant bit. Separate the bit pattern into groups of four bits. Add zeros if the most significant bit group does not fill out the required four bits. Second, assign the hexadecimal equivalent to each 4-bit code. Third, fill in the hexadecimal equivalent of each 4-bit nibble, as illustrated in Figure 4-31.

Binary Value	1001000110110101			
Grouped Binary	1001	0001	1011	0101
Result	9	1	B	5

Figure 4-31 Breaking the 16 bits up into 4-bit groups.

Example 2:

$1101101101111010_2 = $ _____?_____ base 16. Figure 4-32 illustrates the conversion process.

Binary Value	1101101101111010			
Grouped Binary	1101	1011	0111	1010
Result	D	B	7	A

Figure 4-32 Converting 4-bit groups into their respective hexadecimal equivalents.

Example 3:

$101010110111111_2 = $ _____?_____ base 16. Figure 4-33 illustrates the conversion process.

Binary Value	101010110111111			
Grouped Binary	0101	0101	1011	1111
Result	5	5	B	F

Figure 4-33 Converting binary to hexadecimal.

Example 4:

Newer 32-bit PLCs such as the ControlLogix use groups of bits that are 32 bits wide. Converting 10101111000110100011001011101000 to hex is the same basic procedure except that there are more bits. This is illustrated in Figure 4-34.

Binary Value	1010 1111 0001 1011 0011 0010 1110 1000							
Grouped Binary	1010	1111	0001	1011	0011	0010	1110	1000
Result	A	F	1	B	3	2	E	8

Figure 4-34 32-bit binary conversion to hexadecimal.

Hexadecimal-to-Binary Conversion

Conversion of hexadecimal to binary values is easily accomplished by assigning a 4-bit value to each hexadecimal digit. Convert the following:

$$1234_{16} = \underline{\hspace{1cm}?\hspace{1cm}} \text{ base 2}$$

First, separate the hexadecimal digits. Second, convert each separated hexadecimal digit into its 4-bit binary equivalent. Last, combine all 4-bit binary equivalents to form a 16-bit word. The conversion is now complete. Figure 4-35 illustrates the hexadecimal value of 1,234 by breaking it up into 4-bit groups and combining the groups to get the 16-bit binary answer.

Hex Value	1	2	3	4
Grouped Binary	0 0 0 1	0 0 1 0	0 0 1 1	0 1 0 0
Binary Result	0001001000110100_2			

Figure 4-35   Converting hexadecimal to binary.

Example 5:

Convert $92AE_{16}$ to binary. Figure 4-36 illustrates the procedure.

Hex Value	9	2	A	E
Grouped Binary	1 0 0 1	0 0 1 0	1 0 1 0	1 1 1 0
Binary Result	1001001010101110_2			

Figure 4-36   Converting hexadecimal 92AE to binary.

Example 6:

Newer 32-bit PLCs such as the ControlLogix use groups of bits that are 32 bits wide. Converting AF1B32E8 to binary is the same basic procedure as with 16-bit words, except that there are more bits. This is illustrated in Figure 4-37.

Hex Value	A	F	1	B	3	2	E	8
Grouped Binary	1 0 1 0	1 1 1 1	0 0 0 1	1 0 1 1	0 0 1 1	0 0 1 0	1 1 1 0	1 0 0 0
Result	1010 1111 0001 1011 0011 0010 1110 1000							

Figure 4-37   Convert AF1B32E8 to binary.

Figure 4-38 will give you an overview in comparing the different number systems. The left-most column is the decimal number and the other columns give the specific equivalents.

COMPARISON OF NUMBER SYSTEMS				
Decimal	Binary	Octal	BCD	Hexadecimal
1	01	1	0001	1
2	10	2	0010	2
3	11	3	0011	3
4	100	4	0100	4
5	101	5	0101	5
6	110	6	0110	6
7	111	7	0111	7
8	1000	10	1000	8
9	1001	11	1001	9
10	1010	12	0001 0000	A
11	1011	13	0001 0001	B
12	1100	14	0001 0010	C
13	1101	15	0001 0011	D
14	1110	16	0001 0100	E
15	1111	17	0001 0101	F
16	10000	20	0001 0110	10
17	10001	21	0001 0111	11

Figure 4-38   Comparing binary, octal, BCD, and hexadecimal to decimal.

SUMMARY

This chapter introduced the various number systems used in today's computers and programmable controllers. To work with PLCs successfully you need to have a solid grasp of binary fundamentals.

Many of today's industrial systems still have thumbwheel inputs and seven-segment operator displays. In most cases, these systems' inputs and outputs will consist of BCD signals. Since most PLCs have BCD conversion instructions, understanding BCD will come in handy, especially when troubleshooting BCD interfaces. Similarly, a knowledge of the hexadecimal system will come in handy when working with PLC error codes as well as PLC instructions that use hex masks to control data flow. For example, the Allen-Bradley SLC 500, PLC 5, and ControlLogix use hex masks, with the masked move instruction and sequencer output instruction serving as an example. The masked move instruction is used to move data from a source location to a specified destination through a hexadecimal mask. The mask's function is to allow user-selected destination data to be masked out. The sequencer instruction is used to transfer a 16-bit sequence step data to an output status word for the control of sequential machine operations. (We will introduce hex masks later in our lab exercises.)

In some instances you may need to work with a PLC such as the Rockwell Automation PLC 5 that uses octal addressing; in such cases you will need to understand the octal number system to effectively work with the system.

REVIEW QUESTIONS

1. As humans, we understand and use the _____, base _____, number system.
2. Microprocessor-controlled devices and digital computers use the _____, or base _____, number system.
3. Where does the term *bit* come from?
4. What is a bit?
5. Match the following number systems to their corresponding bases:
   A. Decimal   _____ Base 2
   B. Binary   _____ Base 16
   C. Octal   _____ Base 10
   D. Hexadecimal   _____ Base 8
6. 100_{10} is equal to what in decimal?
7. 100_2 is equal to what in decimal?
8. Define LSB.
9. What position in a binary number holds the LSB position?
10. Define MSB.
11. What position in a binary number holds the MSB position?
12. From memory, convert the following binary numbers to decimal:

    1010 = _____   1111 = _____
    1101 = _____   1001 = _____
    0110 = _____   1011 = _____
    1110 = _____   0011 = _____
    0111 = _____   11010 = _____

13. List the binary numbers in order from 1 to 20.
14. List BCD codes from 0 to 20.
15. What is an invalid BCD code?
16. List the 4-bit invalid BCD codes.
17. Convert 110101111_2 into octal.
18. Convert 913 octal into binary.
19. Convert ABC hexadecimal into its decimal equivalent.
20. Convert 1011011101011101 base 2 into _____ base 16.

21. Complete Figure 4-39 by inserting the correct numbers for the corresponding bases as you work across each row.

Decimal	Binary	BCD	Octal	Hexadecimal
10	1010	0001 0000	12	A
				1C
74				
			27	
		0010 0101		
	1011101			

Figure 4-39   Question 21's conversion exercise.

IF PASSWORD IS CORRECT GO TO DRIVE SET UP SCREEN

0022
```
---EQU---
Equal
Source A    N12:21
              7777<
Source B    32767
            32767<
```
B3:10/0
[OSR]

B3:1/15
( )

RESET PASSWORD
ATTEMPTS COUNTER
C5:5
(RES)

0023
```
---GRT---
Greater Than (A>B)
Source A    N12:21
              7777<
Source B        0
                0<
```
```
---NEQ---
Not Equal
Source A    N12:21
              7777<
Source B    32767
            32767<
```
B3:10/1
[OSR]

```
---CTU---
Count Up
Counter     C5:5
              3<
              2<
```
(CU)

(DN)

```
MOV
ce        −1
          −1<
        N12:21
         7777<
```

NOTIFICATION
B3:0/7

0024
] [

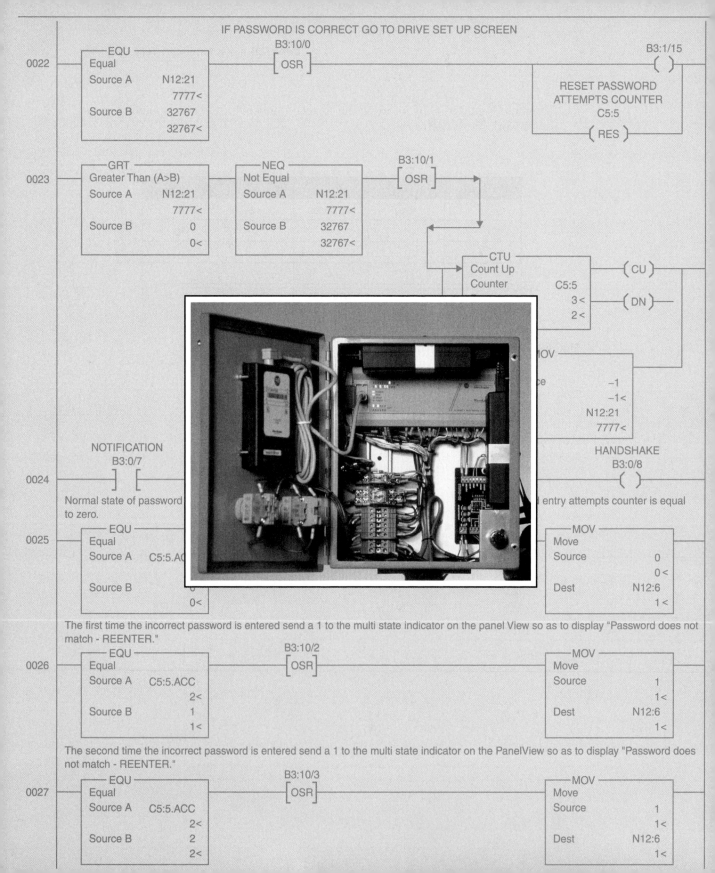

HANDSHAKE
B3:0/8
( )

Normal state of password   entry attempts counter is equal
to zero.

0025
```
---EQU---
Equal
Source A   C5:5.A
Source B       0
               0<
```
```
MOV
Move
Source        0
              0<
Dest      N12:6
              1<
```

The first time the incorrect password is entered send a 1 to the multi state indicator on the panel View so as to display "Password does not
match - REENTER."

0026
```
---EQU---
Equal
Source A   C5:5.ACC
               2<
Source B       1
               1<
```
B3:10/2
[OSR]

```
MOV
Move
Source        1
              1<
Dest      N12:6
              1<
```

The second time the incorrect password is entered send a 1 to the multi state indicator on the PanelView so as to display "Password does
not match - REENTER."

0027
```
---EQU---
Equal
Source A   C5:5.ACC
               2<
Source B       2
               2<
```
B3:10/3
[OSR]

```
MOV
Move
Source        1
              1<
Dest      N12:6
              1<
```

CHAPTER

Introduction to Digital and Analog PLC Interface

OBJECTIVES

After completing the chapter, you should be able to:

- explain how binary information is used to represent input and output in PLCs
- know the difference between digital and analog input and output signals
- observe how digital field device information gets into a PLC
- observe how analog field devices send information into a PLC
- describe how digital and analog I/O data flows from PLC memory to the field device
- understand I/O addresses, and how they are used in a PLC

INTRODUCTION

This chapter will introduce how information flows from input field devices to PLC memory and how it is processed in the user program in PLC memory to control field devices. The PLC is controlled by a microprocessor, which is a digital computer. To program, install, or troubleshoot a PLC effectively, you must understand how the incoming signals and outgoing signals interface with the internal workings of the PLC. These signals fall into two groups, digital and analog.

"Microprocessor" is a term that has become common in our everyday life. The microprocessor has become a common item in many devices we use today. Calculators, video recorders, cameras, microwave ovens, video games, automobile system controls, personal computers, and, indeed, the programmable controller are all controlled by one or more internal microprocessor. Although a microprocessor can be a very complex device, the fundamentals needed to program, operate, and troubleshoot a PLC-based system are easy to understand. Possibly one of your first exposures to a microprocessor-based device was when playing video games on your television. At first, video games were relatively crude, single games available only in black and white. Later came color video games and the ability to change to a different game simply by changing a plug-in cartridge. The cartridge contained the program for the specific game you wanted to play. The cartridge contains an integrated circuit chip, called a memory chip, which stores the program in memory. You then simply change the program by changing the cartridge.

Today modern video games such as Xbox or Playstation use CD-ROMs instead of the older cartridges, but the principle is still the same; load a new game by loading a new program via the CD-ROM.

Using PLCs is very similar. There is a microprocessor in the form of an integrated circuit chip, and other, supporting chips mounted on a printed circuit board. These chips make up the CPU. The microprocessor and its supporting chips make this a digital computer. Figure 5-1 illustrates a PLC block diagram showing the microprocessor, or CPU, reading input signals and controlling output signals.

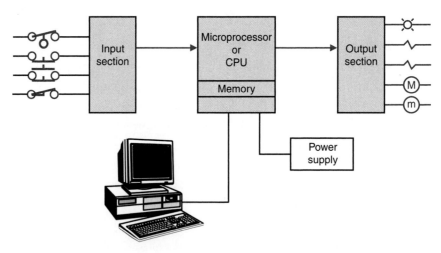

Figure 5-1 Microprocessor controlling input and output circuits in a PLC block diagram.

By changing the program in the PLC, you can instruct the PLC to make a different product. This is very similar to playing a different video game by simply changing a CD-ROM. As a technician, or electrical or maintenance individual, you will need the skills to develop, modify, or troubleshoot the list of instructions that make up the program. Think of the PLC as a large-scale video game and yourself as in control of the program. Figure 5-1 illustrates the main parts of a PLC: the input section, microprocessor, memory, and output section.

Even though a personal computer is used to develop the PLC program, it is not necessary to keep the personal computer connected to the PLC to operate the machine or process. After the personal computer is used to transfer, or download, the PLC program to the PLC processor, the PLC processor is put into run mode. At this point, the personal computer can be removed from the system if desired. This is called going off-line. However, the personal computer could be left connected to allow viewing of the PLC application as it runs. This would be on-line monitoring of the PLC in remote run mode.

A prerequisite to understanding programmable controllers is a working knowledge of how information is represented and worked with by the programmable controller. This chapter will introduce binary and digital concepts. You will learn how binary information is represented and worked with inside the PLC to interface with digital and analog field devices.

THE DIGITAL CONCEPT

Programmable controllers, much like any other microprocessor-based device, operate on **digital** circuits. Digital circuits operate on a simple concept: a circuit is either ON or OFF. A digital circuit is as simple as a light switch. The light switch and a digital circuit have something in common: they can be in one of two possible states, ON or OFF. For instance, a motor can be running or stopped; a valve may be open or closed; and a part may be located in front of a sensor, on a conveyor, or out of range. This two-state concept is the basis of how computer-controlled devices make decisions. It is also the fundamental operating principle of the programmable controller.

A device that provides two-state signals as inputs to, or outputs from, a PLC is called a *discrete* or *digital device*. Another term associated with something that can only exist in one of two predetermined states is the term "binary." Do not let these terms throw you. They are simply another way to identify a two-state signal.

Since microprocessor circuits are also binary in nature, they have only two possible states, either ON or OFF. Each state is represented either by a 1, for an ON condition, or a 0, for an OFF condition. When a valid ON condition is present, it is called a "logical-one." Likewise, a valid OFF condition is commonly called a "logical-zero." Although equipment manufactured in the United States commonly uses the ON-or-OFF convention, most internationally distributed equipment uses 1 for the ON symbol and 0 for the OFF symbol. Figure 5-2 illustrates some common industrial devices that involve two states: discrete, digital, or binary devices.

Consider that a computer only understands (and internally works with) binary information, and it is clear there is a natural fit with input devices that can effectively communicate with the microprocessor in a computer. Since a PLC is a computer, there should be no problem interfacing two-state industrial control devices with a PLC.

To process two-state data it must first be converted to an electronic form that the PLC can understand. This is the job of the input module. Input electrical signals must be converted into usable groups of data. Such a group of data is called a **word.**

BINARY I/O DEVICES		
Two-State Device	**ON**	**OFF**
Alarm bell	Ringing	Silent
Clutch	Engaged	Disengaged
Limit switch	Contacts closed	Contacts open
Motor	Running	Stopped
Pilot light	On	Off
Pressure switch	Contacts closed	Contacts open
Proximity switch	Part Present	No part
Push button	Closed	Open
Temperature switch	Contacts closed	Contacts open
Valve	Closed	Open

Figure 5-2   Common industrial control hardware representing the binary concept.

BINARY DATA REPRESENTATION

In our everyday lives, we communicate to others using groups of letters called words. In a similar fashion, computer data is organized into words and stored in registers in PLC memory. ("Register" is a computer term used by some PLC manufacturers to signify a storage area for a word containing aggregate information.)

The computer uses groups of bits called words. The only differences between words in the English language and computer words are the length and the available characters. English words can be of varying lengths and can contain any of 26 alphabet characters, but computer words are always the same length. Rather than letters making up a computer word, binary bits are used. Unlike in English, when working with computers, there are only two characters available, 1 or 0, each of which is called a bit (BInary digiT). A single bit is the smallest unit of microprocessor data that can be represented.

One measure of a microprocessor's capabilities is the length of the data words on which it can operate. Modern PLCs use 16 bits to represent a word. Figure 5-3 illustrates a 16-bit word. Notice that only 1 bit, either a 1 or 0, is permitted in each of the 16 available positions.

15	14	13	12	11	10	9	8	7	6	5	4	3	2	1	0	◄─Bit Position
1	0	1	0	1	1	1	0	1	1	0	1	0	0	0	1	

1	1	0	1	1	0	0	0	1	1	1	0	1	0	1	1

Figure 5-3   Information represented by arranging the 1s and 0s in different combinations within a 16-bit word.

A 16-bit word is composed of the following parts:

16 single bits

nibble: the lower (**lower nibble**) or upper (**upper nibble**) 4-bit group of a byte

byte: the group of the lower (**lower byte**) or upper (**upper byte**) 8 bits of a 16-bit word

word: a group of 16 bits stored and used together

Each part is illustrated in Figure 5-4.

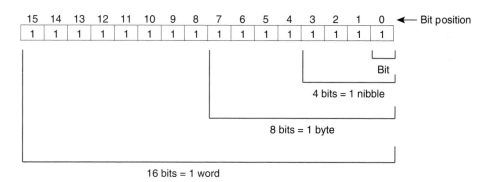

Figure 5-4 Parts of a 16-bit word.

Each 16-bit word is made up of two bytes, an upper and a lower byte. Each **byte** contains two nibbles. Each byte is broken down into an upper and a lower nibble, and each **nibble** is broken down into four individual bits. Figure 5-5 illustrates the breakdown of a 16-bit word.

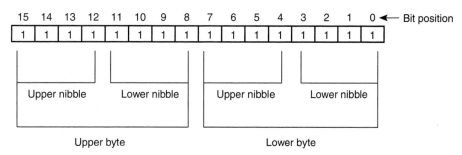

Figure 5-5 A 16-bit word can be broken down into bytes, nibbles, and bits.

Each bit (as you remember from Chapter 4) can have a weighted value depending on its position in the word. As illustrated in Figures 5-3 and 5-4, bit 0 represents the bit of lowest value, or significance. This bit is called the least significant bit (LSB). Bit 15, on the other hand, holds the highest place value, or significance. Bit 15 is called the most significant bit (MSB).

If you think about the decimal number 12,345, the right-most digit position is the lowest value, or 1s place, and the left-most digit position is the highest value.

Even though a 16-bit word can be used to represent numerical data, the 16-bit word can be used to represent other kinds of data within the PLC. Next, we will look at how a 16-bit word is used to represent the ON or OFF status of each screw terminal on a 16-point I/O module.

CORRELATION OF 16-BIT WORDS TO INPUT SIGNALS

As an example, a standard input module might have 16 input points. Each input module corresponds to one 16-bit word of data, and each input point corresponds to 1 bit in the 16-bit word associated with the input module. As illustrated in Figure 5-6, each input screw terminal corresponds to a like-numbered bit position in a 16-bit input word. Each bit position will have either a 1 (if the input is ON) or a 0 (if the input is OFF), as shown in Figure 5-7. A bit position cannot be empty: there is either a 1 or 0 value in each position at all times.

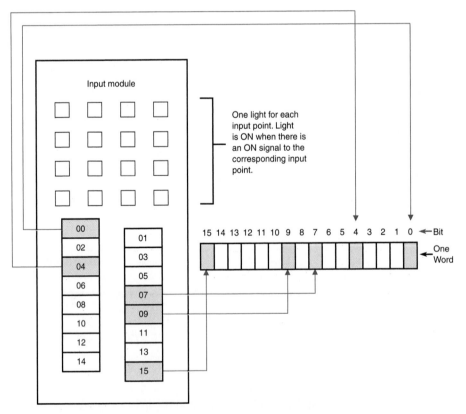

Figure 5-6   Sixteen-point module's I/O points represented in a 16-bit word.

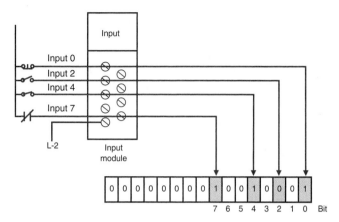

Figure 5-7 Physical input conditions reflected in the corresponding input data word.

In Figure 5-7, input zero is a closed switch going into input screw terminal zero. Since the switch is closed, an ON (1) signal is transferred to the input word, in bit position zero. Input two is an open switch going into input screw terminal two. Since the switch is open, an OFF (0) signal is transferred to the input word, in bit position two. Input four is a closed switch going into input screw terminal four. Since the switch is closed, an ON signal is transferred to the input word, in bit position four. Input seven has a set of closed contacts. The closed contacts input an ON signal into the module, which is reflected as a 1 in bit position seven.

Eight-point input modules are also available. In these modules, only the lower eight bits of the input word are used. Figure 5-8 illustrates an eight-point input module and shows how its input signals are represented in a 16-bit data word.

We have been talking about 1s and 0s stored inside the PLC. Actually, however, there are no real 1s and 0s inside the PLC or any other computer. These ON (1) or OFF (0)

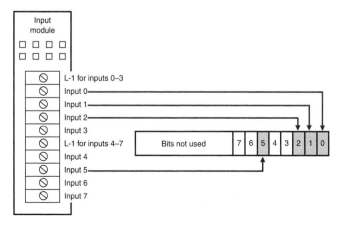

Figure 5-8 An eight-point input module represented in a 16-bit word.

signals are actually representations of voltage levels. A low voltage level represents an OFF condition, and a high level represents an ON condition. Since most computer circuits operate on 5 V dc, a valid ON signal is typically from 2.4 to 5 V dc, while a valid OFF signal is typically from 0 to .8 V dc. The area between .8 V dc, the maximum for a valid OFF signal, and 2.4 V dc, the minimum for a valid ON signal, is undefined, which means that this voltage signal area is not guaranteed as either a valid ON or a valid OFF signal. To ensure valid signals, make sure the input voltage levels are within their respective voltage ranges.

As electronic components became smaller, it became feasible to pack more input points on a single input module. With the arrival of 24- and 32-point modules, it was necessary to represent their input signals with more than one 16-bit data word. Figure 5-9 represents a 24-point input module represented in two 16-bit data words.

15	14	13	12	11	10	09	08	07	06	05	04	03	02	01	00	◄— Bit
																Word 0
These eight bits are not used																Word 1

Figure 5-9   A 24-point I/O module represented in two 16-bit words.

Notice that the first data word in the figure is word zero, while the second is word one. When working with computers, the first bit and first word are always referred to as bit zero and word zero. Since we have only 24 valid input points, two full words are not necessary. As a result, the upper eight bits of data in word one are unused. Input points 0 through 15 are stored in word zero, while input points 16 through 23 are stored as follows:

point 16 is word 1, bit 0

point 17 is word 1, bit 1

point 18 is word 1, bit 2

point 19 is word 1, bit 3

point 20 is word 1, bit 4

point 21 is word 1, bit 5

point 22 is word 1, bit 6

point 23 is word 1, bit 7

Figure 5-10 shows a 32-point input module, in which two full 16-bit words are used for the 32 input points. Word zero contains bits 0 through 15, and word one contains bits 16 through 31.

Now that we have associated each input screw terminal of the input module with a 16-bit data word, how are these data words organized? Let us assume we have a small, modular PLC system with four input modules. Since there are four modules, there will be four associated input data words. These four words need to be organized in some logical manner so that all concerned (including the PLC) will have a clear understanding of which

15	14	13	12	11	10	09	08	07	06	05	04	03	02	01	00	← Bit
																Word 0
																Word 1

Figure 5-10   Two 16-bit words representing the input points of a 32-bit I/O module.

input point is associated with which bit in which word. To simplify the organization, we simply create a data file, organize the words from the first input module as word zero, and list each of the remaining input modules and their associated data words in ascending order. Figure 5-11 illustrates a 16-bit wide by 256-word long data file similar to that found in an SLC 500 PLC.

15	14	13	12	11	10	09	08	07	06	05	04	03	02	01	00	← Bit
1	1	1	0	0	1	1	1	1	1	1	1	1	1	1	1	Word 0
1	1	1	0	0	0	0	0	0	0	1	0	0	0	0	1	Word 1
0	0	0	1	1	1	1	0	1	0	1	1	1	0	1	1	Word 2
0	0	0	1	1	0	1	0	1	1	1	0	1	0	1	1	Word 3
1	0	0	1	0	1	1	0	0	0	0	0	0	0	0	0	Word 4
1	1	1	1	1	1	1	0	0	0	0	0	0	0	0	0	↓
1	0	1	1	0	1	1	0	0	0	0	0	0	0	0	0	
1	0	1	0	1	1	0	1	0	1	1	1	1	1	1	1	Word 254
0	1	1	1	1	1	0	0	0	1	1	1	1	0	0	0	Word 255

Figure 5-11   Sixteen-bit words arranged into a memory data table.

A data file is nothing more than a grouping of words containing like data. The words are associated with the real-world input status of the available input modules in the PLC system. The data file representing a data word for each input module in the system is called the "input status file."

THE INPUT STATUS FILE

Simply put, the input status file is a group of words in memory that are organized to store input data. Data stored in the input status file awaits processing by the CPU when the program is solved. Each **I/O** module in a chassis is "mapped," or assigned a specific word in the input status file. I/O module placement in the PLC chassis is part of the address of input devices connected to that module. Each screw terminal on each input module has one memory location in the input status file for storing of the ON or OFF status of the connected field input device. Figure 5-12 illustrates a five-slot modular PLC, which contains a power supply, CPU module, and four input modules. Notice that each slot of the

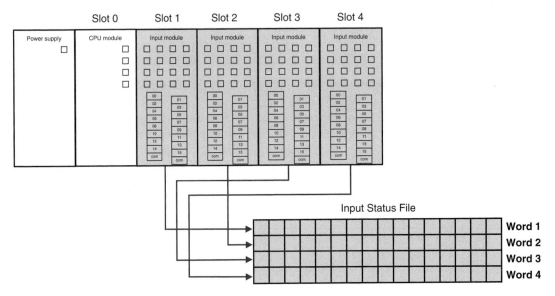

Figure 5-12 Input status file words associated to slots and input modules in a modular PLC.

chassis has a corresponding number. The CPU is in slot zero, while input modules are located in slots one through four. The slot number corresponding to a particular input module identifies in which 16-bit word the input status file will store that particular input module's ON or OFF input data. To simplify tracking modules in the input status file, the file word assigned to the module is the same as the slot in which the module resides in the modular PLC. As a result, since the CPU resides in slot zero on an SLC 500 modular PLC and there are no input words assigned to the CPU, there will be no word zero in the input status file.

THE OUTPUT STATUS FILE

The output status file functions the same as the input status file with the exception that the output status file stores the ON or OFF status of each output. Once the user program has been solved, the resulting ON or OFF state of each output point (and its associated field device) is stored in the output status file. Each 16-point **output module** has one 16-bit word holding the ON or OFF status of each of the module's 16-output screw terminals. This status data is stored in the output status file waiting to be sent to the output module's switching circuitry.

The five-slot modular PLC represented in Figure 5-13 contains a power supply, a CPU module, and four output modules. Notice that each slot of the chassis has a corresponding number. The CPU is in slot zero, while output modules reside in slots one through four. The slot number corresponding to a particular output module identifies in

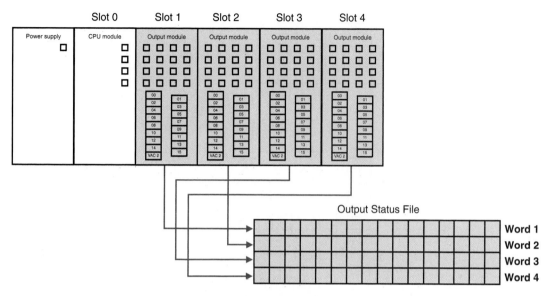

Figure 5-13   Output status file words associated with slots and output modules in a modular PLC.

which 16-bit word the output status file will store that particular module's ON or OFF output data.

When instructed by the CPU, each 16-bit output status file word is sent to its associated output module by way of the backplane. Each output module takes the low-voltage ON or OFF signal (represented as 1 bit in this 16-bit word), isolates it, and turns on or turns off the associated output's switching device. Each output module has one switching device for each output point. These switching devices may be small mechanical relays, as in Figure 5-14, or solid-state devices. The output status file bit's electrical signal is sent from the status file to power the relay coil for output three's relay output point.

Figures 5-12 and 5-13 illustrated PLCs with either all input or all output modules. These examples were designed to help you understand how status file words correlate with input and output modules. Actual applications will have a mix of input and output modules in the same PLC chassis. One primary advantage of a modular PLC is the ability to place input and output modules in the selected chassis in the input or output designation required for a specific application. When using a seven-slot chassis, after the CPU is placed in slot zero, you are free to place any mix of input or output modules you need in the remaining six slots. Figure 5-15 illustrates a seven-slot chassis with input modules in slots 1, 3, 4, and 6 and output modules in slots 2 and 5. Each input module will have an associated word with the same input data file number as the slot in which it resides. Likewise, each output module will have an associated word in the output status file with the same data file word number as the slot in which it resides. These status files are two separate data files. Although they are stored in the PLC's memory, they are located in predefined, separate areas used strictly for the storage of either input or output status data.

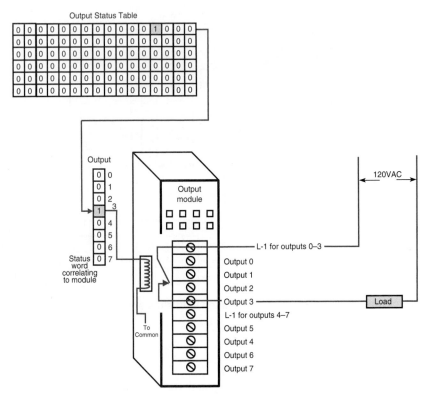

Figure 5-14 Simplified graphical representation of an output status file signal controlling a 16-point relay output module.

Figure 5-15 illustrates an SLC 500 PLC that has reserved four words in the input status file for the four input modules. Notice the gaps in both the input status file's input and the output status word sequence. There are no input status words reserved for nonexistent slots 2 or 5, nor output status words 1, 3, 4, or 6.

The size of a system's input and output status files is dictated by the total number of inputs or outputs it can support. As an example, if the particular PLC can support a total of 4096 inputs and outputs, the input status file will have 4096 bits and the output status file will also have 4096 bits.

When using an SLC 500 modular PLC with a 5/02, 5/03, 5/04, or 5/05 processor the read I/O configuration function creates only the required input and output words in their associated status tables. Because of the age of the PLC 5, the I/O read configuration feature is not available. As a result, the input and output status tables are created with the maximum number of words supported by the specific processor in the chassis. ControlLogix has a ControlNet network as its backplane. The intelligent backplane in conjunction with the intelligent I/O modules will, as a result of the I/O configuration, create only the specific information or tags as required by the specific module residing in each chassis slot.

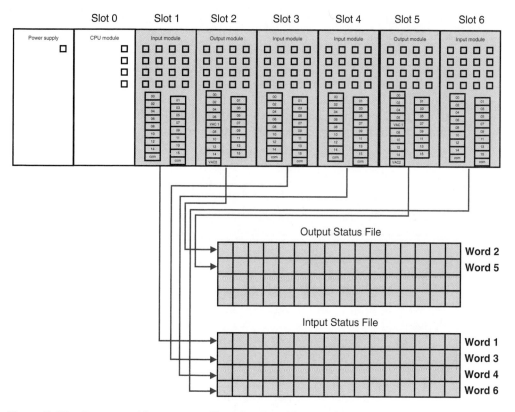

Figure 5-15 Output and input status file reflecting this specific PLC's module mix.

INPUT AND OUTPUT STATUS FILES AND FIXED PLCs

Even though the physical packaging of a fixed I/O PLC is different from the modular PLC, the interaction between inputs and outputs with the input and output status files remains the same. A 16-input fixed PLC has 1 input status file bit that corresponds to each input screw terminal. The typical fixed PLC will have fewer outputs than inputs.

There are three members of the SLC 500 family that are fixed I/O PLCs. The MicroLogix 1000 (see Figure 2-23), MicroLogix 1200 (see Figure 2-30), and MicroLogix 1500 (see Figure 2-24). The MicroLogix 1000 supports up to 20 inputs and 12 outputs. Some MicroLogix 1000 units support up to four analog input channels and one analog output channel. The MicroLogix 1000 I/O is not expandable. The MicroLogix 1200 supports up to 24 discrete inputs and 16 discrete outputs. MicroLogix 1200 I/O can be expanded using up to six Allen-Bradley Bulletin 1762 I/O modules. The MicroLogix 1500 base unit can support up to 16 inputs and 12 outputs. Using Bulletin 1769 Compact I/O expansion modules, the system can be expanded to up to 16 I/O modules.

The CPU unit with the fixed or embedded I/O is called the base unit. The MicroLogix 1200 and 1500 can be expanded by DIN rail mounting and clipping modules directly to the base unit (see Figure 2-29). There is no chassis used to hold the modules. This is called a rackless design.

Unlike a modular PLC, the fixed PLC base unit has all of its input and output points built into the CPU unit. Earlier we learned that the CPU slot number in a modular PLC such as the Allen-Bradley SLC 500 is considered slot zero. To maintain consistency, because fixed I/O points are built into the CPU unit, these I/O points are addressed as slot zero in the input status file and in the output status file. Figure 5-16 illustrates the correlation between the MicroLogix 1000's hardware and the input and output status table in

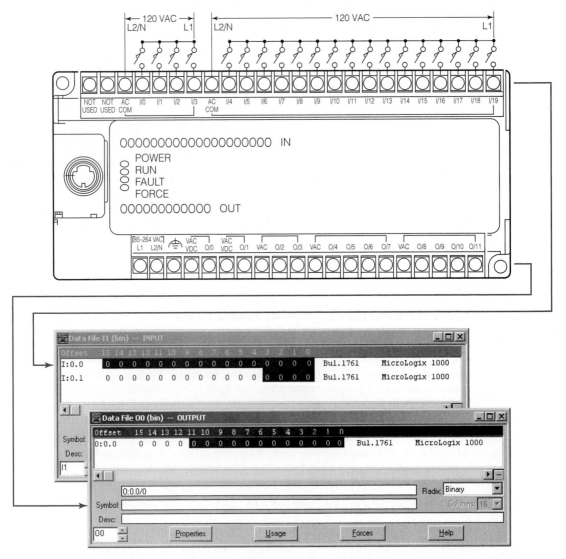

Figure 5-16 Allen-Bradley MicroLogix 1000 Fixed I/O PLC input and output correlation to the RSLogix 500 software input and output status file.

RSLogix 500 software. The shaded portion of each status table identifies the bits used for the illustrated MicroLogix 1000. Because this is a 20-input and 12-output MicroLogix, input bits I:0.0/0 through I:0.0/15 will represent inputs 0 through 15. The address format is as follows. The I identifies an input. An O would be used to designate an output. The colon is a separator. The next placeholder is the slot number the module resides in. Allen-Bradley fixed PLCs like the MicroLogix always identify the built in I/O points as slot 0. The dot is also a separator identifying a word number follows. The slash tells you that a bit number follows. Inputs 16 through 19 will have to go into a second word, I:0.1. Inputs 16 through 19 will be addresses I:0.1/0 through I:0.1/3.

The MicroLogix in Figure 5-16 has 20 inputs and 10 outputs. Since the MicroLogix is a 16-bit computer, the data words will also contain 16 bits. The first word, word zero (I:0.0), will represent the first 16 input bits. The remaining 4 bits will be in word one (I:0.1). Refer to the shaded portion of the input status table in Figure 5-16. The input addresses, as an example, will be I:0.0/0 for input zero and I:0.0/1 for input one. Input 19 will be address I:0.1/3. Inputs 0 through 15 are in word zero. Input 16 will begin with bit 0 of word one, I:0.1/0. When programming RSLogix 500 software, this address could be programmed as I:0.1/0, the formal address, or simply I:0/16.

The MicroLogix 1200 and the MicroLogix 1500 can expand the I/O count by adding additional I/O modules to the base units. When expansion I/O is added, modules are numbered starting with one, two, three, etc. These module numbers are similar to the slots in a modular PLC. Figure 5-17 illustrates an RSLogix 500 software input and output status table for a MicroLogix 1000. The top portion of the figure is the input status table, and the bottom portion is the output status table. Input and output addresses in the base unit are still represented as slot zero starting at I:0.0 and outputs also as slot zero starting at O:0.0. The highlighted bit in the input status table is base address I:0.0/4.

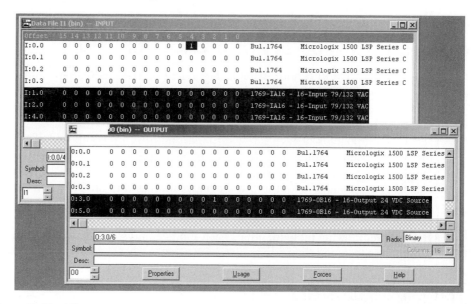

Figure 5-17 MicroLogix 1500 input and output status tables from RSLogix 500 software.

The input status table also contains words I:1.0, I:2.0, and I:4.0. These three words represent three input modules added to the right side of the base unit to expand it. The addresses will be represented as I:1.0/6 for the input point in slot one, terminal or bit position 6. The output status table has words O:3.0 and O:5.0 in addition to the base addresses. These two words represent an expansion output module in slots three and five. The highlighted bit in the output status table in Figure 5-17 refers to O:3.0/6. This is the output module in expansion position three.

Looking at the input and output status tables below, we can determine the I/O modules and addresses available for this specific PLC.

The table below summarizes the information found in Figure 5-17.

I/O Slot	Hardware	I/O Type	I/O Address Range
MicroLogix 1500	Base unit	Up to 12 inputs	I:0.0/0 through I:0.0/11*
MicroLogix 1500	Base unit	Up to 16 outputs	O:0.0/0 through O:0.0/15*
Slot one	1769-IA16	16 inputs	I:1.0/0 through I:1.0/15
Slot two	1769-IA16	16 inputs	I:2.0/0 through I:2.0/15
Slot three	1769-OB16	16 outputs	O:3.0/0 through O:3.1/15
Slot four	1769-IA16	16 inputs	I:4.0/0 through I:4.0/15
Slot five	1769-OB16	16 outputs	O:5.0/0 through O:5.1/15

* Current maximum I/O points for MicroLogix 1500 base unit.

Refer to your lab manual for more information on and exercises regarding the MicroLogix PLCs base unit and expansion addressing and configurations.

SIXTEEN-POINT I/O MODULES WITH DECIMAL ADDRESSING

Modular SLC 500 PLC addressing is dependent on the module type and slot position in which the module resides in the chassis. Figure 5-15 is an SLC 500 Modular PLC in a seven slot chassis with 16-point input modules in slots 1, 3, 4, and 6. 16-point output modules are in slots 2 and 5.

The components of each screw terminal's address are composed of the slot number, the screw terminal number, and a designation as to whether the I/O point is an input or output. If a specific screw terminal is an input, we begin its address with an I. If a particular screw terminal is an output, we begin its address with an O. The next thing we need to identify is what slot the module is in. The input module in slot one will be identified as I:1, the output module in slot two is O:2, and the input module in slot three is identified as I:3. The following table identifies each module and shows how the slot in which each resides is reflected in the address.

CPU is always slot 0   No address

input module in slot 1   I:1

output module in slot 2   O:2

input module in slot 3   I:3

input module in slot 4   I:4

output module in slot 5   O:5

input module in slot 6   I:6

Each screw terminal of a 16-point module is identified as 0 through 15, decimal. If we add the screw terminal designation to the beginning addresses listed in the table, we will have created a complete address for each screw terminal on each module in the chassis. Figure 5-18 illustrates six addressing examples:

Input or Output	Chassis slot	Screw terminal	Address
Input	1	4	I:1/4
Output	2	13	O:2/13
Input	3	2	I:3/2
Input	4	4	I:4/4
Output	5	9	I:5/9
Input	6	0	I:6/0

Figure 5-18   Examples of address structure.

SLC 500 address format consists of three parts (see Figure 5-19):

Part 1: I for input, and a colon to separate the module type from the slot, or O for output, and a colon to separate the module type from the slot

Part 2: The module slot number

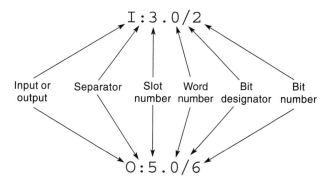

Figure 5-19   I/O address format for SLC 500 family of PLCs.

Part 3: The dot signifies a word follows. This is the word number. If using an 8 or 16 point module only one word is necessary. This is the first word or word 0. When using an 8 or 16 point module, the word number is optional

Part 4: The slash identifies that a bit follows. This is the bit or screw terminal number

Address format—Input or output: slot/screw terminal.

In the Example below, note that the address, I:3/9, signifies the input module in slot three and screw terminal 9.

1. What is the address of screw terminal 9 on the module in slot three?
   Answer: Slot three is an input module, so the address is I:3/9.
2. What is the address of screw terminal 0 on the module in slot two?
   Answer: Slot two is an output module, so the address is O:2/0.
3. What is the address of screw terminal 12 on the module in slot six?
   Answer: I:6/12

I/O INTERACTION WITH THE INPUT AND OUTPUT STATUS FILES

Each PLC manufacturer has its own addressing rules. For this section we will specifically look at the Allen-Bradley SLC 500 PLCs. We will investigate how the previous discussion on fixed and modular addressing applies to the fixed SLC 500 and MicroLogix input and output status files.

SLC 500 Modular I/O Addressing and the Input Status File

The SLC is a 16-bit PLC, thus data is formatted into 16-bit words. When using a 16-point input module, one word of memory contains enough bits for each input screw terminal. However, when using a 32-point input module, a 16-bit word only contains enough bits for half of the input points. As a result, two words will be necessary to represent 32 input points. When using more than 16 bits, multiple words are necessary. The address format introduced in Figure 5-19 changes as we now have a word number identifier included in the address. If only one 16-bit word is necessary to represent input points, the address I:3/2, as an example, would represent the input module in chassis slot 3, screw terminal or bit number 2. This address may also be written as I:3.0/2. This is called the formal address. The .0 is the word number. The first word is always word 0. When using a 32-point input module, word 0 will store input data for the lower 16 input bits. The second word, I:3.1, will represent the upper 16 bits of the 32-point input module. Referring to Figure 5-15, let's assume our SLC 500 modular PLC has the following input modules:

Slot 1   8-point discrete input module
Slot 3   32-point discrete input module
Slot 4   16-point discrete input module
Slot 6   16-point discrete input module

Figure 5-20 illustrates an input status table screen print from Rockwell Software's RSLogix 500 software. RSLogix software is used to program the SLC 500 and MicroLogix family of PLCs. The figure illustrates the words associated with the input modules and their input points.

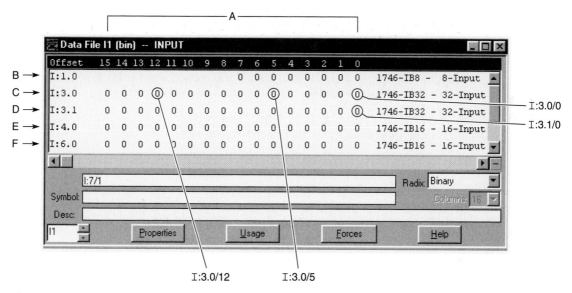

Figure 5-20   RSLogix 500 software input status file screen view.

Following is the analysis of Figure 5-20, the input status table:

A   Bits 0 through 15 correlate to input screw terminals 0 through 15 on the input module.

B   Input word I:1.0 comprises 8 bits for the input module in slot 1.

C   The first 16-bit word for the 32-input point word in slot 3 is word I:3.0.

D   The second 16-bit word for the input module in slot 3 is word I:3.1.

E   I:4.0 is the 16-bit word assigned to the input module in slot 4.

F   Slot 6 input module 16-bit input word.

Slot 1
Eight bits are assigned to the 8-point input module in slot 1. Input addresses range from I:1/0 to I:1/7. Notice the input bits across the top (A in Figure 5-20).

Slot 3
Two input words are assigned to the 32-point input module in slot 3. The first word is I:3.0 and I:3.1 is the second word. The lower 16 input addresses, screw terminals 0 through 15,

are in the first word. These addresses are I:3.0/bits 0 through 15. As an example:

I:3.0/0 is input module in slot 3, bit or screw terminal 0.

I:3.0/5 is input module in slot 3, bit or screw terminal 5.

I:3.0/12 is input module in slot 3, bit or screw terminal 12.

The addresses for the upper 16 inputs on a 32-point module are as follows:

I:3.1/0 is input module in slot 3, bit or screw terminal 16. Since the lower 16-bit word contains input points 0 through 15, the second word contains screw terminals or input points 16 through 31. Remember, the second word is I:3.1, so the first bit is I:3.1/0, or input point 16.

I:3.1/1 is input module in slot 3, bit or screw terminal 17.

I:3.1/2 is input module in slot 3, bit or screw terminal 18.

I:3.1/3 is input module in slot 3, bit or screw terminal 19.

I:3.1/4 is input module in slot 3, bit or screw terminal 20.

Slot 4
One 16-bit word is assigned to the 16-point input module in slot 4. Input addresses range from I:4/0 to I:4/15.

Slot 6
One 16-bit word is assigned to the 16-point input module in slot 6. Input addresses range from I:6/0 to I:6/15.

Modular I/O Addressing and the Output Status File

The output status table is similar in structure to the input status file except that this status table contains output information. Figure 5-21 illustrates an RSLogix 500 software output status file screen print. Our PLC has one output module in slot 2 and one in slot 5.

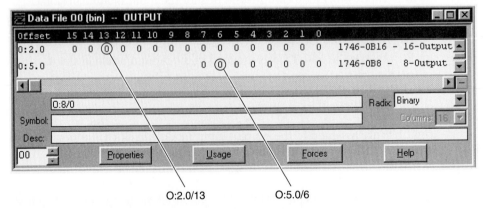

Figure 5-21   RSLogix 500 software output status file screen view.

Likewise, the output modules in slot 2 will have an address like O:2.0/13 to represent output screw terminal 13. O:5.0/6 will represent the output point for the module in slot 5, word 0, bit or screw terminal 6.

Fixed I/O Addressing

The rule for the CPU residing in slot zero holds for the fixed SLC 500 and MicroLogix 1000 PLC, and since there are no modules, all I/O screw terminals must also be slot zero. The addressing for a fixed SLC 500 or MicroLogix PLC is as follows:

I:0/4 for input screw terminal 4

I:0/12 for input screw terminal 12

O:0/7 for output screw terminal 7

O:0/0 for output screw terminal 0

Example 4:

We have a fixed PLC with 16 inputs and 12 outputs. What are the addresses for the following input or output points?

1. Input screw terminal 5
   Answer: I:0/5
2. Output screw terminal 13
   Answer: O:0/13

FIXED PLC I/O ADDRESSING WHEN I/O IS EXPANDED

Only the fixed SLC 500 I/O can be expanded into a special, two-slot expansion chassis. The fixed SLC 500 unit is addressed as slot zero, and the optional two-slot expansion chassis can be clipped on the right side of the fixed unit (see Figure 5-22). The fixed I/O is still addressed as slot zero, while the first slot includes slot one and the second, slot two.

Fixed output terminals, slot 0	Input	Output
	This is slot 1. This slot can be either an input or output module.	This is slot 2. This slot can be either an input or output module.
Fixed input terminals, slot 0		

Figure 5-22   Slot numbering for an Allen-Bradley SLC 500 fixed PLC with two-slot expansion chassis.

The two expansion slots can usually contain any discrete I/O module. Addressing for the two expansion slots is done exactly as if they were modules in a modular PLC. Figure 5-23 illustrates the input status file for the same PLC as in Figure 5-16, which illustrates an input status file with a 16-point input module placed in the first slot of the clip-on expansion chassis. The input module in expansion chassis slot one is addressed the

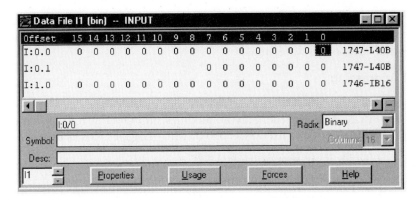

Figure 5-23   RSLogix 500 software input status file for SLC 500 L40B fixed PLC shown in Figure 5-16.

same as a module in slot one of a modular chassis. Input addresses for the input points in expansion slot one range from I:1/0 to I:1/15.

Figure 5-24 illustrates the same expanded PLC with a 16-point output module in slot two of the expansion chassis. The output module in expansion chassis slot two is addressed the same as a module in slot two of a modular chassis. Output addresses for the output points in expansion slot two range from O:2/0 to O:2/15. Notice that the output status file in Figure 5-24 has one word reserved for 0:2. The PLC is smart enough to know that since there is already an input module assigned to slot one, no output module will reside there; hence, the output status file will not assign a 0:1 position in the output status file.

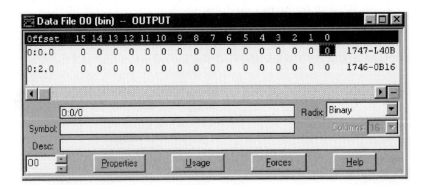

Figure 5-24   RSLogix 500 software output status file for SLC 500 L40B expanded fixed PLC.

THIRTY-TWO-BIT CONTROLLOGIX PLC

The ControlLogix is the latest addition to the Rockwell Automation family of PLCs. ControlLogix is a 32-bit processor as compared to 16-bit processors for earlier PLCs such as the SLC 500 and PLC 5 families of PLCs. ControlLogix stores information in groups of

bits similar to the 16-bit PLC words; however, being a 32-bit processor, ControlLogix'
basic allocation of memory is 32 bits, or a double word (Dword). Figure 5-25 illustrates a
32-bit double word.

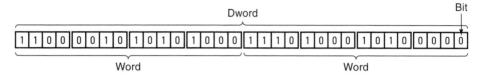

Figure 5-25   ControlLogix input module represented as a double word.

There are no input or output status tables in the ControlLogix PLC as there are in the
SLC 500 or PLC 5 PLC families. The monitor tags window is where input and output status
information is viewed. Figure 5-26 illustrates a portion of the monitor tags window for a
ControlLogix input module. The input address is listed on the left side of the figure as

⊞-Local:2:I.Data	2#0000_0000_0000_0000_0000_0000_0000_0000

Figure 5-26   Analog input signal conversion to binary data.

Local:2:I.Data. The address is composed of the following parts:

Local: = The module input status being viewed is in the local chassis.
2: = This module is in slot two of the chassis.
I. = This module is an input module.
Data = I/O points in the ControlLogix are referred to as data.

The full address including the bit or I/O point number is Local:2:I.Data.3. The dot 3
identifies the bit number of the I/O point.

All information in the ControlLogix is text-based, or simply given a name. This name
is called a tag. Notice the input information is referred to as data rather than inputs (I) or
outputs (O) for other PLCs. The 32 bits representing the input screw terminals are illustrated in the right column of the figure. A 16-point input or output module will use only
the lower 16 bits of the double word. Notice the 2#; this represents the data as being displayed in binary radix, or base two.

We have investigated how discrete field device input and output data moves into and
out of the PLC. Next we need to understand how analog information is represented and
stored inside the PLC.

ANALOG I/O INTERFACE TO THE PLC

Discrete I/O modules allow only discrete, or on/off, devices to be connected to the PLC.
Discrete modules allow on/off control of an application. Continuous process control
applications require the PLC to accept and work with input and output signals that are

constantly changing values such as temperature, pressure, or position. These are not simple two-state signals, but values representing their respective data.

Analog Versus Discrete

Consider a light switch in your home. You flip the switch and the light turns on. You flip the switch again and the light goes off. There is no way with the typical light switch to have the light in any other state than ON or OFF. Making a light switch turn the light halfway on or even three-quarters on is impossible. On the other hand, if you have a light dimmer switch controlling the intensity of your light bulb, you have an *analog* control. By turning the dimmer knob, the brightness of the bulb will vary between totally OFF and full ON. Turning the dimmer knob from OFF to ON varies the voltage to the light bulb in a smooth linear fashion between 0 volts and 115 volts AC.

Analog Signals

A typical PLC analog input or output signal may also be a voltage that may vary between 0 to 10 volts, or current that may vary between 0 to 20 milliamps.

Analog control is employed in applications concerned with continuous process control such as temperature, pressure, humidity, analog valves, load cells, sensing tank levels, variable frequency motor drives, meters, actuators, resolvers, chart recorders, and potentiometers. Analog input modules are used in applications where the input signal is in a continuous and varying form as compared to discrete, strictly ON or OFF, signals. Common Analog Input and Output Voltage and Current Levels analog signal levels are listed below.

$+/- 10$ V dc

0 to 10 V dc

$+/- 5$ V dc

0 to 5 V dc

1 to 5 V dc

0 to 20 mA

4 to 20 mA

$+/- 20$ mA

Temperature is a good example of a signal that continuously changes with time. As temperature varies between 99 and 100 degrees, it does not jump directly from 99 to 100 degrees; there is a steady movement with many different temperature points or values between the two values.

Another example of analog control could be a valve controlling a tank-filling application. The positioning of the valve can be represented by a voltage level of zero volts when the valve is closed, and up to $+10$ V dc for the valve being fully open. The voltage range 0 V dc to 10 V dc would represent 0 to 100 percent open. Since there are many voltage levels between 0 V dc and 10 V dc, the valve could be set to many positioning points to regulate flow. Think of this valve operating in a manner similar to that of the light dimmer mentioned earlier.

Although analog signals like temperature provide a varying input signal to a PLC analog input module, this analog signal must be converted to binary data that can be stored in the data table and used by the PLC when it solves the PLC program.

The analog input module transforms an analog input voltage or current level from a sensor or transmitter into a discrete value. Inside the analog module is an analog-to-digital converter (A/D converter) that makes the conversion. The transformed analog value is the digital equivalent of the analog input signal. As an example, an SLC 500 1746-NI4 analog input module transforms a 0 to 10 V dc signal into a 16-bit binary value with 0 to 32,767 as the resulting value. See Figure 5-27 below.

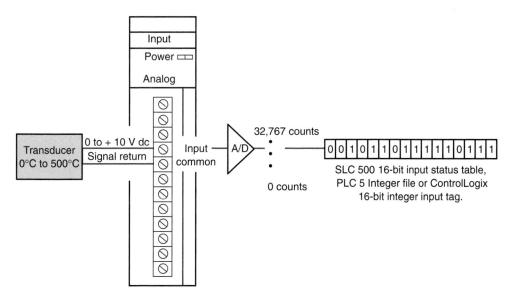

Figure 5-27   SLC 500 four-channel analog input module.

In this example, the PLC's analog input signal was converted into a 16-bit binary equivalent, a 16-bit signed integer. Notice the analog values have been converted into an entire 16-bit word, in comparison to a single bit for a discrete input or output signal. Since analog input and output values are not single digital points, they are not classified as I/O points. Analog values make up an entire data word. Analog input and output module connections are called **channels.** Figure 5-28 shows a four-channel SLC 500 analog input module with four input devices.

The SLC 500 analog input data is automatically stored in the input status table. The PLC 5 requires a set of block transfer instructions to physically transport the data from the analog input module to an integer file location, while the newer ControlLogix PLC automatically stores the converted binary data in the monitor tags window in either integer or floating point form.

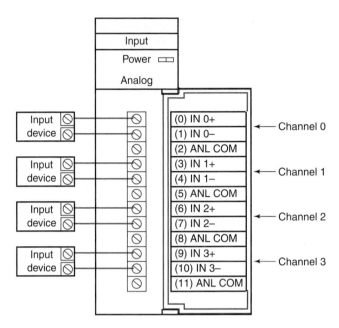

Figure 5-28 SLC 500 four channel analog input module.

Analog Outputs

Another example of analog control could be a valve in a tank-filling application. The positioning of the valve can be represented by a voltage level of zero volts when the valve is closed and up to +10 V dc when the valve is completely open. Any voltage between 0 V dc and 10 V dc would represent how far open the valve would be compared to its fully open state. Table 5-1 illustrates the relationship among the output voltage, the percentage the valve is open, and the decimal equivalent of the digitized binary value for a module that outputs 0 to 32,767 for a 0 to 10 V dc signal.

The PLC program would send a binary value represented by a 16-bit signed integer from the ladder program to the output status table to represent how far to open the valve. The analog output module transforms a digital bit value from the processor's output status table to an analog output voltage or current level. Inside the analog output module, a digital-to-analog converter (D/A converter) is the solid state device that makes the conversion. The transformed binary value is the analog voltage or current equivalent. As an example, an SLC 500 1746-N04V analog output module transforms a 0 to 32,767 digital value into a linear 0 to 10 V dc output signal. In Figure 5-29 on the following page, a digital value of 0 to 32,767 will provide a linear 0 to 10 volt dc signal to open the valve from 0% to 100%.

Module Resolution

The analog-to-digital converter, in the input module, and the digital-to-analog converter, in the output module, perform the signal conversion. A module with a 16-bit converter

Output Volts	Percent valve is open	Digitized Value
0	0	0
1	10	3276
2	20	6554
3	30	9828
4	40	13,106
5	50	16,383
6	60	19,660
7	70	22,936
8	80	26,213
9	90	29,484
10	100	32,767

Table 5-1   Zero to ten volt analog valve values translated into percentages and digitized binary values.

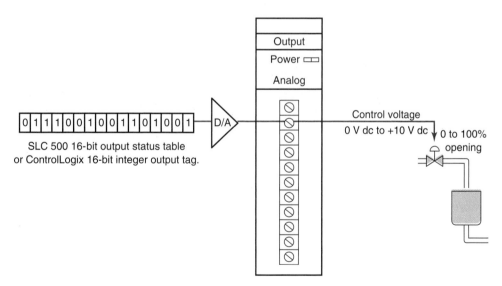

Figure 5-29   Analog output 16-bit binary signal conversion to a control voltage.

can divide the input signal into up to 32,767 bits. This is called 16-bit resolution, as the analog value is converted into a 16-bit signed integer. The number of bits in the digital value that corresponds to the full scale analog value is called the module resolution. Resolution is the smallest amount of change the module can detect. The greater the number of bits, the greater the resolution.

Not all analog input or output modules convert their analog input values into 16 bits. The SLC 500 1746-FIO4I can convert the analog input signal into 12-bit binary values for storage in the processor's input image table. In this case, a 0 to 10 V dc input signal range is digitized into 0 to 4,095, similar to the PLC 5. If a module has 16-bit resolution, the least significant bit (LSB) is equal to approximately 305 microvolts. An analog input module with 12-bit resolution has approximately 2.44 millivolts per the LSB.

This digital representation of the analog input signal is called the raw data. The raw data is typically scaled into the desired engineering units. This data could represent temperature, pressure, or motor speed. As an example, 0 to 32,767 might represent the raw motor speed data coming from a variable frequency drive into a PLC. The raw data needs to be converted to, or scaled to 0 to 1,750 RPM for display on a human interface device. The SLC 500 5/03, 5/04, and 5/05 processors allow analog scaling by programming a Scale with Parameters instruction on the ladder rungs. The MicroLogix, 5/02, 5/03, 5/04 and 5/05 can use the older SCL (Scale Data) instruction. We will work with both scaling instructions in the accompanying programming lab manual. Figure 5-30 illustrates an RSLogix 500 software Scale with Parameters instruction. The Scale with Parameters instruction needs to have the following information provided by the programmer:

Where is the input data coming from? This is the input part of the instruction.

The raw data input minimum value or address.

The raw data input maximum value or address.

The scaled data minimum value or address.

The scaled maximum value or address.

Where to store the scaled data, or the output of the instruction.

Figure 5-30 illustrates a scale with parameters instruction for an SLC 500 PLC where a 0 to 10 volt pot is inputting an analog signal. When the operator turns the pot address I:7.0 from minimum to maximum (0–10 V dc) the PLC will send a signal to a variable frequency drive to run the motor from 0 to 1,750 RPM. The raw data coming in from the analog

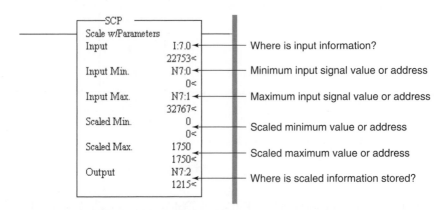

Figure 5-30   SLC 500 scale with parameters instruction.

module is represented as 0 to 32,767. These are the instructions' input minimum and input maximum parameters. This data is stored in N7:0 and N7:1, respectively. The raw data is scaled to 0 to 1,750 for RPM. These are the instructions' scaled minimum and scaled maximum parameters. Notice that these two parameters are not data table addresses, but constants. The instructions' input minimum, input maximum, scaled minimum, and scaled maximum can be programmed either as data table addresses or constants. The scaled data is stored in N7:2 for use in the PLC program. The scaled data stored in N7:2 will be sent to the human interface device to display the desired motor speed.

The PLC 5 and the ControlLogix PLC allow for the raw data to be scaled as part of the I/O configuration process. Your specific PLC will dictate how and where the digital raw input data is scaled for use in the program, and how and where it is scaled before the data is output to the analog module.

Analog Input Addressing and Analog Channel Data Table Representation

Since analog input and output channels are represented as whole words, there must be one word for each analog input channel and output channel in the appropriate status table. Channel representation in the status table not only must represent the slot where the analog module resides, but also the channel word number. Let's assume an SLC 500 1746-NI4, a four-channel analog input module, is in slot three. The input status table must identify the slot and an individual word for each of the four channels. Figure 5-31 illustrates an SLC 500 1746-NI4 channel representation in the input status table.

Notice the analog input module is represented as I:3.0, I:3.1, I:3.2, and I:3.3. The I identifies it as an input module. The three identifies the slot number as three. The channel number is represented by 0 through 3. Each channel's data will be stored in a 16-bit

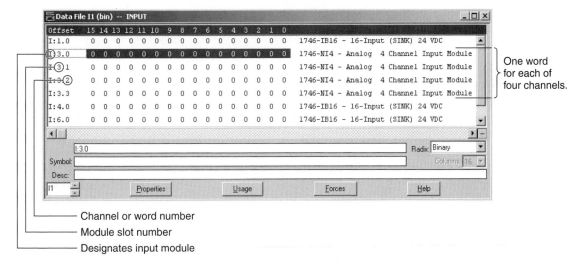

Figure 5-31   Four input channel 1747-NI4 analog module channel representation in an RSLogix 500 input status table.

word. Figure 5-32 illustrates the PLC chassis, analog module, and channels, and the address for channel three.

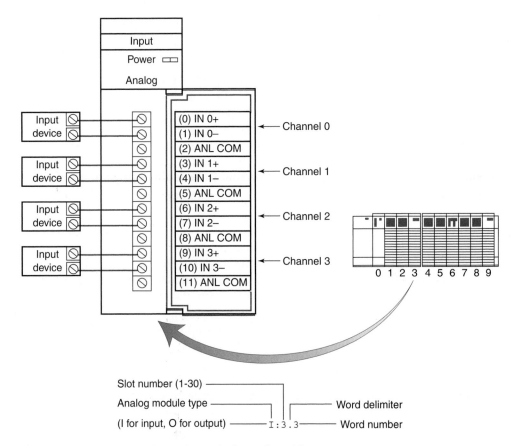

Figure 5-32 Analog input channel correlation to its address.

Analog Output Addressing

Each analog output channel requires one 16-bit word to represent the digitized input information. Each channel's address must identify that the information is an input, the slot where the module resides, and the channel, or word number. Figure 5-33 illustrates a four-channel SLC 500 analog output module. If the module were in chassis slot seven, the address of channel two would be O:7.2—output, slot seven, word two.

Combination Input and Output Analog Modules

The SLC 500 1746-NIO4V analog module has two input channels and two output channels. Two input status table words and two output status table words are necessary to

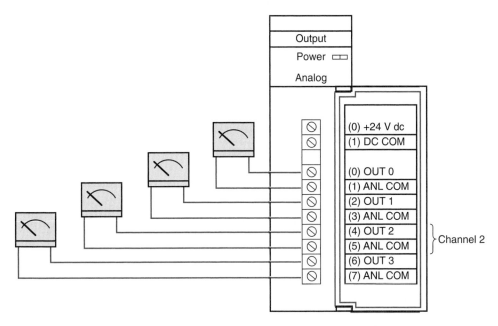

Figure 5-33   Channel two analog output address for module in chassis slot seven.

represent the analog channel data. If this module were in slot five of the chassis, the designation would be I:5.0 and I:5.1 for the input channels, and O:5.0 and O:5.1 for the output channels. I:5.1 in Figure 5-34 shows the input value as a 16-bit signed integer whereas bits 0 through 14 represent the analog input value, currently 32,767. Notice the sign bit is 0, representing a positive 10 V dc input value. Refer back to Chapter 4 for a review of 16-bit signed integers.

The address is listed on the left side, and the 16-bit word representing the value is near the center. The description of the modules and their respective slots is listed on the

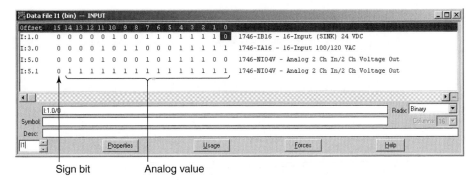

Figure 5-34   A two channel analog input and two channel analog output modules word or channel representation in the status tables for the RSLogix 500 software.

right. Figure 5-35 is the same RSLogix 500 software view of the input status table. Notice that the radix of the display has been changed from binary to decimal. For this module, a 0 to 10 V dc input value is represented by 0 to 32,767. The figure shows a value of 32,767, representing a positive 10 V dc input signal.

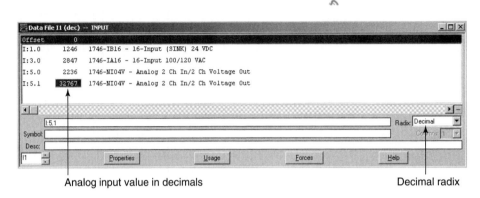

Analog input value in decimals   Decimal radix

Figure 5-35   RSLogix 500 software input status table view displayed in decimal radix.

Figure 5-36 shows the output status table for this SLC 500 PLC. The output addresses are listed on the right side of the output status table. Notice that there are five output words, word O:2, word O:4, word O:5.0, and word 5.1, along with word O:6. The O identifies these as output words. Slots two and four have 16-point discrete modules, and slot five has a two-channel analog module. There are two words, O:5.0 and O:5.1, for the module in slot five. The right side of the window represented in the figure identifies a two-channel analog input and two-channel voltage analog output module in slot five. The two words in the output status file represent the two analog channels for this particular module. In this case, O:5.0 represents channel zero output data, and O:5.1 represents channel one output data. The figure also shows that slot six is an eight-point output module.

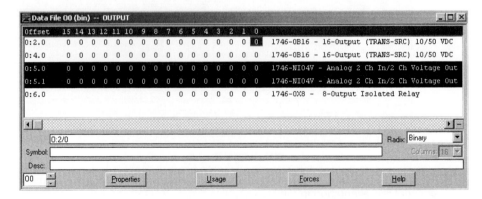

Figure 5-36   Illustrates the output status table from the RSLogix 500 software for the SLC 500.

THE CONTROLLOGIX PLC AND ANALOG

ControlLogix analog input or output data can be either integer or floating point. This is selected by the programmer when the analog module properties are set up in the RSLogix 5000 software I/O configuration. If integer format is selected, input data will be represented as 16-bit, and output data will be either 13- or 16-bit, depending on the specific output module selected. Figure 5-37 illustrates a ControlLogix analog module properties screen in the I/O configuration of the RSLogix 5000 software. For this particular module, floating point or integer format are the basic selections. There are additional, more advanced selections.

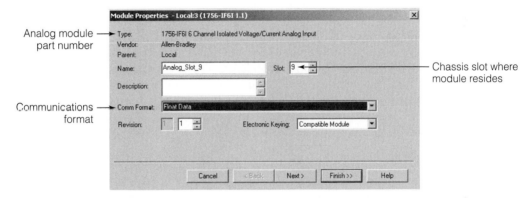

Figure 5-37 ControlLogix 1756-IF6I six-channel analog input module properties screen.

Figure 5-38 illustrates the module properties screen where scaling data is set up. The small boxes across the top left represent each of the six channels for this analog input module. Each channel is set up independently by clicking on the desired box and completing the setup. The input range, top right, is the analog signal coming into the module.

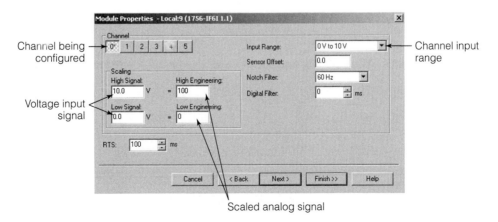

Figure 5-38 ControlLogix 1756-IF6I analog module channel scaling configuration screen.

Notice the scaling section, center left. Scaling is set up in this area. For this example, assume 0 to 10 volts input signal is scaled to 0 to 100.

As an example of ControlLogix analog module resolution, the 1756-IF6I, which is a six-channel isolated voltage/current analog input module, has 16-bit resolution. This module uses all 16 bits to represent a value; thus there are 65,536 counts or values available. Even though the resolution is fixed at 16 bits, the value of each count is determined by the input signal range selected in the properties screen from Figure 5-38, if the input range from Figure 5-38 was 0 to 10 volts. The specifications for the module from the Rockwell literature states that the module's actual input range is 0 to 10.5 volts. Divide the range by the number of counts to determine the resolution or value of each count. In this case, 10.5 volts/65,536 equals about 160 microvolts per count.

ControlLogix input and output data is also stored in the monitor tags window and can be viewed there for each module on a slot-by-slot basis. ControlLogix analog addresses are somewhat similar to the SLC 500, as the I/O module slot number and channel number is part of the address. A ControlLogix analog input or output channel is identified by chassis location, module slot number, type of data input or output, and channel number. I/O information for the ControlLogix is referred to as data. ControlLogix' analog address for channel three of an input module in chassis slot nine, as in the above examples, would look as follows:

Local:9:I.Ch3Data

This address identifies the analog point as in the local chassis, chassis slot nine, an input signal, analog channel three information, or data. Figure 5-39 illustrates part of a ControlLogix monitor tags view showing the analog input addresses and their respective floating point data. The current value in Local:9:1.CH3 Data is 97.43.

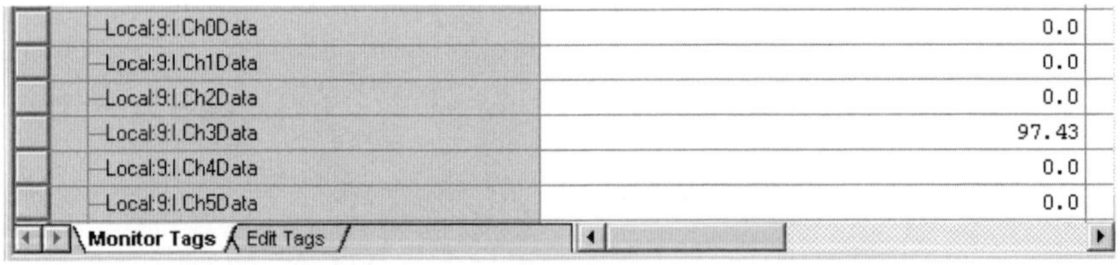

	Local:9:I.Ch0Data	0.0
	Local:9:I.Ch1Data	0.0
	Local:9:I.Ch2Data	0.0
	Local:9:I.Ch3Data	97.43
	Local:9:I.Ch4Data	0.0
	Local:9:I.Ch5Data	0.0
◀ ▶	Monitor Tags ⟨ Edit Tags /	◀ ▶

Figure 5-39   ControlLogix monitor tags window for viewing analog input data.

SUMMARY

Input signals are wired to input module screw terminals, each of which is assigned a specific address. Address assignment is determined, not only by the screw terminal number identified on the module, but also by the position of the module in the PLC chassis.

Input signals coming into the PLC are transformed by the input section electronics to either a valid ON or a valid OFF signal. The PLC will read a +5 V dc signal as ON and a 0 V dc signal as OFF. The ON or OFF status will be stored in the input status file, which represents the actual ON or OFF status of the incoming, or input, signals as seen by the PLC. This input data will be stored in the input status file simply as either a 1 or a 0. These binary 1s and 0s stored in the status file can then be used by the processor to solve the user program.

Analog control is employed in applications concerned with continuous process control such as temperature, pressure, humidity, analog valves, load cells, sensing tank levels, variable frequency motor drives, meters, actuators, resolvers, chart recorders, and potentiometers. Analog input modules are used in applications where the input signal is continuous and varying as compared to discrete, strictly ON or OFF signals. Although analog signals such as temperature provide a varying input signal to a PLC analog input module, this analog signal must be converted to binary data that can be stored in the data table and used by the PLC when it solves the PLC program. Inside the analog module is an analog to digital converter (A/D converter) that makes the conversion. The transformed analog value is the digital 16-bit word equivalent to the analog input signal. This digital representation of the analog input signal is called the raw data. The raw data is typically scaled into the desired engineering units.

On the output side, the PLC program would send a binary value represented as a 16-bit signed integer from the ladder program to the output status table to represent how far to open the valve, as an example. The analog output module transforms a digital bit value from the processor's output status table to an analog output voltage or current level. Inside the analog output module, a digital to analog converter (D/A converter) is the solid state device that makes the conversion. The transformed binary value is the analog voltage or current equivalent.

Because analog input and output values are not single digital points, they are not classified as I/O points. Analog values make up an entire data word; analog input and output module connections are called channels. Analog input and output channels are represented as whole words—there must be one word for each analog input channel and output channel in the appropriate status table. Channel representation in the status table not only must represent the slot where the analog module resides, but also the channel word number.

The user program is sometimes called the logic, which the PLC is said to solve. Your user program will be used to combine the appropriate input signals, and when all appropriate signals combine to cause continuity on a rung in the **ladder diagram,** the associated field output device will be turned ON or OFF. Outputs are controlled through the output status file and the PLC output module.

REVIEW QUESTIONS

1. Explain the difference between a discrete, a binary, and a digital signal.
2. What is the difference between an analog and a binary input signal?
3. The ControlLogix is the latest addition to the Rockwell Automation family of PLCs. ControlLogix is a _____ bit processor as compared to 16-bit processors for earlier PLCs such as the _____ and _____ families of PLCs.

4. ControlLogix stores information in groups of bits similar to the 16-bit PLC words; however, being a 32-bit processor, ControlLogix' basic allocation of memory is _____ bits, or a _____.

5. Explain how a light switch in your home could be either an analog or a digital control.

6. Devices that provide two-state signals as inputs to, or outputs from, a PLC are called _____ or _____ signals.

7. Another term associated with something that can only exist in one of two predetermined states is the term _____.

8. Fill in Figure 5-40 to specify the ON and OFF statuses of the listed discrete field devices.

DISCRETE I/O DEVICES		
Example	OFF—Logical 0	ON—Logical 1
Pilot light		
Limit switch		
Two-state valve		
Motor starter		
Alarm bell		
Proximity sensor		

Figure 5-40   Logical states of discrete field devices.

9. Calculate the resolution for a 1756-IF6I ControlLogix analog module, which is a six-channel isolated voltage/current analog input module with 16-bit resolution. In the RSLogix 5000 software module property screen, the input range was set as 0 to 20 mA. Specifications for the module from Rockwell literature state that the module's actual input range is 0 to 21 mA. What is the module resolution?

10. If you are using a 24-point input module, how can input signals be represented given that there are only 16 bits in a data word?

11. When working with computers, the first bit and first word are always referred to as bit and word _____.

12. A _____ is nothing more than grouping words containing like data.

13. Words that are associated with the real-world input status of the available input modules in your PLC system are stored in a data file called the _____.

14. Most computer circuits operate on 5 V dc. A _____ ON signal typically is from 2.4 V dc to 5 V dc. A valid OFF signal is typically from _____.

15. If a valid ON signal is between 2.4 V dc and 5 V dc and a valid OFF signal is between 0 V dc and .8 V dc, explain how an input signal measuring somewhere between the valid OFF maximum and the valid ON minimum would behave.

16. Illustrate the parts of a 16-bit word in Figure 5-41. Identify how hexadecimal data would be formatted.

Figure 5-41   Hexadecimal data word format.

17. The ControlLogix PLC automatically stores the converted binary data in the monitor tags monitor window in either _____ or _____ form.
18. Digital circuits operate on a simple concept: a circuit is either _____ or _____.
19. To process input data, two-state data must first be converted into an electronic form that the PLC can understand and work with. This is the job of the _____.
20. A digital circuit is as simple as a light switch. The light switch and a digital circuit have one thing in common; they both have two possible states: _____ or _____.
21. Fill in Figure 5-42 with a list of analog field devices.

ANALOG I/O DEVICES	
Inputs	Outputs

Figure 5-42   Analog field devices.

22. Analog input modules are used in applications where the input signal is in a _____ and _____ form as compared to discrete strictly ON or OFF signals.
23. Common Analog Input and Output Voltage and Current Levels analog signal levels include:
    A. + /− 10 V dc
    B. 0 to 10 V dc
    C. + /− 5 V dc
    D. 0 to 5 V dc
    E. 1 to 5 V dc
    F. 0 to 20 mA
    G. 4 to 20 mA

H.   + /− 20 mA

I.   All of the above

J.   only A and C

24. Although analog signals, such as temperature, provide a varying input signal to a PLC analog input module, this analog signal must be converted to _____ data that can be stored in the data table to be used by the PLC when it solves the PLC program.

25. The analog input module transforms an analog input voltage or current level from a sensor or transmitter into a _____ value.

26. A standard input module has:

A.   10 input points

B.   12 input points

C.   16 input points

D.   20 input points

E.   either B or C

F.   any of the above

27. Each input module corresponds to one _____ of data in the input status file.

28. Each bit position will have either a _____, if the input is _____, or a _____, if the input is _____.

29. A _____ is the smallest unit of microprocessor data that can be represented.

A.   single 16-bit word

B.   single nibble

C.   single bit

D.   single byte

E.   either B or C

30. Inside the analog module is an _____ that makes the conversion from a continuous signal to a binary value.

31. The transformed analog value is the digital equivalent to the analog input signal. As an example, an SLC 500 1746-NI4 analog input module transforms a 0 to 10 V dc signal into a 16-bit binary value with 0 to _____ as the resulting value. This is called the _____ data.

32. A PLC's analog input signal was converted into a 16-bit binary equivalent, a _____ integer.

33. Analog values are converted into an entire _____, in comparison to a single bit for a discrete input or output signal.

34. Because analog input and output values are not single digital points, they are not classified as I/O points. Analog input and output module connections are called _____.

35. True or false. All analog input modules' raw input data is always 0 to 32,767.

36. The SLC 500 analog input data is automatically stored in the input status table. The PLC 5 requires a set of _____ instructions to physically transport the data from the analog input module to an integer file location, whereas the newer ControlLogix PLC automatically stores the converted binary data in the _____ window.

37. List ten applications where analog control is employed.

38. The SLC 500 PLC program sends a binary value represented as a 16-bit signed integer from the ladder program to the _____ to represent how far to open an analog valve.

39. The analog output module transforms a digital bit value from the processor's output status table to an _____.

40. Inside the analog output module, a _____ is the solid state device that makes the conversion.

41. The transformed binary value is the analog voltage or current equivalent. As an example, an SLC 500 1746–NI4 analog output module transforms a 0 to _____ digital value into a linear 0 to 10 V dc output signal.

IF PASSWORD IS CORRECT GO TO DRIVE SET UP SCREEN

0022
```
  ┌── EQU ──────────┐        B3:10/0                                    B3:1/15
  │ Equal            │        ┌ OSR ┐                                     ( )
  │ Source A   N12:21│        └─────┘                              ┌──────────────┐
  │              7777<│                                            │RESET PASSWORD│
  │ Source B    32767│                                            │ATTEMPTS COUNTER│
  │             32767<│                                            │    C5:5      │
  └──────────────────┘                                            │   ( RES )    │
```

0023
```
  ┌── GRT ──────────────┐    ┌── NEQ ──────────┐   B3:10/1
  │ Greater Than (A>B)   │    │ Not Equal        │   ┌ OSR ┐
  │ Source A     N12:21  │    │ Source A   N12:21│   └─────┘
  │                7777< │    │             7777<│
  │ Source B         0   │    │ Source B   32767 │
  │                   0< │    │            32767<│
  └─────────────────────┘    └─────────────────┘
```

```
                                      ┌── CTU ──────────┐   ( CU )
                                      │ Count Up         │
                                      │ Counter    C5:5  │
                                      │ Preset        3< │   ( DN )
                                      │ Accum         2< │
                                      └─────────────────┘
```

```
                                              −1
                                              −1<
                                            N12:21
                                             7777<
```

NOTIFICATION   HANDSHAKE
B3:0/7   B3:0/8
0024] [( )

Normal state of passw...   ...try attempts counter is equal
to zero.

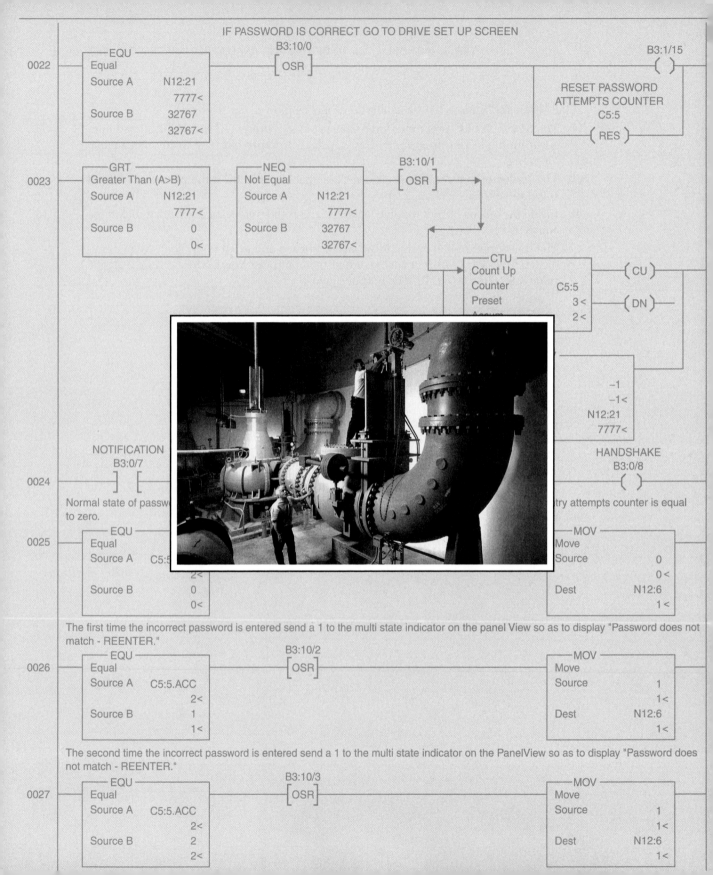

0025
```
  ┌── EQU ──────────┐                                   ┌── MOV ──────────┐
  │ Equal            │                                   │ Move             │
  │ Source A   C5:5..│                                   │ Source        0  │
  │              2<  │                                   │               0< │
  │ Source B     0   │                                   │ Dest    N12:6    │
  │              0<  │                                   │               1< │
  └──────────────────┘                                   └─────────────────┘
```

The first time the incorrect password is entered send a 1 to the multi state indicator on the panel View so as to display "Password does not match - REENTER."

0026
```
  ┌── EQU ──────────┐        B3:10/2                    ┌── MOV ──────────┐
  │ Equal            │        ┌ OSR ┐                    │ Move             │
  │ Source A C5:5.ACC│        └─────┘                    │ Source        1  │
  │              2<  │                                   │               1< │
  │ Source B     1   │                                   │ Dest    N12:6    │
  │              1<  │                                   │               1< │
  └──────────────────┘                                   └─────────────────┘
```

The second time the incorrect password is entered send a 1 to the multi state indicator on the PanelView so as to display "Password does not match - REENTER."

0027
```
  ┌── EQU ──────────┐        B3:10/3                    ┌── MOV ──────────┐
  │ Equal            │        ┌ OSR ┐                    │ Move             │
  │ Source A C5:5.ACC│        └─────┘                    │ Source        1  │
  │              2<  │                                   │               1< │
  │ Source B     2   │                                   │ Dest    N12:6    │
  │              2<  │                                   │               1< │
  └──────────────────┘                                   └─────────────────┘
```

CHAPTER

6

Introduction to Logic

OBJECTIVES

After completing this chapter, you should be able to:

- explain AND logic
- explain OR logic
- explain NOT logic
- develop and fill in a truth table for specified logic
- determine if a PLC ladder rung is true or false under specified conditions
- identify ControlLogix Boolean function blocks

INTRODUCTION

This chapter introduces the concepts of how logic functions are executed when solving a PLC program. We will introduce how a PLC solves a user program using AND, OR, and NOT logic.

Conventional hardwired relay ladder diagrams represent actual hardwired control circuits. In a hardwired circuit, there must be electrical continuity before the load will energize. Even though PLC ladder logic was modeled after the conventional relay ladder, there is no electrical continuity in PLC ladder logic. PLC ladder rungs must have logical continuity before the output will be directed to energize.

Function block diagram programming is a relatively new programming language for American PLCs. Rockwell Automation's ControlLogix family of PLCs is the first Rockwell Automation/Allen-Bradley PLC that supports function block diagram programming. Function block, as it is called, is an additional programming language option where boxes called function blocks replace traditional relay ladder instructions. This chapter will introduce Boolean function blocks. Boolean function blocks replace the traditional normally open and normally closed instructions.

CONVENTIONAL LADDERS VERSUS PLC LADDER LOGIC

The familiar electrical ladder diagram is the traditional method of representing an electrical sequence of operation in hardwired relay circuits. Ladder diagrams are used to represent the interconnection of field devices. Each rung of the ladder clearly illustrates the relationship of turning on one field device and shows how it interacts with the next field device. Due to industrywide acceptance, ladder diagrams became the standard method of providing control information to users and designers of electrical equipment. With the advent of the programmable controller, one of the specifications for this new control device was that it had to be easily programmed. As a result, programming fundamentals were developed directly from the old familiar ladder diagram format, with which electrical maintenance personnel were already familiar.

The difference between a PLC ladder program and relay ladder rungs involves continuity. An electrical schematic rung has electrical continuity when current flows uninterrupted from the left power rail to the right power rail. Electrical continuity, as illustrated in Figure 6-1, is required to energize the load. An electrical current flows from L-1 through SW 1 and onto the load, returning by way of L-2. When current flows, there is electrical continuity to the status table, where it is stored.

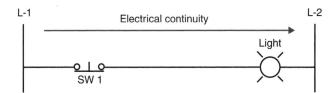

Figure 6-1 Hardwired relay circuit.

Even though a PLC ladder program closely resembles an electrical schematic, PLC ladder symbols represent ladder program instructions. A PLC program is a set of instructions that is stored in memory. These instructions tell the PLC what to do with input signals and then, as a result of following the instructions, where to send the signals.

A PLC relay ladder program uses an electrician's ladder schematic as a model. Even though a PLC ladder program employs familiar terms like "rungs" and "normally open" and "normally closed" contacts, relay ladder logic has no electrical continuity between an input and the controlled output. There is no physical conductor that carries the input signal through to the output. As introduced in Chapter 5, a PLC input signal follows these six steps:

1. The input signal is seen by the input module.
2. The input module isolates and converts the input signal to a low-voltage signal with which the PLC can work.

3. The ON or OFF signal from the input section is sent via the backplane to the input status file, where it is stored.
4. The processor will look at each input's ON or OFF level as it solves the user program.
5. The resulting ON or OFF action, as a result of solving each rung, is sent to the output status file for storage.
6. During the output update portion of the scan, the processor will send the ON or OFF signal from each bit in the output status file to the associated output screw terminal by way of the output module.

Individual ladder-programming symbols are represented as instructions in the CPU section of Figure 6-2. The notations I:01, I:02, and O:01 represent the instructions and their addresses. When programming the PLC, these instructions are entered one by one and stored sequentially in the user program portion of the processor's memory. When the PLC is in run mode, these instructions are combined to arrive at the resulting ON or OFF state of each rung's output.

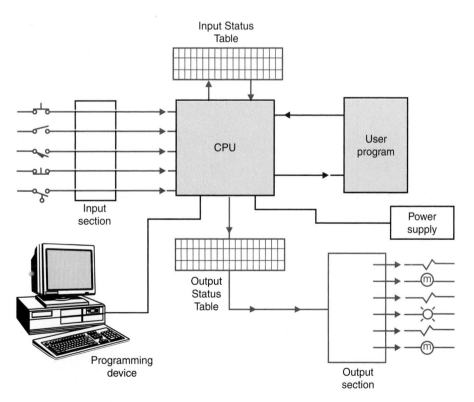

Figure 6-2   Signal flow into and out of a PLC. Notice that there is no electrical continuity between the inputs and the controlled output.

The PLC follows, or executes, the instructions stored in its memory the same way you might follow instructions to make, say, packaged grape drink. The package instructions instruct you to do the following:

1. Get one cup of sugar.
2. Put two quarts of water and the sugar into a container.
3. Add the contents of the package.
4. Mix until uniform.
5. Grape drink is ready to serve.

Following this procedure will provide an end product (grape drink). Likewise, the PLC follows the instructions programmed in its memory to achieve an end product.

WHAT IS LOGIC?

Devices in an electrical schematic diagram are described as being open or closed. PLC ladder instructions are typically referred to as either true or false. When a PLC solves the user program, it is said to be solving the ladder logic.

See Figure 6-3 for a look at a PLC ladder program printout.

```
Rung  2:0
      I:0      I:0       I:0                                              O:0
-+--] [---]/[--+--] [------------------------------------------ ( )--
 |     2     4  |     5                                                    3
 |    I:0       |
 +--] [---------+
       0
Rung 2:1
    I:0    I:0                                                         O:0
--] [---] [----------------------------------------------------- ( )-
    1      5                                                          2
Rung  2:2
     I:0                         I:0       I:0                      O:0
-+--] [-------------+--] [--+--] [--+---------+--( )--+-
 |    3            |     7   |    8  |         |    1  |
 |       I:1       |        I:1      |        O:0     |
 +-+------] [--+-+       +--]/[--+   +--( )--+
 |         6   |              4         5
 |        I:0  |
 +------] [--+
         6
Rung 2:3
    I:0                                                               O:1
--] [----------------------------------------------------------- ( )-
    8                                                                 10
```

Figure 6-3 Printout of SLC 500 ladder logic developed on DOS-based Rockwell Software's A.I. Series PLC 500 Ladder Logistics software.

Each rung is a program statement. A program statement consists of a condition, or conditions, along with some type of action. Inputs are the conditions, and the action, or output, is the result of the conditions. Each PLC ladder rung can be looked at as a problem the processor has to solve. The PLC combines ladder program instructions similar to the physical wiring hardware devices in series or parallel. However, rather than working in series or parallel, the PLC combines instructions logically using logical operators. Logical operations performed by a PLC are based on the fundamental logic operators: AND, OR, and NOT. These operators are used to combine the instructions on a PLC rung so as to make the outcome of each rung either true or false. The symbol that represents the result of solving the input logic on a particular rung is the output.

OVERVIEW OF LOGIC FUNCTIONS

To understand and program programmable controllers, we must understand basic logic. Three logic functions will be introduced here.

One Instruction Combined in Series with Another Is "AND"

You performed a logical operation when you mixed the grape drink. Although you might not realize it, you performed AND logic. Let's rewrite our drink-mixing task to see how it relates to AND logic:

1. Get one cup of sugar.
2. AND put two quarts of water AND the sugar into a container.
3. AND add contents of package.
4. AND mix until uniform.
5. Grape drink is ready to serve.

AND logic is similar to placement in series, as all series devices must pass continuity before the outcome will be allowed to happen. Likewise, when mixing the grape drink, all steps must be performed before a satisfactory beverage is produced.

One Instruction Combined in Parallel with Another Is OR

Our drink-mixing example can be made into OR logic (a parallel operation). When mixing our drink, there may be a choice whereby grape flavoring OR orange flavoring can be used. By following the same instructions but then choosing either grape flavor or orange flavor, OR logic is carried out.

The Opposite of a Normally Open Instruction Is a Normally Closed Instruction

If a set of normally open contacts is represented by a normally open instruction and a set of normally closed contacts is represented by a normally closed instruction, the normally open instruction must be energized to become true, while the normally closed instruction must not be energized to be true. Since there are only two states in digital logic, the normally closed instruction represents the opposite of the normally open instruction.

The term NOT is used for the normally closed instruction, as the input is not energized for the normally closed instruction to be true. NOT logic also refers to the normally closed instruction because the output is the inverse of the input. If the input is true, the instruction will be evaluated as NOT true. If the input is NOT true, the instruction will be evaluated as true. As a result of the input instruction being the opposite, or inverted, in relation to the instruction's status, NOT logic is also referred to as an inverter. Different manufacturers will refer to the normally closed instruction and its logical function with different terminology; however, all terms work in the same manner.

Let's explore these logic functions and see how they relate to PLC ladder programs.

SERIES—THE *AND* LOGIC FUNCTION

The old familiar series circuit can also be referred to as an AND logic function. In the series circuit (Figures 6-4 and 6-5), switch 1 AND switch 2 must be closed to have electrical continuity. When there is electrical continuity, output (light 1) will energize. The key word here is AND.

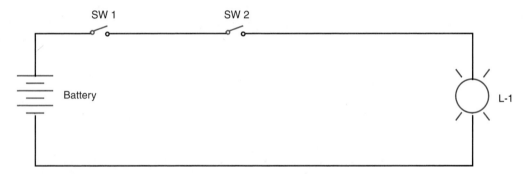

Figure 6-4   Conventional series circuit.

The circuit in Figure 6-4 is represented as a schematic diagram ladder rung in Figure 6-5.

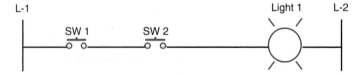

Figure 6-5   Series circuit represented as a conventional ladder rung.

Closing switch 1 and switch 2 will provide power, or electrical continuity, to L-1. This is illustrated in Figure 6-6.

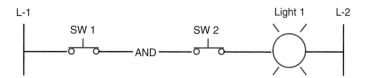

Figure 6-6   Switch 1 AND Switch 2 need to be closed to energize L-1.

Let us look at all the possible combinations that switch 1 (SW 1) and switch 2 (SW 2) can have, and the resulting output signals (Figure 6-7).

POSSIBLE SWITCH CONDITIONS AND RESULTING OUTPUT		
SW 1	SW 2	Light 1
OFF	OFF	OFF
OFF	ON	OFF
ON	OFF	OFF
ON	ON	ON

Figure 6-7   Truth table for AND logic.

Figure 6-7 is called a **truth table.** All possible input configurations for switches 1 and 2 are listed in the two left-hand columns. The expected output signal for light 1 is listed in the right-hand column. From the truth table, you can see that only when switch 1 AND switch 2 are ON will the output (light 1) energize.

Figure 6-8 is an example of the ladder program instructions that would be entered using a handheld programmer and an Allen-Bradley MicroLogix 1000. The ladder rung in Figure 6-8 is identical to that in Figure 6-6 except that the symbols have been changed to PLC ladder format.

```
      SW 1    SW 2              L-1
      I 1     I 2              0 5
 ┤ ├─────┤ ├──────────( )──
```

Figure 6-8   PLC representation of the rung represented in Figure 6-6.

Here is the program listing for the MicroLogix 1000 if you are entering the program with a handheld programmer:

LOAD I1

AND I2

OUT O5

The instructions tell the processor to load input 1 (I1) into memory, AND it with input 2 (I2), and then output the result to output 5 (O5). The resulting output will be determined by the truth table (Figure 6-7). The truth table represents the rules for ANDing two inputs together.

Before we look at the truth table in Figure 6-9, let's update our input status from OFF/ON to a more commonly accepted form. OFF is the same as no power or power not ON, and can be represented by the symbol 0. ON is the same as power ON, which can be represented by the symbol 1. We will use the commonly accepted symbol 1 to represent the presence of a valid signal and the commonly accepted symbol 0 to represent the absence of a valid signal.

Our truth table for the previous example would look as follows (Figure 6-9):

TWO-INPUT *AND* TRUTH TABLE		
SW 1	SW 2	Light 1
0	0	0
0	1	0
1	0	0
1	1	1

Figure 6-9   Truth table for two-input AND logic. This truth table says the same thing as the one in Figure 6-7.

The truth table in Figure 6-9 can also be represented as shown in Figure 6-10.

```
First condition
        FF = ?
        False  AND  False  =  False
                    or
        0 AND 0 = 0
Second condition
        FT = ?
        False AND True = False
                    or
        0 AND 1 = 0
Third condition
        TF = ?
        True AND False = False
                    or
        1 AND 0 = 0
Fourth condition
        TT = ?
        True AND True = True
                    or
        1 AND 1 = 1
```

Figure 6-10   Explanation of Figure 6-9 truth table.

THREE-INPUT *AND* LOGIC

Figure 6-11 has three switches in series controlling the load L-1. The conventional series circuit is shown in Figure 6-12. Figures 6-11 and 6-12 state that switch 1 AND switch 2 AND switch 3 must be energized before output L-1 will occur.

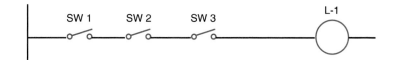

Figure 6-11   Three-input series circuit.

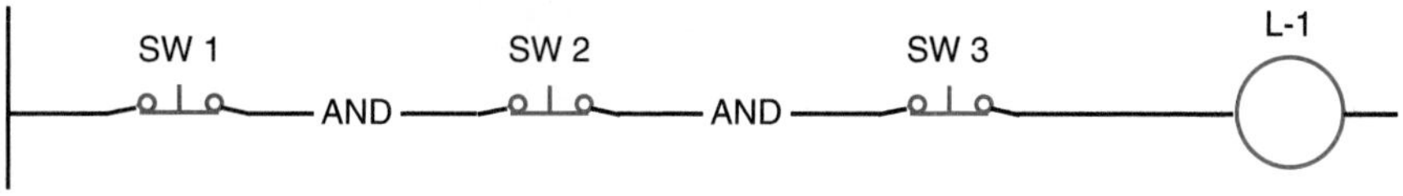

Figure 6-12   Three-input AND circuit.

Figure 6-13 illustrates the converted PLC ladder rungs for the SLC 500 PLC as the top rung. The center rung is for a PLC 5, and the bottom rung is for the ControlLogix PLC. Each rung has its respective input and output addresses.

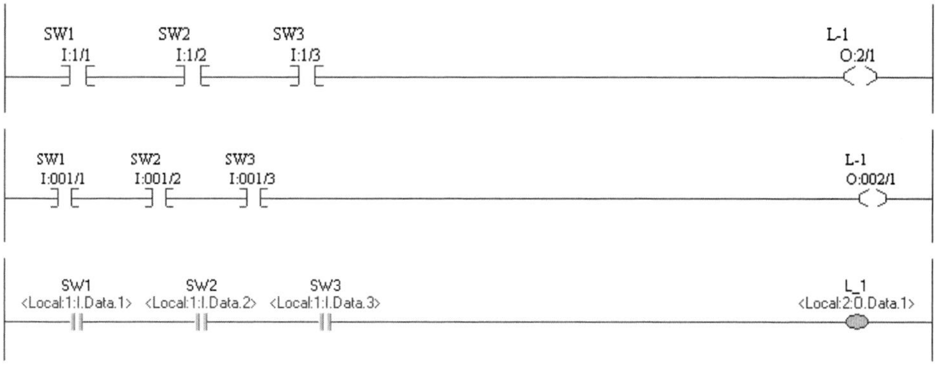

Figure 6-13   PLC three-input AND logic.

The following is the program listing for Figure 6-13 if you are programming with the Allen-Bradley MicroLogix 1000 and entering the program with a handheld programmer:

LOAD I1

AND I2

AND I3

OUT O1

The instructions tell the processor to load input 1 (I1) into memory, AND it with input 2 (I2), AND the result of the previous logical operation with input 3 (I3), and OUTPUT the result to output 1 (O1).

Let's assume that all three inputs are false. If I1, which is false, is loaded into memory and then ANDed with I2, which is also false, the result of this logical AND operation is false. Now we AND I3, which is false, to the result of I1 and I2, which was false. ANDed with I3, the output is also false.

Figure 6-14 is a truth table illustrating the expected outputs for three-input AND logic.

THREE-INPUT *AND* LOGIC			
Switch 1	Switch 2	Switch 3	Light 1
0	0	0	0
0	0	1	0
0	1	0	0
0	1	1	0
1	0	0	0
1	0	1	0
1	1	0	0
1	1	1	1

Figure 6-14   Three-input AND logic truth table.

FUNCTION BLOCK DIAGRAM *AND* LOGIC

Rockwell Automation's ControlLogix PLC family can be programmed in function block diagram language in addition to standard ladder logic. The PLC 5 and SLC 500 families do not support function block.

The principles of combining function block inputs are basically the same as ladder logic. Instead of normally open ladder logic symbols, boxes are used and referred to as function blocks. Figure 6-13 illustrates three rungs, three-input AND ladder logic. A function block diagram will represent the same logic using a Boolean AND (BAND) function block. Figure 6-15 illustrates a BAND function block. Notice the three blocks to the left of the BAND function block; they have Data.1, Data.2, and Data.3 as part of the information inside. These are function block input references. Refer back to Chapter 3 to review basic function block diagram elements. Input references represent the input address where the information is coming from. ControlLogix I/O addressing is a little different than traditional PLC addressing. Input and output addresses for the ControlLogix platform are referred to as tags. The ControlLogix PLC identifies physical inputs and outputs

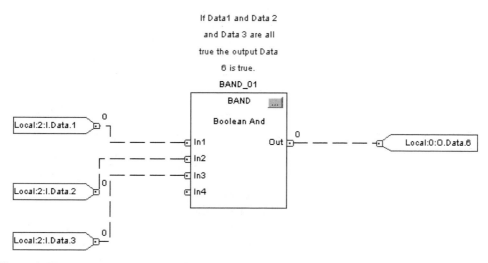

Figure 6-15   Function block BAND.

as data. In this example we are going to AND the three input references together. Here we are looking to see if Data.1 AND Data.2 AND Data.3 are all true. That is the job of the BAND block.

On the left side of the function block are the input points to the block. These are called pins. In 1 through In 4 are shown. This function block can contain up to eight input pins. In this example, since there are only three input references, only three input pins are used. When the inputs represented by Data.1 AND Data.2 AND Data.3 are true the BAND function block will be true. Being true, the output (out pin) of the block will be true. The dotted line or wire connected between the output pin and the output reference symbol identified as Data.6 represents the output data as a bit. Notice the 0 just to the right of the input reference and out pin. This identifies the logical state of the input reference or output pin for the instruction. A zero identifies the reference tag as false, while a one signifies the reference tag is true. In this example each input reference has a zero representing its input state as false. Since we do not have logical continuity, when executed, the BAND block will mark its output as false, a zero. As a result, the associated output reference and its tag will also be false. This function block diagram is equivalent to the ControlLogix rung from Figure 6-13.

Clicking on the View Properties box in the upper right-hand corner of the BAND function block reveals the BAND Properties view as illustrated in Figure 6-16. Notice that there are eight input pins for this function block. They are named In1 through In8. Checking the boxes in the viability (Vis) column turns on the four input pins and displays them on the function block. Since In5 through In8 are not checked, these pins will not be displayed on the function block. This illustrates how the programmer can select the function block options that are specifically needed for this application. Notice that the Out pin is also checked. Other properties and their associated pins that are not checked will not be displayed on the function block.

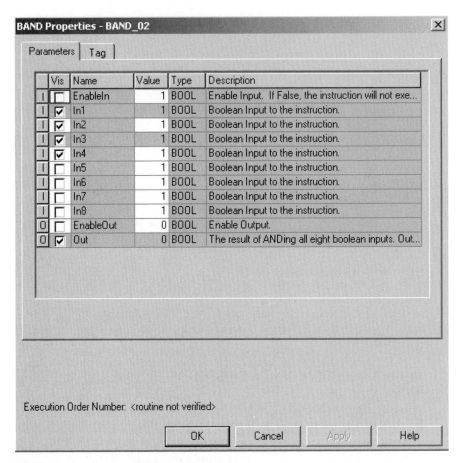

Figure 6-16   Function block BAND properties box.

PARALLEL CIRCUITS—THE *OR* LOGIC FUNCTION

The familiar parallel circuit can also be referred to as the OR logic function. The rule of OR logic is that if any input is true, the output will also be true. OR logic also states that if all inputs are true, the output will be true. In Figure 6-17, if switch 1 OR switch 2 is energized, light 1 will energize. If both SW 1 and SW 2 are true, the output will also be true.

Figure 6-17   Conventional parallel circuit where switch 1 or switch 2 can energize the load, light 1.

Figure 6-18 illustrates Figure 6-17 converted to a PLC ladder rung. Remember, when drawing programmable controller ladder diagrams, do not use the conventional switch symbols such as we employed in the previous examples. A PLC rung of logic will have normally open or normally closed contacts instead of normally open or closed switch symbols. Addresses and instructions are included. Text information in addition to each contact and its address, such as SW 1, SW 2, and L-1, is referred to as instruction comments. Instruction comments can be added from programming software.

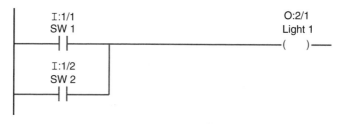

Figure 6-18   Programmable controller ladder diagram symbology. Addressing format is from the Allen-Bradley SLC 500 family of PLCs. Both ladder diagrams, Figures 6-17 and 6-18, are the same circuit, but with different symbols.

A two-input OR truth table representing Figures 6-17 and 6-18 is illustrated in Figure 6-19.

TWO-INPUT *OR* TRUTH TABLE		
SW 1	SW 2	Light 1
0	0	0
0	1	1
1	0	1
1	1	1

Figure 6-19   Two-input OR truth table.

Notice that in the case of the OR circuit, if either switch is ON, the output will be true. In addition, if both switches are ON, the output will be true.

Figure 6-20 shows a three-input parallel circuit, using three-input OR logic.

The following is a PLC program listing from Figure 6-20 for the Allen-Bradley MicroLogix 1000 if you are entering the program with a handheld programmer:

LOAD I1

OR I2

OR I3

OUT O2

The instructions tell the processor to load input 1 (I1) into memory, OR it with input 2 (I2), then OR the resultant of the previous logic operation with input 3 (I3), and then output the result to output 2 (O2).

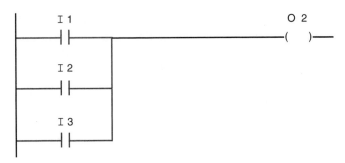

Figure 6-20 Three-input OR logic.

Let's assume that inputs I1 and I3 are false, and input I2 is true. If I1, which is false, is loaded into memory and then ORed with I2, which is true, the result of this logical OR operation is true. Now we OR I3, which is false, with the result of I1 and I2. That resultant was true; ORed with I3, which is false, its output is true. Remember the rule of OR logic: If any or all inputs are true, the output is true.

What are the expected outputs from three-input OR logic for Figure 6-21? Remember that if one or more inputs are true, the output will be true.

THREE-INPUT *OR* LOGIC			
Switch 1	Switch 2	Switch 3	Light 1
0	0	0	0
0	0	1	1
0	1	0	1
0	1	1	1
1	0	0	1
1	0	1	1
1	1	0	1
1	1	1	1

Figure 6-21 Three-input OR logic truth table.

To this point we have been working with normally open inputs. For a normally open input to be true, there must be power to the PLC input screw terminal. This ON signal to the input module will result as a 1 in the associated input status file bit position. A 1 in the input status file will cause the associated normally open PLC ladder instruction to close, or become true.

Typically, when a logical 1 in the input status table is associated with a normally open instruction, something is expected to operate. Likewise, if a logical 0 from the input status table is associated with a normally open instruction, something is expected to become false, or turn off.

The function block Boolean OR (BOR) is illustrated in Figure 6-22. This function block has the same functional components as the BAND. For this example if Data.1 OR Data.2 OR Data.3 are true, the output of the BOR function block will be true, or a 1. As with ladder OR logic, if any combination of input references is true the function block will be true. With the function block true, the output reference tag Data.6 will be true.

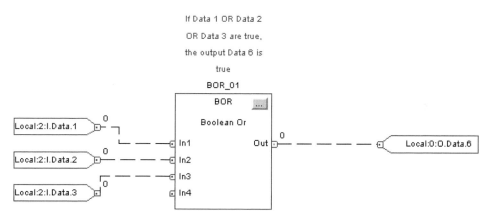

Figure 6-22   Function block BOR.

The NOT logic operator works in a manner opposite from the AND and OR logic with which we have been working. The next section will introduce the normally closed instruction and NOT logic.

NOT LOGIC

A normally closed hardwire relay contact passes power any time the relay coil is not energized. Likewise, the normally closed PLC ladder logic instruction will pass power any time the input status file bit is not a 1. This means that the physical hardware input is not sending an input signal into the PLC's input module. NOT logic is the opposite of a normally open PLC instruction or contact. It can be used in conjunction with AND or OR logic when a logical 0 in the status file is expected to activate some output device. The NOT logic function is used when an input must not be energized for an output to be energized. Likewise, the NOT logic function is used when a logical 1, or true input, is necessary to make the instruction false or deactivate an output device.

The truth table in Figure 6-23 simply states that a normally closed instruction on a PLC ladder rung will be the inverse, or opposite, of the input status table bit associated with the specific instruction. If the input status table bit is a 1, or true, the normally closed instruction will be false. In comparison, when the input table status bit is false, or a 0, the associated normally closed instruction will be true.

The NOT logic function is somewhat difficult to grasp. Let's look a little closer at the relationship between the normally open contact and how it controls the output in comparison to the normally closed contact. Figure 6-24 illustrates two rungs, the first with a normally open instruction and the second with a normally closed instruction.

TWO-STATE LOGIC FUNDAMENTALS		
Input Signal to Input Module	Normally Open PLC Instruction	Normally Closed PLC Instruction
ON	TRUE	FALSE
OFF	FALSE	TRUE

Figure 6-23   Truth table for NOT logic.

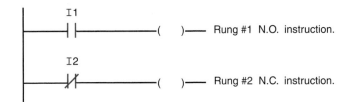

Figure 6-24   Conventional ladder diagram illustrating normally open and normally closed relay contacts controlling load, L-1.

ANALYSIS OF RUNG #1

Instruction I1 will energize the output only when there is a logical 1 in its associated input status file bit. A 1 in this bit position will cause the normally open instruction to become true and change state. In changing state, the instruction will allow logical continuity to pass on to the output instruction and make it true. Instruction I1 is considered true when it passes logical continuity. If there is no valid input signal from the field device attached to I1's screw terminal on the input module, a logical 0 will be placed in the input status file. A logical 0 in the input status file will result in this normally open input instruction becoming false. Being false, the instruction will not pass logical continuity.

ANALYSIS OF RUNG #2

The normally closed instruction works much like the normally closed contacts on a hardware relay. Being normally closed, instruction I2 will energize the output only when there is a logical 0 in its associated input status file bit. Even though there is a logical 0, or false input signal, in the status file, the normally closed instruction is true and passes logical continuity on to the output instruction. If there is a valid ON input signal from the field device attached to I2's input module screw terminal, a logical 1 will be placed in the input status file. A logical 1 in the input status file will cause the normally closed instruction to change state. The normally closed instruction will change from true (closed) to false (open). Being false, the normally closed instruction will not pass logical continuity to the output instruction. Without logical continuity, the output instruction will become false.

The function block Boolean NOT (BNOT) is illustrated in Figure 6-25. If the input reference representing input Data.1 is true, the output pin of the BNOT function block will be false, or a 0. See Figure 6-25. The output is the opposite of the input.

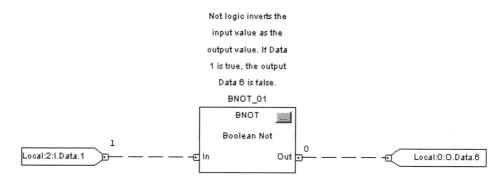

Figure 6-25   Function block NOT.

PARALLEL *NOT* LOGIC

A parallel ladder rung with normally closed inputs is a rung containing OR NOT logic.

Figure 6-26, a PLC ladder rung, has two input instructions, one normally open and one normally closed. This circuit contains parallel NOT logic. This conventional schematic rung will be true under the conditions shown in Figure 6-27. Input 1 must be true OR input 2 must NOT be true to make this rung true and energize output L-1.

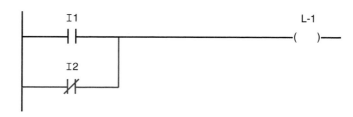

Figure 6-26   Parallel NOT logic.

INPUTS		INPUT STATUS FILE		OUTPUTS
I:1	I:2	I:1	I:2	O:0
0	0	0	1	1
0	1	0	0	0
1	0	1	1	1
1	1	1	0	1

Figure 6-27   NOT logic truth table.

Not all PLC manufacturers use the same terminology to identify the normally open and normally closed contact instruction. Figure 6-28 is a sample of terminology used with different PLCs.

DIFFERENT TERMINOLOGY USED FOR NORMALLY OPEN AND NORMALLY CLOSED INPUT INSTRUCTIONS	
—| |—	Normally open Examine if closed AND
—|/|—	Normally closed Examine if open AND invert OR invert AND NOT

Figure 6-28   Normally open and normally closed instruction identification.

Let's look at the following ladder and program for an Allen-Bradley MicroLogix 1000 (see Figure 6-29). The program listing for Figure 6-25 is a MicroLogix 1000 ladder program when entering it with a handheld programmer. Notice that the normally closed instruction is referred to as "OR invert" (ORI).

LOAD I

OR I2

ORI I3

OUT O2

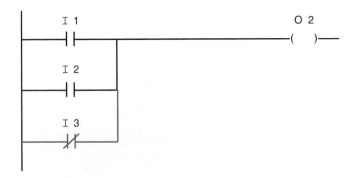

Figure 6-29   Three-input parallel NOT logic.

The instructions tell the processor to load input 1 (I1) into memory, OR it with input 2 (I2), then OR invert (OR a normally closed instruction), with the resultant of the previous logic with input 3 (I3) then being output to output 2 (O2).

OR logic states that when any or all inputs are true, the associated output will be true. Can we develop logic to give a true output if one or the other parallel inputs is true but not both? To solve this problem, we look at exclusive OR logic in the next section.

EXCLUSIVE *OR* LOGIC

Looking at a truth table for a two-input OR logic function, we see that there are three input conditions that will give us an output signal (see Figures 6-30 and 6-31):

1. If input 01 is off and input 02 is on.
2. If input 01 is on and input 02 is off.
3. If input 01 is on and input 02 is on.

Figure 6-30   Two-input OR logic.

TWO-INPUT *OR* TRUTH TABLE		
SW 1	SW 2	Light 1
0	0	0
0	1	1
1	0	1
1	1	1

Figure 6-31   Two-input OR truth table.

The exclusive OR logic function will allow either input 01 OR input 02, but not both together, to control the output (see Figure 6-32).

TRUTH TABLE FOR EXCLUSIVE *OR* LOGIC IN FIGURE 6-32		
I1	I2	O0
0	0	0
0	1	1
1	0	1
1	1	0

Figure 6-32   Exclusive OR logic truth table.

The logic for exclusive OR (sometimes referred to as X-OR) would look as follows (see Figure 6-33).

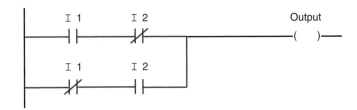

Figure 6-33 Exclusive OR logic.

ANALYSIS OF EXCLUSIVE *OR* LOGIC

If normally closed input I1 is true and input I2 is left as is, the logic on the main rung will become true, thus energizing the output. As for the state of the normally closed I1 instruction on the parallel branch, with the normally open input instruction true or closed, the normally closed contacts for input I1 on the parallel branch will open. With the normally closed contacts from input I1 open on the parallel branch, input I2 cannot control the output.

Input I2's logic will operate in the same manner. If I2's normally open instruction becomes true while I1's instructions remain in their normal state, the parallel branch will become true. With the parallel branch true, the rung will be true. The rung output will become true as there is logical continuity on the parallel branch. With the normally open I2 closed on the parallel branch, the normally closed I2 on the main rung will open and prevent I1 from controlling the output.

If, by chance, both input 1 and input 2 are energized (and therefore change from their normal state), their normally closed counterparts will both open. With an open on the main rung and the parallel branch, there is no way for the rung to become true. Recheck the truth table to verify this. Go through the logic yourself and check that if input 2 on the parallel branch is energized so that it is changed from its normal state and input 1 is left alone, the output will be energized as stated in the truth table.

Figure 6-34 illustrates the Function Block Boolean Exclusive Or (BXOR). If the input reference representing Data.1 OR the input reference representing Data.2 is true, but

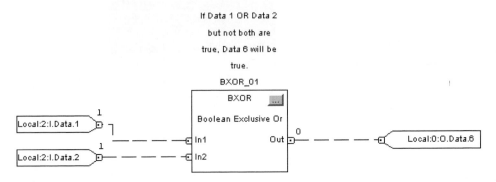

Figure 6-34 Function block XOR.

not both, the BXOR output will be true. The figure illustrates that both input references are true. As a result the output of the BXOR will be false. With the output false, output reference Data.6 will also be false.

COMBINATIONAL LOGIC

Most ladder rungs will include some combination of AND, OR, and NOT logic. No matter what the logic combination or how many logic elements or instructions are on a rung, there must be at least one path of true logical continuity before the output can be made true.

In Figure 6-35, the ladder rung has three logical paths by which it can be true:

1. If I1 AND I2 AND I4 are all true, output O2 will be true.
2. If I1 AND I3 AND I4 are all true, output O2 will be true.
3. If I1 AND I2 OR I3 AND I4 are all true, output O2 will be true.

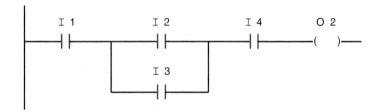

Figure 6-35   Combinational AND-OR logic.

In Figure 6-36, the ladder rung has four logical paths by which it can be true:

1. If I31 AND I10 AND I15 are all true, output O71 will be true.
2. If I4 AND I10 AND I15 are all true, output O71 will be true.
3. If I4 AND I10 AND I7 are all true, output O71 will be true.
4. If I31 AND I10 AND I7 are all true, output O71 will be true.

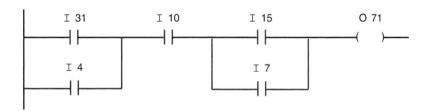

Figure 6-36   Combinational AND-OR logic. Four paths can make this rung logically true.

In Figure 6-37, the ladder rung has many logical paths by which it can be true. Study the figure and see how many paths you can find.

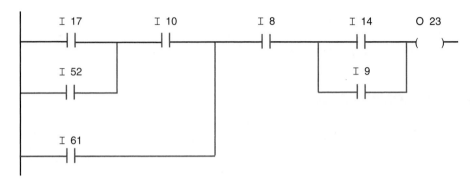

Figure 6-37 Combinational input logic.

PRIORITY OF LOGIC ELEMENTS

The priority of logic is very important in a program. This is especially important when entering a program with a handheld programmer. When using most handheld programmers, you actually enter the list of instructions representing the ladder logic for the current application. Some handheld programmers allow you to enter rungs of logic rather than a list of instructions. If you are able to enter the rungs of logic, the priority of logic concerns is taken care of by the programming device, which places the instructions in the proper position on the terminal's rungs. This is also true when programming PLC ladder rungs on a personal computer. When developing programs on a personal computer, the programmer actually develops the rungs and their associated instructions on the computer monitor's screen. Since the programmer is physically placing the instructions on the rung and in the correct position in relation to surrounding instructions, concerns about logic element priority are eliminated.

Evaluate the following program:

Load I1

AND I2

OR I3

AND I4

Out O7

The ladder rung in Figure 6-38 can be developed from the listed instructions.

The ladder rung in Figure 6-39 can also be developed from the listed instructions. Which is correct?

It should be evident that these two rungs of logic are not equivalent, even though they appear to have the same list of instructions. This is where logic priority becomes important. There need to be rules regarding instruction placement. The PLC was programmed to follow certain programming rules, and as a programmer, you must follow the same programming rules as the PLC. When working together with the same rules, human and

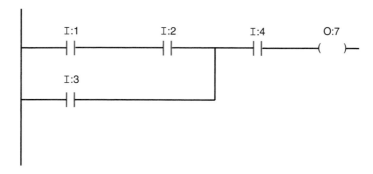

Figure 6-38   This ladder can be developed from the listed instructions.

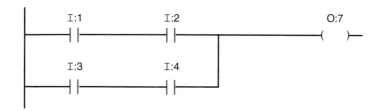

Figure 6-39   This ladder could also be created from the same listed instructions.

machine will understand each other and develop mutually acceptable programs. General programming rules are listed below:

1. Each rung begins at the left power rail.
2. Start each rung with the appropriate beginning instruction. This is typically a "Store" or "Load" instruction.
3. Program the next logic element closest to the one already programmed. In case a series and a parallel logic element are equidistant, always program the parallel logic function first. Continue following this rule until the output instruction is reached.
4. Typically, if two or more instructions are to be programmed on a parallel branch, such as in Figure 6-39, special instructions are used to connect, or group, parallel instructions on that branch. If no grouping instructions are included in the program, the default of one instruction per parallel branch, as illustrated in Figure 6-40, will be programmed.

Figure 6-40 shows a ladder and program for combinational logic programming with an Allen-Bradley MicroLogix 1000.

Rule three identifies the programming sequences when entering an instruction list program, such as in Figure 6-40. Instructions should be entered in the following order: I1, I2, I3, O2.

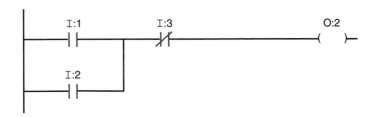

Figure 6-40   Allen-Bradley MicroLogix 1000 combinational logic.

The following is a program listing for a MicroLogix 1000 if entering the program with a handheld programmer. Notice that the normally closed instruction is referred to as an "AND invert" (ANI).

LOAD I:1
OR I:2
AND INVERT I:3
OUT O:2

These instructions tell the processor to load input 1 (I:1) in memory, OR it with input 2 (I:2), AND INVERT the result of previous logic with input 3 (I:3), and OUTPUT the result to output 2 (O:2). The programming rule for the MicroLogix 1000 states that if the programmer does not specify that the I2 AND INVERT I3 instructions are to be grouped together on the parallel branch, the instruction after OR I2 will default back to the main rung. This is illustrated in Figure 6-41.

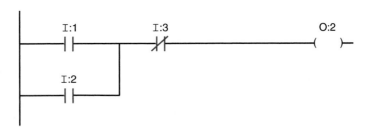

Figure 6-41   MicroLogix instruction list program containing a group of logic on the parallel branch.

Special grouping is accomplished through developing special groups of logic. Changing the programming instructions will direct the processor to evaluate the instructions in the desired manner. As an example, the following program will use the load instruction to separate the I:1 instruction from a second group starting with the LOAD I:2 instruction.

LOAD I:1   GROUP A
LOAD I:2   GROUP 2
AND INVERT I:3
OR Block   Linking Instruction
OUT O:2

The two groups of logic will be linked together using a special linking instruction, the "OR block" instruction. After the groups are linked together in parallel, the output instruction is evaluated (see Figure 6-42).

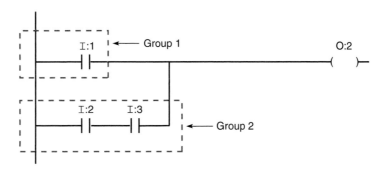

Figure 6-42   The OR block instruction links these two groups in parallel.

FUNCTION BLOCK ORDER OF EXECUTION

The order of function block is typically from input to output. Program your function blocks starting with one or more inputs and add additional blocks from left to right. Figure 6-43 illustrates two input references feeding into a BAND function block. The BAND function block feeds into a timer on-delay with reset (TONR). The TONR block's done output (DN) feeds into the output reference. Here the order of execution is from input to output, or left to right.

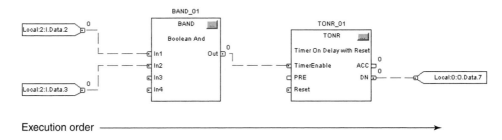

Figure 6-43   Execution order is from input to output in a simple function block diagram.

To view the execution order, go to the View Properties window. Figure 6-44 is a Properties view for the TONR block in our example. Note the execution order notation in the bottom left-hand corner. The execution order of this function block is 2.

Execution order is relative blocks that are wired together. Figure 6-45 illustrates two groups of blocks that are not wired together. The two groups of blocks are not related to each other. Each group of blocks will execute from right to left, or from input to output. Notice how the execution order alternates between groups of blocks.

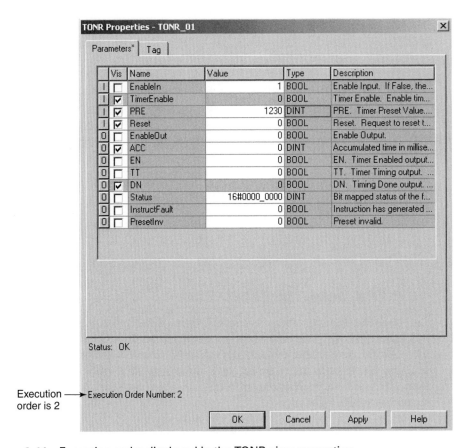

Figure 6-44   Execution order displayed in the TONR view properties.

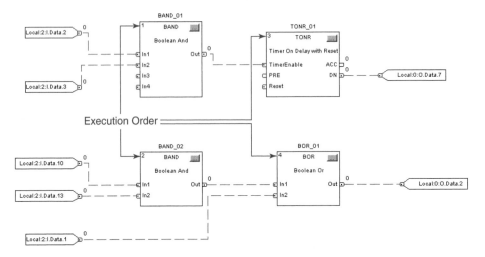

Figure 6-45   Execution order of unrelated groups of function blocks.

When function blocks are in a feedback loop, proper programming is required to identify which block is to be executed first. If not programmed properly, the processor cannot determine which block to execute first. To help the processor understand which block to execute first, place an assume data available indicator on the feedback input pin on the first block as illustrated in Figure 6-46 below.

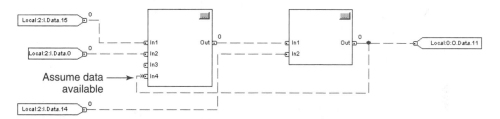

Figure 6-46   Use the assume data available indicator to identify which block the processor is to execute first.

SUMMARY

The main focus of this chapter was to introduce the concepts of how logic functions are executed when solving a PLC program. We looked at how a PLC solves its user program using AND, OR, and NOT logic. PLC ladder logic differs when compared to conventional relay logic in one important aspect: Relay logic has electrical continuity, whereas PLC logic has only logical continuity. Where there is actual current flow in a hardwired relay circuit, there is only logical continuity on a PLC rung.

Even though the PLC accepts the same signals and signal levels as relay circuits, the PLC input signal is transformed to a logical 1 or 0 as it is input to the input status file. The user program uses these logical 1, +5 VDC (ON or true), or logical 0, 0 VDC (OFF or false) signals from the input status file to solve the programmed instructions. The result of solving the user instructions using the basic logic operators, for a particular rung, is the output instruction becoming true or false. The resulting 0 (false) or 1 (true) signal is sent to the output status file, where it is stored. During the output update portion of the scan, the 0 or 1 is sent to the output module as either a 0 VDC level, for false, or +5 VDC level, for true.

REVIEW QUESTIONS

1. PLCs use which of the following logic operators?
   A. AND
   B. OR
   C. NOT
   D. all of the above

2. Each instruction on a ladder diagram has a reference number associated with it, which is:
   A. its instruction comment
   B. its position identifier on the ladder rung

    C. a number referencing the instruction's associated input status table bit address

    D. the instruction's input address.

3. True or false: Contacts shown on PLC ladder rungs can be in any one of the following states—ON, OFF, intermediate.

4. True or false: Programmable controller contacts and relay contacts operate in a similar manner, as both provide either electrical or logical continuity when the contacts are closed.

5. What is the difference between electrical and logical continuity?

6. Fill in the following truth table for NOT logic (Figure 6-47).

NOT LOGIC TRUTH TABLE		
Input Signal to Input Module	Normally Open PLC Instruction	Normally Closed PLC Instruction
ON		
OFF		

Figure 6-47 NOT logic truth table.

7. The logical AND function is similar to:
    A. in parallel
    B. in series
    C. inverted logic
    D. both A and C
    E. depends on the application

8. The logical OR function is similar to:
    A. in parallel
    B. in series
    C. inverted logic
    D. both A and C
    E. depends on the application

9. The logical OR NOT function is similar to:
    A. in parallel
    B. in series
    C. inverted logic
    D. both A and C
    E. depends on the application

10. Illustrate a two-input AND logic PLC ladder rung.

11. Develop a truth table for the answer in 10.

12. Will the following rung be true or false (see Figure 6-48)?

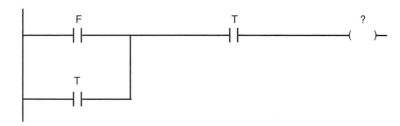

Figure 6-48   Rung for question 12.

13. The BAND function block is similar to:
    A.   series ladder logic
    B.   parallel ladder logic
    C.   NOT ladder logic
    D.   Boolean ladder logic

14. The BAND function block:
    A.   ANDs up to twelve inputs
    B.   ANDs up to eight input references
    C.   logically combines up to eight input references
    D.   performs a logical AND/OR function

15. A BXOR function block:
    A.   is true when all inputs are true
    B.   is true when neither input is true
    C.   is true only when one or the other input pins is true
    D.   is false when only one input is true
    E.   is false when both input references are true
    F.   C and E are correct
    G.   B and D are true

16. A BNOT function block is true:
    A.   when both inputs are true
    B.   when the function block is true
    C.   when the order of execution is true
    D.   when the input is false
    E.   both B and C

17. The order in which a series of function blocks are executed:
    A.   is determined by the programmer
    B.   is determined by the position of the block on the sheet
    C.   typically flows from right to left
    D.   typically flows from inputs to output

18. To resolve a loop:
    A.   the instructions are executed in order from left to right
    B.   the processor uses the assume data available indicator to know when to start
    C.   the processor looks at the output block to determine the state of the feedback
    D.   the processor waits until a start signal is received from the output block

IF PASSWORD IS CORRECT GO TO DRIVE SET UP SCREEN

0022
```
      ┌─── EQU ──────────────┐
      │ Equal                │
      │ Source A      N12:21 │
      │               7777<  │
      │ Source B      32767  │
      │               32767< │
      └──────────────────────┘
```
B3:10/0
[OSR]

B3:1/15
( )

RESET PASSWORD
ATTEMPTS COUNTER
C5:5
(RES)

0023
```
      ┌─── GRT ──────────────┐
      │ Greater Than (A>B)   │
      │ Source A      N12:21 │
      │               7777<  │
      │ Source B          0  │
      │                   0< │
      └──────────────────────┘
```
```
      ┌─── NEQ ──────────────┐
      │ Not Equal            │
      │ Source A      N12:21 │
      │               7777<  │
      │ Source B      32767  │
      │               32767< │
      └──────────────────────┘
```
B3:10/1
[OSR]

(CU)

C5:5
3<   (DN)
2<

−1
−1<
N12:21
7777<

0024
NOTIFICATION
B3:0/7
] [

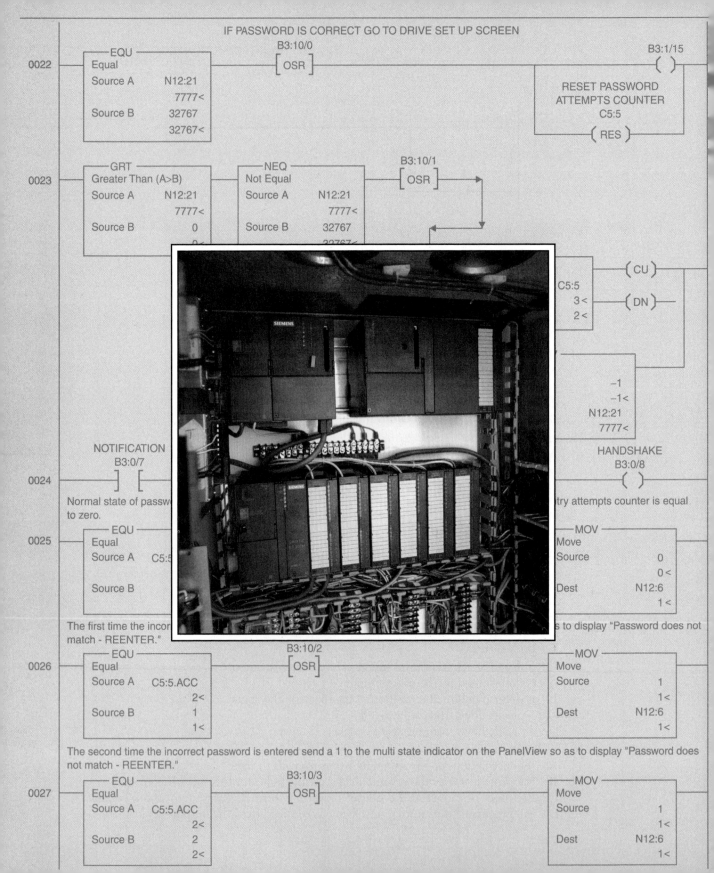

HANDSHAKE
B3:0/8
( )

Normal state of passw... ...try attempts counter is equal to zero.

0025
```
      ┌─── EQU ──────┐
      │ Equal        │
      │ Source A  C5:5│
      │              │
      │ Source B      │
      └──────────────┘
```
```
      ┌─── MOV ──────────────┐
      │ Move                 │
      │ Source            0  │
      │                   0< │
      │ Dest        N12:6    │
      │                   1< │
      └──────────────────────┘
```

The first time the incor... ...s to display "Password does not match - REENTER."

0026
```
      ┌─── EQU ──────────────┐
      │ Equal                │
      │ Source A   C5:5.ACC  │
      │               2<     │
      │ Source B      1      │
      │               1<     │
      └──────────────────────┘
```
B3:10/2
[OSR]
```
      ┌─── MOV ──────────────┐
      │ Move                 │
      │ Source            1  │
      │                   1< │
      │ Dest        N12:6    │
      │                   1< │
      └──────────────────────┘
```

The second time the incorrect password is entered send a 1 to the multi state indicator on the PanelView so as to display "Password does not match - REENTER."

0027
```
      ┌─── EQU ──────────────┐
      │ Equal                │
      │ Source A   C5:5.ACC  │
      │               2<     │
      │ Source B      2      │
      │               2<     │
      └──────────────────────┘
```
B3:10/3
[OSR]
```
      ┌─── MOV ──────────────┐
      │ Move                 │
      │ Source            1  │
      │                   1< │
      │ Dest        N12:6    │
      │                   1< │
      └──────────────────────┘
```

CHAPTER

Input Modules

OBJECTIVES

After completing this chapter, you should be able to:

- describe the available types of input modules
- explain the differences between sinking and sourcing
- explain the correlation between positive and negative logic and sinking and sourcing
- explain how specialty I/O modules enhance a PLC's functionality
- define module and sensor specifications from data sheets
- list the advantages of three-wire solid-state sensors and the disadvantages of using two-wire sensors
- explain the differences between analog and discrete inputs

INTRODUCTION

The input and output section of a PLC system is the physical connection between the outside world and the CPU. A modular PLC uses interchangeable input modules. The modularity feature of a PLC and its ability to interchange modules to meet the immediate interface need makes the PLC so popular and versatile. Input modules are available to accept various voltages and currents and to automatically convert input signals into a logical (typically 5 V dc) signal with which the CPU can work. After solving the user program **logic** and sending the output signal to the output module by way of the output status file, the output module, which is also available in various switching configurations, controls field output devices. This means that a low-voltage (for example, 24 V dc) control signal can be an input controlling 240 VAC output devices.

There are various ways to get information into the CPU from common hardware field devices. First, discrete I/O interface modules provide a method of getting two-state, discrete, or digital signals into and out of the PLC. Second, ever-changing signals such as temperature or pressure (called analog signals) can be interfaced to the PLC using modules specifically designed to accept these variable signals. These modules are called *analog input* or *analog output modules*. There are additional specialized modules that allow the PLC to accept specific variable input signals, including resistance temperature detector (RTD) or thermocouple.

This chapter introduces discrete and analog I/O modules, their operating principles, types of modules available, advantages of using one over the other, and basic interfacing principles.

INPUT MODULES

I/O modules are available as either input only, output only, or a combination of inputs and outputs. Figure 7-1 shows a four-slot Allen-Bradley SLC 500 modular chassis and an I/O module being inserted.

In the figure, the module slides into the rack or chassis on the grooves visible on the inside bottom of the chassis. Also notice the self-locking tabs, which will lock the module

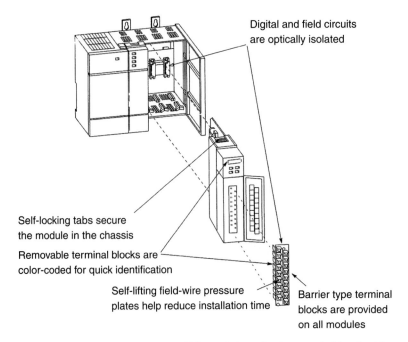

Digital and field circuits
are optically isolated

Self-locking tabs secure
the module in the chassis

Removable terminal blocks are
color-coded for quick identification

Self-lifting field-wire pressure
plates help reduce installation time

Barrier type terminal
blocks are provided
on all modules

Figure 7-1 SLC 500 four-slot chassis with I/O module to be inserted. (Used with permission of Rockwell Automation, Inc.)

into the chassis. An I/O module is made up of the following features: at the very top of the module is identification indicating whether this is an input, output, or combination module. Some manufacturers color-band the identifications portion of the module for further identification. Figure 7-2 illustrates an SLC 500 combination I/O module. Directly below the identification portion lie the status indicators. There will be one status indicator, or light-emitting diode (LED), for each I/O point. These lights alert the operator to the ON or OFF status of each point on that module. Farther down the front of the module are the screw terminals for connecting the I/O wires to the module. Removable terminal blocks make wiring and changing the module easy and avoid the need for rewiring.

Some manufacturers place the I/O connections behind a hinged door, as illustrated in Figure 7-2. There may be an area on the inside of the module door to identify each I/O

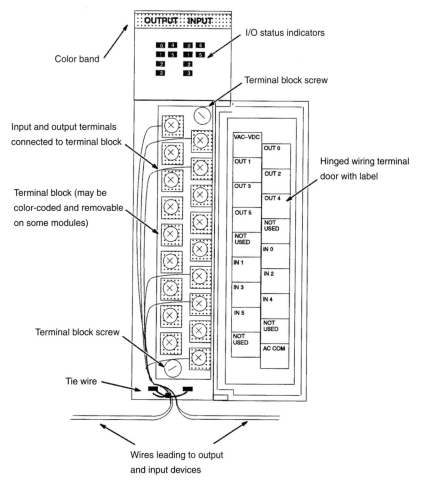

Figure 7-2 SLC 500 combination I/O module. (Used with permission of Rockwell Automation, Inc.)

screw terminal. Notice that each terminal on this combination module is identified as simply OUT 0, OUT 1, OUT 2, . . . and IN 0, IN 1, IN 2, ... This identifies outputs 0 through 5 and inputs 0 through 5 for this module. To build the entire input and output address, the chassis slot in which a module is placed determines the rest of the I/O address for each specific point.

Discrete input modules are selected according to the voltage levels with which they can work. Typical input modules can accept +5 V dc, 10–30 V dc, 120 VAC, 120–240 VAC, and 200–240 VAC levels as inputs.

DISCRETE INPUT MODULE

The discrete input module is the most common input interface used with programmable controllers. Discrete input signals from field devices can be either AC or DC. The most common module types are listed below:

AC INPUT MODULES	DC INPUT MODULES
24 VAC	24 V dc
48 VAC	48 V dc
120 VAC	10–60 V dc
240 VAC	120 V dc
Nonvoltage	230 V dc
120 Volts Isolated	Sink/Source 5–50 V dc
240 Volts Isolated	5 V dc TTL level
24 VAC/DC	5/12 V dc TTL (transistor–transistor logic) level

DISCRETE AC INPUT MODULE

A 120 VAC input module will accept signals between 80 and 135 VAC. Common inputs include limit switches, proximity switches, photoelectric switches, selector switches, relay contacts, and contact closures from other equipment. Figure 7-3 illustrates wiring for a typical 120 VAC input module. Signals from line one through the field input device are wired to input screw terminals on the module. The left module has its commons connected internally. All inputs will have the same voltage. The right module is a 120 or 230 VAC input module. The module has two separate commons, which allows the user to wire two different input voltage levels.

The input module is considered the load for the field input device. The module's job is to convert the 120 VAC, high-voltage signal to the 5 V dc level, with which the PLC can work. The module's job is to verify the input as a valid signal, isolate the high-voltage field device signal from the lower-voltage CPU signal, and send the appropriate ON or OFF signal to the CPU for placement in the input status file. The circuitry contained in an input module is composed of three parts: power file conversion, isolation, and logic, as illustrated in Figure 7-4.

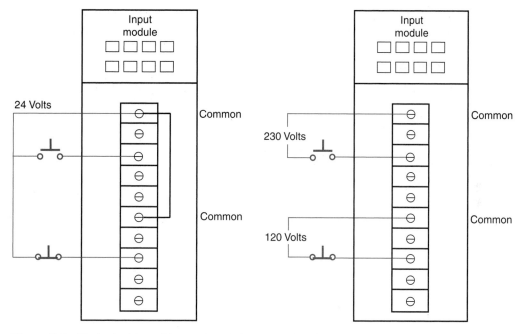

Figure 7-3   Typical wiring of input signals into a 120-VAC eight-point input module.

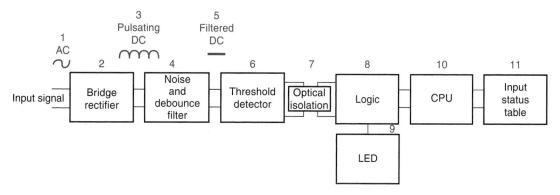

Figure 7-4   Block diagram of a typical AC input circuit.

Alternating current enters the input module (section 1) and then flows to the power conversion section (section 2).

Power Conversion

The power conversion section usually consists of resistors and a bridge rectifier. The bridge rectifier converts the incoming AC signal to a pulsating DC level (see section 3 of Figure 7-4). The DC level is passed through filters and other logic (section 4) to deliver a clean, debounced, DC input signal (section 5). The filtered DC signal goes on to the threshold detector (section 6).

Threshold Detection

Threshold detection circuitry detects if the incoming signal has reached or exceeded a predetermined value for a predetermined time, and whether it should be classified as a valid ON or OFF signal. Module specifications call this the valid ON/OFF state voltage range. A typical valid OFF state is between 0 and 20 or 30 VAC, depending on the module's manufacturer. A valid ON state will be between 80 and 132 VAC, again depending on the manufacturer (see Figure 7-5). The signal area between the upper voltage limit for a valid OFF state (20 volts) and the minimum voltage for a valid ON state (80 volts) is called the undefined, or input state not guaranteed, zone. Signals falling within this undefined area may be ON or OFF, making them unstable and unreliable.

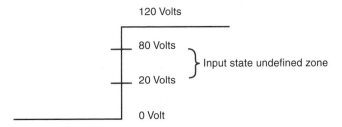

Figure 7-5   Input pulse with valid signal areas defined.

Filtering and time delays are used to filter out possible electrical noise that may be interpreted as a *false* input pulse. To eliminate the possibility of faulty operation due to electrical noise, a valid AC input signal must not only be a specific value, but must be present for a specific amount of time before the input module allows the valid signal to pass to the isolation section.

Isolation

The isolation section of the input circuit (section 7 of Figure 7-4) is usually made up of an optical isolator, or opto-coupler as it is sometimes called. In a 120 VAC input module, isolation separates the high-voltage, 120 VAC input signal from the CPU's low-voltage control logic. The low-voltage control logic signals associated with the CPU will run from 5 to 18 V dc, depending on the module manufacturer and the type of logic employed.

Isolation is accomplished by the input signal energizing a light-emitting diode (LED), which transmits a signal of light energy to a receiver in the form of a photo-conductive diode. Simply put, the LED converts the electrical signal to an optical signal. The receiver, usually a photo-transistor, converts the optical signal back to an electrical signal. An optical isolator works similarly in principle to the sun shining on a solar cell. Think of the LED shining light on the photo-transistor as the sun. The solar cell converts the light into electrical current. There is no actual physical or electrical coupling between the sending LED, its associated input circuitry, and the optical receiver and its low-voltage, associated logic circuitry. The signal is transferred by light from the LED. Figure 7-6 illustrates a simplified optical isolator.

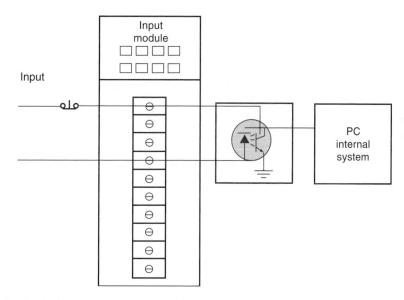

Figure 7-6   Optical isolator to isolate high-voltage incoming signals from the CPU's lower voltage levels.

Optical isolation protects the low-voltage CPU and its associated circuitry by preventing spikes or high-voltage transients on the input circuit from transferring into the low-voltage circuitry.

The Logic Section

DC signals from the opto-coupler are used by the logic section (section 8, Figure 7-4) to pass the input signal to the module's input address LED (section 9) and the CPU (section 10), and then on to the *input status file* via wires on the rack's backplane called the data bus.

AC INPUT MODULE SPECIFICATIONS

Now that we understand how an AC input module conditions, verifies, and passes an input signal on to the CPU, we will look at input module specification data.

Module specifications provide important information during module selection, PLC *hardware* configuration, troubleshooting, and input device selection. Figure 7-7 contains typical AC input specification data. Each specification will be discussed in the following pages.

Voltage	Inputs	Points per Common	Backplane Current Draw at 5 V dc	Maximum Signal Delay	Maximum Off-State Current	Input Current Nominal	Maximum Inrush Current
85 to 132 VAC	4	4	0.035 Amps	ON = 35 ms OFF = 45 ms	2 mA	12 mA at 120 VAC	0.8 A
	8	8	0.050 Amps	ON = 35 ms OFF = 45 ms	2 mA	12 mA at 120 VAC	0.8 A
	16	16	0.085 Amps	ON = 35 ms OFF = 45 ms	2 mA	12 mA at 120 VAC	0.8 A
170 to 265 VAC	4	4	0.035 Amps	ON = 35 ms OFF = 45 ms	2 mA	12 mA at 240 VAC	1.6 A
	8	8	0.050 Amps	ON = 35 ms OFF = 45 ms	2 mA	12 mA at 240 VAC	1.6 A
	16	16	0.085 Amps	ON = 35 ms OFF = 45 ms	2 mA	12 mA at 240 VAC	1.6 A

Figure 7-7 AC input module specifications for Allen-Bradley SLC 500 120-VAC and 240-VAC input modules. (Table compiled from Allen-Bradley Discrete I/O modules data)

Explanation of Figure Headings

Voltage: This is the operating voltage at 47 to 63 hertz (Hz) for the module.

Inputs: This indicates the number of inputs the module has.

Points per Common: This is the number of input points that share the same common connection. As an example, one 16-point input module could have all input points sharing one common, and a different 16-point input module might have two groups of 8 input points. Each group of 8 would have its own separate common. Figure 7-8 illustrates a 16-point input module with two groups of 8 inputs and their respective common terminals.

Backplane Current Draw: Each module takes power from the PLC's power supply to operate the electronics on the module. This specification will be used when calculating power supply loading. We will explore power calculations in Chapter 9. In the lab exercises for Chapter 9, you will calculate power for two PLCs, an SLC 500, and a General Electric Series 90-30.

Maximum Signal Delay: Signal delay is the time it takes for the PLC to pick up the field input signal, digitize it, and store it in memory. This specification is usually listed for a signal turning on and for a signal turning off.

Nominal Input Current: This is the current drawn by an input point at nominal input voltage.

Maximum Inrush Current: This is the maximum inrush current the module can handle.

Maximum Off-State Current: This is the maximum amount of current, typically from leakage from a solid-state input device, that a module can accept while remaining in an OFF state.

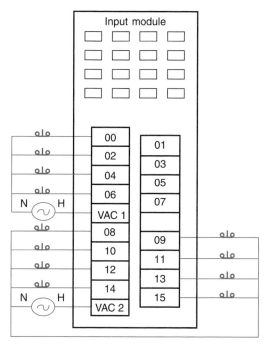

Figure 7-8 Sixteen-point input module with eight points per common.

Additional Module Specifications: In addition to the specifications already mentioned, there are four additional module specifications, sometimes difficult to find, that are important when replacing a mechanical AC limit switch with a solid-state style proximity device. When interfacing an AC, two-wire, solid-state sensor to a PLC, voltage drop, leakage current, minimum load current, and power-up delay are specifications not normally associated with mechanical limit switches, and need to be considered when replacing a mechanical input device with a solid-state device.

The Two Types of Input Devices

There are two types of input devices commonly interfaced to an input module. One type includes the mechanical limit switch, toggle switch, selector switch, push button, and contacts from an electromechanical relay. Every one of these devices has something in common: circuit continuity is either made or broken by physically opening a set of contacts. When open, these contacts have a physical air gap; thus, there is infinite resistance, resulting in zero current flow through the physically open contacts. Since these are mechanical contacts, there is no electrical power required to make the device operate.

The second type, the solid-state proximity device, on the other hand, is an electronic device, which means that it needs power to operate. A small amount of current must continuously flow through the device, even in the OFF state, to keep the internal electronics working so that the switch will be able to sense the presence of an object. Figure 7-9 illustrates a two-wire connection for a solid-state proximity sensor. Notice that there is only a single path for power flow, not only to provide the ON or OFF signal to the load, but also to provide that small amount of current to operate the internal electronics. The current flowing through the sensor to operate the internal electronics is called its "leakage current."

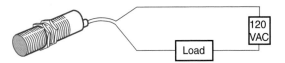

Figure 7-9   Two-wire solid-state sensor. The load usually will be a PLC input module.

This is an interesting concept that we need to investigate: the current leaking from our AC input devices. Excessive leakage current could possibly cause an input module's input point to turn on in error if the module's maximum OFF-state current is exceeded. Figure 7-10 illustrates two inputs going into an input module. Input A is a mechanical limit switch. When the limit switch is open, there exists a physically open circuit; thus, there is zero current flow into input point 0. Input 6 is an inductive proximity switch. Although the

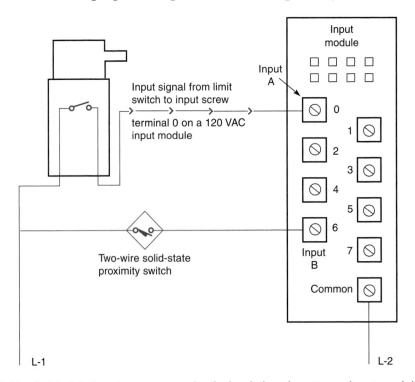

Figure 7-10   Solid-state input versus a mechanical switch as input to an input module.

switch is in the OFF, or deenergized, state and the schematic symbol shows an open circuit, there must be current flow through the electronics to keep the sensor operating so that it will be able to switch from open to closed when a target comes into range.

Leakage current for a two-wire sensor is typically less than 2 milliamps (usually around 1.7 mA). Some larger, high-power sensors can have leakage current as high as 3.5 mA or more. Most sensor manufacturers have a standardized leakage current value of less than 2 mA so that they can interface to most PLCs. A typical AC input module will accept leakage current of less than 2 mA and still read the signal as a valid OFF. Figure 7-11 illustrates a normally open, solid-state input circuit with the sensor seeing no object. The input module screw terminal to which this sensor is wired is in the OFF state. An amp meter shows the circuit's leakage current.

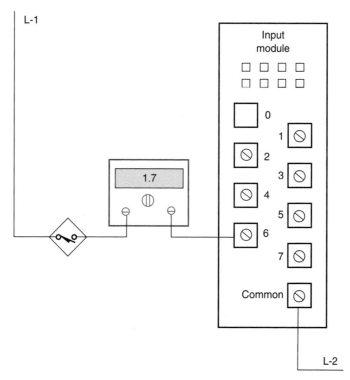

Figure 7-11   Amp meter showing leakage current in a two-wire DC sensor input circuit.

Current multiplied by the amount of voltage equals the amount of operating power. The voltage drop across the input device multiplied by the leakage current equals the voltage available to the device. The higher the voltage drop, the higher the available power. However, the higher the voltage drop across the input device, the less voltage is available for the input module, which is the load in an input circuit. Voltage drop specifications for a specific input device will be found in the sensor's specification sheets. Typically, two-wire, solid-state input devices will have a voltage drop of between 6 and 10 volts. Power supplied to the input module from the sensor must be above the minimum ON-state voltage and minimum ON-state current for the module to see a valid ON signal.

Another consideration is the power-up delay time. Consider the following input circuit, in Figure 7-12.

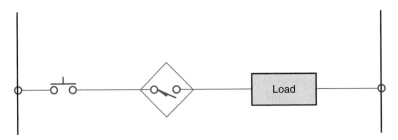

Figure 7-12   Mechanical switch in series with solid-state AC two-wire sensor.

The mechanical switch, in series with the sensor, creates an open circuit to the solid-state sensor. For the solid-state proximity sensor to operate properly, it needs to be powered continuously. Being in an open circuit, the sensor is not operating; thus, it cannot sense an object in the sensing zone. If an object appears in front of the sensor and the push button is depressed, the sensing device will not be available to see the target until the electronics inside the sensor become operational. This delay time is called the "power-up delay" or the "time delay before availability." This time delay can range from 8 milliseconds (ms) up to 100 milliseconds, depending on the sensor and the manufacturer. Because of this phenomenon, wiring any two-wire, solid-state sensor in series with a mechanical switch should be avoided. Figure 7-13 illustrates power-up delay between a push button and a sensor.

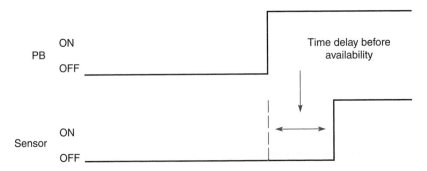

Figure 7-13   Time delay before sensor availability.

There is a solution to the time delay. You can place a resistor in parallel to the mechanical switch to permit enough leakage current to flow to keep the sensor alive for instantaneous operation. The resistor should be sized to provide the leakage current as listed in your sensor specification sheet. Assuming a 115 VAC circuit and a sensor with a 1.7 mA leakage current specification, how do you calculate the value of the resistor?

Resistance (R) = voltage (V) divided by leakage current (I)

R = V/I

R = 115/.0017

R = 67,647

Based on the calculations, you should select a 67 K (67,000) ohm resistor; typically a 1-watt resistor will be used. Figure 7-14 illustrates the input circuit with the proper bypass resistor. The bypass resistor will allow enough current in the circuit to keep the sensor ready.

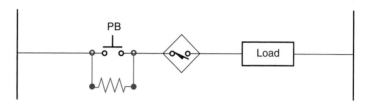

Figure 7-14   Resistor in parallel with input push button to solve time delay before sensor availability.

Minimum Load Current

A two-wire, solid-state sensor requires a minimum current flow to stay alive. If the load has a high impedance, a loading resistor should be placed in parallel with the load to dissipate excessive leakage current.

DC INPUT MODULES

Next, we will investigate DC input modules, operating principles, specifications, selection, operation, and interface. Low-voltage, 24 V dc inputs are commonly used for start/stop control circuitry and sensor interface to the PLC.

DC sensors can drive electromechanical relays, counters, and solenoids in addition to solid-state devices. A DC sensor will interface directly to a PLC without additional interface circuitry when using the proper DC input module. Common industrial sensing applications use discrete, solid-state sensors with transistor outputs for interfacing to PLCs. These sensing devices include inductive proximity sensors, capacitive proximity sensors, and photoelectric sensors, to name a few. Typical industrial sensors that fall into this category are 10–30 V dc sensors.

The sensor's transistor switch controls the signal that is input into a PLC DC input module. Transistors are solid-state switching devices that are available in two different polarities, NPN or PNP. NPN and PNP are descriptions of the two basic types of transistor. By selecting the transistor switching configuration to match the DC polarity of your DC input module, the sensor can be incorporated into any polarity input circuit.

Solid-state input devices with NPN transistors are called "sinking input devices," and input devices with PNP transistors are called "sourcing input devices."

SINKING AND SOURCING

"Sinking" and "sourcing" are terms used to describe current flow through a field device in relation to the power supply and the associated I/O point. Probably the most common PLC interface situation that involves sinking or sourcing occurs when choosing and interfacing a common inductive proximity or photoelectric sensor to a PLC.

The problem with defining sinking and sourcing circuits is that the definition changes depending on whether you are an engineer or a technician. Typically, engineering schools teach one theory, but technical schools may teach another.

The first question we must answer is: In which direction does the current flow? Figure 7-15 illustrates a battery, toggle switch, and light bulb. This is a simple circuit; however, even so simple a circuit can fuel an argument as to which way current is flowing.

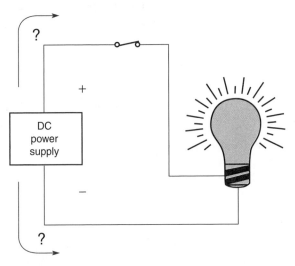

Figure 7-15   Current flow in a simple battery-powered lamp circuit.

On the one hand, many engineering school students and graduates will argue that current flows from the positive side of the battery through the circuit, and returns to the negative side of the battery. On the other hand, many technical school students will argue that current flows from the negative side of the battery through the circuit, and returns by way of the positive terminal of the battery. So what is the correct answer? As a rule of thumb, refer to your particular PLC manufacturer's literature to determine how your PLC handles sinking and sourcing modules and their interface.

There is a commonly accepted definition about what a sinking and sourcing I/O circuit looks like; however, not all manufacturers use the same terminology. Many PLC manufacturers follow the theory that current flows from positive to negative, the theory we will use in this text.

In a DC circuit there must be three pieces: power, a switching device, and the load. The relationship between the switching device, the load, and which one receives current first, defines whether we have a sinking or sourcing circuit. Figure 7-16 illustrates the switch and light circuit we looked at in Figure 7-15. Current flows from the positive terminal of the battery through the switch and onto the light, which is the load. Notice that the switch is the source of current as far as the light is concerned. As a result, the switch is called a sourcing device. The light is then a sinking device, as it sinks the current to ground.

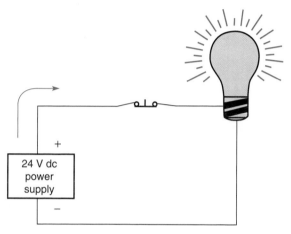

Figure 7-16   Sinking and sourcing DC circuit. The switch is the source of current that the light sinks to ground.

On the flip side is the opposite circuit. Figure 7-17 illustrates the current flow from the positive side of the battery to the light. The light is then the source of the current as it passes it to the switch, which in turn sinks the current to ground.

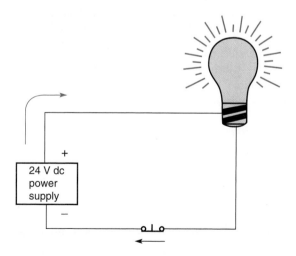

Figure 7-17   Sinking and sourcing DC circuit. The light is the source of current that the switch sinks to ground.

Figure 7-16 illustrates a sourcing switch with a sinking light, whereas Figure 7-17 illustrates a sourcing light with a sinking switch. If we modify Figure 7-17 to include an input module as the load rather than the light, the principles of sinking and sourcing remain the same. Figure 7-18 illustrates a sourcing input module, the load, with a sinking switch.

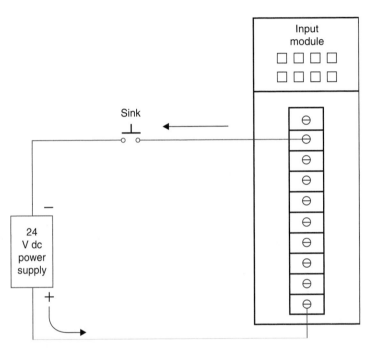

Figure 7-18   Sourcing DC input module with a sinking switch.

Figure 7-19 is a sinking input module. Current flows from the positive side of the battery through the switch, our sourcing input device. The switch is the source of current to the input module.

The circuits in Figures 7-16 and 7-17 may seem as if they would function the same regardless of whether they were hooked up as sinking or sourcing. Neither the toggle switch nor the light is affected by which way current flows through it. Nonetheless, a solid-state transistor device such as an inductive proximity switch is very particular as to how it is hooked up.

The solid-state, switching, inductive proximity sensor uses a transistor as its switching device. For these transistors to operate, they must be wired correctly to the appropriate DC input module. Figure 7-20 illustrates a sourcing (PNP) sensor interfaced to a 24 V dc sinking input module. This configuration has a sourcing sensor (PNP transistor) as an input to a sinking input module.

Figure 7-21 illustrates a sinking (NPN) sensor interfaced to a 24 V dc sourcing input module. The PLC module is also the source of current to the sensor, which makes it a sourcing input.

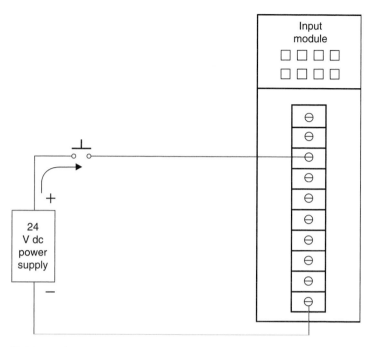

Figure 7-19   Sinking DC input module with a sourcing switch.

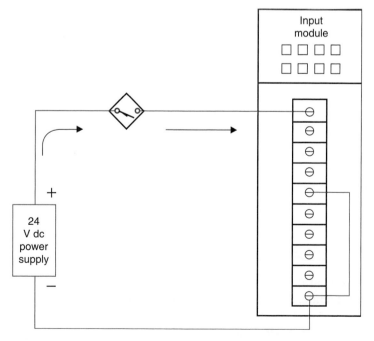

Figure 7-20   Sourcing two-wire inductive proximity sensor interfaced to a sinking 24-V dc input module.

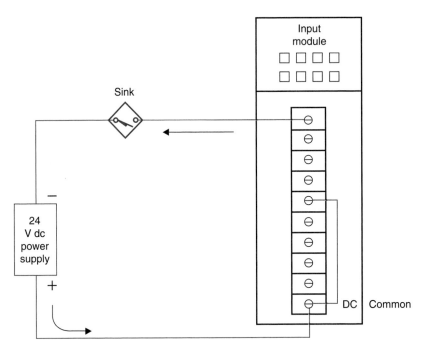

Figure 7-21 Sinking two-wire inductive proximity sensor interfaced to a sourcing 24-V dc input module. This proximity sensor sinks current to ground or the negative side of the power supply. This sensor has an NPN transistor as its switching device.

The following two basic principles pertain to sinking and sourcing circuits:

1. NPN transistors are open-collector, current-sinking devices, which interface to a sourcing input module.
2. PNP transistors are open-collector, current sources, which interface to a sinking input module.

Now that we have defined sinking and sourcing and the two classes of DC input modules, let us look more closely at the modules, operating principles, specifications, and interface.

DC INPUT MODULE OPERATION

Except for the bridge rectifier circuit, the DC input module is very similar to the AC input module. Since the input signal is already DC, no bridge rectifier is necessary. Resistors are used to drop the incoming voltage before passing the signal on to the remaining electronics. Figure 7-22 illustrates a simplified block diagram for a DC input module.

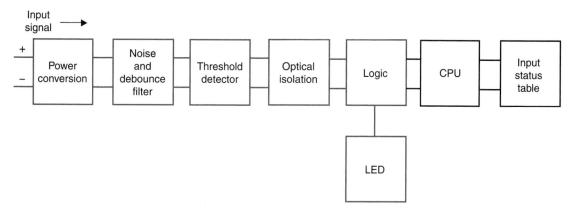

Figure 7-22   Simplified circuitry for a DC input module.

DC SINKING INPUT MODULE SPECIFICATIONS

Now that we understand the basic types, operation, and interface of field devices to a DC input module, let us look into DC input module specifications, as illustrated in Figure 7-23.

Voltage	Inputs	Points per Common	Backplane Current Draw at 5 V dc	Maximum Signal Delay	Maximum Off-State Current	Input Current Nominal	Maximum Off-State Voltage
	8	8	0.050 Amps	ON = 8 ms OFF = 8 ms	1 mA	8 mA at 24 V dc	5 V dc
10 to 30 V dc Sink	16	16	0.085 Amps	ON = 8 ms OFF = 8 ms	1 mA	8 mA at 24 V dc	5 V dc
	32	8	0.106 Amps	ON = 3 ms OFF = 3 ms	1 mA	8 mA at 24 V dc	5 V dc

Figure 7-23   DC input module specifications for Allen-Bradley SLC 500 sinking inputs. (Data compiled from Allen-Bradley DC input module data sheets)

Analysis of New Table Terms

Many of the terms used for DC input module specifications are the same as for AC input modules. Only new terms will be discussed here.

Maximum Off-State Current   This is the maximum amount of leakage current allowed in an input circuit from an input device that will keep the input circuit in an OFF state.

If more leakage current flows through the sensor than the maximum current needed to keep the module's input point (maximum off-state current) in the OFF state, the input module will see a valid ON signal all the time. Figure 7-24 illustrates a typical two-wire connection to its load. When connecting input devices to PLCs, the input point is the load.

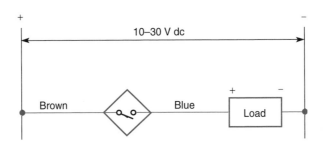

Figure 7-24   Two-wire sensor connected to its load. Notice single path for current to flow.

There are three ways to eliminate the problem of leakage current; first, carefully match the input device and the input module. If the leakage current specification of the sensing device is more than the maximum off-state current the module will accept for a valid OFF condition, choose a different input device.

If using a two-wire sensor is necessary, a bleeder resistor can be inserted into the circuit to bleed current around the load. Each sensor manufacturer has its own resistor selection chart or formula in the sensor data or specification information. Figure 7-25 is an excerpt from an Allen-Bradley bleeder resistor selection chart, in the booklet, "Inductive and Capacitive Sensor and Programmable Controller Interface Manual Preferred Compatibility." For ease of illustration, only one input module has been selected from the available tables. To determine the necessary bleeder resistor, locate the particular SLC 500 input module in the tables in the book. For this example we have chosen the Allen-Bradley SLC 500 input module 1746-IB16. If the sensor's leakage current was 2.0 milliamps, you would locate the proper bleeder resistor at the point where the 2.0 milliamp column intersects with the desired input module. Looking at Figure 7-25, a 1 K ohm resistor is the correct choice. The booklet's text will direct you to select a 3-watt resistor.

ALLEN-BRADLEY BLEEDER RESISTOR SELECTION CHART				
SLC 500 Input Module	Input Module Maximum Off-State Current	Bleeder Resistor Selection per Sensor Leakage Current		
		1.5 milliamps	2.0 milliamps	2.5 milliamps
1746-IB16	1 milliamp	1.5 K ohm	1 K ohm	750 ohm

Figure 7-25   SLC 500 1746-IB16 module bleeder resistor selection data. (Data from Allen-Bradley data tables)

If you need to calculate a bleeder resistor, refer to your sensor manufacturer's calculation method to determine the resistor for your application. The bleeder resistor is installed similar to the illustration in Figure 7-26.

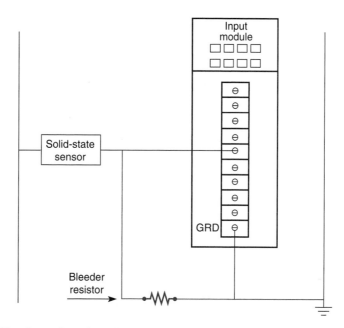

Figure 7-26   Bleeder resistor installation in an input circuit.

The object of the bleeder resistor is to create a parallel circuit so that excess current is shunted away from the input point. As an example, suppose the input impedance of the input module's input point is 1 K ohm. If the leakage current of the sensor is 1.7 milliamps and we put a 1-K ohm resistor in parallel with the input (as illustrated in Figure 7-26), we split the current going into the input module in half. The input module receives .85 milliamps, while the other .85 milliamps are shunted to ground through the bleeder resistor. With the input point receiving .85 milliamps and a maximum off-state current specification of 1 milliamp, there should be no problem with the input signal and the module.

The third option is to choose a three-wire input device. Figure 7-27 illustrates a three-wire inductive proximity switch and three-wire connection to the load. The brown wire provides current flow into the sensor to operate the sensor electronics, along with current for the input signal into the PLC input module. The third wire in Figure 7-27, the blue wire, provides a path for current from the operation of the sensor's internal electronics to the return to ground without affecting the load. The result is a separate circuit for the sensor's internal electronics and a separate signal for the sensor's input signal to the PLC.

We have introduced issues when interfacing a solid-state sensor to DC input modules. Let us look at a practical example of interfacing an actual sensor to a selected manufacturer's 24 V dc input modules.

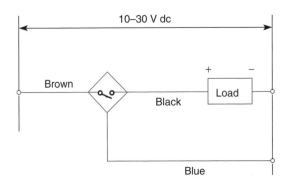

Figure 7-27 Diagram of connection for three-wire sensor.

SOLID-STATE SENSOR INTERFACE TO DC INPUT MODULES

We are going to interface selected two- and three-wire inductive proximity sensors to different 24 V dc input modules. First, let us interface a two-wire Allen-Bradley 871 TM DC inductive proximity sensor to an SLC 500 input module.

Sensor Interface to SLC 500 DC Input Module

Say we have an application where we need to interface a new sensor to an existing SLC 500 PLC system. The application requires sensing a mild steel target at a range of 8 millimeters. We have three input modules available in the SLC 500 chassis. The module part numbers are: 1746-IA16, 1746-IB8, and 1746-IV8. Module specifications are listed in Figure 7-28.

Module Specification	MODULE CATALOG NUMBERS		
	1746-IA16	1746-IB8	1746-IV8
Operating Voltage	85–132 VAC	10–30 V dc Sink	10–30 V dc Source
Number of Inputs	16	0	8
Points per Common	16	8	8
Backplane Current Draw at 5 V dc	.050 A	.085 A	.085 A
Maximum Signal Delay	ON = 35 ms OFF = 45 ms	ON = 8 ms OFF = 8 ms	ON = 8 ms OFF = 8 ms
Maximum Off-State Current	2 mA	1 mA	1 mA
Maximum Off-State Voltage		5 V dc	5 V dc
Nominal Input Current	12 mA at 120 VAC	8 mA at 24 V dc	8 mA at 24 V dc
Maximum Inrush Current	.8 A	NA	NA

Figure 7-28 Selected SLC 500 input module specifications. (Data from Allen-Bradley data tables)

Three Allen-Bradley inductive proximity sensors are available for this application. The available sensors and their specifications are listed in Figure 7-29.

Sensor Specifications	SENSOR CATALOG NUMBERS		
	872C-D8NE-18-A2	871TM-DH4NE-12-A2	871TM-DH8NE-18-A2
Operating Voltage	10–55 V dc Source	10–30 V dc Source	10–30 V dc Source
Barrel Diameter	18 mm	12 mm	18 mm
Sensing Distance	8 mm	4 mm	8 mm
Shielded	No	No	No
Output Configuration	Normally Open	Normally Open	Normally Open
Switching Frequency	500 Hz	75 Hz	60 Hz
Load Current	5–200 mA	<25 mA	<25 mA
Minimum Load Current		2 mA	2 mA
Leakage Current	<1.5 mA	<.9 mA	<.9 mA

Figure 7-29 Selected Allen-Bradley sensor specifications. (Data compiled from Allen-Bradley data tables)

We must select the correct sensor module pair and perform the wiring. We need to ask the following questions:

1. Are the available sensors AC or DC? The three sensors will operate on 10 to 30 V dc.
2. We need to select a DC input module. The 1746-IA16 input module is eliminated as it is an 85–132 VAC input module.
3. We have sinking and sourcing input modules available, but do we have both sinking and sourcing sensors available? No, all available sensors are in the sourcing configuration. From our lessons in the text we have learned that if we have a sourcing sensor, we must choose a sinking input module. For this application we will choose the 1746-IB8, a sinking input module. Always verify wiring diagrams before choosing one. Compare input module wiring with the sensor's wiring diagrams, as in Figure 7-30.
4. Is leakage current from the input sensor a problem? Leakage current from the sensor specifications is listed at <0.9 milliamps for two sensors and <1.5 milliamps for the third. The 1746-IB8 maximum off-state current is listed at 1 milliamp. The sensor with <1.5 milliamp leakage exceeds the module's maximum off-state current. This sensor must be rejected unless a bleeder resistor is to be used. The two remaining sensors should be acceptable from the standpoint of leakage current.
5. What sensor will provide us with the appropriate sensing distance? The 871TM-DH8NE-18-A2 will sense mild steel at an approximate range of 8 millimeters (mm). This sensor will fit our application.

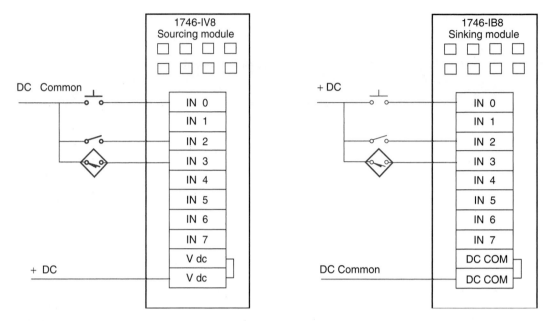

Figure 7-30   Wiring diagrams for the Allen-Bradley SLC 500 1746-IV8 and 1746-IB8. (Used with permission of Rockwell Automation, Inc.)

A FINAL NOTE ON SINKING AND SOURCING

There is one final consideration before leaving the topic of sinking and sourcing. We are going to introduce a possible safety consideration. Consider Figure 7-31. The left, PNP circuit is reportedly more common in Europe, while the right, NPN configuration is more common in the United States. Do you see any safety issues regarding one of these configurations? Looking at the NPN configuration, what would happen if the load developed a short to ground? Would not the load start, or turn on, unintentionally? This may be a consideration when deciding if an input device should be sinking or sourcing.

Before we conclude our study of input modules, we must look at how a PLC interfaces to analog input signals. While discrete signals are simply two-state signals, analog input modules give the PLC the ability to monitor an ever-changing input signal, such as temperature or pressure.

ANALOG INPUTS

Typical analog signals come from temperature, pressure, position, and revolutions per minute (RPM) inputs. Simply, analog input modules convert analog signals to digital words. Analog input modules are selected to accept either a current or a voltage input signal. Input signal levels are usually either 0 to 10 V dc, −10 to +10 V dc, 0 to 5 V dc,

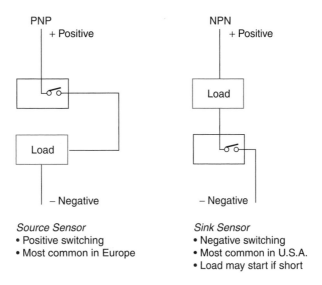

Figure 7-31   PNP versus NPN and safety issues.

1 to 5 V dc, 0 to 20 milliamps, −20 to +20 milliamps, or 4 to 20 milliamps. Analog modules convert the analog input signal through an analog-to-digital converter (A-to-D or A/D converter), and thus to a digital signal. Figure 7-32 illustrates a block diagram of a single-channel input for an IC693ALG222 voltage analog input module for the General Electric Series 90-30 PLC. Notice the A/D converter, which converts the analog input voltage to a digital signal. The digitized signal is passed through optical isolation to an onboard microprocessor and then on to the backplane of the baseplate. The backplane transfers the digitized signal to the CPU and on to the data table for storage.

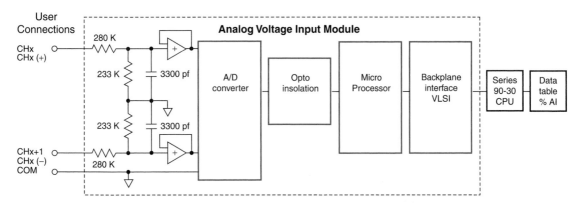

Figure 7-32   Analog voltage input module block diagram for a General Electric IC693ALG222 module. (Courtesy of GE Fanuc Automation)

Input modules come in three configurations: all inputs, all outputs, and combination of inputs and outputs. Figure 7-33 lists the available General Electric (GE) Series 90-30 analog modules. An input to, or output from, an analog module is called a "channel."

GENERAL ELECTRIC SERIES 90-30 ANALOG MODULES		
GE Catalog Number	Module Description	Channels
IC693ALG220	Voltage Analog Input	4
IC693ALG221	Current Analog Input	4
IC693ALG222	Voltage Analog Input	16
IC693ALG223	Current Analog Input	16
IC693ALG390	Analog Output	2
IC693ALG391	Analog Output	2
IC693ALG392	Analog Output	8
IC693ALG422	Combination Analog Module Current/Voltage	4 In 2 Out

Figure 7-33 General Electric Series 90-30 analog modules. (Courtesy of GE Fanuc Automation)

As an example, a PLC can be set up to monitor the temperature in an oven baking cookies. The oven temperature's lower limit could be 340 degrees, while the set point might be 350 degrees. An ever-changing analog temperature value will be input into the PLC. Oven temperature will gradually fall from the set point of 350 degrees to 340 degrees, as illustrated in Figure 7-34. The PLC will be programmed so that when the temperature has fallen to 340 degrees it will turn on the heaters to warm the oven back to its

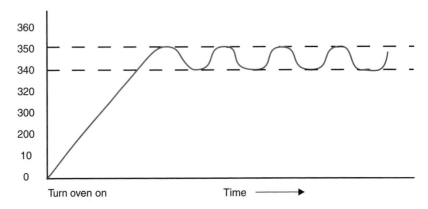

Figure 7-34 Oven temperature correlation to time.

set point. As the PLC continues to monitor the rising oven temperature, the heaters will be turned off when the analog input signal corresponds to 350 degrees.

In another example, let us assume that we have a one-turn potentiometer interfaced to a PLC input module. The potentiometer is used by a process operator to vary the speed of a variable-speed drive controlling a mixer. As the operator turns the potentiometer up from zero to full, the voltage varies in a linear fashion from 0 to 10 volts. The graph in Figure 7-35 illustrates the potentiometer position from 0, which is minimum, to 10, full open, to illustrate the voltage input into our analog input module.

POTENTIOMETER POSITION IN COMPARISON TO INPUT VOLTAGE										
100%										X
90%									X	
80%								X		
70%							X			
60%						X				
50%					X					
40%				X						
30%			X							
20%		X								
10%	X									
0%	1	2	3	4	5	6	7	8	9	10
Input Voltage increasing →										

Figure 7-35   Potentiometer knob position in relation to input voltage signal seen by input analog module.

Let us interface the potentiometer from Figure 7-35 to an Allen-Bradley SLC 500. The input module converts the analog input signal through its internal analog to a digital converter to a 16-bit data word. The digital equivalent of the analog value is sent on to the input status file input word associated with the input slot in which the analog module resides. Each analog input or output connection is called a channel. Figure 7-36 lists Allen-Bradley's SLC 500 analog modules.

Each channel will be represented as one 16-bit word in either the input status or output status files. If an analog input module has two channels, there will be two words of input status file space needed to store the binary representation of the analog voltage or current signal. Likewise, a four-channel input analog module will require four words in the input status file represented by the slot in which the module resides. Considering that a typical discrete input or output module only gets one 16-bit word in the status table, how is analog channel data stored in the status table?

ALLEN-BRADLEY SLC 500 ANALOG MODULES		
Catalog #	Input Channels per Module	Output Channels per Module
1746-NI4	4 differential, voltage or current	None
1746-NI8	8 differential, voltage or current	None
1746-NI16I	16 single-ended, current	None
1746-NI16V	16 single-ended, voltage	None
1746-NIO4I	2 differential, voltage or current	2 current
1746-NIO4V	2 differential, voltage or current	2 voltage
1746-NO4I	None	4 current
1746-NO4V	None	4 voltage
1746-FIO4I	2 differential, voltage or current, selectable	2 current
1746-FIO4V	2 differential, voltage or current, selectable	2 voltage

Figure 7-36 Selected Allen-Bradley SLC 500 analog modules. (Data compiled from Allen-Bradley SLC 500 Analog Modules specification data)

Analog Input Status Word Addressing

A 16-point input module residing in slot three will be addressed as I:3, 0 through 15. A modular SLC 500 with 16-point modules in slots two, three, and four would be represented in the input status table as illustrated in Figure 7-37.

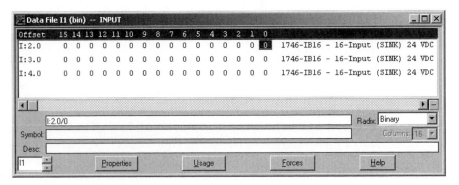

Figure 7-37 RS Logix 500 input status table representation for input modules in slots two, three, and four.

If more data must be stored in the input status table for a particular slot, additional words are available. When only 16 bits of input data need to be stored in the input status table, the CPU will only assign and use one word, word zero. The input address is usually written simply as I:3, followed by the screw terminal number. However, the input address may be written more formally to reflect the word designation. The formal address for an SLC 500 is I:3.0, screw terminal number. Notice the .0 after the slot designation. This designates the word number. The first word is always word zero. When a module needs more than one word, the I/O configuration process for an SLC 500 PLC automatically assigns the required number of input and output status table words. Figure 7-38 illustrates an SLC 500 input status table with a 16-input discrete module in slot two and a 1747-SDN DeviceNet scanner module in slot three. The DeviceNet scanner module is a good example of the principle, as it is assigned 32 input words and 32 output words. The scanner module is used to communicate with a DeviceNet network.

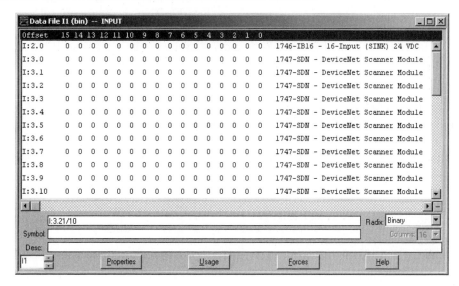

Figure 7-38 SLC 500 input status table showing word zero through word 10 for a 1747-SDN DeviceNet scanner.

A four-channel analog input module would be addressed as illustrated in Figure 7-39. The input status file addresses from Figure 7-39 represent the following:

I:2.0 is the input word for the 16-point module in slot two.

I:3.0 is analog input module, slot three, channel zero.

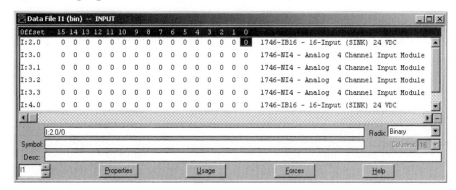

Figure 7-39 RS Logix 500 input status table reflecting words available for slot three.

I:3.1 is analog input module, slot three, channel one.

I:3.2 is analog input module, slot three, channel two.

I:3.3 is analog input module, slot three, channel three.

I:4.0 is the input word for the 16-point module in slot four.

Analog channel data will be stored in its respective words in the input status table. For example, the SLC 500 PLC analog input modules accept current or voltage input signals. The acceptable input signals are listed in Figures 7-40 and 7-41 along with signal specifications. Figure 7-41 lists analog current input specifications for the SLC 500. The far right column of Figures 7-40 and 7-41 refers to least significant bit "resolution." The resolution of an analog module is the weight assigned to the least significant bit.

SLC 500 ANALOG VOLTAGE INPUT SPECIFICATIONS			
Voltage Range	**Decimal Equivalent**	**Significant Bits**	**Least Significant Bit Resolution (in Microvolts)**
−10 V to +10 V	−32,768 to +32,767	16 bits	305.176 uV
0 V to 10 V	0 to +32,767	15 bits	305.176 uV
0 V to 5 V	0 to 16,384	14 bits	305.176 uV
1 V to 5 V	3,277 to +16,384	14 bits	305.176 uV

Figure 7-40 Data compiled from SLC 500 data tables. (Used with permission of Rockwell Automation, Inc.)

SLC 500 ANALOG CURRENT INPUT SPECIFICATIONS			
Current Range	**Decimal Equivalent**	**Significant Bits**	**Least Significant Bit Resolution (in Microvolts)**
−20 mA to +20 mA	−16,384 to +16,384	15 bits	1.22070 uV
0 to +20 mA	0 to 16,384	14 bits	1.22070 uV
4 to +20 mA	3,277 to 16,384	14 bits	1.22070 uV

Figure 7-41 Data compiled from SLC 500 data tables. (Used with permission of Rockwell Automation, Inc.)

Analog Module Resolution

After an analog input signal is sent through the analog-to-digital converter, the digitized word's least significant bit will have a value associated with it. This value is determined by the number of bits into which the analog-to-digital converter breaks the converted signal.

As an example, let's say we have a 0–10 volt analog input signal. Our analog-to-digital converter breaks the signal up into 10 parts. The closest we could measure a voltage input from this module would be to 1/10 of the 10-volt value, or 1 volt. Thus, the resolution would be 1 volt. This coarse resolution would not be acceptable in many applications. What if we represent the digitized value of the analog-to-digital converter with an 8-bit data word? An 8-bit word is equivalent to 255. If we divided a 0–10 volt analog signal into 255 parts, the least significant bit would be 1/255 of the 0–10 volt signal. Ten volts divided into 255 parts is equal to .0392 volts per part. Thus, the least significant bit (the smallest part) is equal to .0392 volts. This would be the resolution of the least significant bit if our input module resolved the 0–10 volt analog input signal through the A-to-D converter into an eight-bit data word.

Figure 7-40 lists the SLC 500 input module with a 0–10 volt analog input signal converted into 15 significant bits. A 15-bit word is equivalent to 32,767 in decimal. That means that a 15-bit word contains 32,767 parts. If 10 volts is divided into 32,767 parts, the least significant bit would be equal to .0003051 volts. This value is the same as the 305.176 microvolts listed in Figure 7-40.

Applying an Analog Input Module

Let us look at an example applying an analog input module. A 0–10 volt analog input signal from the potentiometer described earlier will be an input to channel one of an SLC 500 analog input module. Our analog input module will reside in slot three of an SLC 500 chassis. The converted binary data will be found in the input status table word address I:3.1 (input, slot three, word 1).

A four-channel analog input module would be addressed as illustrated in Figure 7-42. If the potentiometer was inputting 10 volts from a 0–10 volt signal, binary data would be represented as illustrated in the address I:3.1 in Figure 7-42.

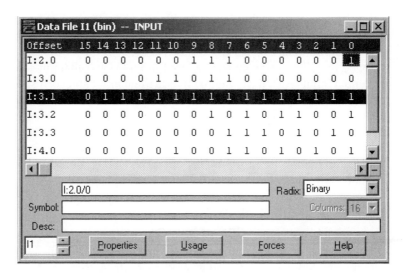

Figure 7-42 RS Logix 500 input status table reflecting words available for slot three.

Figure 7-43 illustrates the decimal value of the binary value stored in the input status file word representing voltages from 0 to 10 volts for that input channel.

ANALOG SIGNAL COMPARISON FOR SAMPLE POTENTIOMETER INPUT		
Potentiometer Position	Voltage Input Signal	Decimal Value Represented in Data PLC's Table
100%	10	32,767
80%	8	26,214
70%	7	22,937
60%	6	19,660
50%	5	16,384
40%	4	13,107
30%	3	9,830
20%	2	6,553
10%	1	3,276
Closed	0	0

Figure 7-43   Percentage of analog potentiometer input correlating to a 0 to 10 V dc analog input signal digitized value.

Differential versus Single-Ended Analog Inputs

Physically connecting analog input signals to an analog input module may be different from using a discrete input module. Analog input modules are classified as either single-ended or differential.

Single-ended inputs to an analog input module have all input commons tied together. Differential inputs each have their own individual input and corresponding common. As a result, a differential connected module will have half as many channels as a single-ended configuration. Figure 7-44 illustrates a typical single-ended analog input interface. Note that shielded cable is used to connect to the input point from the analog transmitter. Shielded cable is used to help keep field noise out of the input signal. The shield should be connected to the chassis ground lug only. The analog source end of the cable should be taped only to insulate the shield from any electrical connection.

Single-Ended versus Differential Inputs

Although there are more types of input channels available with single-ended inputs, they are subject to more potential problems with noise entering the channel, which causes inconsistent input data and ground currents. A converted differential input signal is the difference between the channel's positive input and the channel's negative input. Differential input channels have two input connections per channel. If an equal amount

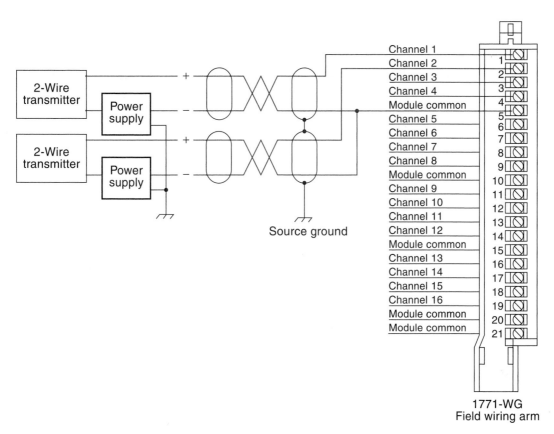

Figure 7-44   Single-ended analog input connections from field devices. (Used with permission of Rockwell Automation, Inc.)

of noise is picked up in the input wiring, the module will look for the difference between the two input lines. With a differential input, noise on both input channel inputs will theoretically be filtered out, leaving only the actual input signal.

Figure 7-45 illustrates differential input connections to an Allen-Bradley PLC 5 analog input module.

This section has only introduced you to the basic principles of analog input modules and their interfaces to a PLC input module. The next section will introduce some input modules that have been developed to allow the PLC to interface in special application situations. These modules provide an interface only for the application for which they were developed. These special modules are called specialty modules.

SPECIALTY MODULES

Specialty modules fall into the category of smart modules. A smart module contains its own microprocessor. Specialty modules are designed to provide a specific function. This section will provide a quick overview of some of the more popular smart input and output modules.

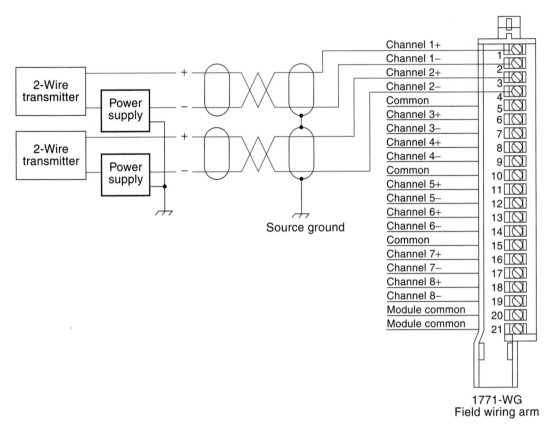

Figure 7-45 Differential analog input connections from field devices to an Allen-Bradley PLC 5, 1771 IFE analog input module. (Used with permission of Rockwell Automation, Inc.)

BASIC Module

A BASIC module is a specialized module that allows the PLC to interface to serial peripheral devices. BASIC modules usually reside in the local chassis, close to the CPU. The operator develops a separate BASIC language program, which is downloaded into the BASIC module. Typically, a program written in the BASIC programming language, or some variation of it, is placed into the BASIC module's memory. Different manufacturers may have their own variations of the BASIC programming language. To receive maximum performance, you must use that manufacturer's software when programming the BASIC module. BASIC modules usually have a programming port that is compatible with RS-232C or RS-422 communication standards. The module will have at least one serial port to interface from the PLC's CPU to printers, ASCII terminals, bar code readers, and asynchronous modems. The BASIC module can also perform computational tasks, thus relieving the burden of the CPU.

Communication Modules

ASCII I/O Modules.   ASCII I/O modules allow the interfacing of bar code readers, meters, printers, and data terminals to a PLC. ASCII modules, which accept only valid ASCII data, are not used as extensively as they once were. Today, the RS-232 module is the module of choice in many applications.

RS-232C Interface Modules.   Communication modules are available that reside in a PLC chassis and enable you to connect a PLC to telephone lines using a modem. PLCs connected to phone lines allow central control room operators to examine ladder programs to modify or edit program operation at remote PLC sites. Today many remote oil, gas, and wastewater applications are unmanned. Remote access by way of phone lines saves maintenance personnel from driving to remote sites each time a PLC encounters a problem or a program change is necessary.

High-Speed Encoder Input Modules

When input pulses come in faster than a discrete input module can handle them, a high-speed input module is used. High-speed counters are also used to interface encoders to a PLC.

Remote I/O Subscanners

When you need an I/O chassis remotely mounted from the base PLC, some PLC systems require a remote I/O subscanner. Simply put, a subscanner resides in the base CPU chassis and relieves the CPU from the burden of scanning the I/O. A subscanner scans the remote I/O chassis and the respective I/O points. After the subscanner has scanned all remote I/O points, their I/O status is stored in a built-in buffer (storage area). At the appointed time in the CPU's scan, the CPU will read the I/O status data stored in the subscanner's buffer.

Resistance Temperature Detector (RTD) Input Modules

A resistance temperature detector (RTD) input module interfaces a PLC to RTD temperature-sensing elements and other types of resistance input devices such as potentiometers. The RTD input module converts analog input signals from a potentiometer or RTD into input signals understood by the PLC. These values are stored in the PLC input table.

Stepper Motor Control Modules

A stepper module is an intelligent module that resides in a PLC chassis and provides a digital output pulse train for microstepping stepper motor applications.

Thermocouple/Millivolt Input Module

The thermocouple/millivolt input module converts inputs from various thermocouple or millivolt devices into values that can be input and stored into PLC data tables. This module greatly enhances the flexibility of a PLC system by interfacing thermocouples, thus eliminating expensive thermocouple transmitters. Using an RTD module, PLCs can thus be used for interface applications requiring temperature and measurement control.

SUMMARY

Early programmable controllers were strictly limited to discrete inputs. Today's programmable controllers are much more sophisticated, thanks to vast advances in microprocessor technology. The increase in sophistication of these programmable controller devices mandates increased abilities and understanding from those individuals who will be applying this new technology to current and new applications. Today's programmable controller can be applied to practically any control problem due to its complete range of discrete, analog, and specialty interface control abilities.

There are four categories of input modules: input modules, combination input and output modules, analog current or voltage input modules, and combination analog input and output modules.

Input modules typically are available with 4, 8, 16, 24, or 32 points. Discrete input modules interface to AC, DC, and +5 DC TTL discrete voltage signals. Analog input modules interface to varying input signals such as temperature and pressure. Analog input modules are available as 2, 4, 8, or 16 channels.

Specialty modules are designed to provide a specific interface solution. Specialty modules give the PLC added functionality to interface modems, high-speed inputs, encoders, stepper motors, thermocouples, and RTDs, to name a few.

In this chapter we explored discrete and analog input modules and how these modules interface signals to real-world devices and the CPU. We also introduced selected specialty modules and described how they enhance a PLC's functionality.

REVIEW QUESTIONS

1. Input modules provide an interface between
   A. Input modules and the CPU
   B. Field equipment and the CPU
   C. Output modules and field devices
   D. The CPU and output modules

2. Discrete I/O input modules accept signals from field devices that are:
   A. Analog devices
   B. Digital devices
   C. Two-state devices
   D. Discrete devices
   E. A , B, and C are correct.
   F. B, C, and D are correct.

3. Name three analog inputs.

4. Name five discrete inputs.

5. What are the five main sections of an AC input module?

6. What is the purpose of the bridge rectifier section of an AC input module?

7. A discrete input signal is considered valid if the signal:
   A. Is a two-state signal
   B. Is successfully changed from AC to DC through the bridge rectifier
   C. Is accepted and passed through the threshold detector
   D. Is successfully debounced
   E. For an AC signal, if it is between 80 and 132 VAC
   F. All of the above

8. Electrical isolation is provided so that there is no physical electrical connection between the CPU and field devices. This isolation is usually accomplished by _____.

9. We introduced thumbwheel switches in Chapter 2. Thumbwheels provide input to the PLC in what format?

10. What are typical applications where we would employ thumbwheel switches?

11. How many wires would be connected to an input module from the three-digit thumbwheel?

12. If we wanted to interface three groups of three-digit thumbwheels to our PLC, how many wires would we have to hook up?

13. Thumbwheels are a state-of-the-art (and very popular) interface medium for modern PLCs—true or false? Explain your answer.

14. Illustrate how you would connect two 115 VAC input devices to an input module. Assume there are different phases on each line.

15. Illustrate a block diagram of a typical AC input module. Identify the blocks.

16. Explain where you would use an isolated I/O module.

17. A sinking input module is:
   A. Always the same no matter who manufactured it
   B. Different among different manufacturers

18. An NPN inductive proximity sensor would interface to what part of an Allen-Bradley SLC 500?
   A. Sinking input module
   B. Sourcing input module

19. A sourcing input module is:
   A. Always the same no matter who manufactured it
   B. Different among different manufacturers

20. A PNP inductive proximity sensor would interface to what part of an Allen-Bradley SLC 500?
   A. Sinking input module
   B. Sourcing input module

IF PASSWORD IS CORRECT GO TO DRIVE SET UP SCREEN

0022	┌─── EQU ───────────────┐ B3:10/0 B3:1/15

```
0022    ┌─── EQU ────────────────┐        B3:10/0                                              B3:1/15
        │ Equal                  │       ─[ OSR ]─                                              ─( )─
        │ Source A      N12:21   │
        │               7777<    │                                              RESET PASSWORD
        │ Source B      32767    │                                              ATTEMPTS COUNTER
        │               32767<   │                                                   C5:5
        └────────────────────────┘                                                  ─( RES )─


0023    ┌─── GRT ────────────────┐    ┌─── NEQ ────────────────┐     B3:10/1
        │ Greater Than (A>B)     │    │ Not Equal              │    ─[ OSR ]─────────────┐
        │ Source A      N12:21   │    │ Source A      N12:21   │                         │
        │               7777<    │    │               7777<    │                         │
        │ Source B        0      │    │ Source B      32767    │                         ▼
        │               0<       │    │               32767<   │
        └────────────────────────┘    └────────────────────────┘          ┌─── CTU ────────────┐
                                                                      ┌────│ Count Up           │──( CU )─
                                                                      │    │ Counter      C5:5  │
                                                                      │    │              3<    │──( DN )─
                                                                      └────│              2<    │
                                                                           └────────────────────┘
```

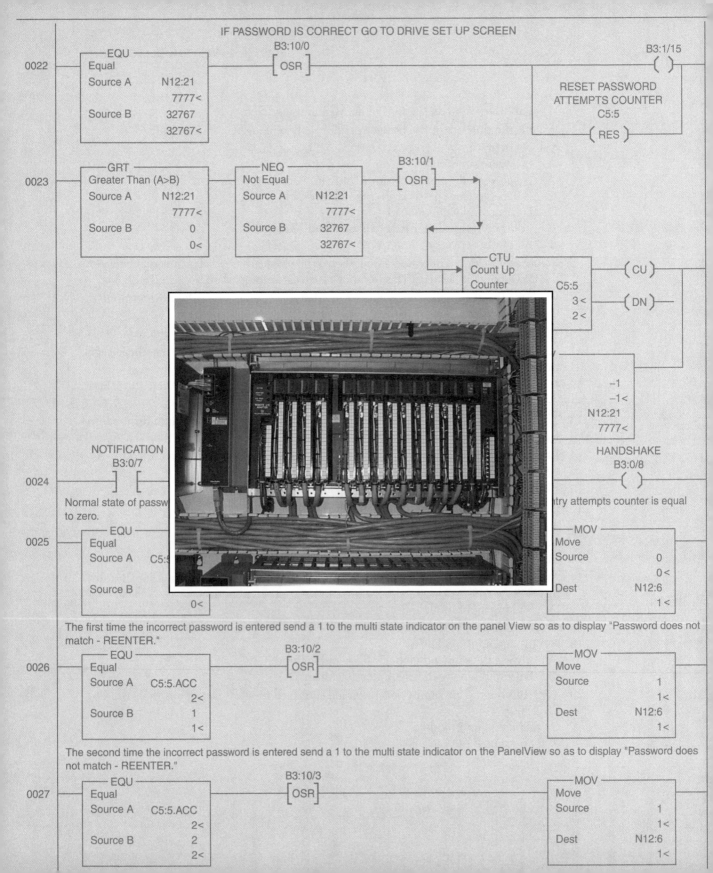

```
                                                                                  ─1
                                                                                  ─1<
                                                                                N12:21
                                                                                7777<


        NOTIFICATION                                                           HANDSHAKE
        B3:0/7                                                                 B3:0/8
0024    ─] [─                                                                   ─( )─

        Normal state of passw                                      try attempts counter is equal
        to zero.

0025    ┌─── EQU ──────┐                                                 ┌─── MOV ──────┐
        │ Equal        │                                                 │ Move         │
        │ Source A  C5:5│                                                │ Source    0  │
        │              │                                                 │           0< │
        │ Source B     │                                                 │ Dest  N12:6  │
        │          0<  │                                                 │           1< │
        └──────────────┘                                                 └──────────────┘
```

The first time the incorrect password is entered send a 1 to the multi state indicator on the panel View so as to display "Password does not match - REENTER."

```
0026    ┌─── EQU ──────────────┐         B3:10/2                     ┌─── MOV ──────┐
        │ Equal                │        ─[ OSR ]─                    │ Move         │
        │ Source A   C5:5.ACC  │                                     │ Source    1  │
        │            2<        │                                     │           1< │
        │ Source B     1       │                                     │ Dest  N12:6  │
        │            1<        │                                     │           1< │
        └──────────────────────┘                                     └──────────────┘
```

The second time the incorrect password is entered send a 1 to the multi state indicator on the PanelView so as to display "Password does not match - REENTER."

```
0027    ┌─── EQU ──────────────┐         B3:10/3                     ┌─── MOV ──────┐
        │ Equal                │        ─[ OSR ]─                    │ Move         │
        │ Source A   C5:5.ACC  │                                     │ Source    1  │
        │            2<        │                                     │           1< │
        │ Source B     2       │                                     │ Dest  N12:6  │
        │            2<        │                                     │           1< │
        └──────────────────────┘                                     └──────────────┘
```

Output Modules

OBJECTIVES

After completing this chapter, you should be able to:

- describe the available types of output modules
- explain applications where analog output modules would be used
- define output module specifications from data sheets
- list the advantages and disadvantages of using relay versus solid-state output switching
- list the advantages and disadvantages of using solid-state versus relay output switching
- explain considerations when applying PLCs in hazardous locations

INTRODUCTION

The output section of a PLC system is the physical connection between the outside world and the central processing unit (CPU). Examples of discrete outputs include motor starter coils, pilot lights, solenoids, alarm bells, valves, fans, other control relays, or input start or stop signals to a variable-speed drive. All these devices have one thing in common: their signals are either ON or OFF, OPEN or CLOSED, YES or NO, or TRUE or FALSE.

Output modules generally fall into three classifications: discrete, analog, or specialty; they are also available in various switching voltages and configurations.

There are various ways to get information from the CPU's output status file to common hardware field devices. First, discrete output interface modules provide a method of getting discrete, or digital, signals out of the PLC to control discrete field devices. Second, ever-changing signals such as temperature control, or the 4–20 milliamp speed reference for variable-frequency drives, can be interfaced to the PLC using analog output modules.

This chapter will introduce discrete and analog output modules, their operating principles, types of modules available, advantages and disadvantages of using one type over the other, and basic interfacing principles.

DISCRETE OUTPUT MODULES

Much like discrete inputs, discrete outputs are the most commonly used type. Discrete output modules simply act as switches to control output field devices. They fall into two classifications: solid-state output switching and relay output switching.

Discrete output modules receive their operating power from the PLC's power supply, which comes from the backplane. Usually, the user must provide the power that the module output point switches to control the field devices. Figure 8-1 illustrates basic field wiring for a discrete 120 VAC output module.

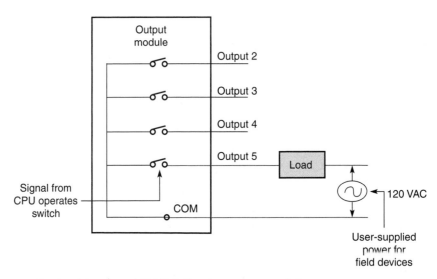

Figure 8-1   Basic wiring for a 120 VAC discrete output module.

Discrete output modules come in various signal levels and specifications. Listed in Figure 8-2 are common types and voltage levels of output modules that are available.

Operating Principles of Output Modules

Output modules are simply switching devices that carry out commands from the CPU. Each output point contains a switching device, which is located inside the output module and is turned on or off according to the bit value residing in that particular output status table address.

DISCRETE OUTPUT MODULES		
Solid-State Outputs		Relay Outputs
AC Output Modules	DC Output Modules	Relay Output Modules
12, 24, 48 VAC	TTL level	Relay Output
120 VAC	12, 24, 48 V dc	Isolated Relay Output
230 VAC	120 V dc	Relay Output
	230 V dc	
	24 V dc, Sink	
	24 V dc, Source	

Figure 8-2   Discrete output module output classifications.

During the portion of the processor operation where outputs are updated, the output status file values will be sent, one 16-bit word at a time, to the respective output modules. On accepting its output status word, the output module will turn each output on or off or will maintain the state of the field device as consistent with the associated bit in the output status file at that specific address. Figure 8-3 illustrates signal flow from the CPU through each step in the control of the field device.

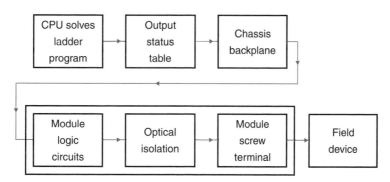

Figure 8-3   Signal flow from the CPU through an output module to the field device.

A typical AC output module contains circuitry as illustrated in Figure 8-4. A single AC output point consists of a latching circuit for the low-voltage (usually VDC or 12-18 V dc)

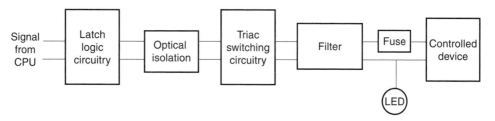

Figure 8-4   Block diagram of a typical output module.

logic signal sent by the CPU from the output status table. The ON or OFF signal represents the logical value of the output. If the output status table bit is a one, the ON signal will be latched into the logic circuitry block. This ON signal will be passed through the optical isolation circuitry to the block containing the switching hardware.

Solid-State Output Module Switching

In a solid-state AC output module, a triac is used to switch the AC high voltage and current controlling the ON or OFF state of the field hardware device. A triac is a solid-state device used to switch AC. The filter block will contain protective devices such as a metal oxide varistor (MOV). The MOV is used to limit peak voltage across the AC switching hardware to a safe level. The output point's LED alerts the operator that the output has been directed by the CPU to turn on. A fuse may also be included on the output line to protect the AC switching device from drawing too much current. If no fuse is provided in the module's circuitry, check your user's manual for instructions on the proper installation of a fuse in your field wiring. Although the PLC's power supply provides power to operate the output module, the power the triac switches to supply to the load must be provided by the user. Figure 8-5 lists specifications for an Omron CQM1-OA221 triac output module.

OMRON CQM1-OA221 TRIAC OUTPUT MODULE SPECIFICATIONS	
Switching Capacity, Maximum	0.4 amp at 100–240 VAC
Leakage Current	1 milliampere (mA) maximum at 100 VAC/ 2 mA maximum at 200 VAC
Residual Voltage	1.5 volts, 0.4 amp
ON Delay	6 milliseconds (ms) maximum
OFF Delay	$1/2$ cycle + 5 ms
Number of Outputs per Common	4
PLC Power Supply Consumption	110 mA at 5 V dc
Fuse	2 amps, one fuse per common. Fuse not user replaceable.

Figure 8-5   Data compiled from Omron CQM1 module specifications.

Switching DC Output Loads

DC output modules are used to control discrete loads by switching a power transistor rather than a triac. The functional operation of a DC output module is similar to the AC output module with the exception of the switching device. Being a solid-state device, the transistor-switching device is susceptible to surge currents and excessive voltages. Care should be taken to ensure that excessive voltages are not applied to the transistor, as they could cause excessive heat and a possible short circuit. Protecting the power transistor with a fast-acting fuse can reduce possible damage from excessive heat.

DC output modules, similar to DC input modules, are available in sinking and sourcing configurations. Figure 8-6 illustrates a sinking output module interfaced to a field device.

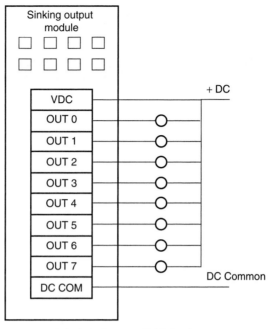

Figure 8-6   Sinking output module interface to field device.

Figure 8-7 illustrates a sourcing output module interfaced to a field device.

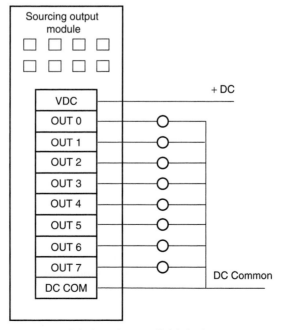

Figure 8-7   Sourcing output module interface to field devices.

Figure 8-8 lists the specifications for the Allen-Bradley SLC 500 sourcing-output module number 1746-OB16.

SLC 500 SOURCING OUTPUT MODULE SPECIFICATIONS	
Operating Voltage	10–50 V dc
Number of Outputs	16
Output Points per Common	16
Backplane Current Draw	0.280 amp at 5 V dc
Maximum Signal Delay (Resistive Load)	ON = 0.1 millisecond OFF = 1.0 millisecond
Maximum OFF-State Leakage	1 milliamp
Minimum Load Current	1 milliamp
Continuous Current per Output Point	.5 amp at 30 degrees centigrade
Continuous Current per Module	8 amps at 30 degrees centigrade
Surge Current per Output Point	3 amps for 10 seconds
ON-State Voltage Drop	1.2 volts at 10 amps

Figure 8-8 Specifications compiled from Allen-Bradley discrete input and output modules product data.

The following terms are associated with output modules (see also Figure 8-8):

continuous current per module: the total of all output currents from all output points on a particular output module

continuous current per point: the maximum continuous output current each output point can supply to a load or field device

surge current: the inrush current to an inductive device

surge current per point: the total inrush current an inductive device can draw out of a single output point

The sinking or sourcing terminology used by different manufacturers for DC input modules also applies to DC output modules. Always check your manufacturer's wiring diagrams and module specifications before applying any module. DC output modules for the General Electric Series 90-30 PLC, for example, are identified as positive or negative logic rather than sinking or sourcing. A positive-logic output module acts as the source of current to the field device. The load is connected between the negative side of the power supply and the module's output screw terminal. A negative-logic output module sinks current from the field device, the load, to the negative side of the power supply. The field device is connected between the positive side of the power supply and the module's output screw terminal.

TRANSISTOR-TRANSISTOR LOGIC (TTL) OUTPUT MODULES

Transistor-transistor logic (TTL) output modules switch 5 V dc signals. A TTL output module allows for interface between the PLC and TTL-comparable devices. An example of a TTL interface would be interfacing a PLC to various 5 V dc field devices including integrated circuits and seven-segment LED displays.

Even though seven-segment displays are still used in some applications, they have been widely replaced with more modern human-machine interface devices that allow customized display screens to be developed with a personal computer and computer software.

RELAY OUTPUT MODULES

Relay output modules are also known as contact outputs or dry contact outputs. Even though relay output modules are used to switch AC or DC loads, usually relay outputs are used to switch small currents at low voltages, to multiplex analog signals, and to interface control signals to variable-speed drives.

Relay output modules use actual mechanical relays, one for each output point, to switch the output signal from the output status file. Figure 8-9 illustrates one point of a relay output module. Notice that the common goes to other relays in the group.

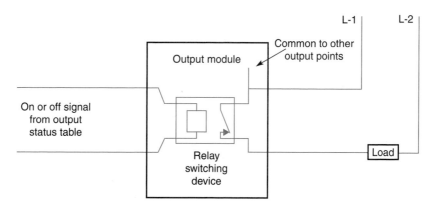

Figure 8-9   Simplified single-point relay output.

Relay output modules are available in three variations, depending on the manufacturer. Most manufacturers offer combination input and relay output modules. Combination relay modules usually come with two, four, or six 120 VAC inputs, and two-, four-, or six-relay outputs.

Figure 8-10 illustrates a 4-point 120 VAC input and a 4-point relay output combination module. Notice that relay outputs accept AC or DC signals.

Relay output modules are available with 8 or 16 outputs. Figure 8-11 illustrates an 8-point relay output module. Notice that the top four outputs have a single common,

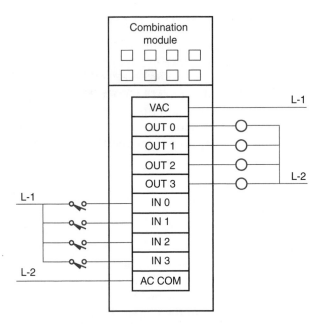

Figure 8-10   Combination 120 VAC input and relay output module.

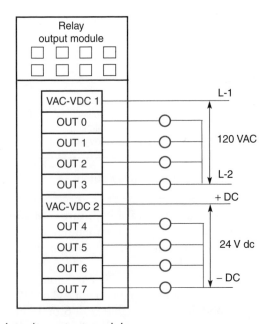

Figure 8-11   Eight-point relay output module.

whereas the bottom four outputs also have their own common. Separating the commons and their associated four outputs into two groups provides users with the flexibility to have one group switch one voltage type or level, say 120 VAC, while the other group of four outputs can switch a different voltage type or level, like 24 V dc.

Figure 8-12 lists module specifications for the General Electric Series 90-30 PLC normally open, relay output module IC693MDL940 (2 amp, 16 point).

GENERAL ELECTRIC RELAY OUTPUT MODULE SPECIFICATIONS FOR IC693MDL940	
Rated Voltage AC DC	120/240 volts 24 volts
Operating Voltage AC DC	5 to 30 volts 5 to 250 volts
Outputs per Module Output Grouping	16 4 per common
Isolation	1,500 volts between field side and logic side 500 volts between groups
Maximum Load, Pilot Duty Maximum Load per Common	2 amps maximum 4 amps maximum
Minimum Load	10 mA
Maximum Inrush	5 amps
ON Response Time	15 ms maximum
OFF Response Time	15 ms maximum
Power Consumed from Power Supply (All outputs on) 5 volts from Backplane 24 volts from Backplane	7 mA 135 mA

Figure 8-12 Compiled from GE Fanuc Automation module specifications data, Series 90-30 Programmable Controller I/O Specifications Manual.

A common application for relay output modules is switching start, stop, and reverse control signals for a variable-speed drive. Figure 8-13 illustrates the wiring of a control wiring terminal block from an Allen-Bradley Bulletin 160 variable-speed drive to an SLC 500 1746-OWA relay output module. This is a similar relay output module to that illustrated in Figure 8-11. Since the drive provides 12 volts of power for the control circuit, the output module only has to provide a contact closure on output four, the reverse signal; output five, the start signal; and output seven, the stop signal.

This drive has eight preset speeds programmed by the operator into the drive's memory. The selection of different speeds is achieved by opening and closing preset switch

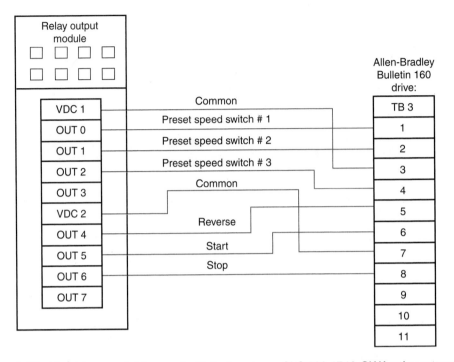

Figure 8-13 Variable-speed drive control interface to an SLC 500 1746-OWA relay output module.

one, output 0; preset switch two, output 1; and preset switch three, output 2, in the proper sequence. Remember from Chapter 4 that three bits can be arranged to reflect eight different binary combinations. These eight binary combinations, when switched by the PLC and input into the drive, select at which of the eight preset speeds the drive will run.

Relay output modules are also available with isolated outputs. In this design, an 8-point, isolated relay output module has a separate incoming line to supply power to the switching portion of each physical relay. Figure 8-14 illustrates an isolated relay output module. Notice the separate IN and OUT signal connections.

Isolated relay output modules have each output point isolated from all the other points. Isolated relay output modules are available with normally open and normally closed form C relays.

The advantage of an isolated relay output module is that since each relay is isolated from the others and has its own common line, each relay and its associated output point can control any voltage level that is compatible with its contacts.

One isolated relay output module could control field devices with 120 VAC, 120 V dc, 230 VAC, 24 V dc, 12 V dc, and 5 V dc, all from one module.

Because these modules use electromechanical relays as their switching devices, there are current limitations and different life expectancies in comparison to solid-state switching

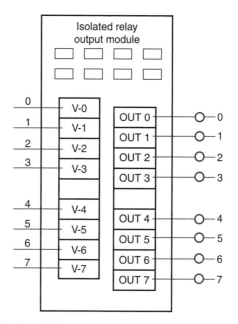

Figure 8-14 Eight-point isolated relay output module.

modules. Figure 8-15 lists contact life and current load specifications for the General Electric IC693MDL940 relay output module.

GENERAL ELECTRIC IC693MDL940 RELAY SPECIFICATIONS			
Voltage	Resistive Load in Amps	Lamp or Inductive Load in Amps	Contact Operations
24 to 120 VAC	2 1 .1	1 .5 .05	300,000 500,000 1,000,000
240 VAC	2 1 .1	1 .5 .05	150,000 200,000 500,000
24 V dc	NA 2 .1	1 .5 .05	300,000 500,000 1,000,000
125 V dc	.2	.1	300,000

Figure 8-15 Compiled from GE Fanuc Automation module specifications data.

Specifications for a Modicon Quantum Series PLC 16-point relay output module are listed in Figure 8-16.

MODICON QUANTUM PLC RELAY OUTPUT MODULE SPECIFICATIONS	
Number of Output Points	10 normally open
LEDs	One green LED for each output point to indicate the point's ON or OFF status
Working Voltage	20–250 VAC 5–30 V dc
Maximum Load Current per Point	2 A max. at 250 VAC or 30 V dc at 60 degrees C ambient resistive load 1 A tungsten lamp load 1 A at a power factor of 0.4 1/8 horsepower (hp) at 125/250 VAC
Maximum Frequency	60 hertz (Hz) resistive loads
Maximum Surge Current per Point	10 A capacitive load at 10 ms
Switching Capability	500 VA resistive load
Response Time	OFF to ON, 10 ms maximum ON to OFF, 20 ms maximum
Relay Contact Life, Mechanical	10,000,000 operations
Relay Contact Life, Electrical	200,000 operations, resistive at maximum voltage and current
Relay Type	Form A
Contact Protection	Internal varistor, 275 volts
Operating Current from Backplane	1,100 mA
Power Dissipation	5.5 watts plus .5 watt per point ON

Figure 8-16   Compiled from Modicon Quantum Automation Hardware Reference Guide.

MODULE SELECTION CONSIDERATIONS

There are several considerations when selecting an output module for your application. The primary issues are the number of operations, leakage current, and safe shutdown in case of a failure.

A solid-state output is used in applications where the output point will be turned on many times per hour or day. Solid-state switching devices theoretically have unlimited life. These devices are rated for hundreds of millions of operations, whereas a mechanical device will switch for only a few hundred thousand times. Switching a solid-state output on and off every few seconds for 24 hours a day every day will result in a probable life in the 20- to 30-year range. On the other hand, the same application with a mechanical output module would require replacement every few months.

By their very nature, solid-state devices leak current. As a result, leakage current may be adequate to turn on low-level actuators, or even turn on and keep on certain LED pilot lights or display devices. A relay output module may be required for these applications. The mechanical relay in a relay or contact output module is the only output circuit with no leakage current.

Even though today's PLCs are rugged and dependable devices, where safety is a primary concern, you should not depend on solid-state output modules. In many cases, solid-state switching devices will fail in a shorted, rather than an open, condition. By failing in a shorted (or ON) condition, such devices pose an added safety hazard. On the other hand, where safe shutdown is a necessity, an electromechanical relay can be jumpered or wired to fail in either an open or a closed state.

Figure 8-17 lists the specifications for a General Electric Series 90-30 solid-state output module.

GENERAL ELECTRIC SERIES 90-30, 120/240 VAC OUTPUT MODULE SPECIFICATIONS	
Number of Output Points	8, two groups of four
Output Voltage Range	85–264 VAC
Output Current	2 amps maximum per point 4 amps maximum per group (ambient dependent)
Inrush Current	20 amps maximum for one cycle
Maximum Load Current	100 mA
Output Leakage Current	3 mA maximum at 120 VAC 6 mA maximum at 240 VAC
Response Time	ON, 1 ms maximum OFF, ½ cycle maximum
Operating Current from Backplane	160 mA at 5 V dc, all outputs ON
Power Dissipation	5.5 watts plus .5 watt per point ON

Figure 8-17   Series 90-30 output module specifications.

CHOOSING THE PROPER OUTPUT MODULE

Take into consideration the advantages and disadvantages of each type of output module as you go through the selection process.

Advantages to Solid-State Switching

1. fast switching speeds
2. high reliability and almost infinite life
3. low power required to energize
4. no contact arcing
5. little or no switching noise

6. positive switching and no contact bounce
7. can be hermetically sealed—good for hostile environments

Disadvantages to Solid-State Switching

1. solid-state switch may be destroyed by an overload
2. solid-state devices tend to fail in the ON state
3. heat dissipation
4. expensive to purchase
5. possibility of false trips from electrical noise

Advantages to Relay Output Switching

1. contacts forgiving to a temporary overload
2. immune to false trips from electrical noise
3. little voltage drop across contacts
4. no restrictions when connecting in series or parallel configurations
5. definite ON or OFF state, with contacts physically open
6. no leakage
7. contacts generate little heat
8. inexpensive to purchase

Disadvantages to Relay Output Switching

1. mechanical switching is slow
2. mechanical life is limited by demands of the load and the contacts
3. require 50 milliamps or more to energize
4. subject to contact arcing or welding
5. subject to contact bounce
6. cannot be completely sealed

Solid-state output modules, like relay output modules, are available in isolated outputs, too.

ISOLATED OUTPUT MODULES

Isolated output modules operate the same as other discrete output modules with the exception that the isolated module has its own separate common. Isolated output modules are available as either AC output modules or relay output modules. Isolated output modules are also similar to isolated input modules in that they allow for interfacing output devices powered by different sources, different phases, or different grounds. Such output modules allow a single module to interface many different output voltage levels. Isolated output modules have fewer output points, as there are only two points per output address, one for the signal in and one for the isolated, or separate, return. Figure 8-18 illustrates

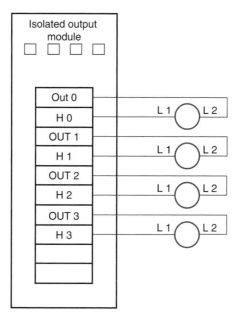

Figure 8-18   AC isolated output module.

an AC isolated output module. Notice that three different phases are being used to control three different output field devices.

Now that we have investigated the basic types of output modules and the advantages and disadvantages of both, let us see what you can do in cases where an output module is not capable of switching the load.

INTERPOSING RELAYS

After looking at output module specifications, we see that most output modules switch between one-half and four amps. For the PLC to switch higher current loads, a mechanical relay needs to be placed between the high current load and the output module. The output module will switch the relay, and the relay will switch the load. This relay is called an "interposing relay." Figure 8-19 illustrates an output module switching an interposing relay coil, CR 1. Relay contacts CR 1-1, in turn, switch the motor starter coil.

SURGE SUPPRESSION AND OUTPUT MODULES

Electrical noise and microprocessor equipment like PLCs do not work well together. Electrical noise is generated by many common industrial devices found in today's industrial environment. Electrical noise can cause microprocessor equipment to malfunction or lock up. It can be picked up on low-voltage communication cabling transmitting data

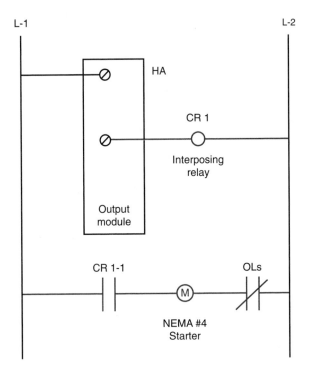

Figure 8-19 Interposing relay switching a load that exceeds output module's switching capability.

between PLCs or between a PLC and other hardware field devices. When noise is picked up on a communication cable, data can become corrupted. Corrupted data can bring a PLC network to a screeching halt.

Common devices such as motor starters, relays, contactors, solenoids, clutches and brakes, variable-frequency drives, and welding equipment are the primary culprits. Noise disturbances on the power line are generally caused by devices that have coils. These devices are called inductive devices. Noise is created when inductive devices are switched off.

Transmitted noise is created by devices that create radio frequency (RF) noise. Transmitted noise is generally caused from high-current applications such as welders and variable-frequency drives. Variable-frequency drives with newer Insulated Gate Bipolar Transistors (IGBT) switching transistors are known for generating large amounts of noise. In many cases, noise problems can be overcome by proper grounding, shielded cabling, inductors or chokes on communication lines, chokes on variable-frequency drive motor leads, and suppression devices on inductive devices such as motor starters or contractors.

Although most PLC output modules have built-in surge suppression to reduce the possibility of damage from high-voltage transients, it is a good idea to install additional surge suppression to any output module circuit switching an inductive load. Surge suppression should always be installed on a PLC output circuit where there is a set of hard contacts controlling an inductive field device. Install the surge suppressor, also called a snubber, across the coil of the field device. The surge suppressor will not only reduce the effects of voltage transients caused by interrupting the current to the inductive field device, it will also prevent electrical noise from entering the system wiring. Surge suppression will also help prolong the life of your output module's switching device.

Depending on the type of output module and its switching device, the suppression device may vary. As an example, the typical choice for a surge suppressor for a triac output module would be a metal oxide varistor (MOV). In some cases, a resistor-capacitor (RC) network is used for surge suppression. Surge suppressors must be carefully selected to suppress the switching transient of the particular inductive device. As an example, Allen-Bradley SLC 500 literature recommends a Harris MOV, part number V220MA2A, for 120 VAC inductive loads. MOVs are usually purchased from a local electronics supply house. Always consult your specific manufacturer's output module specification data sheet for selection of the correct suppressor. Note that most industrial control suppliers offer surge suppressors that are intended for installation on motor starters in conventional control circuits. In many cases these surge suppressors contain RC networks. RC network surge suppressors are typically not recommended for output modules with triac switching. DC output circuits, on the other hand, can usually use a diode for inductive load device suppression. In some cases, a diode such as a 1N4004, is just as acceptable as an MOV. Again, consult your module manufacturer's data sheets and always use the recommended surge suppressors.

HIGH-DENSITY I/O INTERFACE

High-density 32-point DC I/O modules are available for reducing panel space requirements by interfacing 32 input or output points to a single-slot module. Wiring 32 I/O points to the terminal block on the front of the module can become quite a challenge due to the high density of screw terminals and limited space. Figure 8-20 illustrates one example of wiring of high-density modules using a prewired cable that plugs into the front of the I/O module and connects to a remote DIN rail-mounted terminal block.

ANALOG OUTPUTS

Analog output modules accept a 16-bit output status word, which they convert to an analog value through a digital-to-analog converter. The converter (called a D-to-A converter) is part of the electronics inside the analog output module. Typical analog signals are 0 to 10 V dc, -10 to $+10$ V dc, 0 to 5 V dc, 1 to 5 V dc, 0 to 20 milliamps, -20 to $+20$ milliamps, or 4 to 20 milliamps. Analog output modules are selected to send out either a varying current or voltage signal.

An analog output could send a 4 to 20 milliamp signal to a variable-speed drive. The drive will control the speed of a motor in proportion to the analog signal received from the analog output module.

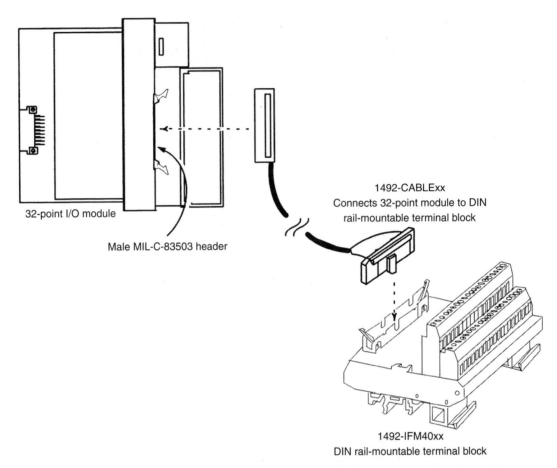

32-point I/O module

Male MIL-C-83503 header

1492-CABLExx
Connects 32-point module to DIN
rail-mountable terminal block

1492-IFM40xx
DIN rail-mountable terminal block

Figure 8-20   High-density I/O module wiring simplified by cabling I/O points to a nearby DIN rail-mounted terminal block. (Used with permission of Rockwell Automation, Inc.)

Another analog application could be controlling how far an analog water valve opens. A discrete valve is either fully open or fully closed. In many applications, more precise control is required than either a fully open or fully closed valve can provide. An analog valve can provide this precise control. An analog output module could output a 0 to 10 volt signal to an analog valve to provide the needed control. As an example, the output signal can be divided into 32,767 increments and represented in a 16-bit word. To achieve precision in controlling the valve, the 0 to 10 volt signal will be split into 32,767 steps, or possible output values, which correspond to 32,767 possible voltages between 0 and 10 volts. This will result in 32,767 possible valve positions. The voltage value output to the valve will control how far the valve will open or close. Figure 8-21 illustrates the valve position correlation to the module's output voltage. Column 3 represents the decimal value (between 0 and 32,767) of how far the program will direct the valve to open. Since the output

ANALOG SIGNAL COMPARISON FOR SAMPLE ANALOG VALUE OUTPUT		
Valve Position	Voltage Output Signal	Decimal Valve Output to Output Section
Full Open	10	32,767
80%	8	26,214
70%	7	22,937
60%	6	19,660
50%	5	16,384
40%	4	13,107
30%	3	9,830
20%	2	6,553
10%	1	3,276
Closed	0	0

Figure 8-21   Analog voltage correlation to one analog output module's data range.

module automatically converts the 16-bit output word to the proper analog voltage, the programmer only has to output the desired decimal integer value to the output status file. The value of the step will be directly correlated to the current signal output from the analog output module to the valve motor.

EMERGENCY-STOP SWITCHES AND PLC APPLICATIONS

Even though the typical emergency-stop (E-stop) palm button or switch may be thought of as a PLC input device, this is not the correct method of applying emergency-stop switches. There are important considerations regarding connecting E-stop switches in your control circuitry and the control of PLC-controlled field output devices in an emergency.

Emergency-stop switches should never be programmed in your PLC user program ladder diagram. Any emergency-stop switch needs to turn off all machine power. This is typically accomplished by a hardwired master control relay. Do not confuse this hardwired master control relay for E-stop control and the master control relay instruction programmed in a ladder program. The ladder program master control relay instruction should never be used in place of a hardwired master control relay. A hardwired master control relay provides a convenient means to shut down a system in an emergency. Typically, emergency-stop push and palm buttons, over travel limit switches, and system-stop switches are all normally closed devices wired in series with a hardware master control relay. The master control relay is wired to cut power to all input and output circuits. See Figure 8-22 for a typical master control wiring configuration. Remember always to

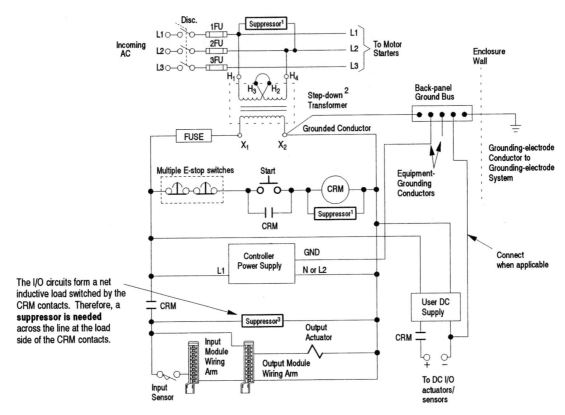

Figure 8-22 Typical master control relay wiring for emergency stop control. (Used with permission of Rockwell Automation, Inc.)

observe all local codes concerning the placement and labeling of E-stop switches and master relay control application for E-stop circuits.

When a master control relay is used to remove power from external PLC I/O circuits, power continues to be provided for the PLC controller itself (see Figure 8-22). Continuing power to the PLC's power supply enables the troubleshooter to use the PLC's diagnostic indicators. The troubleshooter can attach to the processor and monitor the PLC program and its status file's diagnostic information.

Important Rules Regarding Emergency-Stop Switches

1. Never program E-stop switches in the PLC program.
2. Always use a hardwired, hardware master control relay to cut power to all input and output circuits.

3. Do not control the master control relay instruction with the PLC.
4. The E-stop switch must turn off all machine power by turning off the master control relay.
5. Make sure E-stop switch contacts have sufficient current break ratings to handle breaking the circuit.
6. E-stop switches must be easy to reach.
7. *Always* check and follow all applicable codes concerning placement and labeling of E-stop switches and their associated master control relay.

I/O MODULES IN HAZARDOUS LOCATIONS

Many PLC manufacturers provide I/O modules that are rated for installation in hazardous locations. First, we need to define a hazardous area. A hazardous area or environment is one that contains amounts of explosive gases or dust. These hazardous materials may be present under normal operating conditions, or only in the event of equipment breakdown or an accident. The types of materials and the circumstances of their presence determine how the area is classified. Hazardous materials are divided into three classes, as defined by Underwriters Laboratories (UL) and the Canadian Standards Association (CSA). Class I includes flammable gases or liquid vapors that may ignite or explode. Class II is for combustible dust. Class III covers ignitable fibers or materials that produce combustible flyings (airborne debris from manufacturing processes such as cutting or grinding).

Classes are further broken down into divisions. The division identifies the conditions in which the hazard is present. Division 1 is an area where the hazard specified in the class can exist under normal operating conditions. Division 2 specifies that the hazardous materials as specified in the class are handled, processed, or used, but under normal operating conditions, these materials are either in closed containers or closed systems. Under Division 2, the hazardous materials are only present if they escape during equipment failure or accident.

Last, hazardous materials are classified by groups. Materials are grouped by their common range of ignition temperatures or explosion pressures.

A number of PLC manufacturers offer PLC hardware that is certified as Class I; Groups A, B, C, or D; Division 2. This certification means that the PLC can operate in an environment where volatile flammable liquids or gases are handled, processed, or used. Under normal conditions the hazard is either contained or removed by mechanical ventilation. Also included in Class I, Division 2, are manufacturing areas that are next to Class I, Division 1, areas and where hazardous gases or vapors might occasionally be present. (Refer to the National Electrical Code for the full text on hazardous classified locations.)

Potential Class I hazardous locations include petrochemical plants, petroleum refining or distribution facilities, dry-cleaning plants, dip tanks, solvent extraction plants, spray paint or finishing areas, aircraft hangars, and fueling areas.

HARDWARE CONSIDERATIONS BEFORE INSTALLING PLCs IN CLASS I, DIVISION 2, AREAS

1. Ensure that the PLC and its associated hardware are all certified as Class I, Division 2.
2. Verify that all electrical equipment installed in a Class I, Division 2, area falls within operating temperature codes.
3. All devices interfacing with the PLC must be Class I, Division 2, certified. This includes push buttons and pilot lights.
4. Follow National Electrical Code wiring methods.
5. Follow the manufacturer's installation instructions on all hardware devices.
6. Never install, remove, or operate any device that could cause an electrical arc while the circuit is alive.
7. Always check and follow local codes for the installation and operation of any equipment in a hazardous location.

Always carefully check your hardware against any hazardous rating standards before installation.

SUMMARY

Early programmable controllers were strictly limited to discrete inputs and outputs. Today's programmable controllers are much more sophisticated, thanks to vast advances in microprocessor technology. The increase in sophistication of these programmable controller devices mandates increased abilities and understanding from those individuals who will be applying this new technology to current and new applications. Today's programmable controllers can be applied to practically any control problem due to their complete range of discrete and analog interface control advancements and abilities.

There are output modules available along with combination input and output modules. I/O modules typically are available as 4-, 8-, 16-, 24-, or 32-point modules. Discrete output modules interface to AC, DC, and +5 DC TTL voltage signals to control field hardware devices. Output modules are available in two configurations: AC or DC solid-state output switching or mechanical-relay output switching.

In this chapter we explored discrete and analog output modules and how these modules interface signals to real-world devices from a CPU and the output status file. We also explored the basic techniques of interfacing analog output modules to the programmable controller.

REVIEW QUESTIONS

1. Our system of I/O provides an interface between
   A. input modules and the CPU
   B. field equipment and the CPU

    C. output modules and field devices
    D. the CPU and output modules
    E. input modules and output modules

2. Discrete output interface modules accept signals from field devices that are:
    A. analog devices
    B. digital devices
    C. two-state devices
    D. discrete devices
    E. A and B
    F. B and C
    G. B, C, and D

3. Discrete output modules act as a _____ to control output field devices.

4. Output modules fall into two classifications, _____ output switching or _____ output switching.

5. Most PLCs have solid-state output modules that can switch outputs with:
    A. +5 V dc
    B. 10–30 V dc
    C. 120 VAC
    D. 120–240 VAC
    E. 200–240 VAC
    F. All of the above are correct.

6. Relay outputs accept a wide range of voltages, usually _____ VAC and _____ V dc.

7. Typical discrete output devices include:
    A. motor starters
    B. alarm bells
    C. inductive proximity switches
    D. pilot lights
    E. solenoids
    F. A, B, D, and E
    G. all of the above

8. Discrete output modules receive their operating power from the PLC's power supply by way of the _____.

9. Power that the module output switches to control field devices must be provided by:
    A. the CPU
    B. the chassis backplane
    C. the user
    D. the master control relay
    E. A and B
    F. all of the above

10. Name five analog outputs.

11. What are the five main sections of an AC output module?

12. What is the purpose of the hardwired master control relay in your control circuit?
    A. The master control relay is the master controller of the PLC.
    B. The master control relay is the master control of input and output circuit power only.
    C. The master control relay is the master control of input and output circuit power and power to the CPU's power supply.
    D. The master control relay controls power to all local and remote PLC chassis.
    E. A and D are correct.
    F. B and D are correct.

13. Examples of discrete outputs include:
    A. motor starter coils
    B. pilot lights
    C. solenoids
    D. start or stop signals to a variable-speed drive
    E. A, B, and C
    F. all of the above

14. All discrete outputs have one thing in common; their signals are either:
    A. ON or OFF
    B. OPEN or CLOSED
    C. YES or NO
    D. TRUE or FALSE.
    E. A, B, and C
    F. all of the above

15. Analog output control includes:
    A. pressure regulation
    B. temperature regulation
    C. analog valve positioning
    D. variable-frequency drive speed reference
    E. A, B, and C
    F. all of the above

16. The output section of a PLC system is the physical connection between the outside world and the _____.

17. Output modules generally fall into three classifications: _____, _____, or _____.

18. Electrical isolation is provided so that there is no physical electrical connection between the CPU and field devices. How is isolation usually accomplished?

19. Analog output signals can be interfaced to the PLC using modules specifically designed to accept these variable signals. These modules are called _____ output modules.

20. A _____ PLC's ability to interchange modules to meet the immediate interface need makes it popular and versatile.

21. Output modules are available in various switching _____ and configurations.

22. Each output module accepts its associated 16-bit output status word from the _____ during the output update portion of the CPU operating cycle.

23. The output module's internal electronics automatically converts the output status word into a separate ON or OFF signal for each output _____ terminal to control its associated field device.

24. Discrete output interface modules provide a method of getting two-state, _____, or _____ signals out of the PLC to control field devices.

25. Changing signals like temperature control or the 4 to 20 milliamp speed reference for a variable-frequency drive are called _____ signals.

IF PASSWORD IS CORRECT GO TO DRIVE SET UP SCREEN
B3:10/0

0022

B3:1/15
—( )—

RESET PASSWORD
ATTEMPTS COUNTER
C5:5
—(RES)—

0023

B3:10/1
[OSR]

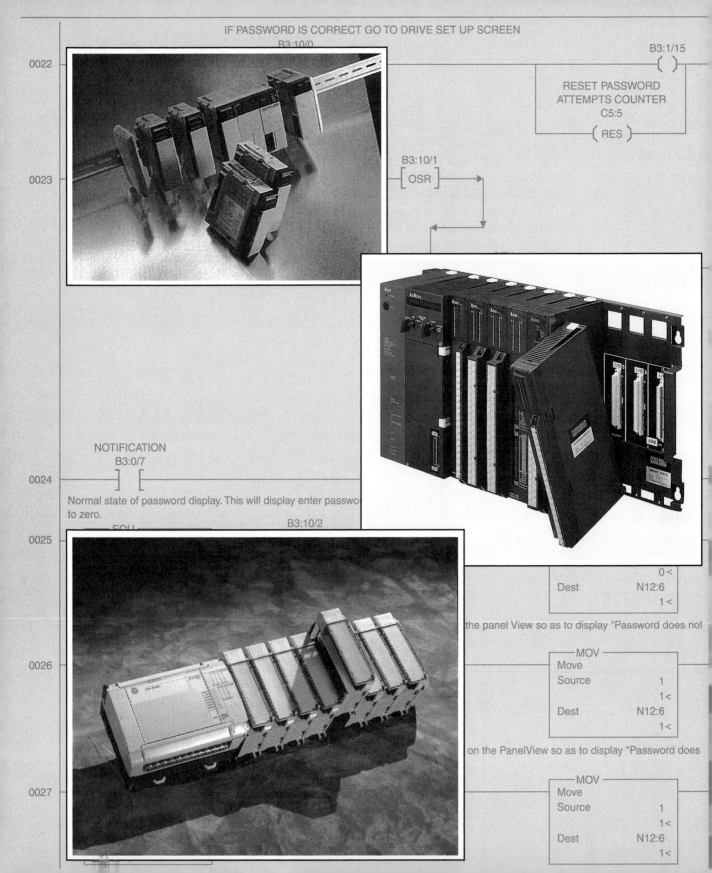

0024 NOTIFICATION
B3:0/7
—] [—

Normal state of password display. This will display enter passwo
to zero.

0025 B3:10/2

Dest   N12:6
              1<

the panel View so as to display "Password does not

0026
—— MOV ——
Move
Source   1
                1<
Dest   N12:6
                1<

on the PanelView so as to display "Password does

0027
—— MOV ——
Move
Source   1
                1<
Dest   N12:6
                1<

0 <

CHAPTER

9

Putting Together a Modular PLC

OBJECTIVES

After completing this chapter, you should be able to:

- define rack, chassis, and baseplate and tell how or why they differ
- select the proper type of I/O to interface a specific input signal
- explain why power supply loading must be determined as a PLC system is configured
- explain why power supply loading should be recalculated in some troubleshooting situations
- calculate power supply loading on selected PLCs
- determine the proper power supply for selected PLCs

INTRODUCTION

PLCs come in two styles, fixed and modular. The I/O of a fixed PLC is built-in and not changeable. A modular PLC consists of user-selected I/O modules, a CPU, a power supply, and a chassis. Figure 9-1 illustrates a Modicon TSX Quantum modular PLC in a ten-slot backplane.

The chassis or backplate is used to hold all the parts together. The Modicon Quantum PLC in Figure 9-1 consists of the ten-slot backplane, a power supply, a controller (CPU) and I/O modules. Since modular PLCs are application-specific, a modular PLC cannot be purchased preconfigured from a catalog. The user must determine and select all the pieces to build the PLC to fit the specific application. This chapter will address considerations when selecting and configuring a modular PLC.

Figure 9-1 AEG Schneider Automation's Quantum PLC. (Courtesy of AEG Schneider Automation/Square D-Modicon)

RACKS, CHASSIS, OR BASEPLATES

Depending on the modular PLC manufacturer, the term used to identify the hardware device that holds all the modules together may vary. One manufacturer may use the term rack, base unit, or backplane, another may use chassis, while yet another uses the word baseplate. The rack, chassis, or baseplate holds the power supply, CPU, and I/O modules together in an easy-to-mount package. Figure 9-2 illustrates an Allen-Bradley MicroLogix

Figure 9-2 Allen-Bradley MicroLogix 1500 programmable controller. (Used with permission of Rockwell Automation, Inc.)

1500 modular PLC with an I/O module being inserted. The MicroLogix 1500 is a rackless-design modular PLC. Being rackless, the MicroLogix 1500 I/O modules slide into the tongue-and-groove slots in adjacent modules. The rackless modular PLC can be either mounted to the panel or to the DIN rail. Many newer PLCs are transitioning to a rackless design.

A single power supply is usually mounted on one end of each rack, backplane, baseplate, or chassis. The power supply provides power to operate the CPU, along with the electronics inside each I/O module. Most often, the power supply does not provide power to operate input or output field devices. The PLC 5 allows a power supply to be inserted into the chassis or an external power supply fastened to the left side of the chassis. Power supplies mounted inside a PLC 5 chassis can consume either one or two slots, depending on chassis power requirements. The SLC 500 allows only for an external left-mounted power supply. In both PLCs the processor is installed in the left-most slot of the chassis. The remaining chassis slots are for I/O or communication modules. Chassis are classified by the number of slots. There are four chassis available for the modular SLC 500. SLC 500 modular chassis are available with four, seven, ten, or thirteen slots. The ControlLogix chassis sizes are four-, seven-, ten-, thirteen-, or seventeen-slot. Slots are numbered starting at slot zero, the left-most slot, and increment moving to the right. A ControlLogix processor can be placed in any chassis slot. ControlLogix also provides the user the flexibility to insert multiple processors in a chassis if additional processor power is required in the application. Figure 9-3 shows a ControlLogix PLC as a ten-slot chassis. The power supply is on the left followed by ten slots for modules. There are two processors, one in slot three and one in slot six.

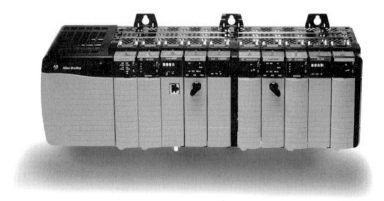

Figure 9-3 ControlLogix PLC in a ten-slot chassis with multiple processors. (Used with permission of Rockwell Automation, Inc.)

Figure 9-4 illustrates the Mitsubishi A3NCPU modular PLC with an I/O module being installed in the base unit. The module is hooked on to the bottom of the base unit and then pushed back into place. A clip at the top of the module holds the module firmly in place on the base unit.

Figure 9-4 Securing an I/O module to a Mitsubishi A3NCPU base unit. (Courtesy of Mitsubishi Electric Automation Inc.)

Figure 9-5 illustrates a completely configured General Electric Series 90-30 Modular PLC Model 331 CPU with a ten-slot baseplate, with power supply and I/O modules installed. The Series 90-30 is available with either five- or ten-slot CPU baseplates. Figure 9-6 shows a five- and a ten-slot baseplate.

Figure 9-5 General Electric Series 90-30 Modular PLC. (Courtesy of GE Fanuc Automation)

To install or remove a module is as simple as pressing the release lever at the bottom of the module and swinging the module upward using the pivot hook as a hinge. Disengage the pivot hook at the top rear of the module by moving the module up and away from the baseplate as illustrated in Figure 9-7. To install an I/O module, simply reverse the removal procedure.

Before removal or after installation of a module, the module's terminal board needs to be either installed or removed. Refer to Figure 9-8. Installation of a terminal board consists of the following steps:

1. Hook the pivot hook, number 1 in Figure 9-8, located on the bottom of the terminal board, to the lower slot of the module.

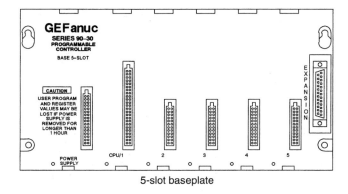

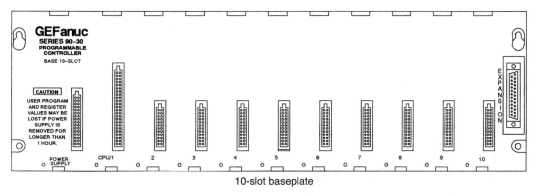

Figure 9-6 Series 90-30 PLC Model 331 CPU five- and ten-slot baseplates. (Courtesy of GE Fanuc Automation)

2. Push the terminal board toward the module, illustrated as number 2, until it snaps into place.
3. Open the terminal board cover, number 3, and ensure that the latch on the module is holding the terminal board in place.

The Omron CQM1 does not have a physical rack or chassis. After selecting the needed CPU and I/O modules, the PLC is assembled in a connect-and-lock fashion. The CQM1 will support the power supply, CPU, and up to eleven I/O modules connected together. The assembled PLC is either DIN rail or panel-mounted. Figure 9-9 illustrates the Omron CQM1 PLC.

Figure 9-10 illustrates the procedure for putting together an Omron CQM1 PLC.

Although these PLCs may look different, they all perform the same function. They hold the power supply, CPU, and selected I/O modules together to form a modular PLC.

Rather than list all variations of terms when referring to the rack, chassis, backplane, or baseplate, we will refer to these hardware devices, generally, as a chassis. When discussing a specific manufacturer's product, the appropriate terms associated with that product will be used.

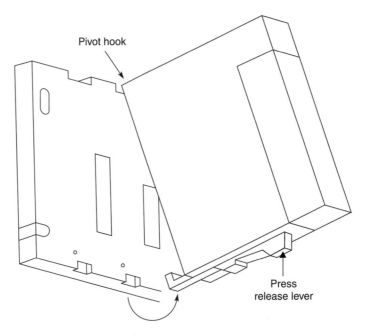

Figure 9-7 Removing a General Electric Series 90-30 I/O module. (Courtesy of GE Fanuc Automation)

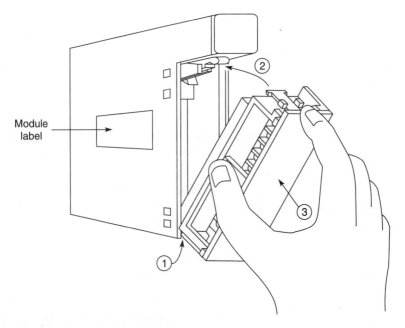

Figure 9-8 Installing a Series 90-30 I/O module's terminal board. (Courtesy of GE Fanuc Automation)

Figure 9-9   Omron CQM1 Modular PLC. (Courtesy of Omron Electronics, Inc.)

This diagram shows the connection of two modules that make up a
CQM1 PLC. Join the modules so that the connectors fit exactly.

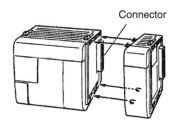

The yellow locking tabs at the top and bottom of each module
lock the modules together. Slide these locking tabs toward the back
of the modules as shown here.

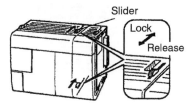

Attach the end cover to the module on the far right side of the PLC.

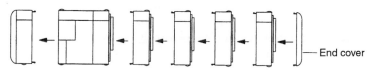

Figure 9-10   Omron CQM1 PLC "connect and lock" packaging design. (Courtesy of Omron
Electronics, Inc.)

THE LOCAL, OR BASE, PLC

A modular PLC in a single chassis can be a stand-alone system where the PLC is dedicated to controlling one machine or process. This type of configuration is usually called a base rack, resident, or local chassis. In applications such as this, each machine or process may have its own dedicated, fixed, or modular PLC. The General Electric Series 90-30 PLC in Figure 9-3 contains all the pieces necessary to control a machine or process. This local PLC will only control the machine to which it is dedicated.

However, a PLC chassis can have multiple uses beyond being a stand-alone local PLC. Let us explore these other uses.

LOCAL EXPANSION

A single modular PLC chassis can be used as the base for expanding other chassis to make a larger system. This is accomplished by installing I/O and a power supply into a second chassis. The expansion chassis is connected to the base PLC with a communication cable. The communication cable carries I/O status signals between the CPU, in the base chassis, and the expanded chassis and its I/O. Each expansion chassis has a power supply as necessary for the particular application. Look ahead at Figure 9-30, which illustrates a General Electric Model 331 CPU and local expansion. Figure 9-11 illustrates the Allen-Bradley SLC 500 PLC expanded into three chassis. Note concerning the expansion cables that the 1746-C9 communication cable is 36 inches long and the 1746-C7 communication cable is 6 inches long. These are the only two cables available for expansion.

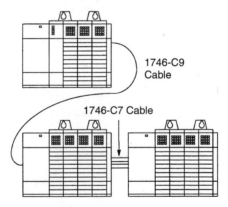

Figure 9-11   Allen-Bradley SLC 500 three-chassis local expansion. (Used with permission of Rockwell Automation, Inc.)

As an example, let us configure an Allen-Bradley SLC 500 where 18 I/O modules are required. Since there is no chassis available that will hold the CPU along with 18 I/O modules, expansion into a second chassis is necessary. Two SLC 500 ten-slot chassis can hold up to 19 I/O modules and the processor. The electrical signals to communicate I/O status between the local chassis and the locally expanded chassis will be communicated through a cable connected between the two.

REMOTE I/O EXPANSION

A PLC chassis can also be the base for expanding and distributing I/O around the factory floor to the manufacturing locations where this I/O is needed. This is called remote I/O. Rather than scattering separate PLCs around the factory or running numerous wires back from remote field devices, remote I/O expansion makes it possible for one base, or local, PLC to house the CPU and control several remotely distributed racks of I/O through a communication link. The advantage of remote I/O is that there is one central CPU supervising all I/O points. The differences between local and remote I/O expansion are:

- the communication method, parallel for local, serial for remote I/O
- number of chassis that can be used
- the distance the remote chassis may be away from the base chassis

Figure 9-12 shows an Allen-Bradley PLC 5 with four chassis, three of which are connected as remote I/O from the resident chassis on the top left.

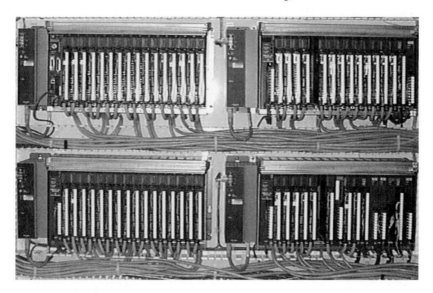

Figure 9-12   Four Allen-Bradley PLC 5 chassis connected via remote I/O. (Used with permission of Rockwell Automation, Inc.)

Even though the picture shows the three remote chassis close to the processor, remote I/O chassis in a PLC 5 system could be mounted up to 10,000 feet away from the local chassis.

The processor resides in the resident chassis and communicates to the three remote chassis through a two-wire communication cable. Connected to the immediate left of each chassis is its power supply. The left-most module in each of the remote chassis is the 1771-ASB communication module that sets up communication between the remote chassis and the resident chassis.

The SLC 500 remote I/O also uses special communication modules, one in the base, or local, chassis and one in each remote chassis. The 1747-SN module resides in any slot in the local SLC chassis except the processor slot. The 1747-SN is an asynchronous scanner that sends and receives information to the remote chassis. After the SN scanner receives information for the remote chassis, information is then shared with the processor. Information from the processor is passed on to remote I/O chassis by the SN scanner. Each remote chassis has a communication module in the left-most slot of that chassis. This is a 1747-ASB module. The ASB module is the other end of the communication link between the processor and each remote I/O chassis. The ASB is placed in the left-most slot of each remote I/O chassis. A different communication cable is used for remote I/O versus local I/O. The PLC 5 and the SLC 500, both Allen-Bradley products, use the same "blue hose" cable for remote I/O communication cabling. Maximum cable length for SLC 500 remote I/O is 10,000 cable feet. The maximum number of I/O points an SLC 500 can support including remote I/O is 4096 inputs and 4096 outputs. Figure 9-13 provides an overview of an SLC 500 remote I/O configuration. The left-most SLC 500 is the local chassis, which is the only chassis that contains a processor. Notice each of three remote chassis and their associated 1747-ASB modules. Also notice there is a PanelView operator interface terminal on this remote I/O link. Many different devices, including operator interface terminals and

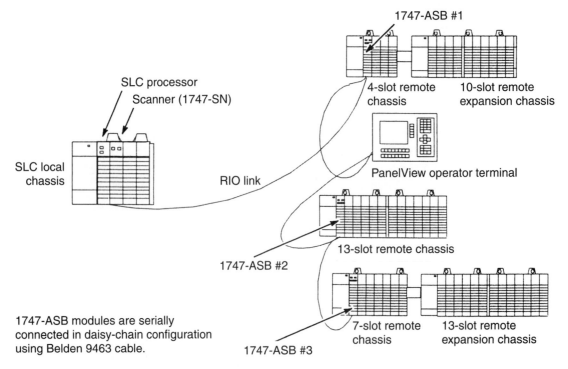

Figure 9-13   SLC 500 local chassis connected to remotely mounted chassis and a PanelView operator terminal by way of remote I/O. (Used with permission of Rockwell Automation, Inc.)

variable-frequency drives, may be placed on a remote I/O network. Information can then be transferred between each of these devices and the local processor.

PLC NETWORKING

A single-chassis PLC can be a part of a larger, factory-wide communication **network.** Each PLC on the network will have its own CPU. The network is the medium where data can be shared between several PLCs. As illustrated in Figure 9-14, each PLC would be a "node" on the network.

Now that we have looked at the uses for a PLC chassis, let us look into selecting the proper module for specific types of inputs and outputs.

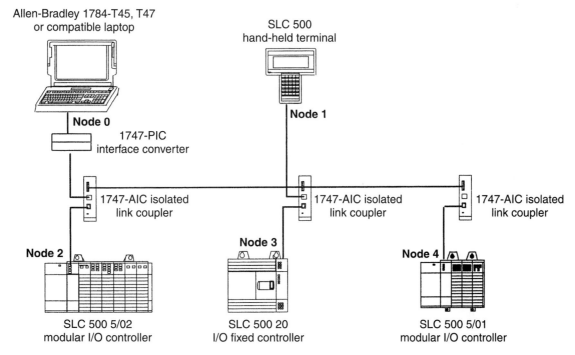

Figure 9-14   An Allen-Bradley SLC 500 Data Highway-485 network. (Used with permission of Rockwell Automation, Inc.)

SELECTION AND PLACEMENT OF I/O MODULES

Configuring a modular programmable logic controller consists of more than simply collecting the desired input and output modules and a processor, plugging them into a chassis with a power supply, connecting the power, and developing a program.

First, the input and output signals must be considered. The appropriate input or output module is selected by determining incoming or outgoing signals and matching these signals to an I/O module. Use Figure 9-15 to help you decide what module family to choose.

DETERMINING MODULE FAMILY	
If you have this type of signal to interface to your PLC	Select this family of modules
Either AC or DC, two-state, ON or OFF, OPEN or CLOSED incoming signals	Discrete input module
Output signals controlling either AC or DC, two-state devices that are either ON or OFF, OPEN or CLOSED	Discrete output module
Continuously changing voltage or current incoming signals like 4 to 20 mA, +10 to −10 V dc, 0 to 10 V dc.	Analog input module
Continuously changing voltage or current incoming signals like 4 to 20 mA, +10 to −10 V dc, 0 to 10 V dc. Signals could be a speed reference going into a drive or the signal to control an (analog) variable opening value.	Analog output module
SPECIALTY MODULES	
High-speed discrete inputs from an incremental encoder	High-speed counter specialty module
Network communication, weigh scales, modem, bar code, RS-232 communication, motion control or stepper motor control, local or remote I/O communication, certain display devices, encoders, resolvers, certain electronic operator interface products	There are typically specialty modules. Many are controlled by their own internal microprocessors. These advanced devices will not be covered in this text.

Figure 9-15   Table for determining input and output types and their associated module family.

Figure 9-16 will help you select the proper I/O module type if you know the type of field hardware device you need to interface to your PLC.

Care must be taken when placing modules in the chassis in relation to other modules and the CPU. Although this may not be obvious, high-power and noise-generating I/O modules need to be separated from low-voltage and noise-susceptible modules and from specialty and communication modules. Many of these modules are low voltage and very sensitive to electromagnetic noise.

Module placement rules are straightforward. Starting with the CPU, place the modules as follows:

Low- to high-voltage DC input modules, in ascending order away from the CPU.

Low- to high-voltage DC output modules, in ascending order away from the CPU.

Low- to high-voltage AC input modules, in ascending order away from the CPU.

Low- to high-voltage AC output modules, in ascending order away from the CPU.

Now that you can identify I/O signals and select the appropriate module for the application, let us look into selecting a power supply.

I/O MODULE SELECTION	
If you have these hardware field devices to interface to the PLC	**Select this type of module**
Push buttons; selector switches; motor starter auxiliary contacts; level switches; two-state pressure or temperature switches; inductive, capacitive, or ultrasonic, two-state proximity sensors; two-state photoelectric sensors; relay contacts; limit switches	Discrete input module with the appropriate input voltage
Alarms, pilot lights, motor starters, solenoids, fans, relays, two-state valves	Discrete output module with the appropriate switching voltage
5 V dc TTL signals from other solid-state or microprocessor-controlled devices	Discrete 5 V dc TTL input module
5 V dc TTL signals to be output to other solid-state or microprocessor-controlled devices	Discrete 5 V dc TTL output module
Temperature, pressure, humidity, load cell, or flow transducers; analog photo switches; analog proximity switches; potentiometers	Analog input module
Analog valves or meters, speed control into a drive	Analog output module

Figure 9-16   I/O module selection from input or output type.

POWER SUPPLY SELECTION

Each power supply is designed to handle a specific load. When configuring a modular PLC, there must be a separate power supply in each chassis. The power needed to power the internal electronics of each module, handheld programming device, personal computer interface hardware, and other peripherals is drawn through the rack backplane from the chassis power supply. When multiple power supplies are available for a particular PLC, a power supply must be chosen for each rack based on calculations of the power needed to operate the specific hardware associated with that chassis. Often, modular PLC hardware is selected, purchased, and assembled into a working PLC with no concern to power supply loading on the grounds that it is a bother to calculate power consumption for all those modules.

Why Verify Power Supply Loading?

When configuring a system or working as a troubleshooter attempting to find phantom or inconsistent problems, power supply loading should be verified for each modular chassis. Calculate the loads that each module, the CPU, the programmer, and any peripheral devices put on the power supply. For most modular programmable logic controllers, the load, or current draw, is represented as 5 volts and 24 volts. Five volts DC and 24 volts DC are used to power the CPU and modules in the chassis in addition to any 24 V dc user power taken from the power supply. The power consumed by each piece of hardware powered from a particular chassis is simply totaled for the 5 volt and 24 volt loads placed on the power supply. The totals are used to compare the total load placed on the power supply with the available power. If the available power exceeds the power consumed by all the chassis-associated hardware, the configuration is acceptable and should produce acceptable system performance.

However, if the calculated power consumed by the chassis hardware exceeds the total available power from the power supply, adjustments must be made before putting this system into service. Listed below are a few suggestions.

1. If your PLC manufacturer offers multiple sizes of power supplies, select a larger power supply.
2. Rearrange the module mix so as to satisfy power demand versus power availability in order to eliminate potential problems.
3. If you must use the modules that were selected, then consider local I/O expansion into a second chassis. By dividing the modules between two chassis and two power supplies, you can usually place one-half the load on each power supply.

Note that power calculations are based on the worst case—all I/O points on all input and output modules, specialty modules, and some peripheral devices are on, not just those actually being energized at any specific moment.

Procedure for Determining Proper Power Supply

To decide which power supply is correct for your specific application:

1. Determine what CPU is needed for this specific application.
2. Determine which modules are required for the input hardware to be interfaced to the PLC.
   a. how many 24 V dc sourcing inputs?
   b. how many 24 V dc sinking inputs?
   c. how many 120 VAC inputs?
   d. how many 240 VAC inputs?
   e. how many 5 V dc inputs?
3. Determine which modules are required for the output hardware to be interfaced to the PLC.
   a. how many 24 V dc sourcing outputs?
   b. how many 24 V dc sinking outputs?
   c. how many 120–240 VAC outputs?
   d. how many 5 V dc outputs?
   e. how many relay (hard-contact) outputs?
4. List the modules you will need.
5. Now that you know the total number of modules, determine the chassis needed for the application.
6. Determine the placement of modules in each rack, as described earlier in this chapter.

7. Perform power-loading calculations.
8. Verify power-loading calculations.
9. Choose correct power supply for each chassis.

Depending on the PLC programming software you are using, you may or may not have to manually calculate power-supply loading for your PLC. As an example, newer software such as Allen-Bradley's RSLogix 500 software will calculate the power-supply loading for you. You will find power-supply load calculations under the I/O configuration folder. Figure 9-17 illustrates the SLC 500 software, RSLogix 500 I/O configuration window, left, and power-supply loading view, right.

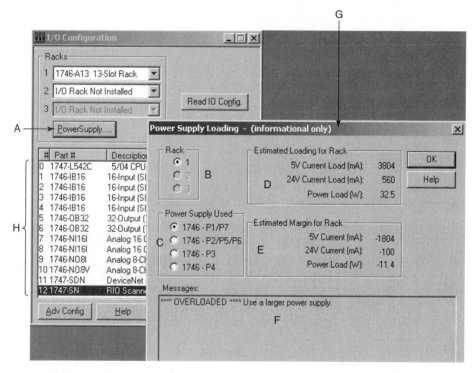

Figure 9-17   RSLogix 500 software I/O configuration and power supply loading view.

Notice the following main points from Figure 9-17:

   A. From the I/O configuration screen, click here to view the power-supply loading screen.
   B. Select the chassis or rack for which you want power supply loading information.
   C. Select the power supply currently installed on this chassis.
   D. Calculated estimated loading for the selected rack or the chassis.
   E. Estimated margin for rack.
   F. Messages as the result of the calculations.

G. This screen only does the calculations from the information you have provided.

H. This I/O configuration information is where the calculated information comes from.

Keep the following important points in mind. First, the software cannot determine the power supply that is actually installed on the chassis. The software will always default to the P1 or smallest power supply. Second, power-supply loading is calculated using the I/O configuration information from H in Figure 9-17. Finally, if you need to manually calculate power-supply loading, the following section steps you through manual calculations.

Let us look at three examples of calculating a power load and choosing a power supply for a specific group of I/O modules. We will look at two different manufacturers, Allen-Bradley's SLC 500 and General Electric's Series 90-30.

SLC 500 Power Supply Selection, Loading, and Installation

Like other PLCs, the Allen-Bradley SLC 500 modular chassis, modules, and power supply must be carefully selected to configure a system that will provide safe and satisfactory operation. Each chassis of a modular system requires a power supply. Excessive power-supply loading can cause inconsistent results or even power-supply shutdown. Figure 9-18 lists the seven power supplies available.

After selecting the correct power supply, install it on the modular chassis. Figure 9-19 illustrates SLC 500 power-supply installation on a four-slot modular chassis.

Description	Catalog Number						
	1746-P1	1746-P2	1746-P3	1746-P4*	1746-P5	1746-P6	1746-P7
Incoming Line Voltage	120 or 220 VAC	120 or 220 VAC	19.2 to 28.8 V dc	120 or 220 V dc	90 to 140 V dc	30 to 60 V dc	10 to 30 V dc
5 V dc Current Capacity	2 Amps	5 Amps	3.6 Amps	10 Amps*	5 Amps	5 Amps	12 V dc input: 2 Amps 24 V dc input: 3.6 Amps
24 V dc Current Capacity	.46 Amps	.96 Amps	.87 Amps	2.88 Amps*	.96 Amps	.96 Amps	12 V dc input: 3.6 Amps 24 V dc input: .87 Amps
User Current Capacity	200 mA at 18 to 30 V dc	200 mA at 18 to 30 V dc	NA	1 Amp at 20.4 to 27.6 V dc*	200 mA at 18 to 30 V dc	200 mA at 18 to 30 V dc	NA
*Note additional considerations for P4 power supply later in this chapter.							

Figure 9-18 SLC 500 power-supply specifications. (Table data compiled from Allen-Bradley power-supply data)

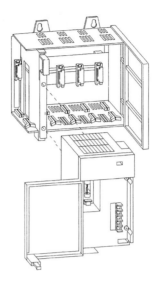

Figure 9-19 The SLC 500 power supply slides onto the left end of the chassis. When the power supply is in place, two screws are tightened to secure the unit. (Used with permission of Rockwell Automation, Inc.)

Figure 9-20 lists a sample of available SLC discrete I/O modules and their power draw specifications for the chassis power supply.

SLC 500 Processor Specifications.   Figure 9-21 lists the power requirements for the available Allen-Bradley SLC 500 modular processors. Use these power-supply loading values in your calculations.

Example of Power Supply Selection

For this example, let us determine the proper power supply for the following SLC 500 modular PLC hardware:

1. 1747-L511 CPU
2. seven-slot chassis
3. two #1746-IA Input Modules
4. two #1746-OW16 Relay Output Modules
5. one #1746-IG16 Input Module
6. one #1746-OV16 Output Module

Performing the Calculations.   Performing the calculations is as simple as:

1. List the selected hardware on a chart like the one illustrated in Figure 9-22.
2. Look up the 5 V dc and 24 V dc loading current values from Figure 9-20 for each piece of hardware selected. Write the appropriate loading value in the proper position on the chart.

SLC 500 DISCRETE INPUT MODULE POWER			SLC 500 DISCRETE OUTPUT MODULE POWER		
Module	+5 Volts	+24 Volts	Module	+5 Volts	+24 Volts
120 VAC Input Modules			**120–240 VAC Output Modules**		
1746-IA4	.035	0	1746-OA8	.185	
1746-IA8	.050	0	1746-OA16	.370	0
1746-IA16	.085	0	1746-OAP12	.370	0
24 V dc Input Modules			**24 V dc Output Modules**		
1746-IB8	.050	0	1746-OB8	.135	0
1746-IB16	.085	0	1746-OB16	.280	0
1746-IB32	.106	0	1746-OBP16	.250	0
1746-ITB16	.085	0	1746-OB32	.452	0
1746-IV8	.050	0	1746-OV8	.135	0
1746-IV16	.085	0	1746-OV16	.270	0
1746-ITV16	.085	0	1746-OV32	.452	0
1746-IV32	.106	0			
200–240 VAC Input Modules			**Relay AC/DC Output Modules**		
1746-IM4	.035	0	1746-OW4	.045	.045
1746-IM8	.050	0	1746-OW8	.085	.090
1746-IM16	.085	0	1746-OW16	.170	.180
			1746-OX8	.085	.090
5 V dc TTL Input Module			**5 V dc TTL Output Module**		
1746-IG16	.140	0	1746-OG16	.180	0

Figure 9-20   Selected Allen-Bradley SLC 500 I/O module loading specifications. (Chart data used with permission of Rockwell Automation, Inc.)

PROCESSOR SPECIFICATIONS		POWER SUPPLY LOADING IN AMPS	
Processor	User Instructions	+5 V dc	+24 V dc
5/01	1 K	.350	.105
5/01	4 K	.350	.105
5/02	4 K	.350	.105
5/03	12 K	.500	.105
5/04	12 K	1.00	.200
5/04	28 K	1.00	.200
5/04	60 K	1.00	.200

Figure 9-21   SLC 500 processor power supply loading specifications. (Chart data used with permission of Rockwell Automation, Inc.)

SLC 500 MODULAR POWER CALCULATIONS			
Slot	Hardware Description and Catalog Number	5 V dc Loading Current	24 V dc Loading Current
0	1747-L511 CPU	0.350	.105
1	1746-IA 16 Input Module 120 VAC	0.085	0
2	1746-IA 16 Input Module 120 VAC	0.085	0
3	1746-OW16 Relay Output	0.170	.180
4	1746-OW16 Relay Output	0.170	.180
5	1746-IG16 5 V dc Input Module	0.140	0
6	1746-OV16 Output Module DC Sink	0.270	0

Figure 9-22   Example of calculating power for Allen-Bradley SLC 500 PLC.

3. Add up values in the 5 V dc loading current column. Place the total at the bottom in the total area.
4. Add up the values in the 24 V dc loading current column. Place the total at the bottom in the total area.
5. Compare the totals with the power-supply specifications in Figure 9-18.
6. Select the appropriate power supply. See the example in Figure 9-22.

Analysis of Our Calculations.   In our example, the total current requirement is 1.27 amps at 5 V dc. This is within range of the 1746-P1 power supply (see Figure 9-18), which is rated at 2 amps. However, with 32 relay output points, we are over the maximum 24 V dc allowed on the 1746-P1 power supply; that is, .460. Considering future possible module mix considerations, it may be wise to select the 1746-P2 power supply in case you need to add an additional relay output module to this rack in the future.

Special Considerations When Using the 1746-P4 Power Supply.   When calculating power for the 1746-P4 power supply, one additional step needs to be completed. After determining the total current for the 5-volt and 24-volt columns, the total watts, or power, must be calculated. The total power-supply loading current of the chassis must not exceed 70 watts. As an example, let us assume we are using the P4 power supply in the previous example. Completed calculations are in Figure 9-23.

Now that we have looked at an example for the Allen-Bradley SLC 500 PLC, our second example will calculate power and configure a General Electric Series 90-30 programmable logic controller.

Series 90-30 Power Supplies and Specification

There must be a power supply installed in the left-most slot of all baseplates, each of which, whether a CPU, expansion, or remote baseplate, requires a power supply. General

CALCULATING TOTAL POWER FOR THE 1746-P4		
Total Current	**Multiplied by Volts**	**Watts**
1.27	5 Volts	6.35 Watts
.465	24 Volts	11.16 Watts
0	24 Volts User Current	0 Watts
TOTAL POWER (Must Not Exceed 70 Watts)		17.51 Watts

Figure 9-23 Additional power supply calculations for the 1746-P4.

Electric offers three power supplies for the Series 90-30 PLC. Power supply specifications are listed in Figure 9-24.

Catalog Number	Load Capacity	Input Voltages	Output Capacities		
IC693PWR321	30 Watts	100–240 VAC	+5 V dc	+24 V dc Isolated	+24 V dc Relay
		100–125 V dc	15 Watts	15 Watts	15 Watts
IC693PWR330	30 Watts	100–240 VAC	+5 V dc	+24 V dc Isolated	+24 V dc Relay
		100–125 V dc	30 Watts	20 Watts	15 Watts
IC693PWR322	30 Watts	24 or 48 V dc	+5 V dc	+24 V dc Isolated	+24 V dc Relay
			15 Watts	20 Watts	15 Watts

Figure 9-24 Series 90-30 power supply specifications. (Table data courtesy of GE Fanuc Automation)

The General Electric IC693PWR321 power supply is illustrated in Figure 9-25. Note the following features of the 90-30 power supply:

1. system status indicators: "PWR" indicates that the power supply is operating properly. If the green LED is OFF, either power is not being applied or the power supply is faulted. "OK" indicates that the PLC is operating properly. If the "OK" LED is off, a problem has been detected in the PLC. "RUN," when steady on, indicates that the PLC is in run mode. "BATT" is on when the memory backup battery voltage is too low to maintain memory through a power loss. Do not remove line power to the power supply without replacing battery.
2. RS-485 compatible serial port: the connection point for either the handheld programmer or the personal computer, interface converter #IC690ACC900.
3. lithium backup battery: for memory retention when line power to the power supply is disconnected or there is an unexpected power failure. The battery status is shown by the "BATT" system status indicator on the top of the power supply.
4. power connection terminals: connections for 100 to 240 VAC line power and 24 V dc internally supplied, isolated, 24 V dc user power.

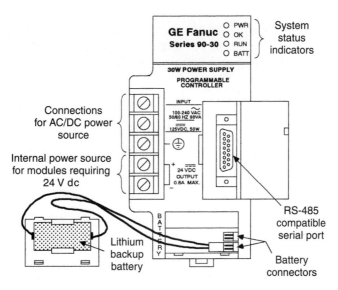

Figure 9-25   Series 90-30 power supply for 100 to 240 VAC or 100 to 125 V dc. (Courtesy of GE Fanuc Automation)

Selected Series 90-30 Hardware Load Requirements.   Figure 9-26 lists selected General Electric Series 90-30 PLC hardware and the associated power supply loading specifications.

Data in Figure 9-24, power supply specifications, and Figure 9-26, module load requirements, provide the necessary data to configure a Series 90-30 PLC. Now we will apply our knowledge as we configure a General Electric Series 90-30 PLC.

Configuring a Series 90-30 PLC

We will configure the following ten-slot General Electric Series 90-30 PLC (Figure 9-27).

The list shows us what hardware is required for our application. Next we need to calculate the total power all hardware items will draw from our power supply. To do this:

1. List all hardware in a chart similar to Figure 9-28.
2. Fill in the power each piece of hardware will draw from the power supply.
3. Add each column and put the sum of the column in the "Total Milliamps" space. Do this for each of the three voltage columns. The total of each column will be in milliamps.
4. Convert each total to watts. To convert milliamps to watts, multiply milliamps by the voltage for that column.

Hint: One milliamp is equal to .001 amp, so 1,542 milliamps is equal to 1.542 amps. Converting makes completing the calculations (by multiplying 1.542 amps by 5 volts to get 7.71 watts) easier. Do the same with the 240 milliamp value for the +24 isolated

SERIES 90-30 HARDWARE LOAD REQUIREMENTS				
Catalog Number	Description	+5 V dc (mA)	+24 V dc Isolated (mA)	+24 V dc Relay (mA)
IC693CPU311	Model 311 Five-Slot Baseplate	410		
IC693CPU321	Model 311 Ten-Slot Baseplate	410		
IC693CPU331	Model 331 CPU	350		
IC693CPU340	Model 340 CPU	490		
IC693CPU341	Model 341 CPU	490		
IC693CPU351	Model 351 CPU	890		
IC693CHS397	Model 331 Five-Slot CPU Baseplate	270		
IC693CHS391	Model 331 Ten-Slot CPU Baseplate	250		
IC693CHS398	Model 331 Five-Slot Expansion Baseplate	170		
IC693CHS392	Model 331 Ten-Slot Expansion Baseplate	150		
IC693ACC307	Expansion Bus Termination Pack	72		
IC690ACC900	RS-422/RS-485 to RS-232 Converter (PLC to Personal Computer Converter)	170		
IC693ACC300	Input Simulator	120		
IC693MDL230	120 VAC Isolated, 8-Point Input	60		
IC693MDL231	240 VAC Isolated, 8-Point Input	60		
IC693MDL240	120 VAC, 16-Point Input	90		
IC693MDL310	120 VAC, .05A, 12-Point Output	210		
IC693MDL340	120 VAC, .05A, 16-Point Output	315		
IC693MAR390	24 VAC Input, Relay Output, 8 In/8 Out	80		70
IC693MAR590	120 VAC Input, Relay Output, 8 In/8 Out	80		70
IC693MDL640	24 V dc Positive Logic, 16-Point Input	5	120	
IC693MDL646	24 V dc Fast, 16-Point Input	80	125	
IC693MDL740	12/24 V dc 16-Point Positive Logic Output	110		
IC693MDL930	8-Point Isolated Relay Output	6		70
IC693MDL940	Relay Output Module, 16-Point	7		135
IC693ALG220	Analog Input, Four Channels, Voltage	27	98	
IC693ALG221	Analog Input Module, Current	25	100	
IC693ALG390	Analog Input, Two Channels, Voltage	32	120	
IC693ALG391	Analog Output Module, Current	30	215	

Figure 9-26 Load requirements for selected Series 90-30 modules. (Table data courtesy of GE Fanuc Automation)

Quantity	Part Number	Hardware Description
1	IC693CPU331	Model 311 CPU
1	IC693CHS391	Ten-Slot CPU Baseplate for Model 331
1	IC693ACC300	Input Simulator
1	IC693ACC300	Input Simulator
1	IC693MDL240	16-Point Input Module, 120 VAC
1	IC693MDL240	16-Point Input Module, 120 VAC
1	IC693MDL640	16-Point 24 V dc, Positive Logic Inputs
1	IC693MDL640	16-Point 24 V dc, Positive Logic Inputs
1	IC693MDL940	16-Point Relay Output Module
1	IC693MDL940	16-Point Relay Output Module
1	IC693MDL310	12-Point Output Module, 120 VAC
1	IC693MDL740	16-Point 12/24 V dc, Positive Logic Outputs
1	IC690ACC900	Interface Converter from PLC to PC

Figure 9-27 Hardware list for Series 90-30 PLC configuration. (Chart data courtesy of GE Fanuc Automation)

GENERAL ELECTRIC SERIES 90-30 POWER SUPPLY LOADING Series 90-30 Model 331 Ten-Slot Baseplate			
Hardware	+5 Volts in mA	+24 Isolated in mA	+24 Relay in mA
Model 331 CPU	350		
Ten-Slot CPU Baseplate for Model 331	250		
Input Simulator	120		
Input Simulator	120		
16-Point Input Module	120 VAC	90	
16-Point Input Module	120 VAC	90	
16-Point 24 V dc, Positive Logic Inputs	5	120	
16-Point 24 V dc, Positive Logic Inputs	5	120	
16-Point Relay Output Module	7		135
16-Point Relay Output Module	7		135
12-Point Output Module, 120 VAC	210		
16-Point 12/24 V dc, Positive Logic Outputs	118		
Interface Converter from PLC to PC	170		
TOTAL MILLIAMPS	1542	240	270
TOTAL WATTS (volts multiplied by milliamps = watts)	7.71	5.76	6.48

Figure 9-28 Completing the power supply loading calculations for our example.

.240 amps multiplied by 24 volts (equals 5.76 watts). Likewise, .270 milliamps times 24 volts equals 6.48 watts. Keep this principle in mind when working through these exercises.

Check to see that none of the three voltages has totaled up to be more than 15 watts. If any column alone totals more than 15 watts, you must change your module mix. Either replace a module with another module with less power draw or add an expansion chassis. We will look at adding an expansion chassis and redistributing modules later in this chapter.

Total up the watts for each of the three voltages as illustrated in the calculations table in Figure 9-29. Remember, the total of any one voltage cannot exceed 15 watts, and the total of all three voltages must not exceed 30 watts.

CALCULATIONS	
7.71	5 Volts DC (15 watts Max.)
5.76	24 Volts Isolated (15 watts Max.)
6.48	24 Volts Relay (15 watts Max.)
19.95	Total (30 watts Max.)

Figure 9-29   Totaling individual voltage loads in watts to get total load on power supply.

Figure 9-28 illustrates the process of determining the milliamp loading value calculation for each of the three power-supply output voltages.

Analysis of Our Calculations.   The load capacity of a Series 90-30 power supply is the sum of the internal loads placed on it by all hardware components associated with that baseplate. The total power output of the power supply is 30 watts and the totals of each of the three voltages must not exceed 15 watts for any one voltage. Our totals for this application were 7.71 watts at 5 V dc, 5.76 watts at 24 V dc isolated, and 6.48 watts at 24 V dc relay. By combining the three watts totals, we arrive at a grand total of 19.95 watts loading on the power supply, well within the power-supply ranges.

I/O RACKS AND EXPANSION

Consider a manufacturing process where a product is manufactured as a conveyor moves it through many assembly stations. Each assembly station would have its own dedicated, or local, PLC. This local PLC would only control the activities within that station. The disadvantage with a system such as this is that there is no communication between assembly stations. Thus, there is no way to tell what is really happening or method to track the assembly process. How can we direct, control, or communicate to other workstations if one cell fails or runs out of parts? Each local PLC does its task with no interaction with its neighbor. Of course, if the assembly process continues without any problems, this could be a workable solution.

Think of the control advantage if we could distribute needed I/O at each workstation. One CPU will control all blocks of I/O. As a result, the CPU could oversee each I/O point for the entire process. Configuring a PLC system in this manner is a second use for the PLC chassis.

Racks, baseplates, and chassis have multiple uses, including use as a PLC local chassis. They can also be used to expand I/O to locations close by as an expansion chassis off the local chassis. We have identified the **local I/O** chassis holding the processor, selected I/O modules, and the associated power supply as a base chassis. We will now use this local chassis as the base from which to expand our system into using other chassis containing only a power supply and I/O modules. Expanding the I/O of a PLC system to locations in close proximity to the local chassis is called local I/O expansion. A chassis containing additional I/O is typically called an expansion chassis. This ability to expand I/O into additional chassis allows us to add to the PLC's I/O while enhancing control under the supervision of a single processor. Local I/O expansion is a good alternative for small PLC systems with I/O distributed close to the local chassis. Different manufacturers handle I/O expansion differently. Let us look at how the Series 90-30 expands off the local PLC baseplate.

The General Electric (GE) Series 90-30 uses bases rather than a rack or chassis. I/O modules clip onto a flat base about three-quarters of an inch thick rather than slide into a rack. The GE Series 90-30 employs a special expansion baseplate (illustrated in Figure 9-30), into which modules for the remote I/O locations are plugged. Extending the backplane of our CPU baseplate is accomplished by using a short parallel cable to connect the CPU baseplate to the first expansion baseplate. The parallel cable provides parallel data transfer between the CPU baseplate and the locally expanded I/O baseplate.

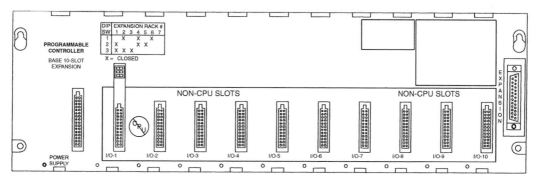

Figure 9-30   Series 90-30 ten-slot expansion baseplate. (Courtesy of GE Fanuc Automation)

Parallel cable connections between two baseplates makes data transfer seamless. Communication occurs as if the two backplanes were physically one. Expanding the I/O off the Series 90-30 CPU baseplate, as illustrated in Figure 9-31, is accomplished by selecting the special expansion baseplates (pictured in Figure 9-30) to hold the expanded I/O modules and the necessary interconnect cables. There is no CPU in any expansion baseplate; however, each expansion baseplate must nonetheless have a power supply. Power-supply loading should be calculated for each expansion baseplate's modules and the load that will be placed on each baseplate's associated power supply. In Figure 9-31,

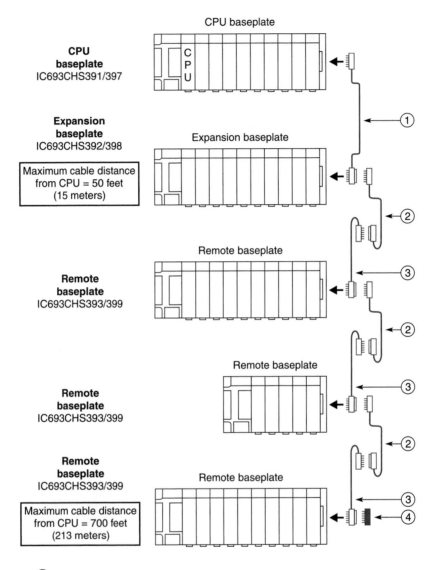

① Standard wye cable

② Custom-built point-to-point cable

③ IC693CBL300 standard wye cable, used as wye jumper

④ IC693ACC307 bus terminator

Figure 9-31   Series 90-30 expansion and remote baseplate connections. (Courtesy of GE Fanuc Automation)

note the I/O bus termination resistor plug placed at the end of the farthest cable. Let's look at an example of configuring a General Electric 90-30 PLC with local expansion.

Configuring a Series 90-30 PLC with Local Expanded I/O

The second GE Series 90-30 configuration example requires the hardware shown in Figure 9-32. Additional hardware needed to configure this particular PLC is shown in Figure 9-33.

Quantity	Part Number	Hardware Description
1	IC693CPU331	Model 331 CPU
1	IC693MDL753	12/24 V dc, 32-Point Output
2	IC693ALG391	Analog Output
2	IC693ALG221	Analog Input
1	IC693MAR590	120 VAC, 8 Inputs and 8 Relay Outputs
1	IC693MDL930	8-Point Relay Output
1	IC693MDL231	240 VAC, 8-Point Input
1	IC693MDL646	24 V dc Fast, 16-Point Output

Figure 9-32   Hardware list for Series 90-30 PLC configuration. (Chart data courtesy of GE Fanuc Automation)

Quantity	Part Number	Description
1	IC693CHS391	Ten-Slot Baseplate
1	IC690ACC900	RS-422/RS485 to RS-232 Converter

Figure 9-33   Additional hardware components needed to complete configuration.

Looking at the column totals for Figure 9-34, notice that the +24 V dc isolated column totals to 18.6 watts. The maximum wattage for any of the three voltage columns cannot exceed 15 watts. Since the isolated 24 V dc loading on this power supply exceeds the maximum allowable watts, this configuration cannot be reliably applied. Let us analyze why our module mix has exceeded the maximum value and what options we have to correct the problem.

There are five I/O modules that contribute to the problem, two IC693ALG391 analog outputs, two IC693ALG221 analog inputs, and one IC693MDL646 24 V dc, fast, 16-point input module. One option is to see if any of these problem modules can be replaced with another module that requires less power. Let us assume we must keep the analog modules. However, we could possibly remove the IC693MDL646 module and replace it with

GENERAL ELECTRIC SERIES 90-30 POWER SUPPLY LOADING Series 90-30 Model 331 Ten-slot Baseplate			
Hardware	+5 Volts	+24 Isolated	+24 Relay
Model 331 CPU	350		
Ten-slot Baseplate	250		
12/24 V dc, 32-Point Output	260		
Analog Output	30	215	
Analog Input	30	215	
Analog Output	30	110	
Analog Input	30	110	
120 VAC, 8 Inputs, and 8 Relay Outputs	80		70
240 VAC, 8-Point Input	60		
24 V dc FAST, 16-Point Input	80	125	
120 VAC, 16-Point Output	315		
RS-422/RS485 to RS-232 Converter	170		
TOTAL MILLIAMPS	1685	775	70
CONVERT MILLIAMPS TO WATTS (Volts multiplied by milliamps = watts); 15 watts maximum for any column	8.42	18.6	1.68

Figure 9-34   Calculating 90-30 power.

something else. Would this solve our power-supply loading problem? The power required to operate the four remaining analog modules is:

$$IC693ALG391, \text{ analog output module} = 215 \text{ milliamps}$$
$$IC693ALG391, \text{ analog output module} = 215 \text{ milliamps}$$
$$IC693ALG221, \text{ analog input module} = 110 \text{ milliamps}$$
$$IC693ALG221, \text{ analog input module} = 110 \text{ milliamps}$$
$$\text{Total} = 650 \text{ milliamps}$$

Calculate the total watts from total milliamps:

Total milliamps multiplied by voltage   =   total watts

Convert 650 milliamps to .650 amps

.650 milliamps multiplied by 24 volts   =   15.6 total watts

The total watts required from the power supply is 15.6 for the four remaining analog modules. Is this an acceptable hardware configuration? Since the maximum is 15.0 watts, our configuration is still not acceptable. We must find another solution. Substitution of the 24 V dc, fast, 16-point input module with another module will not solve our problem.

A second option is to reconfigure our system onto two five-slot baseplates in a local expansion configuration. By splitting the system onto two baseplates, there will be a separate power supply for each baseplate and the resident five hardware devices. As a result, we will have twice the total power to operate the same nine modules and CPU. Since the analog modules create the power loading problem, they will be divided between the two five-slot baseplates. Our new system configuration is illustrated in Figure 9-35.

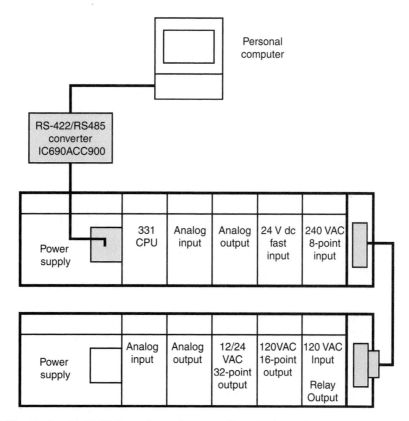

Figure 9-35   Series 90-30 PLC configured on two five-slot baseplates.

Our system has two separate baseplates and power supplies. Modules have been placed on each baseplate as suggested in the module placement section earlier in this chapter. This lab exercise will be found in your lab manual.

A few things to keep in mind regarding the calculations:

1. You will have two separate baseplates.
2. Each baseplate will have its own power supply.

3. The power supply in one baseplate has no relationship to power in the other. The cable between the local baseplate and the remote baseplate is for communication for input and output signals to and from modules in the second baseplate.
4. Only the first baseplate has a CPU. The second baseplate is simply an extension of the first.
5. The first baseplate will power the RS422/RS485 converter to the personal computer.
6. Baseplate number two will provide power to the resistor terminal plug, part number IC693ACC307. This is the expansion bus termination pack. Include 72 milliamps in your +5 V dc calculation for this device.

Summary of Series 90-30 Local Expansion. Figure 9-35 illustrates a typical General Electric Series 90-30 local expansion application. Notice that each expansion baseplate is specifically used for expansion from the CPU baseplate. While each expansion baseplate has a power supply, there is only one CPU, which resides in the CPU baseplate. The CPU controls all of the I/O in the entire system. The maximum distance from the CPU baseplate and the last expansion baseplate is 50 cable feet. This distance is made up by using any of the three available cables (3 feet, 6 feet, and 50 feet). The system cannot be expanded more than 50 cable feet without going to a remote expansion configuration. Last, notice the I/O bus terminator plug on the last expansion chassis. This terminator is required at the end of the communication cable to eliminate signal reflection problems (commonly also known as ringing).

ROCKWELL AUTOMATION CONTROLLOGIX FAMILY OF PLCs

The newest members of the Rockwell Automation PLC family are the ControlLogix. The modular ControlLogix PLC can support up to 128,000 inputs and outputs, or 4,000 analog. Modular processors have from 750 Kbytes of user memory up to 8 Mbytes. Modular ControlLogix processors come with an integrated serial port. Separate communication modules such as Ethernet, Data Highway Plus, or Remote I/O, ControlNet, and DeviceNet are available for insertion into the ControlLogix chassis. These modules are used for upload, download, or going on-line with the processor, or to configure a gateway to bridge data between different networks. In Figure 9-36, the ControlLogix PLC is top center. Directly below the ControlLogix is the FlexLogix PLC, and below that is the CompactLogix PLC. The FlexLogix and CompactLogix are smaller versions of the ControlLogix modular PLC in memory size, in the number of I/O supported, and in software functionality. As an example, FlexLogix processors come with either 64 Kbytes or 512 Kbytes of user memory. Up to two communication cards can be inserted into the FlexLogix processor module to provide for Ethernet, ControlNet, or DeviceNet network connectivity. FlexLogix can support up to 512 I/O in two banks of Flex I/O blocks and terminal bases. Extended local I/O can be connected using a communication cable from one to three meters long. Banks can also be mounted up to ten feet away with the appropriate extender cables. The top shows a VersaView 1500P industrial computer running SoftLogix. There is a PowerFlex 700S variable-frequency systems drive displayed at the left. The 700S is a system or supervisory drive. The equivalent of a ControlLogix processor can be installed in the drive by opening the door on the left front

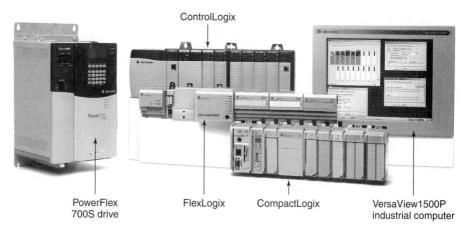

ControlLogix

PowerFlex
700S drive

FlexLogix   CompactLogix

VersaView1500P
industrial computer

Figure 9-36 Rockwell Automation ControlLogix family members. (Used with permission of Rockwell Automation, Inc.)

and installing the interface card in the appropriate slot in the drive. Figure 9-36 shows the members of the ControlLogix family. RSLogix 5000 software can be used to program all members of the ControlLogix family. The ControlLogix platforms share the same PLC instruction set as the Allen-Bradley PLC 5.

Figure 9-37 below shows a CompactLogix PLC with the major hardware features identified. CompactLogix can be a small to medium modular size standalone PLC. A total

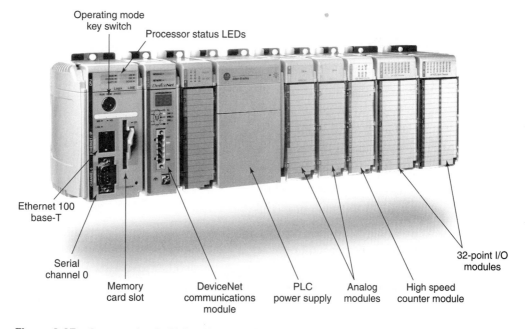

Operating mode
key switch
Processor status LEDs

Ethernet 100
base-T

Serial
channel 0

Memory
card slot

DeviceNet
communications
module

PLC
power supply

Analog
modules

High speed
counter module

32-point I/O
modules

Figure 9-37 CompactLogix PLC with major features identified. (Used with permission of Rockwell Automation, Inc.)

of up to 30 I/O modules can be configured in three banks of modules for a total of 1,024 I/O points. Processors are available starting with 64 Kbytes up to 1.5 Mbytes of user memory.

SUMMARY

PLCs come in two styles, fixed and modular. Some fixed PLCs can be expanded beyond their total built-in I/O points, and all modular PLCs can be expanded beyond their local chassis in one way or another. Some manufacturers allow a single part-numbered chassis to be used for local, extended local, and remote I/O. Another manufacturer may require a special chassis or baseplate for local expansion and a different one for remote I/O expansion. Though there may be no advantage to one method over the other, you must be aware of what hardware is necessary to expand your I/O into one or more expansion chassis.

I/O modules are characterized as either discrete (two-state), analog, or specialty. Discrete I/O modules are simply devices that either accept ON/OFF inputs or provide ON/OFF output switching. Analog modules input or output signals that vary between two defined limits. They either accept or provide voltage or current signals. Temperature or pressure can be input as either an analog or discrete signal to a PLC. If a temperature or pressure switch with a two-state contact block is to be interfaced to a PLC, a discrete input module is used. On the other hand, if a temperature or pressure input device provides a continuously varying signal (such as 4 to 20 milliamps or 1 to 5 volts), an analog input module will be selected.

REVIEW QUESTIONS

1. Place the following modules in a ten-slot chassis in their proper position relative to the CPU:

    Two relay output modules

    One 120 VAC input module

    Two 120 VAC output modules

    One 5 V dc input module

    Two 24 V dc input modules

    One 24 V dc output module

2. As a troubleshooter, why would you consider recalculating a troublesome PLC chassis's module power requirements in relation to power supply output?

3. One manufacturer may use the term *rack*, another _____, while to another the same component is a _____.

4. A modular PLC in a single chassis can be a stand-alone system where the PLC is dedicated to controlling one machine or process. This configuration is usually called a base rack or _____.

5. Selecting the appropriate input or output module is accomplished by determining incoming or outgoing signals and matching these signals to an _____ module.

6. A _____ is usually mounted on one end of each rack, baseplate, or chassis.

7. An _____ is connected to the base PLC with a communication cable. The communication cable carries I/O status signals between the CPU, in the base chassis, and to the expanded chassis and its I/O.

8. Two ten-slot chassis can hold up to _____ I/O modules and the processor.

9. The _____ provides power to operate the CPU, along with the electronics inside each I/O module.

10. In regard to local expansion requirements:
    A. Placement of a CPU is required only in the first, base, or local chassis.
    B. A CPU is needed in both the local chassis and the expansion chassis.
    C. The expansion chassis will hold only I/O modules and a power supply.
    D. The expansion chassis receive their operating power from the base chassis.
    E. A and D are correct.
    F. B and D are correct.

11. The power supply does not usually provide power to operate input or output _____.

12. When connecting separate PLCs on a network, each base PLC will be a _____ on the network.

13. A PLC chassis can also be the base for expanding and distributing I/O around the factory floor to manufacturing locations where it is needed. This is called _____ I/O.

14. Rather than scattering separate PLCs around the factory, remote I/O expansion makes it possible for one base, or local, PLC to house the CPU and control several _____ racks of I/O.

15. The advantage of _____ is that one, central CPU can supervise all I/O points.

16. Suppose you need to select a module to interface AC or DC, two-state, ON/OFF, or OPEN/CLOSED incoming signals. Which will you choose:
    A. analog input module
    B. analog output module
    C. discrete input module
    D. discrete output module
    E. A or C
    F. B or D
    G. none of the above

17. Suppose you need to select a module to interface outgoing AC or DC, two-state, ON/OFF, or OPEN/CLOSED outgoing signals. Which will you choose:
    A. analog input module
    B. analog output module
    C. discrete input module
    D. discrete output module
    E. A or C

F. B or D

G. none of the above

18. Suppose you need to select a module to interface a continuously changing incoming voltage or current signal. Which will you choose:

A. analog input module

B. analog output module

C. discrete input module

D. discrete output module

E. A or C

F. B or D

G. none of the above

19. A discrete input module can interface:

A. push buttons

B. selector switches

C. two-state pressure switches

D. pilot lights

E. inductive proximity switches

F. motor starters

G. A, B, C, and E

H. D and F

20. A discrete output module can interface:

A. push buttons

B. selector switches

C. two-state pressure switches

D. pilot lights

E. inductive proximity switches

F. motor starters

G. A, B, C, and E

H. D and F

21. An analog output module can control:

A. pilot light

B. valves

C. speed of variable-speed drive

D. temperature

E. B, C, and D

F. all of the above

22. Module placement rules are straightforward. Starting with the CPU, modules are placed as follows (list in order from first to fourth):

_____ low- to high-voltage AC output modules, in ascending order away from the CPU

_____ low- to high-voltage DC output modules, in ascending order away from the CPU

_____ low- to high-voltage DC input modules, in ascending order away from the CPU

_____ low- to high-voltage AC input modules, in ascending order away from the CPU

23. The power needed to power _____ is drawn through the rack backplane from the chassis power supply.
    A. the internal electronics of each module
    B. any handheld programming device
    C. interface hardware
    D. expansion chassis
    E. output field devices
    F. A, B, and C
    G. A, B, C, and D

24. When multiple power supplies are available for a particular PLC:
    A. always select the cheapest power supply available
    B. always select the largest power supply available
    C. select the power supply for each rack based on calculations of the power needed to operate the specific hardware associated with that chassis
    D. choose your power-supply based on calculations for the inrush current for all field devices connected to the PLC
    E. a single power supply must be chosen for all local racks based on calculations of the total power needed to operate all the hardware in all the associated chassis

25. When configuring a system or troubleshooting phantom or inconsistent problems:
    A. replace the power supply as it is probably bad
    B. replace the CPU as it is probably bad
    C. replace the rack or chassis as the backplane is not allowing power to flow to all modules
    D. verify power-supply loading for each modular chassis
    E. determine which remote rack is draining power from the base rack's power supply

IF PASSWORD IS CORRECT GO TO DRIVE SET UP SCREEN

```
                        B3:10/0                                                    B3:1/15
0022 ──┤ EQU ├─────────[ OSR ]──────────────────────────────────────────────────────( )──
       │ Equal      │                                                          RESET PASSWORD
       │ Source A   N12:21│                                                   ATTEMPTS COUNTER
       │            7777< │                                                         C5:5
       │ Source B   32767 │                                                       ─( RES )─
       │            32767<│
       └────────────┘

0023 ──┤ GRT ├────────────┤ NEQ ├────────────  B3:10/1
       │ Greater Than(A>B)│ Not Equal        │ [ OSR ]──────┐
       │ Source A   N12:21│ Source A   N12:21│              │
       │            7777< │            7777< │              ↓
       │ Source B   0     │ Source B   32767 │         ┌───←┘
       │            0<    │            32767<│
       └──────────────────┘
```

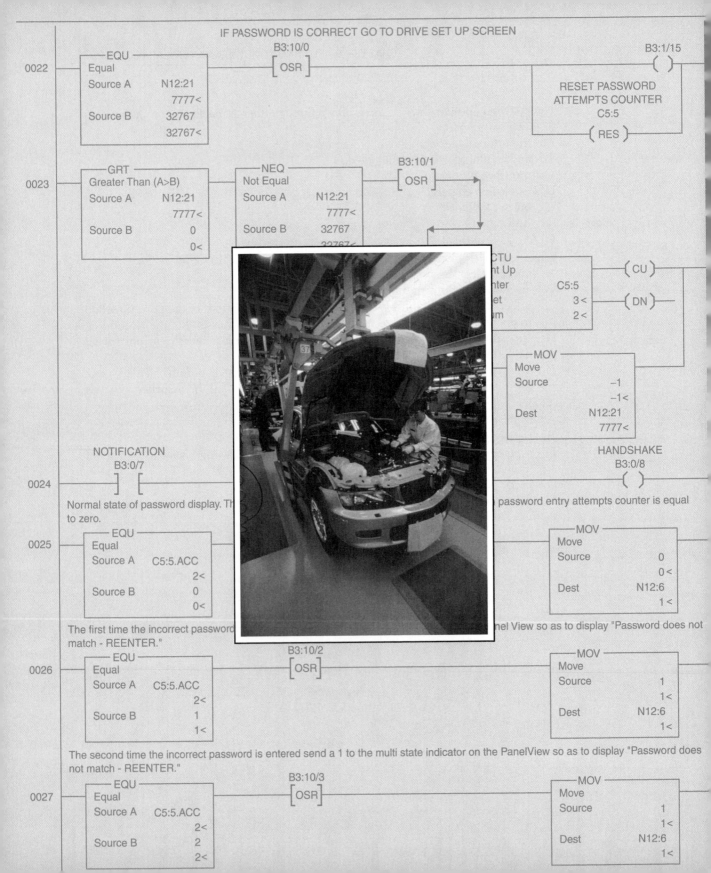

```
                                                         ┌ CTU ─────────
                                                         │ ...nt Up        ─( CU )─
                                                         │ ...nter   C5:5  
                                                         │ ...et       3<  ─( DN )─
                                                         │ ...m        2<
                                                         └───────────────

                                                         ┌ MOV ─────────
                                                         │ Move
                                                         │ Source      −1
                                                         │             −1<
                                                         │ Dest    N12:21
                                                         │          7777<
                                                         └───────────────
```

```
       NOTIFICATION                                                        HANDSHAKE
       B3:0/7                                                              B3:0/8
0024 ──┤ / ├──────────────────────────────────────────────────────────────( )──
```

Normal state of password display. Th... ...password entry attempts counter is equal to zero.

```
0025 ──┤ EQU ├─────────────────────────────────────────────┌ MOV ─────────
       │ Equal       │                                      │ Move
       │ Source A  C5:5.ACC│                                │ Source       0
       │            2<  │                                    │             0<
       │ Source B   0   │                                    │ Dest    N12:6
       │            0<  │                                    │          1<
       └───────────────┘                                    └───────────────
```

The first time the incorrect password... ...nel View so as to display "Password does not match - REENTER."

```
                        B3:10/2                           ┌ MOV ─────────
0026 ──┤ EQU ├─────────[ OSR ]───────────────────────────│ Move
       │ Equal       │                                    │ Source       1
       │ Source A  C5:5.ACC│                              │             1<
       │            2<  │                                  │ Dest    N12:6
       │ Source B   1   │                                  │          1<
       │            1<  │                                  └───────────────
       └───────────────┘
```

The second time the incorrect password is entered send a 1 to the multi state indicator on the PanelView so as to display "Password does not match - REENTER."

```
                        B3:10/3                           ┌ MOV ─────────
0027 ──┤ EQU ├─────────[ OSR ]───────────────────────────│ Move
       │ Equal       │                                    │ Source       1
       │ Source A  C5:5.ACC│                              │             1<
       │            2<  │                                  │ Dest    N12:6
       │ Source B   2   │                                  │          1<
       │            2<  │
       └───────────────┘
```

CHAPTER

10

PLC Processors

OBJECTIVES

After completing this chapter, you should be able to:

- describe the function of the processor
- describe processor operating modes
- explain the function of the watchdog timer
- explain how PLC user program updates can be accomplished in the field when using a nonvolatile memory chip on your processor
- explain on-line editing
- explain the differences between off-line and on-line editing
- discuss password protection of a user program, various types of password protection levels, and the advantages and disadvantages of using passwords

INTRODUCTION

Programmable controllers, their processors, and their programming instructions have been evolving since their inception in the late 1960s and early 1970s. The early PLC as a relay replacer has become more and more sophisticated as advances in microprocessor technology have increased the capabilities of the processor's computer power. As microprocessor computing and data processing capabilities increased, so did the instructions and complexity of tasks performed. In this chapter we will explore PLC processors, introduce basic operating principles, and look at selection criteria.

THE PROCESSOR'S FUNCTION

The PLC processor is a microprocessor with a memory system, circuits to store and retrieve information from memory, and other circuits for communicating with outside devices like a personal or industrial computer programming terminal. Devices external to the personal computer (PC) or PLC are called "peripheral devices." The processor, or central processing unit (CPU), is built into the fixed PLC chassis or else is a self-contained modular device that plugs into a chassis or clips onto a baseplate. The microprocessor is the decision maker in a PLC system.

THE PROCESSOR'S OPERATING SYSTEM

A processor's operating system is simply the processor's personality and capabilities stored on a nonvolatile memory chip. More specifically, an operating system is a set of software instructions, typically stored in nonvolatile memory, that is responsible for directing system activities. The operating system directs the allocation of system resources, including memory allocation for program and data storage, input and output distribution, processor timing, and communication with peripheral devices. The operating system also contains a list of programming instructions used to develop and solve user ladder programs, directly execute the user program, and perform associated housekeeping chores. When a series of instructions, such as an operating system, is stored in nonvolatile memory, it is called **firmware** (FRN).

Older PLC processors stored their operating systems on programmable, read-only memory (PROM) chips. When the processor manufacturer made changes to the operating system, the PROM chips had to be replaced. More modern PLC processors stored their operating systems on electrically erasable, programmable, read-only memory (EEPROM) chips. If the processor manufacturer made changes or upgrades to the operating system, in some cases, the EEPROM chips (firmware) had to be replaced. Replacing EEPROM chips for firmware upgrading meant that the processor had to be disassembled. Firmware upgrades requiring processor disassembly could not always be completed in the field. Sometimes either the processor had to be sent in for repair or a service engineer had to make a service call to replace EEPROM chips. Clearly, there was a need for a better, easier way to upgrade a processor's firmware.

Formerly, upgrading an EEPROM chip operating system in some PLCs meant that the manufacturer supplied a floppy disk containing the new operating system. The EEPROM chips were removed from the processor and placed in a device used to erase the old operating system and download or write the new operating system from personal computer memory to the EEPROM chips. This device was called an "EEPROM burner."

An easier way to upgrade an operating system was developed whereby no disassembly or chip removal was necessary. Newer PLC processors use nonvolatile memory that can be reprogrammed without processor disassembly and chip removal. This newer type of memory is called "flash memory."

Flash Memory

Newer processors, along with other electronic microprocessor-based equipment such as modern operator interface devices, have their operating systems installed on the newer, flash memory EEPROMs in place of the old-style EEPROMs. Flash EEPROMs eliminate the need to replace the EEPROM chips physically when making a firmware upgrade. Flash memory is a special form of nonvolatile memory that can be erased and reprogrammed with signal levels commonly found inside the computer device in which they reside.

There are two methods commonly used to upgrade a system's flash EEPROM–based operating system. The steps to complete the operating system upgrade will be determined by the hardware vendor.

Probably the easiest way to upgrade an operating system is when the new operating system is supplied from the manufacturer on a floppy disk. The floppy is simply loaded into a personal or industrial computer, a communication cable is hooked up, and by following the manufacturer's instructions, the operating system upgrade is completed. Another method is for the hardware manufacturer to offer a firmware upgrade kit. The kit will usually contain a plug-in integrated chip and a set of instructions. To illustrate how this method works, we will look at upgrading an Allen-Bradley SLC 500 5/03 or 5/04 processor. Be aware that the user program will be cleared and all communication ports will be reset to factory defaults. An overview of the procedure is listed below. Refer to Figure 10-1 as you review the procedure.

1. Remove power.
2. Remove the processor from its chassis.
3. Change Jumper J4 on the processor's circuit board, as directed on the instruction sheet. This is the operating system write-protect jumper. The jumper will be moved to the unprotected position.
4. Plug the firmware upgrade pack into the memory module socket on the processor circuit board.
5. Replace the processor in the chassis.
6. Connect power.
7. The processor will automatically do the upgrade. The download process takes about 45 seconds.
8. The instruction sheet will indicate which LEDs on the processor will be lit up when the procedure has been successfully completed.
9. Remove power.
10. Remove the processor, remove the upgrade pack, and replace the operating system write-protect jumper to the protected position.
11. Reassemble the processor.
12. The operation is complete. Restore the user program and reset communication channels, if necessary.

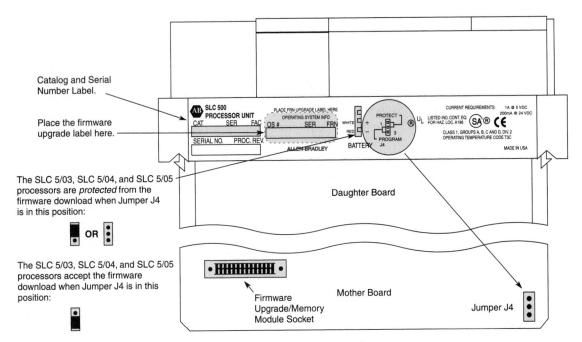

Figure 10-1 SLC 500 processor side view showing Jumper J4, firmware upgrade/memory module socket, and firmware upgrade label placement. (Used with permission of Rockwell Automation, Inc.)

The upgrade kit can be used to upgrade many processor-operating systems by following the indicated steps. On newer processors like the Allen-Bradley SLC 5/05 and ControlLogix, the operating system can be upgraded or "flashed" directly from a file downloaded from the Internet.

Each operating system upgrade is used to fix problems found in earlier versions, along with adding new features, instructions, and functionality. As an example, the Allen-Bradley SLC 500 operating system identification is as follows. Firmware that goes with the 5/03 processor is identified as OS 300 (OS stands for operating system), Series A, FRN release number 1. The first operating system that came with the first 5/03 processor would be the OS 300, Series A, FRN 1. These numbers will increment as additional operating system upgrades are offered.

The list below provides only a sample of some major enhancements contained in OS 302 operating system upgrades:

1. addition of trigonometric math functions
2. addition of the compute instruction
3. enhanced analog instruction functionality
4. added indexed addressing
5. remote I/O enhancements
6. network enhancements

The Operating System

The operating system defines how the processor will react to internal and external signals. Thus, the operating system defines the processor's operating cycle. During the typical operating cycle, the processor reads inputs and stores the input data in the input status file. While solving the user program logic, the processor uses input status file data to solve the programmed logic. The ultimate result of solving the processor logic is an ON or OFF state to be passed on to discrete outputs or binary words representing analog outputs. The output statuses are passed on to the *output status file* awaiting transmission to each output point. The process of reading outputs, solving logic, and updating outputs is called the **processor scan** or **processor sweep.**

PROCESSOR SCANNING

The processor in any PLC is designed to perform specific duties in a specific sequence and then continuously repeat the sequence. The operating cycle is also called the processor scan or sweep. A scan consists of a series of sequential operations that include housekeeping, data input, program execution, data output, servicing or updating the programming device, system communications, and diagnostics. Each of these steps will be described in further detail. This scan cycle is performed sequentially and repeatedly when the processor is in run mode. The processor could be in run mode for hours, days, weeks, even months. An example of a standard processor scan or sweep is shown in Figure 10-2. This is a typical processor scan or sweep for a General Electric Series 90-30 PLC.

The user program will execute from the top, or rung zero, straight through to the last rung, continuously, unless altered by an instruction specifically designed to alter the flow of the program. Program flow instructions direct the flow of instructions and the execution of instructions within a ladder program. Program flow is altered when using instructions like "jump" and "branch."

Input Scan

During the input scan, the CPU scans each input module for the ON or OFF states of each of the associated input points. The ON or OFF input states are stored in the input status file.

Program Scan

After the inputs are read and stored in the input status file, the processor will use this information to solve the user ladder program. The processor scans the user program starting at rung zero at the left power rail, working left to right, and evaluating one instruction at a time until the output instruction is reached. The output status is the logical resultant of the solved input logic for that rung. The logical one or zero output status is placed in the output status file. When rung zero is completed, the processor goes on to rung one, rung two, rung three, and so on, sequentially, to the last rung. After the last rung of ladder logic is executed, there is one additional rung in the program. This last rung is automatically

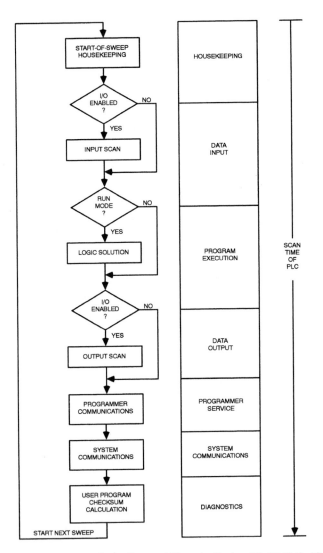

Figure 10-2 PLC program sweep for a General Electric Series 90-30 PLC. (Courtesy of GE Fanuc Automation)

inserted by the software. The last rung is the "end rung." The end rung alerts the CPU that it has reached the end of the ladder program. Most PLC CPU's solve the user logic in this manner. The time it takes for the CPU to scan the user program varies depending on which instructions are used and the ON or OFF status of the instruction. With some PLCs, if they determine a rung has no chance of becoming true during the current scan, they will skip over the remaining instructions on that particular rung and evaluate it as false. This procedure is used to speed up the processor scan time.

After the inputs are read and stored in the input status file, the user program is solved, the processor sends (writes) the resulting logical ON or OFF state for each output to the output status file, and the output scan is executed.

Output Scan

The output scan is the function where the CPU writes the ON or OFF status, one word at a time, to the associated output module. Each output status word is made up of ON or OFF electrical signals; there is one ON or OFF signal for each output point. Each module output point latches its ON or OFF signal into its electronic hardware to keep the output in the proper status until the next output scan sends an update. The time it takes to read an input, solve the user program, and turn on or off the corresponding output is called the system throughput.

Service Communications

After updated output data has been written to the output modules, the CPU services communications. The servicing of communications includes updating the handheld or personal computer's monitor screen and sending communications to other PLCs on a network or on operator interface devices.

Housekeeping and Overhead

Housekeeping and overhead are the part of the scan cycle in which the CPU takes care of memory management, updating timers and counters, internal time base, the processor status file, and other internal registers.

OMRON CQM1 OPERATING CYCLE

Figure 10-3 illustrates the sequence of CPU operation in a flowchart format for Omron's CQM1 programmable controllers. The Omron CQM1 scan or sweep is called a cycle. The CQM1 cycle consists of six segments: overseeing, program execution, cycle time calculation, I/O refresh, RS-232C port servicing, and peripheral port servicing. We will investigate each segment of the cycle.

Overseeing

During the overseeing portion of the cycle, the watchdog timer is set, I/O bus is checked, the clock is refreshed, and the bits assigned to new functions are also refreshed. The watchdog timer is a safeguard to ensure that the processor does not become stuck while scanning the user program or become, for another reason, unable to complete the current scan.

Program Execution

During the program execution portion of the CQM1 cycle, the user program is executed. The time of this portion of the cycle varies according to the mix of instructions in the user program.

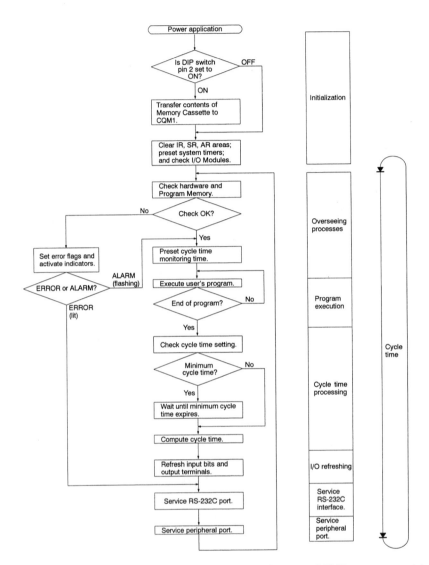

Figure 10-3 Flowchart of CPU operation for an Omron Sysmac CQM1 programmable controller. (Courtesy of Omron Electronics, Inc.)

Cycle Time Calculation

Cycle time calculation is based on the time it takes for inputs to be read and output bits to be written, plus program execution times and time needed for servicing peripherals such as the RS-232C and peripheral ports. The calculated cycle time is stored in an internal register. If the cycle time is less than the minimum cycle time, execution will wait until the minimum time has expired. The minimum cycle time can be used to standardize the cycle time and eliminate variations in I/O response time.

I/O Refresh

The I/O refresh portion of the operating cycle is used to read input module status information and write output status information to the output modules.

Port Servicing

During the port-servicing portion of the operating cycle, the processor services the RS-232C port and the peripheral port.

THE WATCHDOG TIMER

During normal operation, the processor scan is completed and started again in an orderly, predictable cycle. Inputs must be read and outputs written to, or updated, predictably. A control system with unpredictable control of I/O would be unreliable and unsafe. To ensure system predictability, a **watchdog timer** is used to ensure that the processor completes each scan in a timely manner.

The watchdog timer is typically a hardware timer incorporated into the CPU's circuitry that monitors the cyclical process, or scan, of the processor or CPU. The processor must complete the input scan, program execution, output scan, communication update, housekeeping, and processor overhead before the watchdog timer times out. The watchdog timer is a safeguard that verifies that the processor does not become stuck while scanning the user program, or for some other reason become unable to complete the current scan. Being unable to complete the scan means that the processor has no control over real-world outputs. Real-world outputs without control might be left running when they were to be stopped. Field hardware devices that run without control of the processor could spell disaster.

The watchdog timer is reset at the conclusion of each scan cycle by the processor when the scan time is less than the watchdog timer's preset time. If the processor scan is not completed and the watchdog timer not reset before the watchdog timer times out, the processor will fault. If the processor faults, it enters fault mode and all outputs will be turned off.

A program that faults due to the watchdog timer timing out may have become caught in an endless loop within the program. This could be the result of programming problems, such as the use of too many uncontrolled backward jumps where there is no defined jump out of the loop. Using subroutines where too much time is spent within one or multiple subroutines could also result in the **program scan** time exceeding the watchdog timer time value. In some instances, lengthening the watchdog timer's preset value can solve the problem.

Some PLCs have watchdog timers that have fixed time intervals, while others are adjustable within specific limits. A typical default time of 100 or 200 milliseconds is standard for many PLCs with either fixed or variable watchdog timing cycles. The General Electric Series 90-30 watchdog timer is fixed at 200 milliseconds. In the Series 90-70, the big brother to the Series 90-30 PLC, the watchdog timer default value is 200 milliseconds. The Series 90-70 watchdog timer value is adjustable by the user in the programming software. The valid range for a Series 90-70 PLC watchdog timer is from 10 milliseconds (ms) to 2.55 seconds (s).

The Allen-Bradley SLC 500 PLC's watchdog timer default value is 100 milliseconds. The SLC 500 stores the watchdog timer value in a specific word in the processor status file. The watchdog scan time byte can be modified by the user from a minimum of 20 ms to a maximum of 2.5 s. A user would adjust the watchdog timer when running longer, more complex programs.

As an example, in the SLC 500 processor status file, address S:3, the high byte of the 16-bit word is the variable watchdog time value. The watchdog scan time byte can be modified by the user from a minimum of 20 ms to a maximum of 2.5 s. Changing the watchdog timer value for an SLC 500 is as easy as double-clicking on the processor status folder in the RSLogix 500 project tree and selecting the scan times tab. Figure 10-4 illustrates the scan time tab in the processor status file.

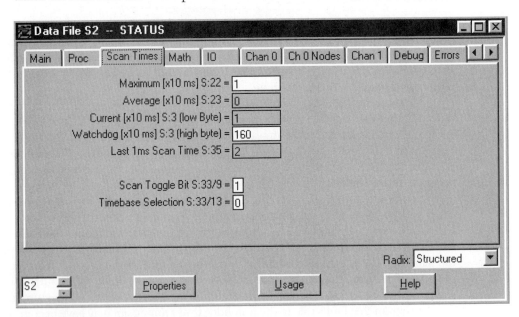

Figure 10-4   RSLogix 500 software processor status file screen printout.

The notation of the high byte of word S:3 is typically listed as S:3 (high byte). This signifies the upper eight bits of status word S:3, which is the current value of the watchdog timer in milliseconds. The low byte of S:3 is signified as S:3 (low byte). This signifies that the lower eight bits of status word S:3 contain the last or current processor scan time (see Figure 10-4).

The value contained in S:3 (low byte) is the elapsed program scan time. At the end of each scan, this byte's value is zeroed by the processor immediately preceding execution of rung zero of the main program. As long as the processor is in run mode, this byte is incremented every 10 milliseconds. If this value equals the value in S:3 (high byte) and the processor has not completed its scan and reset the watchdog timer value to zero, the processor will fault. To remove the fault, go to the processor status screen and the errors tab. Figure 10-5 illustrates the error tab in the status file. Notice the processor FAULTED

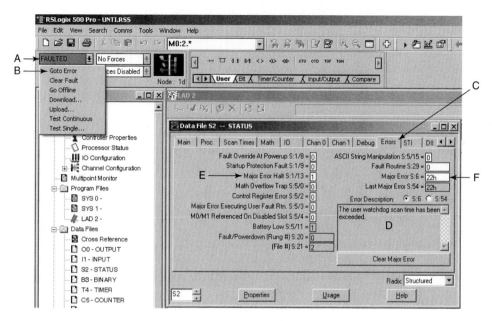

Figure 10-5   SLC 500 processor status file errors tab.

display in the on-line toolbar, A in Figure 10-5. Click on the down arrow to the right of the word FAULTED and the drop-down menu, B, will appear. Click on "Goto Error" and the errors tab in data file S2 will display, C. Notice the Error Description Window (D) and the message it contains. Notice that the Major Error Halt bit, S:1/13 (E), is set to a one, and a Major Error code is displayed (F). The address where the error code is stored is S:6. Click on the Clear Major Error button directly below the error description box, to clear the error and return the processor to program mode.

PROCESSOR PORTS

The physical connection between a processor and the outside world is called a "communication port." Think of this port like a port where ships unload and load cargo. The communication port is simply the point where communication signals enter or leave a computer device such as a personal computer, industrial computer, or PLC. Figure 10-6 illustrates the communication ports on three SLC 500 processors, the 5/03, 5/04, and 5/05; one PLC 5 processor; and a ControlLogix processor.

The SLC 5/03 processor has a serial port plus a Data-Highway 485 (DH-485) port. DH 485 is the native SLC 500 network. The SLC 5/04 processor has a serial port and a Data Highway Plus (DH+) port. DH+ is the native PLC 5 network. The 5/04 processor can be directly connected to a PLC 5 system and its DH+ network. The SLC 500 5/05 processor has an Ethernet port in addition to the serial port. The modular ControlLogix processor only has a serial port. Other network communications to the ControlLogix will be made through a separate communication module inserted into the ControlLogix chassis. The specific PLC 5 processor selected will dictate the specific processor communication ports.

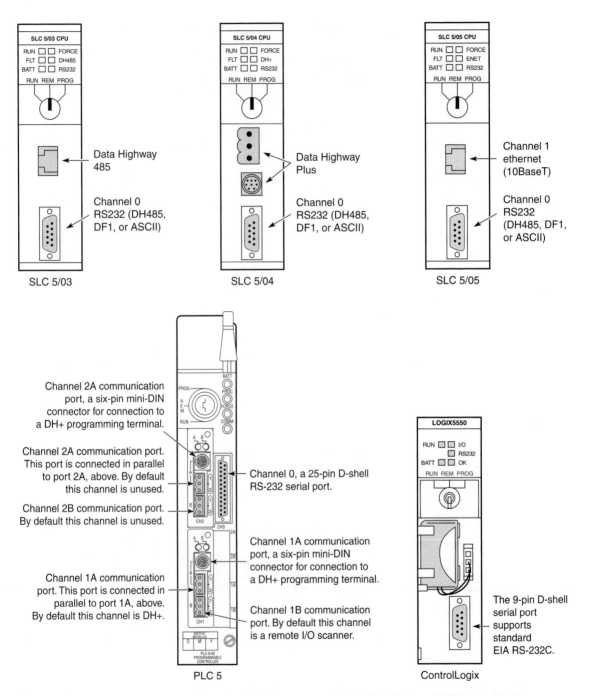

Figure 10-6 Communication ports on SLC 500 processors, the 5/03, 5/04, and 5/05; one PLC 5 processor; and a ControlLogix processor. (Used with permission of Rockwell Automation, Inc.)

The PLC 5 processor illustrated would represent a PLC 5/40, PLC 5/60, or PLC 5/80. If a ControlNet PLC 5 processor were selected, such as a PLC 5/40C, channels 2A and 2B would be replaced with ControlNet redundant bayonet connectors, whereas an Ethernet PLC 5 processor such as a PLC 5/40E would have an Ethernet connection.

For two devices to communicate with each other, they must be connected so that electrical signals transmitted by one device are received by the other. Communication between two devices is accomplished either by direct connection or by indirect connections through intermediate communication devices. Direct connection between two devices is achieved by directly running a cable from one device to the other.

There are two types of communication hardware: data terminal equipment (DTE) and data communication equipment (DCE). Data terminal equipment consists of terminals, computers, PLC processors, and other devices that serve as the source or final destination of data communicated between two or more pieces of hardware. Data communication equipment, on the other hand, usually consists of two modems that are intermediate communication devices between two pieces of data terminal equipment. The intermediate devices usually convert communication signals from the communicating device to a form that can be transmitted over the selected transmission medium. A typical intermediate communication device is the modem (MOdulator DEModulator), which is used to convert the signals from the data terminal equipment into signals suitable for transmission over phone lines, fiber-optic cables, or radio frequencies if transmitting via radio frequency (RF) modems. The receiving end of the phone line or radio frequency receiver will include another modem of the same type for converting the transmitted signals back to signals that the data terminal equipment can understand. For different manufacturer's devices to communicate, there need to be commonly accepted communication standards.

THE RS-232C COMMUNICATION STANDARD

Standards have evolved to ensure that equipment from different manufacturers will be able to communicate. The most widely used is the RS-232C standard. Originally, the RS-232C standard (usually called simply RS-232) was devised to specify connections between terminals, data terminal equipment, and modems and other data communication equipment. The standard was established by the Electrical Industries Association to define electrical, functional, and mechanical characteristics for asynchronous communication transmissions between a computer (DTE) and a peripheral device (DCE).

The letters RS stand for "recommended standard." The C portion of RS-232C identifies the third revision of the standard. Asynchronous communication transmissions are those in which characters may be sent randomly or at unequal time intervals. Each character will also have start and stop bits used for transmission control.

With the evolution of microcomputers, printers, and other microprocessor equipment like PLCs, the RS-232 standard has been carried forward to include these devices, too. The standard specifies electrical characteristics of the communication link between two hardware devices. It also gives names and numbers to the wires used in the cable for joining the two communicating devices. RS-232C uses either a 25-pin or a 9-pin D-shell connector. (In many applications the 9-pin connector is most common.) Figure 10-7 illustrates the typical RS-232 9-pin connections and identifies the pins.

Figure 10-7   DB-9 pin connector showing typical RS-232 connections for IBM personal computers.

RS-232 DB-9 Pin-outs

The wires in a nine-pin connector are identified by the numbers assigned to each of the connector's pins. Rather than identifying wires, the pins in the connector are used for identification. This identification is referred to as the cable's pin-outs. Figure 10-7 illustrates the pin identification methodology. The function of each pin is identified as follows:

> Pin 1 is the Carrier Detect signal.
>
> Pin 2 is the Data Set Ready signal.
>
> Pin 3 is TXD (transmitted data).
>
> Pin 4 is the Data Terminal Ready signal.
>
> Pin 5 is the signal ground.
>
> Pin 6 is Data Set Ready.
>
> Pin 7 is the Request to Send signal.
>
> Pin 8 is the Clear to Send signal.
>
> Pin 9 is the Ring Indicator signal.

Each of these signals and its functions will be introduced as we proceed through the chapter.

One of the simplest methods to communicate between a PLC processor and a programming terminal is to connect the proper cable directly to the COM 1 port of the programming computer terminal and the other end to the RS-232 serial communication port. Not all PLC processors have an RS-232 communication interface, however.

SERIAL COMMUNICATION BETWEEN A PERSONAL OR INDUSTRIAL COMPUTER AND THE PLC

Many newer PLC processors come with RS-232 serial communication ports. When communicating between a personal computer or industrial computer and a PLC, we need the capability to download information from the computer terminal to the PLC processor. We also need to upload programs and program-related data from the PLC and the computer terminal. When data is transferred in two directions, this is called two-way communication.

It might seem that any PLC processor with an RS-232 serial port could communicate with any other RS-232 port. This is not true, however. When configuring a communication link between two devices, such as a computer and a PLC processor, there are two important aspects of the communication link. First is the communication standard.

The R-232 Communication Standard

The RS-232 communication standard defines only the physical cable connections, the use for each of the nine wires inside the standard communication cable, and their associated connector pins. Remember, when referring to a communication cable, the pin's connector pins are used for identification. The standard does not define how many pins and wires must be used.

The minimum configuration for two-way communication only requires three pins in the 9-pin D-shell connector. In a typical RS-232 connection, the computer (DTE) device uses pin 2 for data output and peripheral (DCE) equipment uses pin 2 for data input. For sending data back from the peripheral (DCE) device to the computer (DTE) device, pin 3 is data output from the peripheral device, while pin 3 is data input on the computer device. Pin 7 is used as the ground. Figure 10-8 illustrates the minimum connections between the computer and peripheral equipment. These wires go directly between the two devices, pin for pin; this is referred to as a straight-through connection.

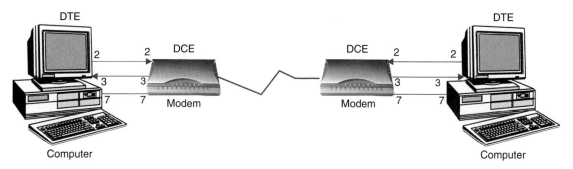

Figure 10-8 Straight-through cabling between computer and peripheral equipment.

For ease of connection the RS-232 standard specifies that computer devices have male connectors, whereas peripheral equipment has female connectors. When communicating between a personal or industrial computer terminal and a PLC processor (also a computer) there is no intervening peripheral equipment. If the same straight-through cable were used to connect the computer terminal to the PLC processor (refer to Figure 10-8), we would be connecting pin 2 of one computer to pin 2 of the other computer, as illustrated in Figure 10-9.

A computer's pin 2 is for outgoing data. Figure 10-9 shows that both computer devices are sending output data out pin 2 and attempting to use each other's pin 2 as an input for data. Both computer devices are also looking for input data on pin 3. In this cabling configuration, pin 3 from each computer device is connected to the other. This type of connection will obviously not allow any communication between the two devices. We need to change our cabling so the output of one computer, pin 2, is connected to the input, pin 3, of the opposite computer. To do this, pin 2, the output from each computer device, must cross and connect to the input, pin 3, of the opposite computer device. Now

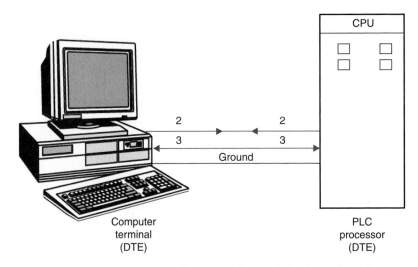

Figure 10-9 Connecting two computer devices with a straight-through cable.

communication sent by one computer device can be received by the other. Data sent back by the receiving computer device can be received by the originator of the transmission. Figure 10-10 illustrates the necessary minimum connection to communicate between two computer (DTE) devices. (Remember that the RS-232 standard specifies that computer connections are male.) The common name for the communication cable illustrated in Figure 10-10, where wires 2 and 3 are crossed, is a "null-modem cable."

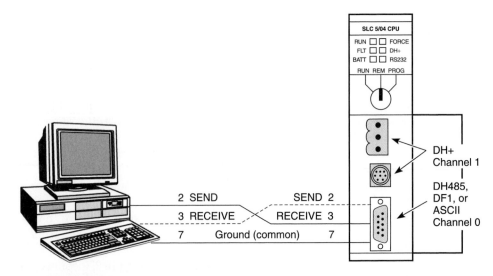

Figure 10-10 Wiring connections between a personal computer and a 5/04 processor using a null-modem cable.

The null-modem cable will have female 9-pin or 25-pin D-shell connectors on each end. This cable is called a null-modem cable because it replaces two modems.

Handshaking

Some manufacturers use more than three pins in their RS-232 communication connection. Additional pins and their respective signals are used to get verifiable communication between devices. Verifiable communication between devices is possible when the two devices cooperate with each other in the exchange of data. This cooperation is called "handshaking." Figure 10-10 has no handshaking. Data is simply sent out pin 2, without verification that the data is being received. Since there are no signals coming back from the receiving device, the sending device sends what it has and assumes the information was received. What if the receiving device had been turned off or broken, or the communication cable was bad or disconnected? What if the receiving device was not ready to accept data? To circumvent these problems, information is sent back from the receiving device to the sending device to indicate when it is ready to accept data, when data has been received, and when it is clear to send additional data. Handshaking is sometimes called "flow control."

There are two types of handshaking, hardware and software. Software handshaking consists of special characters transmitted across the data lines, pins 2 and 3, rather than separate handshaking lines. Software handshaking is popular as no additional wires are necessary.

One of the more popular software handshaking protocols is XON/XOFF. This is an asynchronous communication protocol whereby the receiving device controls the flow of data from the transmitting device by sending control characters. When the receiving device's buffer is almost full, an XOFF (ASCII DC3) character will be sent to the transmitting device and the transmitting device will stop sending data. The receiving device will send an XON (ASCII DC1) character when the transmitting device is to start sending data again. These characters that control the flow of data are called "control characters."

Hardware handshaking uses dedicated handshaking lines to control the transmission of data. Peripheral equipment uses pin 6, "Data Set Ready," as the primary handshaking line to tell the computer that it is ready to control transmissions. Pin 5, "Clear to Send," is used to control transmissions from a computer device. Computer equipment uses pin 20, "Data Terminal Ready" (DTR) as its main handshaking line. DTR simply tells the peripheral that it is ready to receive data. DTR also uses pin 4, which serves as a "Request to Send" handshaking line. In addition to the basic communication lines illustrated in Figure 10-10, handshaking lines are illustrated in Figure 10-11.

When using modems, two additional lines are used for communication. Pin 8 is the "Carrier Detect" signal line. Carrier detect is used to indicate the presence of a carrier signal. Pin 22 is the ring indicator and is used to indicate that the modem is being called by a remote device over phone lines and would be ringing if it were indeed a telephone.

How Handshaking Works

Electrical, or hardware, handshaking involves signals being exchanged on additional wires to communicate the status of each device to the other. Basically, handshaking works as follows.

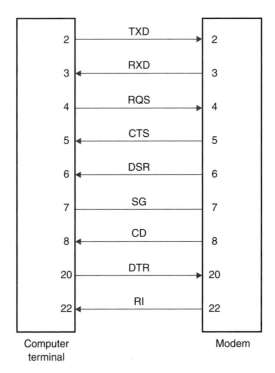

Figure 10-11   The nine RS-232 standard connections. Nine-pin D-shell connectors are typically used on each end.

The sending computer sends a message to the receiving device informing the receiving device that it has a message to send. This is done by the sending computer setting pin 4 high. This is called the request to send signal.

If the receiving device is ready to accept the data that the sending computer wants to send, the receiving device sets the clear to send pin (pin 5) high. Receiving this signal, the sending computer now knows that the receiving device is there, the cable is good, and it is OK to transmit the message or data. Other handshaking lines are used as secondary lines and are not always used.

Communication Protocols

Not all PLC processors support RS-232 communication. Moreover, even if two different PLC processors both support RS-232 communication, there is one other consideration when connecting two devices together: the protocol. The protocol is a set of rules that govern the way that data is formatted and timed as it is transmitted between the sending and receiving devices. Each manufacturer designs a protocol that defines data format, timing, sequence, and error checking. As a result, one processor from one PLC manufacturer will probably not be able to talk to a processor from another manufacturer, even if they both support RS-232 communication standards, as their protocols will differ.

Even processors from the same manufacturer but different PLC families may not be able to communicate because they have different protocols. As an example, an Allen-Bradley PLC 5 processor (with the Data Highway Plus network) cannot talk directly to an Allen-Bradley SLC 500 processor (with the Data Highway DH-485 network), as they have different protocols. Protocol converters are available in some cases to allow different devices with different protocols to communicate. A protocol converter is called a "bridge" because it connects two dissimilar networks.

Rockwell Automation's ControlLogix PLC can be used either as a PLC, a gateway, or a combination of both in the same chassis.

A gateway is a protocol translator. The ControlLogix backplane is a ControlNet network, also called a ControlBus. The backplane acts as a gateway that translates the communications from one communication module to either the controller or out a different communication module to its respective network. As an example, one might wish to use Ethernet communications for upload, download, or going on-line to the ControlLogix processor. This will require an Ethernet communication module to be inserted into a chassis slot. This communication module is called a bridge module.

A bridge module in a ControlLogix PLC is a communication module, in this case, Ethernet, when inserted into the chassis to "bridge" from Ethernet to the backplane of the ControlLogix chassis, which is a ControlNet network. The backplane translates the data from one network to a different network.

Separate communication modules are available for installation in the ControlLogix chassis to configure a gateway or bridge to route information between different networks. Messages are sent directly from one communication module to another across the backplane. Current ControlLogix communication, or bridge modules, include ControlNet, Data Highway Plus—Remote I/O, DeviceNet, or Ethernet/IP. There is also a linking device for DH-485 and Foundation Fieldbus.

Since a ControlLogix modular processor contains one port, a serial port, any other network that needs to be interfaced to the ControlLogix processor will require a bridge module in the chassis. The bridge module can be used to interface to a personal or industrial computer for programming or monitoring.

Now that we have looked into communicating between a personal or industrial computer and a PLC processor, we need to investigate processor-operating modes and how these modes correlate with communication between a computer and the PLC processor.

PROCESSOR-OPERATING MODES

A processor has two modes of operation. Either the processor is in program mode or it is in some variation of run mode. Different processors and different manufacturers will have different operating modes. The most common modes are *run mode* and *program mode*. With the ability to connect to a personal computer for programming and program monitoring came **remote run mode** and remote programming mode. In addition, some processors have test mode and single-step mode.

Allen-Bradley SLC 500, PLC 5, and ControlLogix Processor-Operating Modes

The operating modes of the Allen-Bradley PLCs are explained in the following paragraphs.

Program Mode
When the processor is in program mode, it is accepting instructions, either as a new program or as changes to instructions (adding instructions, or rungs, or deleting instructions, or rungs, of an existing program). The processor, or CPU, must be in program mode. Changing or adding instructions or rungs to an existing program is called "editing." Most processors cannot edit a program and have the program running at the same time.

Run Mode
After all instructions have been entered in a new program or all editing has been completed to an existing program, the processor is put in run mode. While in run mode, a processor is executing the instructions programmed on the rungs of ladder logic that make up the user program. In run mode, the processor or CPU is in its operating cycle, which is called the processor scan or sweep.

Remote Run Mode
Some processors have keyswitches built on the front of the processor module. This keyswitch can be used to put the processor in run or program mode. Using the keyswitch to change modes is called operating in "local mode." If a PLC processor is put in run mode when programming or monitoring a PLC program from an industrial or personal computer, the PLC is in remote run mode.

Remote Program Mode
If the processor is put in program mode from a remote programming device, it is said to be in "remote program mode." Any programming operations can be accomplished from a computer while it is connected to the processor that is in remote program mode.

Test Mode
After developing a user program or when done editing a user program, it is a good idea to test program execution before allowing the PLC to operate the actual hardware. Most test modes operate much like run mode with the exception of actually energizing real-world outputs. The processor still reads inputs, executes the ladder program, and updates the output status tables, but without energizing output circuits. As an example, newer Allen-Bradley SLC 500 processors have the following options available for testing programs before they are put into production.

Single-Step Test Mode
Single-step test mode is used to verify ladder operation. This feature is typically used for checking new program development for proper operation or troubleshooting. A single-step test directs the processor to execute a selected single rung or group of rungs. The output points are disabled.

Single-Scan Test Mode

Single-scan test mode executes a single processor operating scan or cycle. Inputs are read, the ladder program is executed, and the output status table or file is updated. The output circuits are not energized.

Continuous-Scan Test Mode

The continuous-scan test mode is used to continuously run the program for program checkout or troubleshooting. The only difference between normal run mode and continuous-scan test mode is that in the latter, although the output status table or file is updated, the output points are not energized.

As of this writing, ControlLogix has one test mode that is similar to continuous test mode in a PLC 5 or SLC 500. ControlLogix test mode is different than traditional Rockwell Automation PLCs. In test mode, each ControlLogix output point state is determined by how the output state during program mode parameter is set up in the RSLogix 5000 I/O configuration for the PLC project. Selections are on, off, or hold, and are configured by the user as part of the software I/O configuration.

General Electric Series 90-70 CPU Operating Modes

The General Electric Series 90-70 has the following operating modes.

Run with Outputs Enabled

With the processor in this operating mode, inputs are read, the ladder program is executed, the output status table is updated, and outputs are turned on or off. This is the normal processor-running mode.

Run with Outputs Disabled

This operating mode is used to continuously run the program for program checkout or troubleshooting. The only difference between normal run mode and the run with outputs disabled mode is that in the latter, although the output status table or file is updated, the output points are held in their default state. Thus, the outputs are not changed as a result of the program.

Stop Mode

In stop mode, the processor will only communicate with the programmer, with devices connected to the serial port, and with faulted I/O modules (for recovering from a fault condition).

Stop and I/O Scan Mode

Stop and I/O scan mode operates the same as the stop-operating mode, but with the I/O scan running. The output status table will not be updated as a result of the ladder program being solved. Outputs will be frozen at the ON or OFF states that were in the output status table when the transition to stop and I/O scan occurred.

Run Mode Store Function

The run mode store function provides the ability to transfer program logic, system configurations, and reference table data to a running PLC processor. When executing this function, General Electric's Logicmaster software switches the processor into stop and I/O scan mode to make the transfer. When the transfer is in process, the processor scan, or sweep, is stopped. Typical transfer time is approximately one to ten seconds, depending on program size and the communication baud rate. When the transfer is complete, the software automatically switches the processor back into run mode.

ON-LINE EDITING

If you wanted to edit the program currently running in the processor to which your computer is connected, you would have to go off-line and edit the PLC ladder program in your computer's memory. When completed editing, you would go on-line, put the receiving processor in program mode, and download the edited version to the processor from the computer's hard drive. When the program had been downloaded, the processor would be put back into run mode.

The problem with **off-line** editing is that the process the PLC is controlling has to be shut down while the edited program is downloaded into the processor. In today's manufacturing environment, shutting down an operating manufacturing line to make a couple of edits to the existing program is an expensive and time-consuming proposition.

Many newer processors allow the operator to edit an executing program while on-line with the processor in run mode. This is called **on-line** editing. Simply put, on-line editing allows the operator the ability to edit a copy of the current processor's program while the processor is still running. When editing is completed, the edited program is loaded into the running processor "on-the-fly," without removing the processor from run mode.

BATTERY BACKUP FOR THE PROCESSOR'S VOLATILE MEMORY

After the user program has been loaded into the processor's main memory, the processor is put into run mode. But what about the permanence of the read/write memory? We have learned from previous lessons that read/write (RAM) memory is volatile and not permanent through a power failure or shutdown of the PLC system. Almost all processors have battery backup provisions for their volatile memory.

The read/write memory (RAM) in your processor makes it possible to enter and change your user program, and to store data in memory. This memory is referred to as volatile memory, as there must always be power applied to the memory's integrated circuit chips. As soon as power is removed, or lost through a power failure, volatile memory chips forget what was stored in them. A PLC would be of little value if every time you shut off the disconnect or there was a power failure (from either blown fuses or a tripped circuit breaker) or power interruptions (from the power utility or weather-related power outages), the operator had to reenter the user program and any data stored in the data tables. To circumvent memory failure problems, backup power for PLC read/write memory is provided by either capacitor backup with an optional battery (for many fixed PLCs) or a replaceable battery (for modular units).

Newer memory chips are much more energy-efficient. Since these new memory chips use less power, many of the newer, fixed PLCs have built-in capacitors with optional battery backup for memory retention. When the PLC is powered by line voltage, the capacitor charges up. When power is lost, the capacitor begins discharging, providing power to the volatile memory chips. As an example, the SLC 500 fixed units have a built-in capacitor for memory retention. Battery installation in these fixed PLCs is optional. Figure 10-12 illustrates battery placement on a fixed SLC 500 PLC.

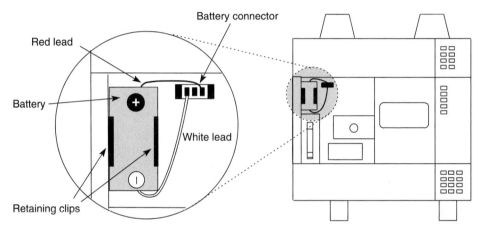

Figure 10-12   Battery placement in an Allen-Bradley SLC 500 fixed PLC. (Used with permission of Rockwell Automation, Inc.)

Capacitor backup time is dependant on the ambient temperature along with the size of the capacitor and resistive element and the age of the unit. If the ambient temperature is 25 degrees C (77 degrees F), the capacitor will hold memory for about 30 days. If it is 60 degrees C (140 degrees F), memory retention will be degraded to about 5 days. If this is not adequate, the optional battery may be installed. Typical battery life is about five years.

The Allen-Bradley MicroLogix 1000, a fixed PLC, uses a built-in EEPROM for memory retention; thus, no battery or backup capacitor is required.

Today many PLC manufacturers use lithium batteries to provide backup power for volatile memory. Battery life varies between processors and is usually determined by the amount of memory that is backed up and the amount of time the battery is actually providing power to the memory chips. As a rule, if the PLC is being powered by line power, the battery is not being used to supply power to memory. Typically, lithium batteries will last between one and five years.

Most processors have a battery condition status indicator light (an LED). Normally this light is in the OFF condition because the battery is not in use. When the battery voltage falls below the acceptable threshold, the battery condition status indicator light will come on. When the indicating light comes on, you have a few days to change the battery before there is a danger of memory loss. Check your particular processor's specifications for the amount of time after which there is the danger of memory loss. To change the battery, consult your manufacturer's documentation and procedures.

Changing the battery on an Allen-Bradley SLC 5/01 or 5/02 processor is as simple as unplugging the battery behind the door on the front of the processor and installing a new one. The PLC need not be powered down. Battery replacement on an Allen-Bradley SLC 5/03, 5/04, or 5/05 processor is a bit more tricky. The battery on these processors is located on the processor's printed circuit board. Power down the PLC and remove the processor from the chassis. The battery is found on the circuit board in the upper left-hand corner. Simply disconnect the plug and remove the old battery from the retaining clips. Then install the new battery in the reverse order. There is a capacitor built into the processor that will hold the memory for up to 30 minutes while you change the battery, so there is no danger of memory loss during battery removal and installation. A battery will last about two years when installed on a 5/03, 5/04, or 5/05 processor.

LITHIUM BATTERY HANDLING AND DISPOSAL

Used lithium batteries are considered hazardous waste. When lithium batteries are used there are special precautions necessary with their handling, transporting, and disposal.

Follow the following guidelines in handling or disposing of lithium batteries:

1. These batteries are not rechargeable. Attempting to recharge a lithium battery could cause the battery to overheat and possibly explode.
2. Do not attempt to open, puncture, or crush. The possibility of an explosion and/or contact with corrosive, toxic, or flammable liquids could result.
3. Do not incinerate batteries. High temperatures could cause the battery to explode.
4. Severe burns could result if the positive or negative terminals are shorted together.
5. The U.S. Department of Transportation (DOT) regulates the transportation of materials such as lithium batteries within the United States. As a general rule, lithium batteries cannot be transported on a passenger aircraft. They may be transported via motor vehicle, rail freight, cargo vessel, or cargo aircraft. Before shipping lithium batteries, check on the proper procedures, packaging, and hazard identification.

As a general rule, for disposal procedures, contact your city or county recycling authority for the correct methods of disposal in your area.

In many situations, a machine builder does not want to have to worry if the end customer will be responsible for replacing a low battery. In addition, if the end customer is negligent in replacing the battery and power to the PLC is lost or interrupted, causing the PLC program to be lost, who will pay for the service call to reload the PLC program? Because most machine builders consider their user programs proprietary, they would not wish to mail the end user a floppy disk to reload the program. To circumvent these types of problems, an optional, nonvolatile memory chip is available for most processors.

NONVOLATILE MEMORY: EEPROM

If nonvolatile memory is an option for your PLC, there will be an empty memory chip holder on your fixed PLC or on the processor module. EEPROM chips are available from many PLC manufacturers as a field-installable option. Figure 10-13 illustrates an EEPROM memory chip mounted on a plastic carrier. This assembly is called a memory module.

Figure 10-13 EEPROM memory module for SLC 500 5/01 and 5/02 processors. (Used with permission of Rockwell Automation, Inc.)

There are two types of nonvolatile memory chips available for PLCs, EEPROMs and PROMS. Many newer PLCs allow the EEPROM installed on the processor to be programmed and reprogrammed by simply selecting the proper function from the programming software and downloading the new or edited program into the EEPROM chip. The disadvantage to this is that machine builders must password-protect the program on the EEPROM to avoid the possibility of the end user writing over it. Figure 10-14 illustrates installation of an EEPROM memory module on a fixed SLC 500 PLC.

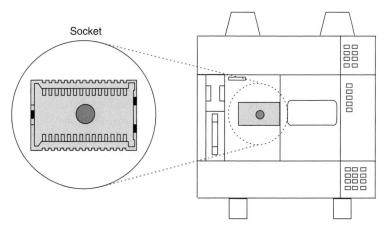

Socket

Figure 10-14 Placement of EEPROM on a fixed SLC 500 PLC. (Used with permission of Rockwell Automation, Inc.)

PROMs have the advantage of not being able to have the program written over in the field. A machine builder can feel relatively confident that the end user will not be able to write over the user program as the PROM chip is not electrically erasable. Thus, the stored program, although nonvolatile, cannot be changed by simply downloading a new program from the programming software. Some PROMs can be reburnt and loaded with a new program. To reprogram a PROM, the chip needs to be removed from the processor and placed in a PROM eraser. When the PROM has been erased, a new program can be loaded into the PROM chip using a PROM burner (a special device that is specifically used to write, or burn, a program into a PROM chip).

There are many advantages to using a nonvolatile memory chip:

1. PROMs or EEPROMS are nonvolatile memory chips. There are no battery problems since a battery is not necessary for memory retention.

2. If a machine builder needs to upgrade the program contained on a nonvolatile memory chip, it is simple to produce new chips with the current version of the program on it and send them out to the end user. The end user simply shuts down the PLC, removes the old memory chip(s) from the processor, installs the new memory chip(s), and powers up the unit. If programmed to do so, the processor will automatically upload the new program from the EEPROM or PROM into processor memory and go into run mode. By receiving the old EEPROM or PROM chip, the machine builder can feel confident that the program was successfully upgraded and can use the EEPROM or PROM chip over for another upgrade.

3. These chips offer automatic upload on start-up or RAM memory error.

A major advantage when using nonvolatile memory in a PLC is the option for automatic program load on start-up or RAM memory error. If the PLC is in a remote location such as an unmanned facility, an automatic program load can be a real benefit. This could eliminate a field service engineer from making a service call to reload memory and start the PLC. A machine builder's remotely installed, PLC-controlled equipment can automatically load nonvolatile memory into the processor's user memory under the conditions listed below.

These automatic memory load functions can be set up in the SLC 500 status file by setting the correct bits in the proper words when the user program is developed, as illustrated in Figure 10-15.

1. when a processor memory error or memory corruption is detected, S:1/10

2. whenever processor power is cycled and the processor status is dependent on mode before power down, S:1/11 (see SLC 500 Reference Manual)

3. whenever processor power is cycled and the processor automatically goes into run mode, S:1/12

Some newer ControlLogix processors have the option to use the newer industrial compact flash cards for nonvolatile memory storage. ControlLogix processors require a minimum firmware version of eleven to use the compact flash card. In the case of a 1756-L61, L62, or L63 processor, a 64 Mbyte CompactFlash card can be installed as illustrated in Figure 10-16.

The top portion of the figure shows the processor on its side and a wire locking clip that must be unlocked before the compact flash card can be installed. The bottom portion illustrates insertion of the compact flash card. Once inserted, the wire locking clip is moved downward until it snaps over the card, locking it into position. Refer to Figure 10-17 for a picture of a 64 MB compact flash card.

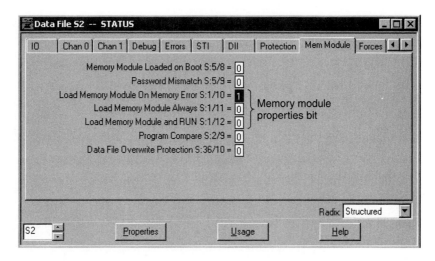

Figure 10-15   RSLogix 500 software properties screen for memory module setup.

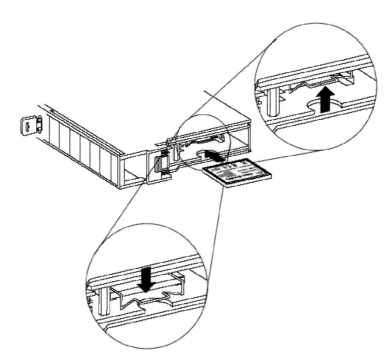

Figure 10-16   ControlLogix industrial CompactFlash nonvolatile memory card installation. (Used with permission of Rockwell Automation, Inc.)

Figure 10-17 Rockwell Automation 64MB capacity 1784–CF64 Industrial CompactFlash card. (Used with permission of Rockwell Automation, Inc.)

Setting up the properties of the card is similar to the process for the SLC 500. The nonvolatile memory tab in the RSLogix 5000 software Properties view provides the user three selections:

Load from card at power up

Load from card on corrupt memory

User initiated memory load

PASSWORDS

A machine builder will usually want to protect proprietary programs from modification or being copied by another party. A "password" is used to limit or deny access to a PLC program and data files. Passwords are placed on a program to deny access by outside individuals. Some manufacturers offer different access levels of password protection that can be placed on their processors and the resident programs.

Typically a company will design and build a machine for a specific purpose. It will develop the user program and load it into the PLC as part of the build stage. After the machine and program have been tested for proper operation, the machine is shipped to the end customer. In some instances, the end customer may decide to improve the original equipment manufacturer's PLC user program. If the end user was able to make PLC program changes and someone was hurt as a result, who is responsible? Even though the end user modified the original program, chances are that the original manufacturer will end up in court. Passwords are placed on PLC programs for just this reason.

PLC passwords only restrict access to the PLC processor and its resident program and associated data tables or files, not the programming software on your personal computer.

Passwords are usually placed on a user program after the program has been developed, tested, and verified for proper operation. After password protection has been enabled, the level of password protection is selected. Older PLCs had passwords that restricted anyone without the password access to any part of the user program or data tables. As processors increased in functionality, however, access levels were added to password protection.

Some manufacturers offer passwords and master passwords. In large PLC systems there will be many processors. With many processors, a master password can be put on all processors while each processor also has a separate password of its own. The master password will override a separate processor password. A master password allows an authorized individual to access any processor without remembering all the individual passwords.

Password access levels are provided so that a machine builder can limit access to certain parts of the program, such as modification of the user program, data file modification, access to memory module overwrite, or ability to force I/O or reconfigure communication channels.

As an example, General Electric Series 90-30 PLC passwords have four access levels. Each successive access level includes all the privileges of the lower levels. Level four is the default if there are no passwords assigned. Series 90-30 password access levels are listed below:[1]

Access level four	Write to all configuration logic. Configuration may only be written in stop mode; logic may be written in stop or run mode. Display, set, or delete passwords for any level.
Access level three	Write to any configuration or logic, including word-for-word changes, the addition or deletion of program logic, and the overriding of discrete I/O.
Access level two	Write to any data memory, except overriding discrete I/O. The PLC can be started or stopped. PLC and I/O fault tables can be cleared.
Access level one	Read any PLC data except passwords; no PLC memory may be changed.

Forgotten Passwords

The reason passwords are placed on PLC programs is to limit access to the programs and data tables. As a result, it is not easy to get around a password. In many instances if you forget a password on a program that you legitimately have a right to edit, you will have to contact the manufacturer. The manufacturer may send out a representative to use a backdoor password to enable access to your program, assuming you can verify need and ownership. Another possible option is to start over and download a new copy of the program over the password-protected one.

1. Information from General Electric, *Logicmaster 90-30/20/ Micro Programming Software User's Manual*, Publication Number GKF-0466F (General Electric, October 1994), p. 5-4.

THE ALLEN-BRADLEY SLC 500 MODULAR PROCESSORS

We have looked at the general characteristics of PLC processors. Next we will look at physical features and specifications for selected PLC processors.

The Allen-Bradley SLC 500 family of modular PLC processors includes the 5/01, 5/02, 5/03, 5/04, and 5/05 processors. Each processor has enhanced capabilities over its predecessor.

The SLC 5/01 modular processor has the same instruction set as the SLC 500 fixed controller. SLC 5/01 specifications are listed in Figure 10-18.

ALLEN-BRADLEY SLC 5/01 PROCESSOR SPECIFICATIONS		
Part Number	1747-L511	1747-L514
Program Memory	1 K	4 K
Local I/O Capacity	256 discrete	480 discrete
Remote I/O	None	None
Local Chassis	3 chassis	3 chassis
Total Modular Slots	30	30
Typical Scan Time	8 milliseconds per K (thousand) instructions	8 milliseconds per K (thousand) instructions
Programming Instructions	52	52
Memory Backup	EEPROM UVPROM	EEPROM UVPROM
RAM Backup	Capacitor, up to 2 weeks	Lithium battery, up to 2 years
Programming Methods	Handheld terminal or software	Handheld terminal or software

Figure 10-18 Overview of SLC 5/01 processor specifications. (Data compiled from Allen-Bradley data sheets)

The SLC 5/02 modular processor has an expanded instruction set, increased diagnostics, additional memory, and additional I/O, along with additional communication options as compared to the SLC 500 fixed controller and SLC 5/01. An SLC 5/02 and 5/03 processor specification overview is listed in Figure 10-19.

Figure 10-20 illustrates a front and side view of the SLC 5/02 processor series B and series C. Notice the memory module socket for nonvolatile memory chip installation, battery for memory retention, and DH-485 channel communication port.

Allen-Bradley SLC 5/03 Processor

Allen-Bradley's SLC 5/03 processor made a major leap in PLC processor functionality. This processor's memory was expanded up to 16,000 (16 K) words of program memory. With this increase in memory size also came an increase in processor scan speed. A 1 K

ALLEN-BRADLEY SLC 5/02 and 5/03 PROCESSOR SPECIFICATIONS		
	SLC 5/02	**SLC 5/03**
Part Number	1747-L524	1747-L531 1747-L532
Program Memory	4 K	L531 = 8 K L532 = 16 K
Communications	DH-485	Channel 0: RS-232 (DF1, ASCII, DH-485 Protocols) Channel 1: DH-485
Local I/O Capacity	480 discrete	960 discrete
Remote I/O	4096 inputs 4096 outputs	4096 inputs 4096 outputs
Local Chassis	3 chassis	3 chassis
Total Modular Slots	30	30
Typical Scan Time	4.8 milliseconds per K (thousand) instructions	1 milliseconds per K (thousand) instructions
Programming Instructions	71	107
Memory Backup	EEPROM UVPROM	Flash EEPROM
RAM Backup	Lithium battery, up to 2 years	Lithium battery, up to 2 years
Programming Methods	Handheld terminal or software	Software only

Figure 10-19   Overview of SLC 5/02 processor specifications. (Data compiled from Allen-Bradley data sheets)

program containing simple ladder logic can be scanned in as little as one millisecond. In addition to 36 additional instructions, the following features were added:

- built-in real-time clock
- Flash EEPROM for firmware upgrades
- on-line editing
- floating-point math
- keyswitch for selecting processor operating modes
- a built-in RS-232 port

Another important change in the 5/03 processor is that it was the first processor that cannot be programmed with the handheld terminal. The only means for programming this processor, the 5/04, and the 5/05 is an IBM-compatible personal computer or industrial computer.

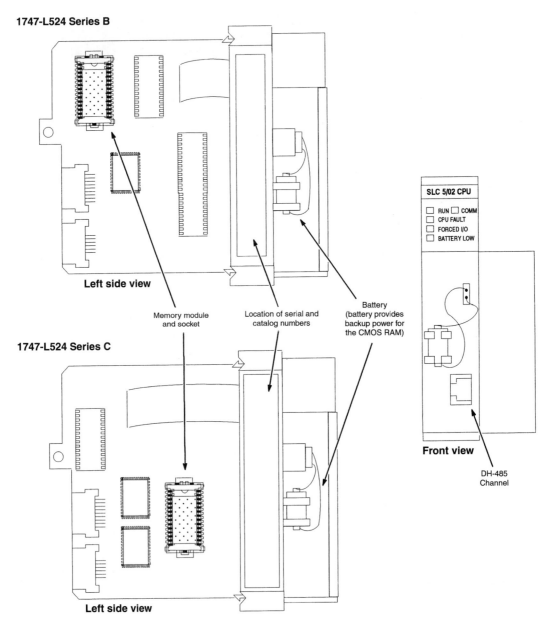

Figure 10-20 Allen-Bradley SLC 5/02-processor, series B and series C. (Used with permission of Rockwell Automation, Inc.)

The RS-232 port is a first for an SLC processor. This RS-232 channel provides a direct connection to an IBM-compatible programming device, in addition to other enhanced communication options.

Looking at Figure 10-21, notice the memory module connector position, battery position, key switch, LEDs on the top front, channel 0 and its standard D-shell connector for RS-232 connectivity, and channel 1 for Data Highway connection. Also, notice the memory module download protection jumper in the bottom left-hand corner of the processor's circuit board. This jumper is used for flash memory firmware or operating system upgrades. To perform an operating system upgrade, plug the flash EEPROM upgrade kit's plug-in module into the memory module connector. Next, move the operating system memory module download protection jumper, as directed in the instructions. Reinsert the processor into the chassis and follow the instructions to complete the operating system upgrade.

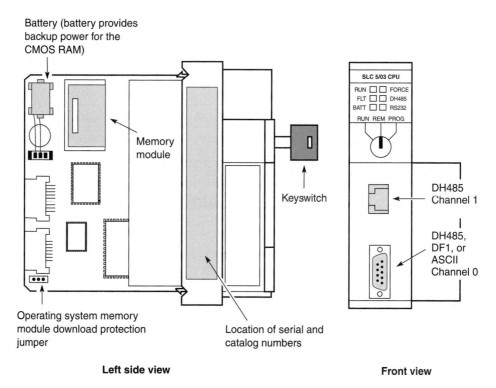

Figure 10-21   Main features of the SLC 5/03 processor. (Used with permission of Rockwell Automation, Inc.)

Allen-Bradley SLC 5/04 Processor

Major enhancements built into the SLC 5/04 processor include memory expansion up to 64 K of program memory and built-in Data Highway Plus network connectivity. The 5/04 processor is the first SLC processor with the ability to directly connect to the Allen-Bradley PLC 5's Data Highway Plus (DH+) network. Figure 10-22 lists the 5/04's specifications.

ALLEN-BRADLEY SLC 5/04 PROCESSOR SPECIFICATIONS			
Part Number	1747-L541	1747-L542	1747-L543
Program Memory	16 K instructions	32 K instructions	64 K instructions
Communications	Channel 0: RS-232 (DF1, ASCII, DH-485 Protocols) Channel 1: DH Plus (PLC 5 Network)		
Local I/O Capacity	960 discrete	960 discrete	960 discrete
Remote I/O	4096 inputs 4096 outputs	4096 inputs 4096 outputs	4096 inputs 4096 outputs
Local Chassis	3 chassis	3 chassis	3 chassis
Total Modular Slots	30	30	30
Typical Scan Time	.9 milliseconds per K (thousand) instructions	.9 milliseconds per K (thousand) instructions	.9 millisecond per K (thousand) instructions
Programming Instructions	107	107	107
Memory Backup	Flash EPROM	Flash EPROM	Flash EPROM
RAM Backup	Lithium battery, up to 2 years	Lithium battery, up to 2 years	Lithium battery, up to 2 years
Programming Methods	Software only	Software only	Software only

Figure 10-22   Overview of SLC 5/04 processor specifications. (Data compiled from Allen-Bradley data sheets)

Figure 10-23 shows a side and front view of the SLC 5/04 processor. Notice that the features are much the same as those of the 5/03 processor. The major physical change is the elimination of the Data Highway 485 network port. The 5/04 processor was designed specifically to interface with the PLC 5 DH+ network. Channel 1 on the front of the 5/04 processor has two different connections. The bottom round connector connects from 1784 KTX desktop personal or industrial computer interface card, or a 1784 PCMK notebook computer interface card to provide communication between a personal computer and the 5/04 processor. The top three-pin connector is used to connect to the Data Highway Plus (DH+) network "blue hose." With the blue hose, the 5/04 processor can connect directly to other DH+ network processors or personal computers. These two connections are connected together inside the processor. As a result, one could connect to the processor from a PCMK interface card using the round connector and then out to the network through the blue hose.

Allen-Bradley SLC 5/05 Processor

The SLC 5/05 processor shares the same features, memory options, and functionality as the 5/04 processor with the exception of the DH+ port on the 5/04 and the Ethernet port

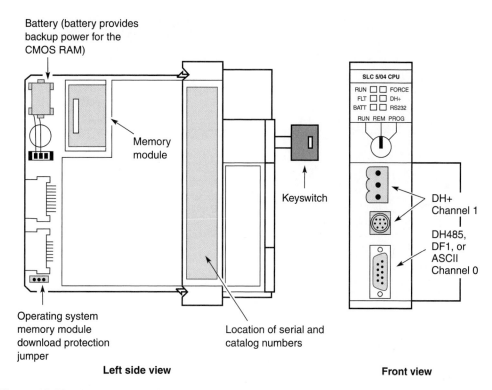

Battery (battery provides backup power for the CMOS RAM)

Memory module

Keyswitch

SLC 5/04 CPU

RUN ☐☐ FORCE
FLT ☐☐ DH+
BATT ☐☐ RS232

RUN REM PROG

DH+
Channel 1

DH485, DF1, or ASCII Channel 0

Operating system memory module download protection jumper

Location of serial and catalog numbers

Left side view

Front view

Figure 10-23   Allen-Bradley SLC 5/04 processor. (Used with permission of Rockwell Automation, Inc.)

on the 5/05 processor. The 5/05 processor is used primarily to attach to a TCP/IP Ethernet network. Ethernet is a high-speed local area network used to exchange information between computers, PLCs, and other similar devices at a rate from 10 Mbps to 100 Mbps (million bits per second). Using an Ethernet network, you have the capability to communicate between many pieces of equipment from different vendors even over the Internet. Figure 10-24 lists the 5/05 processor's specifications.

Figure 10-25 shows a side and front view of the SLC 5/05 processor. Notice that the features are much the same as for the 5/04 processor. The major feature change is the Ethernet port. The 5/05 processor was designed specifically to interface with an Ethernet network. Channel zero on the front of the 5/05 is the same RS-232 port found on the 5/03 and 5/04 processors.

SLC 500 Processor LEDs

Each processor has a number of LEDs on the front near the top of the processor module. The LEDs vary as to the particular SLC processor. We will introduce a general overview of the LEDs' functions. Refer to Figure 10-26 along with the SLC 500 hardware manual for specifics on the LED functionality for your particular SLC 500 processor.

ALLEN-BRADLEY SLC 5/05 PROCESSOR SPECIFICATIONS			
Part Number	1747-L551	1747-L552	1747-L553
Program Memory	16 K	32 K	64 K
Communications	Channel 0: RS-232 (DF1, ASCII, DH-485 Protocols) Channel 1: Ethernet (10BaseT)		
Local I/O Capacity	960 discrete		
Remote I/O	4096 inputs 4096 outputs		
Local Chassis	3 chassis		
Total Modular Slots	30		
Typical Scan Time	.9 milliseconds per K (thousand) instructions		
Programming Instructions	107		
Memory Backup	Flash EPROM		
RAM Backup	Lithium battery, up to 2 years		
Programming Methods	Software only		

Figure 10-24 Overview of SLC 5/05 processor specifications. (Data compiled from Allen-Bradley data sheets)

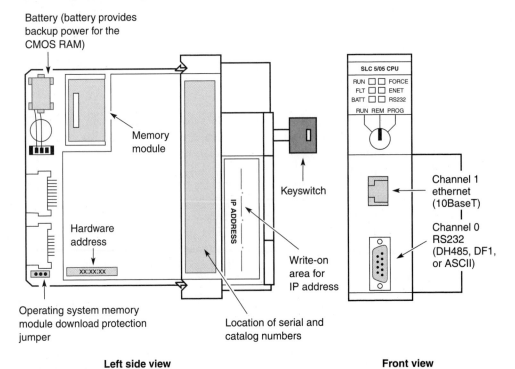

Left side view   **Front view**

Figure 10-25 Allen-Bradley SLC 5/05 modular processor. (Used with permission of Rockwell Automation, Inc.)

ALLEN-BRADLEY SLC 500 PROCESSOR LED FUNCTIONALITY			
LED	Color	LED State	LED Indication
RUN	Green	ON steady	Processor in RUN mode.
		Flashing	Transferring program from RAM to memory module.
		OFF	Processor other than run mode.
FLT (Fault)	Red	Flashing at power-up	Processor not configured.
		Flashing during operation	Processor detects major error.
		ON steady	Fatal error, no communication.
		OFF	No faults.
BATT (Battery)	Red	ON steady	Low battery or battery jumper problem.
		OFF	Battery is OK.
FORCE	Amber	Flashing	I/O has been forced ON or OFF, but not enabled.
		ON steady	Forces have been enabled.
		OFF	No forces present or enabled.
DH485	Green	ON steady	Processor communicating on network.
		Flashing	No active nodes on network.
		OFF	Fatal error, no communication.
RS232	Green	ON flashing (DF1 Mode)	Processor transmitting on network.
		OFF (DF1 Mode)	Processor not transmitting.
		ON steady (DH 485 Mode)	Processor communicating on the network.
		Flashing (DH 485 Mode)	Processor trying to establish communication but there are no active nodes on the network.
		OFF (DH 485 Mode)	No communication, fatal error.
DH+		ON steady	Processor communicating on the network.
		Flashing green	Processor trying to establish communication but there are no active nodes on the network.
		Flashing red	Duplicate nodes on network.

Figure 10-26 Overview of SLC 500 processor LED functionality. (Data compiled from Allen-Bradley data sheets)

ROCKWELL AUTOMATION'S LOGIX 5550 CONTROLLER

The ControlLogix PLC is Rockwell Automation's next generation of PLC. The heart of the ControlLogix PLC is the highly functional 32-bit industrial processor. ControlLogix architecture provides a state-of-the art control platform that allows combining current separate technologies such as PLC ladder logic, motion control, variable-frequency drives, and a communication gateway, along with process control into one modular PLC system. Figure 10-27 lists the Logix 5550 processor's specifications.

SELECTED CONTROLLOGIX PROCESSOR SPECIFICATIONS						
Part Number	1756-L55M22	1756-L55M23	1756-L55M24	1756-L61	1756-L62	1756-L63
Data and Logic Memory	750 Kbytes	1.5 Mbytes	3.5 Mbytes	2 Mbytes	4 Mbytes	8 Mbytes
I/O Memory	208 Kbytes	208 Kbytes	208 Kbytes	478 Kbytes	478 Kbytes	478 Kbytes
Nonvolatile Memory	Yes	Yes	Yes	CompactFlash Card	CompactFlash Card	CompactFlash Card
Communication	Ethernet, ControlNet, DeviceNet, Foundation Fieldbus, AutoMax, Serial, DH+/RIO					
Discrete I/O	Up to 128,000					
Analog I/O	Up to 4,000					
Remote I/O	Up to 32 logical racks using 1756-DHRIO (Data Highway Plus and Remote I/O Interface Module)					
Chassis Available	4, 7, 10, 13, and 17 slot					
RAM Backup	Lithium battery					
Programming Options	Ethernet, ControlNet, DH+, Serial processor port					
Programming Port	Nine-pin D-shell RS-232 serial					
Programming Methods	RS Logix 5000 Software					

Figure 10-27 Overview of Rockwell ControlLogix 5550 processor specifications. (Data compiled from Allen-Bradley data sheets)

The Logix 5550 processor's physical features include an RS-232 port on the processor that provides a local programming connection in addition to communication to the processor's backplane for communication bridging to other devices in the control system. The Logix 5550 controller's modular memory allows user insertion of one of three high-speed battery-backed static memory cards, enabling memory expansion from the standard 160 Kbyte memory to 512 Kbytes and up to 2 Mbytes. The processor's operating system is IEC1131-compliant, providing a true preemptive multitasking environment. By installing multiple processors in a single chassis, system applications and processing applications can be shared between controllers. Figure 10-28 shows the Logix 5550 processor features.

Let us move on to look at the processors available from General Electric and the Series 90-30 PLC family.

THE GENERAL ELECTRIC SERIES 90-30 PROCESSORS

The 90-30 PLC family has numerous processors which differ in I/O capacity, processing speed, instruction set, user memory size, and remote I/O capabilities. The processors, or CPUs, are divided into two groups, embedded and modular. Embedded CPUs are built into the baseplate called an embedded CPU baseplate. Modular CPUs are separate modules which clip onto a different baseplate called a CPU baseplate. In both cases user

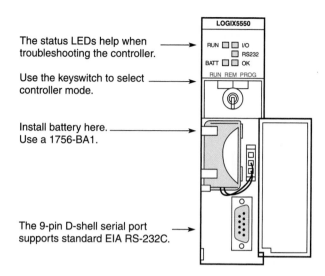

The status LEDs help when troubleshooting the controller.

Use the keyswitch to select controller mode.

Install battery here. Use a 1756-BA1.

The 9-pin D-shell serial port supports standard EIA RS-232C.

Figure 10-28   Rockwell Automation ControlLogix 5550 processor. (Used with permission of Rockwell Automation, Inc.)

selected I/O modules clip onto the remaining baseplate slots. Figure 10-29 illustrates a five-slot and a ten-slot embedded CPU baseplate.

Power Supply and Serial Port Connection

The power supply clips onto the left-most slot of each baseplate. The power supply contains the serial port for CPUs other than models 351, 352, and 363. The serial port is RS-485 compatible and uses Series Ninety Protocol (SNP) slave mode only. The serial port connector is functional only on baseplates that contain a processor. Figure 10-30 illustrates the position of the serial port on the power supply module.

Embedded CPUs

Embedded CPUs are built, or embedded, into an embedded CPU baseplate. Embedded CPUs are basic CPUs and do not have the PLC power, features, and I/O capabilities of their modular cousins. There are three models of embedded CPUs: Model 311, Model 313, and Model 323. Embedded CPUs all share the following features:

- The CPU is built into the baseplate and cannot be upgraded without changing to another baseplate.
- Since the embedded CPU baseplates do not require a modular CPU, all slots are available for I/O modules.
- The largest baseplate, the Model 323, supports ten slots. If additional I/O slots are required, a modular CPU system and baseplates will have to be used.
- Embedded CPUs do not support expansion or remote baseplates.
- Embedded CPUs do not support floating-point math or Flash user memory, and do not have a time of day clock.

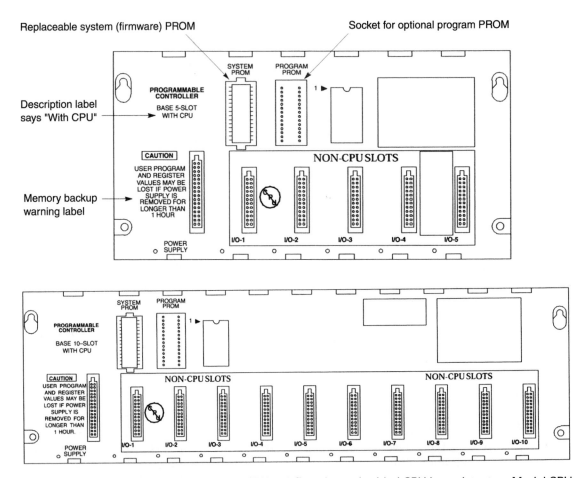

Figure 10-29 Model CPU 311 and Model CPU 313 five-slot embedded CPU baseplate, top. Model CPU 323 ten-slot embedded CPU baseplate, bottom. (Courtesy of GE Fanuc Automation)

Figure 10-31 provides a quick overview of the embedded CPU specifications. For complete specifications refer to the *General Electric Series 90-30 Hardware Installation Manual.*

Modular CPUs

A modular CPU baseplate has the CPU as a separate clip-on module. Starting with Model 331 and above, the CPUs are modular pieces of hardware. Being separate from the baseplate, a modular CPU can easily be changed to a more powerful model if the system needs change. Figure 10-32 illustrates a five-slot modular baseplate. Refer to the numbers below for identification of the baseplate features:

1. Module retainer
2. Baseplate mounting holes

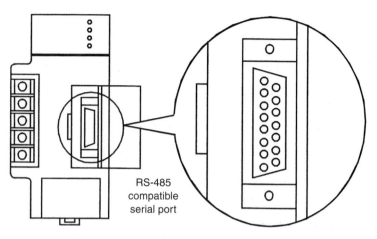

RS-485 compatible serial port

Figure 10-30 CPU serial port on the Series 90-30 power supply. (Courtesy of GE Fanuc Automation)

GENERAL ELECTRIC SERIES 90-30 EMBEDDED CPU SPECIFICATIONS			
Part Number	Model 311	Model 313	Model 323
CPU	Five-slot baseplate with embedded CPU		Ten-slot baseplate with embedded CPU
Base Plates	Five-slot	Five- or Ten-slot	Five- or Ten-slot
Total Baseplates per System	1		
Typical Scan Rate	18 milliseconds per 1 K of logic (Boolean contacts)	0.6 millisecond per 1 K of logic (Boolean contacts)	
User Program Memory (maximum)	6 Kbytes	12 Kbytes (6 Kbytes prior to release 7)	
Analog Inputs—% AI	64 words		
Analog Outputs—% AQ	32 words		
Communications	LAN-supports Multidrop, Ethernet, FIP, Profibus, GBC, GCM+ option modules.		
Memory Storage	RAM and optional EPROM or EEPROM		
I/O Capacity	160 total combined inputs and outputs		320 total combined inputs and outputs
Expansion or Remote Baseplates	None		
Programming Methods	Handheld programmer; software, Ladder Logic or SFC		

Figure 10-31 Overview of General Electric Series 90-30 embedded CPU baseplate specifications. (Data compiled from General Electric data sheets)

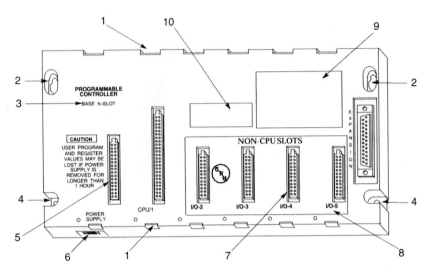

Figure 10-32 Five-slot modular CPU baseplate. (Courtesy of GE Fanuc Automation)

3. Description of baseplate
4. Baseplate lower mounting holes
5. Power supply connector
6. Serial number label
7. I/O module connection points. This is a five-slot baseplate. Notice that slot CPU/1 is reserved for the CPU. I/O slots two through five are reserved for I/O modules.
8. Slot identifier labels
9. Compliance label
10. Catalog number label

Modular CPUs share the following features:

- CPU slot one is reserved for the CPU. This slot is labeled as CPU/1. Only the CPU or special option modules can be plugged into this slot.
- Modular CPUs can be easily changed as system requirements change.
- The modular CPU baseplate is always identified as rack 0.
- Only one CPU and CPU baseplate are allowed per system. Additional baseplates of I/O expansion will be either expansion or remote baseplates.
- Modular CPU baseplates have a 25-pin D-shell type expansion connector (refer to Figure 10-32) for connection to expansion or remote baseplates.
- Modular CPUs have a time of day clock.
- Most modular CPUs support floating-point math and flash EEPROM.

Figure 10-33 provides an overview of selected specifications for modular CPU models 331, 340, 341, 350, and 360. Refer to General Electric documentation for complete specifications.

GENERAL ELECTRIC SERIES 90-30 MODULAR CPU SPECIFICATIONS				
Part Number	**Model 331**	**Models 340, 341**	**Model 350**	**Model 360**
CPU	Single-slot modular	Single-slot modular	Single-slot modular	Single-slot modular
Baseplates	5 or 10 slot	5 or 10 slot	5 or 10 slot	5 or 10 slot
Program Memory	16 K instructions	340: 23 K instructions 341: 80 K instructions	32 K instructions	Up to 240 K instructions
Register Memory	4 Kbytes	19.9 Kbytes	9,999 words	9,999 words
I/O Capacity	512 inputs 512 outputs	512 inputs 512 outputs	2,048 inputs 2,048 outputs	2,048 inputs 2,048 outputs
Expansion Baseplates	Up to 4	340 up to 4 341 up to 5	Up to 7	Up to 7
Built-In Ports	1 (uses connector on PLC power supply). Supports SNP/SNPX slave protocols. Requires CCM module for SNP/SNPX master, CCM, or RTU slave support; PCM module for RTU master support.			
PCM/CCM Compatibility	Yes			
Typical Scan Time	.4 milliseconds per K (thousand) instructions	.3 milliseconds per K (thousand) instructions	.22 milliseconds per K (thousand) instructions	.22 milliseconds per K (thousand) instructions
Programming Methods	Handheld Programmer; software, Ladder Logic and SFC		Ladder Logic, C, and SFC	

Figure 10-33   Overview of General Electric Series 90-30 Modular CPU specifications. (Data compiled from General Electric data sheets)

Models CSE 311, CSE 323, CSE 331, and CSE 340

The CSE model CPUs can be programmed using state logic. State logic is an alternative to traditional programming languages such as ladder. State logic provides the programmer the option to break the process into tasks and steps, similar to a flowchart. This type of programming is especially useful when programming complex machine systems and processes.

Series 90-30 CPU 350 through CPU 364

The CPUs in the range from 350 to 364 are high-end CPUs. These CPUs support additional I/O, larger user memory, and faster processing speeds, along with an additional programming option, programming in C. This group of CPUs provides the following features:

- Limited compatibility with the handheld programmer. (See Hardware Installation Manual.)
- Configurable memory. Configurable memory provides the programmer additional flexibility in assigning analog and register memory.

- Starting with firmware version 9.0 the 351–364 family of CPUs has up to 240 K of user configurable memory.
- Floating-point math.
- Flash memory for nonvolatile storage of CPU firmware, program, configuration, and register memory.
- Additional serial ports (CPU 351, CPU 352, CPU 363) mounted on the CPU rather than accessing the serial port on the power supply by way of the backplane. This direct access improves system performance.
- Keyswitch on processor for On/Run and Off/Stop. See documentation for processor specific functionality.
- Sequential event recorder for troubleshooting and debugging.
- Embedded Ethernet Interface on CPU 364 only. This provides built-in, one-slot interface to the Ethernet network.

Figure 10-34 illustrates the CPU 363 and the CPU 351.

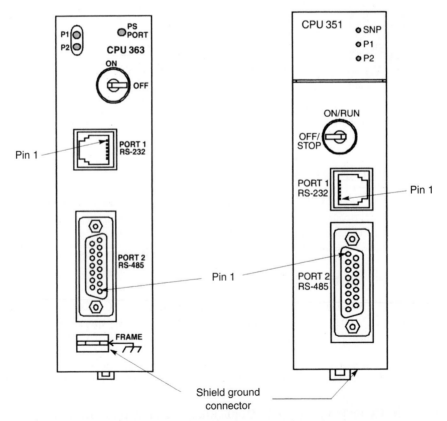

Figure 10-34   Modular CPU 351 and CPU 363. (Courtesy of GE Fanuc Automation)

CPU 364 Physical Features

There are four LED indicators on the model 364 CPU. Three relate to the Ethernet interface and the fourth, PSPORT, is the status of the CPU's serial port. The EOK, LAN, and STAT LED indicators provide information as to the status of the Ethernet port and communications. These indicators may be steady on, flashing slow, flashing fast, or in different combinations depending on the status of the port. Refer to the *Ethernet Communications* manual for additional information. Refer to Figure 10-35 for CPU 364 features:

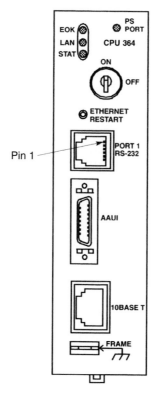

Figure 10-35   Modular CPU 364. (Courtesy of GE Fanuc Automation)

- The Ethernet restart button has four functions: Test the LEDs, restart, restart and enter software load state, and restart and enter maintenance state.
- Port 1, RS-232 has two functions. Port 1 is used to connect to a terminal or terminal emulator to access the Ethernet Station Manager software. Also Port 1 can be used to connect to a personal or industrial computer to update the Ethernet interface firmware.
- The AAUI port is a 14-pin port for connecting to an external Ethernet-compatible transceiver for either 10BaseT or 10Base2 connectivity.

- The 10BaseT port is an 8-pin RJ-45 connection that provides direct connection to a 10BaseT Ethernet network.
- Shield ground connection tab is used to connect the module's shield ground connection.

Figure 10-36 lists selected specifications for model 351, 352, 363, and 364 CPUs.

GENERAL ELECTRIC SERIES 90-30 MODULAR CPU SPECIFICATIONS				
Part Number	**Model 351**	**Model 352**	**Model 363**	**Model 374**
CPU	Single-slot CPU module			
Baseplates	CPU plus 7 expansion and remote			
Program Memory	Firmware prior to 9.0 = 80 Kbytes Firmware 9.0 and above = 240 Kbytes		240 Kbytes	240 Kbytes
Register Memory	Up to 32,640 words. See specifications.			
I/O Capacity	2,048 inputs 2,048 outputs			
Typical Scan Time	.22 milliseconds per K (thousand) instructions			.15 msec per 1K
Three Built-In Serial Ports	Supports: SNP/SNPX slave and RTU slave SNP/SNPX master/slave Serial I/O on ports 1 and 2 Requires CMM module for CCM or PCM module for RTU master support			None
Communications	LAN—supports Mulitdrop, Ethernet, FIP, Profibus, GBC, GCM, GCM+ option modules			
Programming Methods	Ladder Logic, C, and SFC			
Interrupt Support	Periodic Subroutine Feature			
Memory Storage	RAM and Flash			
Floating-Point Math Support	Firmware-based with firmware Release 9.0 and later	Hardware-based with built-in math co-processor	Firmware-based with firmware Release 9.0 and later	Hardware-based

Figure 10-36   Overview of General Electric Series 90-30 Modular CPU specifications. (Data compiled from General Electric data sheets)

Modular CPUs have either five- or ten-slot baseplates available. The power supply, which is modular, must be installed in the left-most position. Slot one is for the CPU. The four or nine remaining slots are for I/O modules. Figure 10-37 illustrates a ten-slot modular CPU baseplate.

Up to four expansion baseplates (seven for the model 351 CPU and above) can be connected to the CPU baseplate. The baseplates contain either five or ten slots. Again, the left-most slot is reserved for the power supply. There are no expansion modules necessary

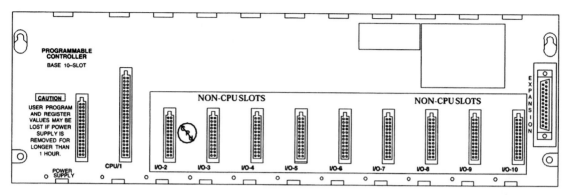

Figure 10-37   Series 90-30 ten-slot modular CPU baseplate. (Courtesy of GE Fanuc Automation)

in an expansion baseplate. Each expansion baseplate must be address configured by selecting DIP switch positions residing directly above slot one. There are two types of baseplates: expansion baseplates, for local expansion, and remote baseplates, for remote placement.

Local expansion consists of using either a five- or a ten-slot expansion baseplate, as illustrated in Figure 10-38. Note the expansion connector, a 25-pin female D-type connector, on the CPU baseplate and also on the expansion baseplate. The total length of cable connecting all expansion baseplates cannot be more than 50 feet.

Remote baseplates are used when distances greater than 50 feet are necessary between the CPU baseplate and the last baseplate. Remote baseplates provide extended expansion capability for the Series 90-30 Model 331, 340, 341, and 351 programmable

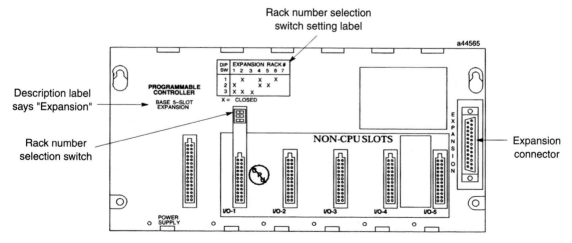

Figure 10-38   Series 90-30 five-slot expansion baseplate. (Courtesy of GE Fanuc Automation)

controllers. Remote baseplates are available as five or ten slots. Figure 10-39 illustrates a ten-slot remote baseplate. Remote baseplates are the same size and use the same power supply and I/O modules as CPU baseplates and expansion baseplates. Remote baseplates allow for up to 700 feet of cable from the CPU baseplate and the last remote baseplate.

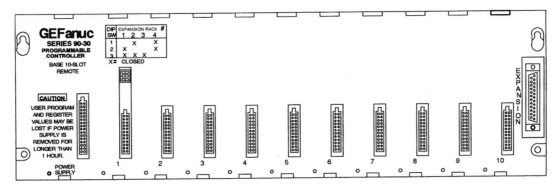

Figure 10-39   Ten-slot remote baseplate. (Courtesy of GE Fanuc Automation)

CHOOSING THE CORRECT PROCESSOR FOR AN APPLICATION

A perplexing question for many of us is, "How do I choose the proper PLC processor for my application?" In some situations you have to look into the future as you select your PLC processor. Depending on your particular control situation, choosing a PLC processor may be very straightforward or quite complex.

Take, for example, a situation in which you work for an original equipment manufacturer (OEM). If you are building a machine that will go to the field to do a specific task with little chance of this task being modified or expanded in the future, choosing the proper processor can be a simple exercise. Simply make a list of the immediate needs regarding instructions, I/O count, and memory size. If, on the other hand, you are working in a manufacturing environment where the PLC is going to be used for many years through expansion and equipment upgrades and modifications, selecting the correct processor in some instances becomes an exercise of predicting the future. The following questions will help you in deciding which particular processor to choose from the many selections your particular manufacturer offers:

1. How much memory will be needed for this application and future program additions?
2. How many I/O are needed?
3. Do I need only discrete I/O?
4. Do I need specialty I/O modules?

5. Do I need analog capabilities?
6. Will I need to interface my PLC to a network (today or in the future)?
7. Will I need to interface my PLC to a human-machine interface device?
8. Will I need to program with a handheld programming terminal or software?
9. Will I need passwords? What does the manufacturer offer?
10. Will I need to use an EEPROM for nonvolatile memory backup? Is this available?
11. What instructions are needed?
12. Is scan time critical?
13. What processor ports are needed?
14. What port is needed to interface my programming terminal to the processor?
15. Will I need a special interface card or device to interface my programming terminal to the processor?
16. Do I need local I/O expansion? How much expansion might be needed? How far will my PLC allow me to place the expansion chassis away from the CPU chassis?
17. Do I need remote I/O expansion? How much remote I/O expansion is needed? How far will my PLC allow me to place a remote chassis away from the base CPU chassis?

SUMMARY

This chapter introduced the hows and whys of programmable controller processor operation. We randomly chose two of the major players in the programmable controller market, the Allen-Bradley SLC 500 family of processors and the Series 90-30 family of General Electric processors, to investigate the available processors and their specifications. We looked into the scan, sweep, or cycle time of a processor as a way to understand what happens internally when a processor is in run mode. Nonvolatile memory backup was introduced as a method to ensure user program permanence, especially for machine builders who do not want to worry about field maintenance individuals monitoring and replacing batteries. We also looked at nonvolatile memory chips as a way to autoload user programs after possible memory corruption or failure in remote field locations. LEDs are included on processors to assist field personnel in monitoring operations and troubleshooting PLCs. Last, we looked at questions that need to be considered when selecting a processor for a particular application.

REVIEW QUESTIONS

1. The early PLC as a relay replacer has become more and more sophisticated as advances in _____ increased the capabilities of the processor's computer power.
2. As microprocessor computing and data processing capabilities increased, so did the _____ and _____ performed by the PLC.

3. One K of memory is:
   A. 1,000 bits
   B. 1,000 words
   C. 1,020 bits
   D. 1,020 words
   E. 1,024 bits
   F. 1,024 words

4. Volatile memory is:
   A. RAM
   B. ROM
   C. PROM
   D. EEPROM
   E. Flash EEPROM

5. The operating system directs the allocation of system resources, including:
   A. memory allocation for program storage
   B. memory allocation for data storage
   C. input and output distribution
   D. processor timing
   E. processor communication with peripheral devices
   F. all of the above

6. Nonvolatile memory is:
   A. RAM
   B. ROM
   C. PROM
   D. EEPROM
   E. A, B, and C
   F. B, C, and E

7. The watchdog timer is:
   A. a hardware timer to ensure the input scan is completed successfully
   B. a hardware timer to ensure the program scan is completed successfully
   C. a hardware timer to ensure the subroutine scan is completed successfully
   D. a hardware timer to ensure the output update is completed successfully
   E. a timer in the user program that ensures the program completes on time
   F. a timer in the user program that ensures that the predetermined throughput time is not exceeded

8. _____ eliminates the need to physically replace the EEPROM chips when making a firmware upgrade.

9. Flash memory is a special form of _____ memory that can be erased and reprogrammed with signal levels commonly found inside the computer device in which it is located.

10. The processor in any PLC is designed to perform specific duties in a specific sequence and then continuously repeat the sequence. The operating cycle is called the _____ or _____ .

11. The scan consists of a series of sequential operations, which include housekeeping, reading _____ , execution, _____ output data, servicing or updating programming device, system communications, and diagnostics.

12. The servicing of communications includes updating:
    A. your handheld
    B. personal computer monitor screen
    C. the watchdog timer
    D. sending communications to other PLCs or devices on a network
    E. A, B, and C
    F. A, B, and D

13. The _____ is reset at the conclusion of each scan cycle by the processor when the scan time is less than the watchdog timer's preset time.

14. If the processor scan is not completed and the _____ reset before the watchdog timer times out, the processor will be _____ .

15. The physical connection between a processor and the outside world is called a communication:
    A. connection
    B. input
    C. output
    D. tie point
    E. port
    F. status point

16. For two devices to communicate with each other, they must be connected so electrical signals transmitted by one device are _____ by the other.

17. Communication between two devices is accomplished either by direct connection or indirect connections, through _____ communication devices.

18. Direct connection between two devices is achieved by directly running a cable from one device to another. This is called a _____ connection.

19. There are two types of communication hardware: _____ and _____ equipment.

20. Data terminal equipment consists of:
    A. terminals
    B. computers
    C. PLC processors
    D. modems
    E. other devices that are the source or final destination of data
    F. all of the above

21. In order for different manufacturer's devices to communicate, there need to be commonly accepted communication _____ .

22. The way the wires are identified in a 9-pin connector is by the numbers assigned to each of the _____.

23. Verifiable communication between devices is possible when the two devices cooperate with each other in the exchange of data. This cooperation is called _____.

24. Even if the RS-232 wiring is standard between two processors of different manufacturers, the protocol will probably be different. The _____ is a set of rules that governs the way that data is formatted and timed as it is transmitted between the sending and receiving device.

25. What type of battery is typically used in PLC processors?
    A. carbon
    B. alkaline
    C. any battery that will physically fit
    D. lithium
    E. silver-carbon

26. Are there any special precautions you need to be aware of when transporting PLC batteries?
    A. You only need to wrap them securely and place them in a sturdy cardboard box before shipping.
    B. You must place the batteries in a static bag before packaging and shipping.
    C. You must place batteries in a metal container before wrapping.
    D. You must place batteries in a plastic container supplied by the manufacturer.
    E. A and B.
    F. B and D.

27. Are there any special precautions you need to be aware of when disposing of PLC batteries?
    A. No, simply throw them in the trash unless directed differently by local ordinance.
    B. Wrap them securely before throwing them in the trash.
    C. They must be shipped to a federal depository.
    D. They must be shipped back to the manufacturer for proper disposal.
    E. A and B.
    F. C and D.

28. Batteries are used in a PLC's processor to:
    A. keep the processor alive to make one final housekeeping scan before it shuts down after power failure
    B. maintain memory in volatile memory when line power is removed from the processor
    C. maintain memory in nonvolatile memory when line power is removed from the processor
    D. operate the LEDs on the front of the CPU
    E. maintain outputs through a power failure

29. PLC power interruption where memory battery backup is necessary could be caused by which of the following?
    A. shutoff of the disconnect to the PLC
    B. power failure from blown fuses
    C. tripped circuit breaker
    D. power interruptions from the power utility
    E. weather-related power outages
    F. any of the above

30. Many PLC manufacturers use _____ batteries to provide backup power for volatile memory.

IF PASSWORD IS CORRECT GO TO DRIVE SET UP SCREEN

B3:10/0

0022
```
──── EQU ────
     Equal
     Source A    N12:21
                  7777<
     Source B    32767
                 32767<
```
──[OSR]──

B3:1/15
──( )──

RESET PASSWORD
ATTEMPTS COUNTER
C5:5
──(RES)──

0023
```
──── GRT ────
     Greater Than (A>B)
     Source A    N12:21
                  7777<
     Source B      0
```
```
──── NEQ ────
     Not Equal
     Source A    N12:21
                  7777<
     Source B    32767
```
B3:10/1
──[OSR]──

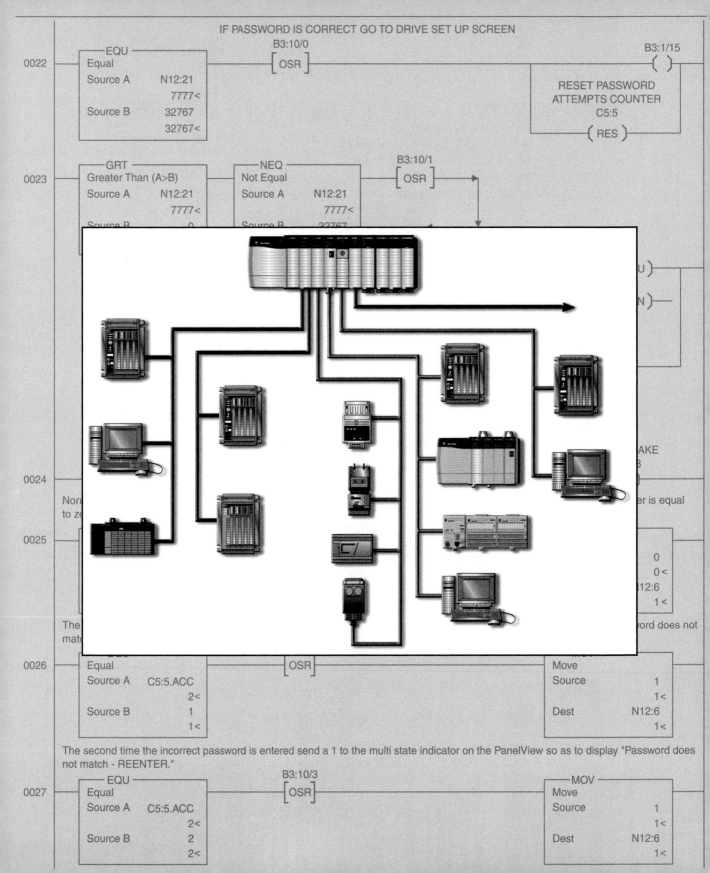

0024

Nor...

to ze...

0025

```
                    0
                   0<
                 12:6
                   1<
```

The ... ...ord does not
mat...

0026
```
     Equal
     Source A   C5:5.ACC
                   2<
     Source B      1
                   1<
```
──[OSR]──
```
──── MOV ────
     Move
     Source       1
                  1<
     Dest       N12:6
                  1<
```

The second time the incorrect password is entered send a 1 to the multi state indicator on the PanelView so as to display "Password does not match - REENTER."

B3:10/3

0027
```
──── EQU ────
     Equal
     Source A   C5:5.ACC
                   2<
     Source B      2
                   2<
```
──[OSR]──
```
──── MOV ────
     Move
     Source       1
                  1<
     Dest       N12:6
                  1<
```

Introduction to ControlNet and DeviceNet

OBJECTIVES

After completing the chapter you should be able to:

- explain the differences between ControlNet and DeviceNet networks
- decide when a ControlNet or DeviceNet network might be selected
- calculate maximum trunk line length for a ControlNet segment
- determine when a repeater should be used on a ControlNet network
- determine maximum cable lengths of a DeviceNet network
- define basic ControlNet terms
- understand the function of RSNetWorx for ControlNet and DeviceNet

INTRODUCTION

Networks are becoming more and more popular every day. The days of stand-alone PLCs appear to be fading fast, only to be replaced by networks. In some cases, older networks such as Data Highway 485 and Data Highway Plus are being upgraded to faster, more efficient, networks offering more deterministic data transfer. Installing a device-based network such as DeviceNet provides the important feature of reducing field wiring costs. Rather than running large wiring bundles back to the PLC from distant I/O field devices, a single network cable can provide device operating power and data transfer between many devices and the PLC processor. Another advantage of a network is the option to remotely mount a PLC chassis containing I/O modules close to the field devices, thus shortening I/O wires, and then using a single cable or fiber-optic cable to transfer data between the remote chassis and the local processor. PLC processors connected together afford a programmer or maintenance individual the ability to access any PLC processor from a single computer via a network connection for program downloads, editing, or monitoring. PLC processors can also communicate time-critical or interlocking information between themselves over the network.

As networks evolve, they become more and more complex to install, modify, and maintain. This chapter is intended only to introduce you to the basics of ControlNet and DeviceNet networks.

THE CONTROLNET NETWORK

ControlNet is an open-control network used for real-time data transfer of time-critical and non-time-critical data between processors or I/O on the same link. The ControlNet data transfer rate is five million bits per second, and up to 99 nodes can be connected on a ControlNet network, depending on how much data needs to be transferred and how efficiently the bandwidth of the network is set up. ControlNet is basically a combination of Data Highway Plus and Remote I/O. The network provides high-speed peer-to-peer information exchange and I/O networking on the same network.

CONTROLNET NETWORK APPLICATIONS

Sharing data between a processor and I/O in a remote ControlLogix chassis is done over a ControlNet network. ControlNet is typically used in the following applications:

- ControlLogix processor, SLC 500, or PLC 5 processor-to-processor scheduled data exchange
- Local ControlLogix or PLC 5 chassis connection to a remote chassis for high-speed remote I/O connectivity
- Interlocking or synchronization of two nodes such as starting two variable-frequency drives
- Connecting two Data Highway Plus networks
- Connecting multiple DeviceNet networks

ControlNet Network Makeup

A ControlNet network is made up of a number of nodes (1–99), each having a ControlNet communication interface. ControlNet nodes could include ControlLogix, PLC 5, or SLC 500 PLCs; variable-frequency drives; operator interface, etc. Figure 11-1 shows a ControlNet trunkline/dropline network with eight devices, or nodes. This view is taken from RSNetWorx for ControlNet software. ControlNet is an open network managed by ControlNet International (www.controlnet.org). Being an open network, hardware from many different vendors can be purchased and connected as a node to the network. The network is a combination of Data Highway Plus (DH+) and remote I/O. Processor-to-processor information can be exchanged as scheduled or unscheduled information and as control of I/O in remote chassis or other pieces of hardware. Scheduled or time-critical information is that which needs to transfer on a repeatable and deterministic timed interval. Unscheduled or

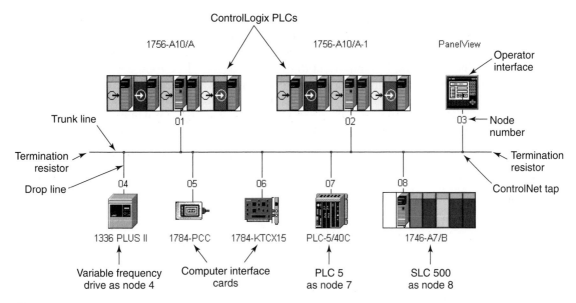

Figure 11-1   ControlNet trunk line/drop line topology.

non-time-critical information is associated with a message instruction programmed on a ladder rung. The message instruction is triggered by the desired ladder logic only when the information needs to be transferred. Figure 11-2 illustrates three possible nodes on a ControlNet network.

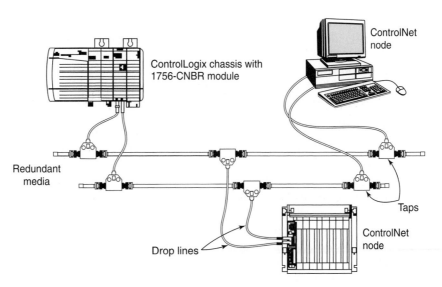

Figure 11-2   Three nodes on a Redundant ControlNet network. (Used with permission of Rockwell Automation, Inc.)

When configuring a ControlNet network, you can use two communication cables to get redundancy. With redundancy, if one cable fails or is damaged, the second cable can pick up communications seamlessly. Refer to Figure 11-2. When designing a redundant ControlNet network, it is important that redundant cables be routed separately, so if one cable is damaged the other cable remains intact and can continue communications. Notice that the connections on the cables from the main cable, called the trunkline, are taps and drop lines. The taps are required to connect from the trunkline to each individual node. Maximum drop-line length is one meter (39.4 inches). Taps are available in many different configurations as illustrated in Figure 11-3. From left to right, the taps are identified as: straight T, straight Y, right-angle, and right-angle Y taps.

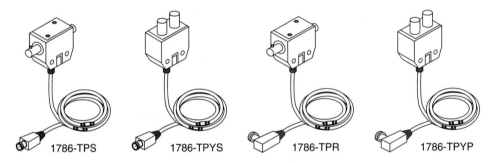

Figure 11-3   ControlNet taps. (Used with permission of Rockwell Automation, Inc.)

SLC 500 ControlNet Communications Modules

The SLC 500 modular PLCs have two ControlNet communication modules for connection to the ControlNet network. The 1747-SCNR is shown in Figure 11-4. The scanner is used to provide scheduled and unscheduled messaging between devices on ControlNet and the SLC 500. We will discuss scheduled and unscheduled messages later in this chapter. The module's features include the node address and status display. Directly below are three status indicators, one for Channel A, one for Channel B, and one for general module status. The ControlNet network access port, referred to simply as the NAP, is an RJ45 connection point for personal computer interface. The bottom of the module has two BNC connectors for coax network connectivity. The connections or ports are identified as Channel A and Channel B. Channel A is on the right. If redundancy is desired, both ports must be used.

The second ControlNet communication module for the SLC 5/03, 5/04, and 5/05 processors is a bridge between the ControlNet network and the processor's RS-232 serial port. This module allows the SLC processors to receive or send unscheduled ControlNet messages, along with monitoring or editing of the ladder program. The 1747-KFC module is typically installed in the chassis next to the SLC processor. A short serial cable is connected between the module and the SLC processor's serial port. The KFC module has the same basic features as the SCNR module, with the addition of the 9-pin D-shell serial connection. Figure 11-5 illustrates the KFC module.

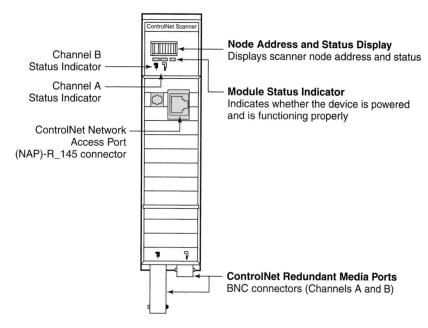

Channel B
Status Indicator

Channel A
Status Indicator

ControlNet Network
Access Port
(NAP)-R_145 connector

Node Address and Status Display
Displays scanner node address and status

Module Status Indicator
Indicates whether the device is powered
and is functioning properly

ControlNet Redundant Media Ports
BNC connectors (Channels A and B)

Figure 11-4   SLC 500 1747-SCNR. (Used with permission of Rockwell Automation, Inc.)

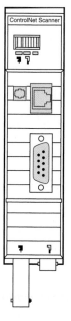

Figure 11-5   1747-KFC15 ControlNet interface to SLC processor communications module.
(Used with permission of Rockwell Automation, Inc.)

Figure 11-6 shows the 1747-KFC module installed in a SLC 500 chassis. Note the serial cable connected between the KFC module and the processor's serial port. The ControlNet coax cable is connected to the BNC connector on the bottom of the module.

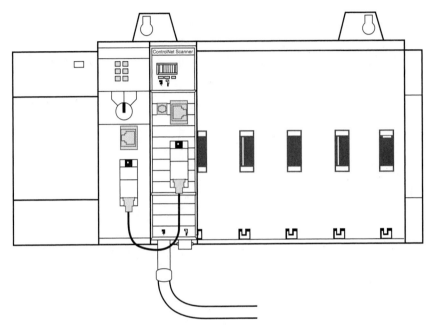

Figure 11-6   1747-KFC module installed in an SLC 500 chassis. (Used with permission of Rockwell Automation, Inc.)

CONTROLLOGIX CONTROLNET INTERFACE MODULE

The 1756-CNB or 1756-CNBR are the ControlLogix PLC interface modules to the ControlNet network. CNB stands for ControlNet bridge module. The CNB has only one network connection, whereas the CNBR contains two connectors for network redundancy. Figure 11-7 illustrates the main features of a 1756-CNBR module. Notice the two BNC connections on the bottom of the module. One is Channel A, the other Channel B. The Channel B connector is present only on a CNBR module and is used for redundancy. The network access port (NAP) is an RJ-45 connector used as an interface between a notebook personal computer and the network, typically while working on the factory floor. A maintenance or electrical individual can walk up to any CNB or CNBR module with a personal computer and the appropriate interface card and cable, and connect to the network at any network access port.

ControlNet channel status indicators and the module status indicator are used to view the health of the module, and typically are used for troubleshooting. Also used for troubleshooting is the module status alphanumeric display. Here, module information such as firmware level can be viewed as the module is powered up. The module's node address

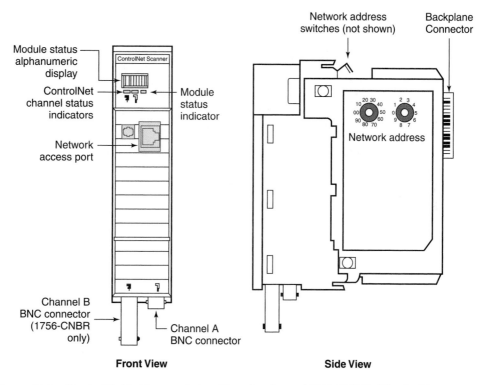

Figure 11-7 ControlNet bridge module with redundancy (1756-CNBR). (Used with permission of Rockwell Automation, Inc.)

and troubleshooting messages are displayed as the module is running. Located on the top of the module are the network address switches. These two rotary switches are used to set the ControlNet network node address for this module.

When distributing I/O around the plant floor, there are three choices available. First, a dedicated PLC can be placed at the location where the I/O is required. This PLC will have its own program running in the processor to manage the I/O at that location. The downside to this method is that there is no communication back to a main or supervisory PLC or other PLCs in the plant. A second option would be to purchase a PLC chassis or other form of I/O, such as Flex I/O, and connect it to a communication module via a network connection in a PLC chassis. This method will allow communication back to the supervisory PLC processor so that information can be shared. If you want intelligence at the point of the remote I/O, a processor could be installed at the site of the remote I/O and communication set up back to a supervisory PLC. Figure 11-8 shows a FlexLogix processor connected to Flex I/O blocks. Instead of having a communication module connected to the Flex I/O blocks as in the second option, here we have a 32-bit ControlLogix processor running RSLogix 5000 software at the remote site using the same Flex I/O blocks. FlexLogix, a member of the ControlLogix Family, uses standard Flex I/O blocks and bases for its I/O connectivity. A Flex I/O block (A) and terminal strip base (B) can be seen

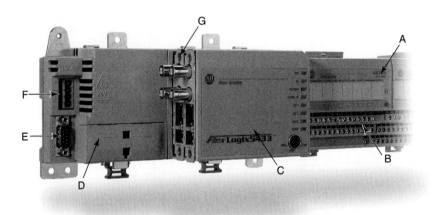

Figure 11-8 FlexLogix PLC with two ControlNet daughter cards installed providing ControlNet network connectivity. (Used with permission of Rockwell Automation, Inc.)

at the right end of Figure 11-8. The FlexLogix 5433 processor is in the center of the figure (C), and the power supply is to the left. The battery is directly under the door on the power supply (D). The nine-pin D-shell connector (E) is for serial communication. F is the 24-volt power connection. Notice that the left side of the processor has four ControlNet bayonet connectors. This particular picture has two ControlNet communication cards (G) (called daughter cards) installed in the FlexLogix processor.

The two connectors on each card provide redundancy. Now communication between this remote PLC and the supervisory PLC can take place across a ControlNet network. There are two daughter-card slots in the FlexLogix processor. In this particular picture, two ControlNet cards are installed. They could be any combination of Ethernet/IP, ControlNet, or DeviceNet.

DESKTOP OR INDUSTRIAL COMPUTER INTERFACE CARD

A circuit-board interface card needs to be installed in an expansion slot in your desktop or industrial computer to provide an interface between your personal computer and the ControlNet network. Figure 11-9 is a KTCX15 interface card. Note the network access port. Channel A and Channel B are BNC, for coax network connectivity and redundancy, if desired.

Figure 11-10 is an example of using the KTCX15 interface card to connect through the BNC connectors to an SLC 500 1747-SCNR module on a redundant ControlNet network.

If using a notebook personal computer, a PCMCIA-type card is used in place of expansion cards such as the KTCX15. A 1784-PCC card is a PCMCIA credit-card-sized interface card for insertion into a notebook computer's PCMCIA slot to interface to the ControlNet network. Figure 11-11 shows ControlNet interface using a notebook

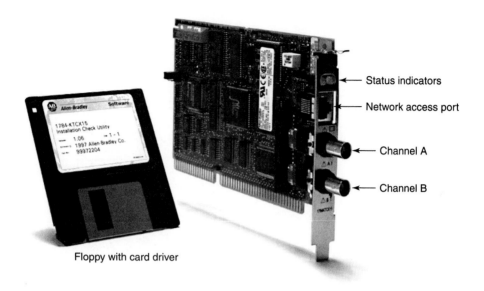

Figure 11-9   Personal or industrial computer ControlNet KTCX15 ControlNet interface card. (Used with permission of Rockwell Automation, Inc.)

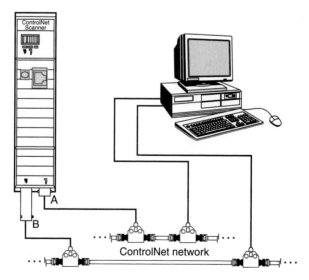

Figure 11-10   Personal computer interface to 1747-SCNR using a KTCX15 interface card connecting via coax. (Used with permission of Rockwell Automation, Inc.)

computer, with the interface cable connected to a 1747-SCNR network access port (NAP). Note network redundancy, with Channel A and Channel B taps connected to the network coax cable trunk line.

Figure 11-12 is a KTCX15 interface card and a desktop personal computer connected to the network access port of a PLC 5 ControlNet processor.

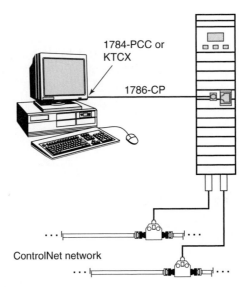

Figure 11-11 Personal computer interface to 1747-SCNR using either 1784-PCC or 1784-KTCX15 interface card connecting to NAP. (Used with permission of Rockwell Automation, Inc.)

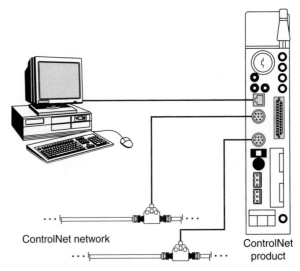

Figure 11-12 KTCX15 ControlNet interface card connected to a PLC 5. (Used with permission of Rockwell Automation, Inc.)

CONTROLNET CABLE SYSTEM

The trunk cable is the bus or backbone of the network. The trunk cable may contain one or multiple sections called segments. Up to 48 nodes may be on a segment before a repeater must be used. If using fiber and the appropriate repeaters, a ControlNet network could be up to 18.6 miles long.

The trunk line is either quad-shielded RG-6 coax, fiber, or special-use cable depending on the environment in which your cable will be installed. Cable can range from standard quad-shielded RG-6 coax, to armored, to fiber, and also to special cables such as flood burial, high-flex, or a mixture of fiber and cable.

A tap is required as the drop line off the trunk cable to each field device or node. Taps come in many configurations. Refer back to Figure 11-3. The drop line length is fixed at 39.6 inches. A node is a physical device on the network; up to 99 nodes are permitted on a ControlNet network. Each node must have a unique decimal node address within the range of 1 to 99. There is no node 0 in ControlNet.

CONTROLNET SEGMENT

A segment consists of a number of sections of trunk cable separated by taps and the two required termination resistors. The maximum length of a segment is defined as 1,000 meters (3,280 feet) of standard RG-6 coax with two taps and one termination resistor on each end. Figure 11-13 illustrates a basic segment with three taps. Note the termination resistors on each end. If you started with two taps and 1,000 meters of coax, as additional taps are added to the network the cable must be shortened by 16.3 meters (53 feet) for each additional node added.

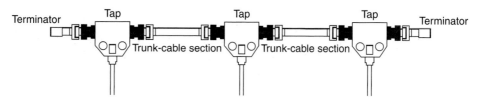

Figure 11-13 Basic ControlNet segment with RG-6 coax cable. (Used with permission of Rockwell Automation, Inc.)

Up to 48 nodes can be placed on any one segment. The total cable length for a network segment with 48 nodes is 250 meters (820 feet). There is no minimum trunk line distance between taps, so taps can be connected directly. The following cable calculation formula is used to determine the total length of cable.

1,000 meters − [16.3 meters × (the number of taps − 2)]

As an example, a segment of standard RG-6 coax requires 15 nodes. To calculate the maximum segment length, use the formula:

1,000 − [16.3 × (15 − 2)]

1,000 − [16.3 × (13)]

1,000 − 211.9

788.1 meters is the maximum trunk line length for this network using standard or light industrial R6-6 coax.

Other considerations and calculations are required for special cables such as high-flex. Refer to the ControlNet Coax Media Planning and Installation Guide for additional assistance for determining trunk cable section lengths. For fiber calculation and installation information, refer to the fiber installation manuals.

If more than 48 nodes are required a repeater must be used to connect a second segment. Figure 11-14 illustrates a repeater used to connect two segments. Two segments connected by a repeater is called a link.

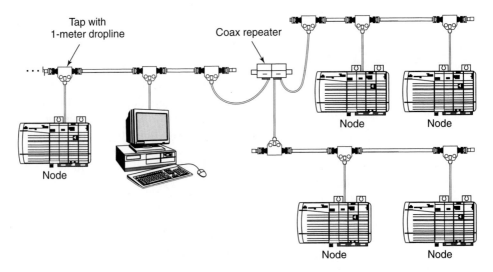

Figure 11-14   Three ControlNet segments connected by a repeater to create a star topology.

Figure 11-15 is a ring topology. Notice this network is a combination of coax and fiber and four repeaters.

Repeaters

A repeater is required if additional nodes are needed after either the maximum number of nodes or cable length has been reached. Two modules are required to build a repeater. A 1786-RPA or repeater adapter supplies power and coordinates communication. Figure 11-16 illustrates a repeater adapter module. Notice the power connections, lower left, and LED status indicators and BNC connector, lower right.

The other part required is the 1786-RPCD module. These two modules are connected together on DIN-rail to give us a repeater. Up to four 1786-RPCD modules can be connected to a single repeater adapter module. The copper repeater module 1786-RPCD, in conjunction with a repeater adapter, 1786-RPA, allows multiple 1,000-meter sections to be connected. See Figure 11-17 for a 1786-RPCD module and coax connection.

Figure 11-18 illustrates two RPA-RPCD repeater pairs connecting two ControlNet segments. Notice a PLC 5, upper left, connected to the network. This network is designed

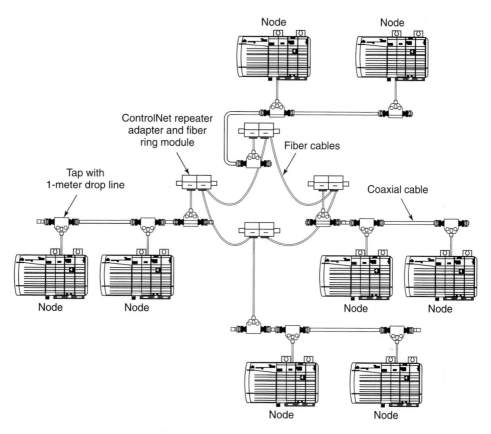

Figure 11-15   ControlNet ring topology using repeaters and a combination of fiber and coax.

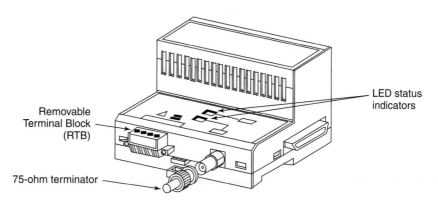

Figure 11-16   1786 repeater adapter module. (Used with permission of Rockwell Automation, Inc.)

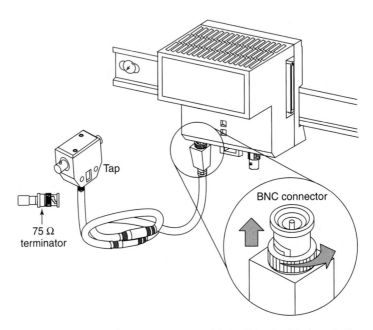

Figure 11-17   1786-RPCD dual copper repeater module. (Used with permission of Rockwell Automation, Inc.)

so the PLC 5 can communicate to two remotely mounted blocks of I/O across a Control-Net network; such as the two 1794-ACNR15 communication modules, one on Channel 1 (CH 1) and the other on Channel two (CH 2), bottom center. These two devices are on separate segments, so this network has three segments. The ACNR15s are DIN-rail-mounted communication modules that allow for up to eight modular Flex I/O digital or analog blocks to be connected. This was one of our distributed I/O options discussed in conjunction with the FlexLogix processor.

Figure 11-15 illustrated a fiber ring topology using fiber repeaters in conjunction with coax cable. Figure 11-19 shows a fiber repeater. This would be the right half of the fiber repeaters illustrated in Figure 11-15.

CONTROLNET AND SCHEDULED AND UNSCHEDULED COMMUNICATIONS

When working with ControlNet, the programmer needs to separate data into two categories, information that is time-critical and information that is not. The programmer must determine what is real time for this specific application. In many cases programmers wish they could have everything instantly; unfortunately, network bandwidth is not unlimited. As an example, if you had a tank that took four hours to fill, why would the level be needed every ten milliseconds? In many cases this could be lengthened to be

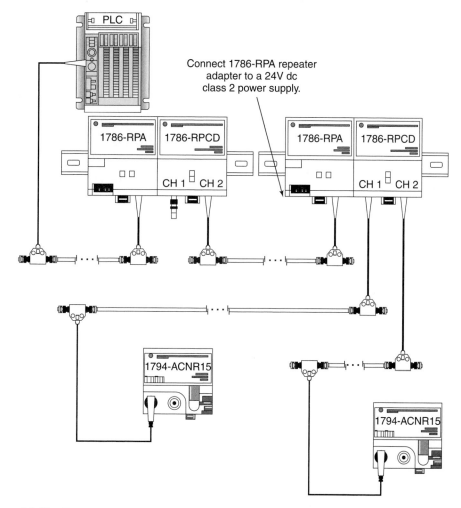

Figure 11-18   Repeaters connecting three segments. (Used with permission of Rockwell Automation, Inc.)

more realistic, thus allowing valuable bandwidth to be conserved for something more important, such as interlocking the starting of two variable-frequency drives. In some cases the tank-level information could be moved to the non-time-critical portion of the bandwidth. Non-time-critical data is transferred by triggering a message instruction on the ladder rung.

ControlNet communication falls into two general categories, scheduled and unscheduled. When determining communications on your system, determine which data is time-critical and must be transferred in a repeatable, deterministic fashion, and which data can be categorized as unscheduled, and be transferred as time allows. Scheduled data typically

Figure 11-19   ControlNet fiber repeater.

will have a requested packet interval, or RPI, as part of the setup. In many cases these will be produced and consumed tags created in a ControlLogix application, as an example. Unscheduled information typically is transferred by triggering a message instruction on a PLC ladder rung.

ControlNet has three sections of the network bandwidth:

Scheduled

Time-critical data would be classified as scheduled. Each node on the ControlNet network would be guaranteed an opportunity to transmit information for each rotation of the network.

Unscheduled

Non-time-critical information would be unscheduled. Unscheduled information would be transferred when time became available within the bandwidth. The amount of time available within the bandwidth is determined not only by how the network is set up, but, more important, by the amount of scheduled traffic on the network. More scheduled traffic means less time for unscheduled traffic.

Nodes transmit information from the lowest to the highest node address on a rotating basis. When the unscheduled portion of the bandwidth is active, the current node whose turn it is to transmit will transmit its data. If there is additional time in the unscheduled portion of the bandwidth, then the next node in sequence will be allowed to transmit its unscheduled data. This is continued until the unscheduled time expires. The amount of scheduled network traffic will determine how much time is available for unscheduled service. Heavy scheduled network traffic could leave little time for unscheduled service. ControlNet guarantees that one node will be given the opportunity to transmit per rotation of the network. Examples of what would fall under unscheduled service include:

Message instruction triggered from logic on a ladder rung

Uploads and downloads

I/O and communication modules' connection establishment

Maintenance or Guard Band

The maintenance or guard band is fixed and is reserved for network maintenance duties. The lowest node number should be a module such as a CNB. This module is called the keeper and is typically node 1. The keeper is kind of like a traffic cop. It transmits information to keep all nodes synchronized during the maintenance portion of the bandwidth.

NETWORK UPDATE TIME

Scheduled information is transferred with each rotation of the network, called the network update time, or NUT. The network update time is the sum of scheduled service, unscheduled service, and the maintenance band. Network update time is determined by the network designer in conjunction with the amount of scheduled service required. As the network expands over time, maintenance workers or electricians might have to adjust the network update time. The network update time and other network parameters are set up in RSNetWorx Software for ControlNet. As the number of scheduled messages increases, the network update time will need to be extended. The downside of extending the network, allowing for more scheduled traffic, is that network updates will be longer. Figure 11-20 illustrates the network update time and the three pieces to the bandwidth-scheduled service, unscheduled service, and the maintenance band.

If the network is properly designed and installed, all scheduled information is guaranteed to transfer with each network update. By design, a minimum of one packet of up to 500 bytes of unscheduled data is transferred each NUT. If there is remaining time in the NUT after the initial node has transferred its information, additional nodes may transfer data until the NUT has expired. The maintenance, or guard band, time is fixed and is used for network housekeeping duties.

SCHEDULED MAXIMUM NODE

The scheduled maximum node (SMAX) that will be transferring data must be at or below the SMAX setting in RSNetworx for ControlNet. Network nodes with scheduled

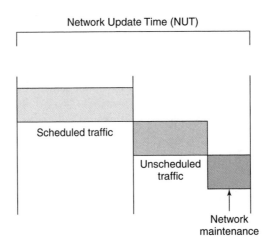

Figure 11-20   ControlNet Network update time bandwidth pieces.

information to transfer should all be grouped together as lower sequential node numbers. For maximum network efficiency, there should be very few unused node numbers, starting with the lowest node number, usually node 1, up to the last node number to have scheduled data to transfer. Figure 11-21 shows the network properties screen in RSNetworx for ControlNet software. Notice that the network update time is being changed to 10 milliseconds in the pending column and node 15 in this example is set as SMAX. As a result, no node greater than 15 can send scheduled data across the network, the maximum scheduled address or SMAX.

Caution must be used when configuring a new network or adding nodes to an existing network. If there is a node wanting to transfer scheduled data whose node address is above SMAX, its data will not be transferred, as the node is above the maximum scheduled and is not permitted to transfer data. Keep in mind that a node can transmit both scheduled and unscheduled data.

UNSCHEDULED MAXIMUM NODE

Nodes transferring unscheduled data should be grouped together sequentially, starting at the next node sequentially above SMAX, and leaving a minimum of open unused node addresses. Another consideration during the design stage should be: Is anyone going to connect to a NAP with a personal computer in the future to gain access to the network for PLC program monitoring or editing? If so, then what node addresses need to be left unused for future PLC connectivity? Remember, PLC uploads and downloads, including monitoring or on-line editing via ControlNet, is unscheduled service. There should be one or two unused node addresses left open for computer connectivity.

When configuring RSNetWorx, an unscheduled maximum node parameter will be set to identify the maximum node number that will be permitted to transmit unscheduled

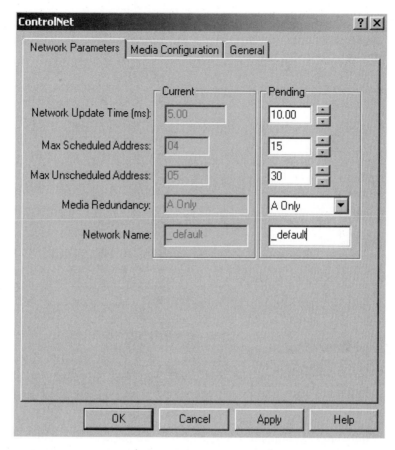

Figure 11-21   RSNetWorx ControlNet network view showing NUT and SMAX parameters being edited. (Used with permission of Rockwell Automation, Inc.)

data. This is called the UMAX. If any node is assigned an address above UMAX it will not be permitted to transmit data. Refer to Figure 11-21 to see that the UMAX parameter for this example is being set to 30. Communication for anyone who might connect a desktop computer or a notebook computer to the network to upload, download, or monitor the devices on the network, would need to know the available node addresses for computer connectivity so as to be at or below UMAX, but they would have to be careful to not create a duplicate node situation. If someone attempted to connect to the network with a personal computer at node 31 or above, they would be outside the UMAX range and would be unable to communicate on the network.

Care must also be exercised when setting up communication drivers for personal computer communication to the network. RSLinx software is used to configure the driver necessary for a personal computer to communicate. As an example, if using a notebook personal computer and a PCMCIA-type ControlNet interface card such as a

1784-PCC card, the default network node address for the personal computer would be set at 99. Hopefully, it is obvious that a UMAX of 30 will not allow network communication to a personal computer with a node address of 99.

SLOT TIME

When allocating node addresses for scheduled and unscheduled service, we had mentioned earlier that only one or two node addresses should be unused for future connectivity to the network or for NAP connectivity with a personal computer. All scheduled nodes should be grouped together below SMAX, and all unscheduled nodes should be grouped together at or below UMAX, so as to minimize the effect of slot time on network performance. Slot time is the time the network waits for an unused network address or missing node. The actual time depends on the physical attributes of the network. Variables such as trunk line cable length and the number of repeaters will affect the slot time. Sloppy network design, such as high values in SMAX and UMAX, and a large number of unused node addresses scattered over the range of these two parameters, will result in an excessive amount of slot time. Excessive slot time means more time waiting for nodes that do not exist and results in a very slow, inefficient network.

RSNETWORX FOR CONTROLNET

Before scheduled connections can be used on the ControlNet network, the network must be scheduled. Also, any time any scheduled connection is added to, modified, or removed from the network, the network will have to be rescheduled. To schedule a ControlNet network, you must use RSNetWorx for ControlNet software. The software is used to configure the network parameters and schedule the network components, and to calculate the bandwidth of each device and the bandwidth of the entire network. After the network parameters are set up in RSNetWorx, the network configuration is saved. Part of the save includes optimizing and rewriting the network connections. This is called scheduling. Basically, scheduling is downloading the network configuration to the ControlNet communication module such as a 1756-CNB, called the "keeper." The keeper is typically the lowest node number, and the suggested node number for the keeper is node 1. The keeper acts as a "traffic cop" and directs traffic on the network.

Figure 11-22 shows a screen from RSNetWorx for ControlNet. Notice the right pane; this is a graphical representation of the ControlNet network. This is created when the user goes on-line with the network using RSNetWorx. There are two ten-slot ControlLogix chassis on the left at nodes 1 and 2. Node 5, on the right, shows the personal computer-to-network interface as a 1784-PCC card.

If RSLinx is not set up, or not communicating with a ControlNet network, going on-line to read the network configuration with RSNetWorx is not an option. The left pane in Figure 11-22, entitled hardware, can be used to configure a network manually.

Near the top of the screen in Figure 11-22 is the section entitled Network Usage. This is where editing is enabled and network usage information can be viewed.

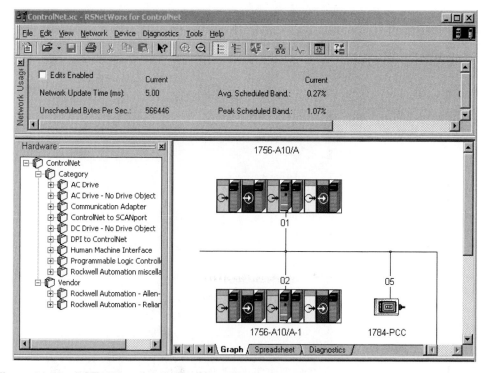

Figure 11-22   RSNetWorx for ControlNet software screen.

INTRODUCTION TO THE DEVICENET NETWORK

DeviceNet is also an open network intended to link low-level devices such as sensors, push-button stations, distributed I/O blocks, intelligent motor starter overload devices, and variable frequency drives to higher-level devices such as PLCs. DeviceNet, similar to ControlNet, is managed by an independent group called the Open DeviceNet Vendors Association, or ODVA. Their Web address is http://www.odva.org. Because it is an open network, devices or nodes from many different manufacturers can be nodes on it. A DeviceNet network is made of the following components:

Trunk line

Drop lines

PLC scanner

Nodes

Minimum of one power supply

Two termination resistors

Figure 11-23 provides a small sample of some of the DeviceNet media components available. A key on the next page accompanies the figure. Keep in mind that there is an entire catalog of media, switches, distributed I/O devices, and encoders, etc.

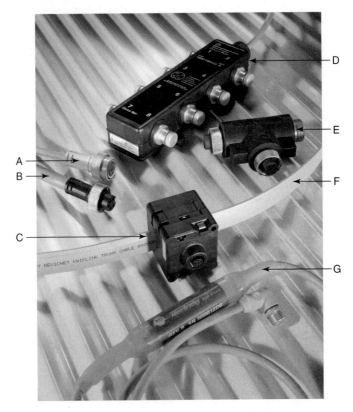

Figure 11-23 Samples of some DeviceNet media components. (Used with permission of Rockwell Automation, Inc.)

A. Thick round drop line cable.

B. KwikLink™ drop line cable.

C. KwikLink™ flat trunk line cable insulation displacement connector. The insulation displacement connector's connection teeth bite through the flat cable insulation as it is clamped to the flat cable. No tools other than a screwdriver are required to install a drop line off the flat trunk cable. See Figure 11-24. Note the round connector on the bottom of the insulation displacement connector. A drop cable similar to B would be connected here.

D. A DevicePort™. Devices are passive four- or eight-point taps that are connected to the trunk line using a drop cable. Figure 11-25 shows a cable connected to the top of the DevicePort™. This is the drop cable that would connect back to the trunk line. The picture also illustrates an eight-port device. Notice the connection points on either side. Nodes are connected to these points by way of another drop cable.

E. The T-Port is used to connect a drop line off the trunk line. As shown, the trunk line connects to the right and left sides of the T-Port while the drop line cable (similar to A in this figure) is screwed to the bottom connection. T-Ports are rated NEMA 6P.

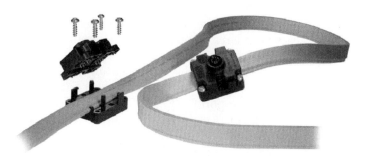

Figure 11-24   KwikLink™ DeviceNet Connection. (Used with permission of Rockwell Automation, Inc.)

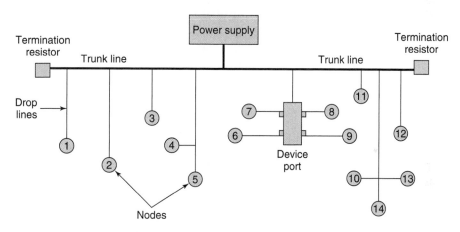

Figure 11-25   DeviceNet network components.

F. KwikLink™ flat trunk line cable.

G. DeviceLink™ is an adapter that will allow interfacing of standard non DeviceNet two- or three-wire 24v sensors with open collector sourcing outputs, mechanical devices such as limit switches, or any device with relay contacts to the DeviceNet network. One DeviceLink is required for each node.

Flat cable is typically used for non-washdown conveyor applications. Flat cable can be run on the inside rail of a conveyor. No conduit is required. Nodes can be easily connected to the flat cable by insulation displacement connectors. Insulation displacement connectors are hinged units that have teeth that, when closed, bite through the insulation to mask contact with the conductors. These are simply placed over the cable and closed. Two screws hold each connector in place. Refer to Figure 11-24. There is no minimum spacing between connectors, and new nodes, after they are properly commissioned, can be attached to a running network.

Figure 11-25 illustrates the basic components of a simple DeviceNet network. Notice the trunk line. Each end is required to have one 121-ohm, 1/4-watt terminating resistor. For this example we are using the thick cable for the trunk line and thin cable for the lines dropping from the trunk. These are called drop lines. The circles represent nodes on the network, and each node has a number called a node address. Duplicate node numbers are not allowed. A DeviceNet network could have up to 64 nodes.

There are many devices available for connecting nodes or I/O points to the network. As an example, a DevicePort™ is being used to connect nodes 6, 7, 8, and 9 to the trunk line. The figure illustrates a four-port DevicePort™. The DevicePort™ is available in either four or eight ports. A picture of an eight-port DevicePort™ is included in Figure 11-23 D.

The software used to configure a DeviceNet network is RSNetWorx for DeviceNet. This is not the same software as RSNetWorx for ControlNet or RSNetWorx for Ethernet/IP. Each network has its own specific software. Figure 11-26 illustrates an RSNetWorx for DeviceNet screen. The icons show an actual picture of the device. The node address is listed directly above or below the node object. Node 00 is a ControlLogix DeviceNet Communication Module, also called a DeviceNet Bridge (DNB). This module is the interface between the ControlLogix PLC and the DeviceNet network. Node 01 is an SLC 500 1747-SDN. The SDN is the SLC 500 communication module interface between the SLC 500 and DeviceNet. These communication modules are also called scanners. All other nodes are devices on the DeviceNet network.

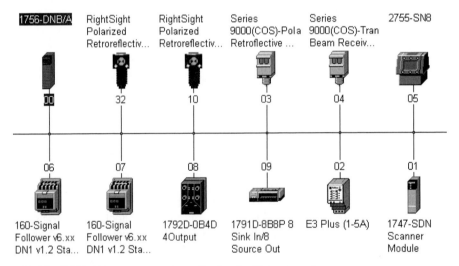

Figure 11-26 RSNetWorx screen view of a DeviceNet network.

A DeviceNet network can support up to 64 decimal nodes, depending on the total amount of data transferred between the nodes. Keep in mind that the DeviceNet network was intended for low-level devices that exchange small amounts of data, usually either a few bits or bytes. DeviceNet was not intended to support operator interface, such as a number of PanelViews, each exchanging 100 words each. Hardware such as this should be placed on Ethernet. If data needs to be shared between a DeviceNet node and the Ethernet operator interface, a pair of Ethernet and DeviceNet bridge modules can be used in a ControlLogix chassis.

DeviceNet network is popular, as connecting devices together on a network can save wiring costs. All nodes attach to one cable called the trunk line. The cable has four wires, two for signal and two for power. The round cable also has a drain wire. There is also specialty cabling, depending on the environment where you are running your cable. Thick round and thin round can both be used for the trunk line of the network. While thin round is in many cases used for drop lines, it can also be used as the trunk line in small applications with low current draw. Selection of either cable is determined by the length of the backbone and total current draw.

NETWORK POWER SUPPLIES

A DeviceNet network will have one, possibly more, power supplies depending on the design and current requirements of the network. When designing a new network or adding additional devices to the network, power-supply calculations must be completed to ensure proper power supply selection and network operation. Even though there are several considerations when selecting, installing, and sizing a power supply for your network, the calculations are rather straightforward—just add up the total current required by all the devices on the network. This is the minimum nameplate current rating of the power supply you select. On the other hand, if you need to add one or more new devices to the network, ask if the current power supply will handle the present requirements for current and new devices. Power-supply calculations need to be rechecked. If the calculations show the current supply is undersized, either the existing power supply will need to be replaced with a larger one, or the situation might require adding a second power supply to the network. It is very important that a quality ODVA-compliant power supply be used to ensure proper network operation. Refer to the DeviceNet Media Design and Installation Guide, which can be found at http://www.ab.com. Select Publications Library, then Manuals On-line, to access manuals and documentation.

MAXIMUM TRUNK LINE DISTANCE

The maximum trunk line distance is defined as the maximum cable distance between any two nodes. This does not necessarily mean the actual length of the trunk line, as we will discuss. The maximum length of the cable is determined by the cable type and the network baud rates. The table below describes baud rates, sample cable types, and their respective maximum trunk line distances.

Maximum Trunk Line Cable			
Baud Rate in bits per second	Thick Round	Thin Round	Flat Cable
125K	1,640 ft (420m)	328 ft (100m)	1,378 ft (420m)
250K	820 ft (250m)	328 ft (100m)	656 ft (200m)
500K	328 ft (100m)	328 ft (100m)	246 ft (75m)

Example One

Referring to Figure 11-27, let's assume thick round cable trunk line:

1. The distance from the left terminating resistor and the node 1 drop line is 12 feet.
2. The drop line length to node 1 is 2 feet.
3. The length from the right terminating resistor and the drop to node 12 is also 12 feet.
4. The drop line length to node 12 is 2 feet.
5. The distance from the node 1 drop line to the node 12 drop line is 800 feet.

For this example, the total trunk line length is a total length of cable between the terminating resistors, which is 12 + 800 + 12 = 824 feet. Referring to the table using the round thick cable, what are our baud rate options? Since we have over 820 feet of trunk line, our only option is 125K. The figure below highlights the cable lengths used in the calculations.

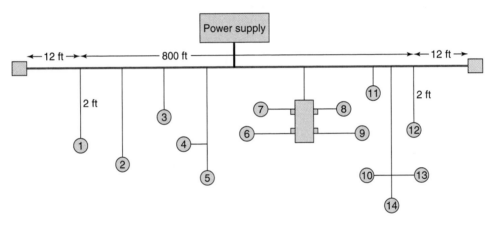

Figure 11-27   Determining maximum trunk line cable distance.

Example Two

Refer to Figure 11-28:

1. The distance from the left terminating resistor and the node 1 drop line is 12 feet.
2. The length of the drop line to node 1 is 20 feet.
3. The length from the right terminating resistor and the drop to node 12 is 8 feet.
4. The length of the drop line to node 12 is 15 feet.
5. The distance from the node 1 drop line to the node 12 drop line is 800 feet.

For this example, the total trunk line length is the *maximum* length of cable between node 1 and node 12, which is 20 + 800 + 15 = 835 feet. Referring to the table above, using the round thick cable, what are our baud rate options? Since we have over 820 feet, our only option is 125K. The figure below highlights the cable used for the calculations. The main point here is that you need to determine the maximum length between two nodes, or the terminating resistor(s)—which is longer? In this case, the drop lines were longer than the distance from the drop line to the terminating resistor, so the drop line lengths must be included in the calculation.

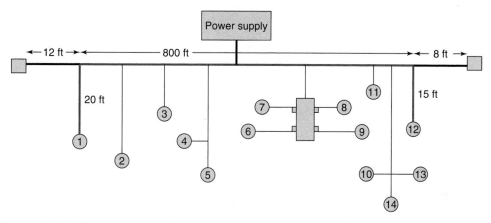

Figure 11-28   Maximum cable distance between two nodes used for calculation.

Example Three

Refer to Figure 11-29:

1. The distance from the left terminating resistor and the node 1 drop line is 20 feet.
2. The length of the drop line to node 1 is 6 feet.
3. The length from the right terminating resistor and the drop to node 12 is 2 feet.
4. The length of the drop line to node 12 is 8 feet.
5. The distance from the node 1 drop line to the node 14 drop line is 300 feet.
6. The distance between the node 14 drop line and the node 12 drop line is 3 feet.
7. Node 14's drop line is 12 feet.

For this example, the total trunk line length is a maximum length of cable between the left terminating resistor and node 14, which is 20 + 300 + 12 = 332 feet. The longest cable distance is between the left terminating resistor and node 14. Referring to the table above, using the round thick cable for our trunk line, what are our baud rate options? Since we

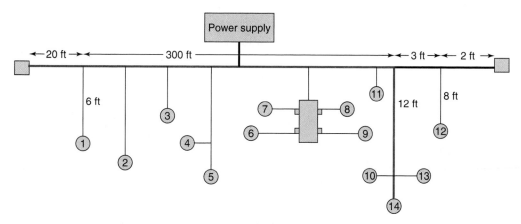

Figure 11-29   Determining maximum truck line distance.

have over 328 feet, our options are 250K or 125K. The main point here is that you need to determine the maximum length between two nodes, or the terminating resistor(s). In this case, the drop line to node 14 was longer than the distance from node 14's drop line to the right terminating resistor, so only one drop line length must be included in the calculation. Figure 11-29 highlights the network used for the calculation.

CUMULATIVE DROP LINE LENGTH

Our next consideration is to determine the cumulative drop line length. This is simply the sum of all drop line cables. The table below lists the maximum cumulative drop line lengths for each network baud rate.

DeviceNet Cumulative Drop Line Length	
Baud rate in bits per second	Cumulative length
125K	512 feet (156m)
250K	256 feet (78m)
500K	128 feet (39m)

Figure 11-30 below illustrates the same DeviceNet network we have been working with, now with drop line lengths identified. To determine the cumulative length, simply add all the lengths together.

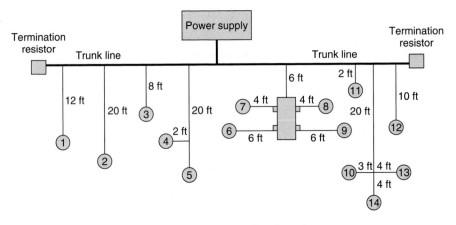

Figure 11-30   DeviceNet network cumulative drop line length.

The total is 131 feet. Referring to the table above, what is the maximum baud rate allowable for this network?

There is another rule, however; the maximum length for any one drop is 20 feet. Refer to Figure 11-30. Do you see a problem? Node 14 is a 24-foot drop. This must be fixed. Let's assume you shortened node 14's drop to 20 feet. Does this have any affect on your maximum allowable baud rate?

DEVICENET INTERFACE

The type of personal computer you will use to attach to the DeviceNet network will determine the interface card and cable required for your particular application. As with other network connections, either a circuit board, for installation into a desktop or industrial computer, or a PCMCIA card, for installation into a notebook personal computer, are needed to communicate to the network.

If using a notebook personal computer, the 1784-PCD, a PCMCIA-type card, could be installed in the computer for interfacing to DeviceNet. There is also a gray plastic interface box known as a 1770-KFD that can be connected to a computer's serial port to interface to DeviceNet. In all cases listed above, the appropriate RSLinx driver will need to be configured. If using a ControlLogix PLC, most people would use Ethernet to upload, download, and do on-line editing of their PLC project. When using DeviceNet, there would be a DeviceNet bridge module in the ControlLogix chassis to access the DeviceNet network. Because of this, the easiest way to access the DeviceNet network would be through Ethernet communications to the chassis, and simply bridging across the backplane to DeviceNet. Communicating in this manner eliminates the need to purchase a DeviceNet interface card for your computer.

To interface to your PLC, a DeviceNet communication module is required in the chassis. Figure 11-31 illustrates a DeviceNet daughter card for a FlexLogix (refer back to Figure 11-8). Notice the top five-pin connector (A); this is the DeviceNet network cable connector connection point. Directly below this is the switch for setting up the interface card's baud rate (B). There is some text just above the network cable connector that reads 0-125K, 1-250K, 2-500K, and PGM. Set the switch to the network's baud rate. PGM means the baud rate is set up in software. Directly below this switch are some status LEDs (C). Switches on the very bottom of the card (D) set up the card's node address, which currently appears to be 00. If communicating with an SLC 500, a 1747-SDN scanner module is needed. A PLC 5 needs a 1771-SDN, whereas ControlLogix requires a 1756-DNB (DeviceNet Bridge) module. Figure 11-32 is that of a ControlLogix DeviceNet bridge module. Notice the network connection near the bottom of the module. There is a square window near the top left of the module where information such as error codes and module information will be displayed. Directly below the window are three LEDs providing information on the module and network. The window and LEDs are used to view the overall health of the network, and any error codes currently on the network.

NODE COMMISSIONING

Node commissioning consists of taking a new device and setting up its node address and baud rate, and programming its parameters. This can be accomplished in one of two ways—connecting to a live network, and using RSNetWorx to complete the setup. Keep in mind that some devices can be hardware-commissioned, meaning they have switches for setting up node address and baud rate. Some devices need to be commissioned using

Figure 11-31 DeviceNet interface card that would be installed in a FlexLogix PLC. (Used with permission of Rockwell Automation, Inc.)

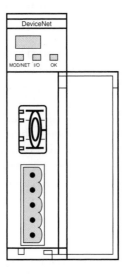

Figure 11-32 ControlLogix DeviceNet Bridge module. (Used with permission of Rockwell Automation, Inc.)

software, but others can be commissioned either way. Remember the program (PGM) switch on the interface card in Figure 11-31? One very important consideration for on-network live commissioning is that duplicate node addresses are not allowed. Many devices come from the factory as node 63, but some newer hardware- and software-configurable devices are shipped with an address of 99. Be sure to review the manufacturer's installation instructions before attempting to commission a new network device. Second, if you put a new device on a live network at the wrong baud rate, the network will crash immediately. Note that some newer devices come with an auto baud selection switch; again, consult the manufacturer's documentation before proceeding.

A safer way to commission a new device is to have a separate node commissioning station. This will allow setup of the device off the network, reducing possible baud rate and duplicate node problems. A node commissioning station would consist of a personal computer with RSLinx and RSNetWorx, the proper DeviceNet interface card and cables, a DeviceNet-compliant 24-volt power supply, and the device to commission. Once completed, simply install the device on the network.

RSNETWORX FOR DEVICENET SOFTWARE

RSNetWorx software is used for either off-line or on-line configuration of the network. The left pane of Figure 11-33 shows a list of categories of network devices. To create a new off-line configuration, find the device you wish to put on the network and double-click on it, and it will be displayed on the right pane on the network. Now set up the properties of

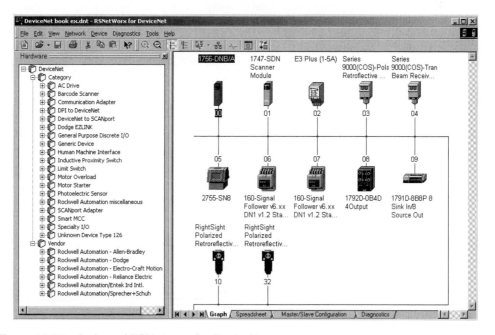

Figure 11-33   A view of RSNetworx for DeviceNet.

the device. By going on-line with the network, you can have the scanner and software go out and read the network. All devices on the network will be displayed on the right pane. The Scanlist and device properties setup will be discussed next.

DeviceNet Scan List

After all devices have been installed or the network modified, RSNetWorx software is used to go on-line with the network. Going on-line allows the software to scan the network and identify each node. Nodes will be displayed in the right pane of RSNetWorx software. Refer to Figure 11-33. Once the DeviceNet scanner knows what nodes are out there, they are placed in the available devices view under the Scanlist tab for the scanner module in RSNetWorx. These are the nodes on the network that are available; however, they are not scanned by the DeviceNet scanner module until they are put into the Scanlist. Refer to Figure 11-34 below and note the following:

    A. Available devices view. These devices were seen by the scanner and software when going on-line with the network.

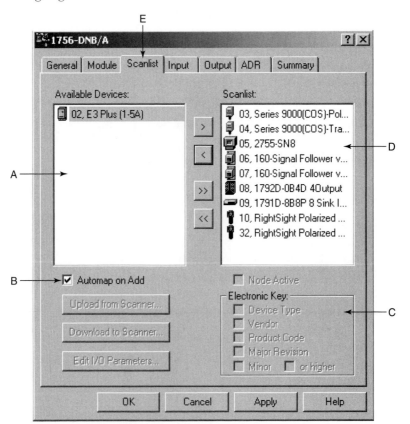

Figure 11-34   Scan List for the ControlLogix DNB module.

B. Do you want the available devices auto-mapped when they are added to the scan list? We will explore mapping in a minute.

C. Electronic keying. How close a match does a node replaced, say due to failure, need to be to the original? Check the boxes you desire. Typically the device type, vendor, product code, and, possibly, major revision are checked.

D. This is the Scanlist, or the devices that will be scanned on the network. If a device or node is not in the Scanlist, as far as the scanner is concerned it does not exist. Notice the single and double arrows between the available devices view (A) and the Scanlist view (D), click on a device to put on or remove from the Scanlist, click on the single arrow pointing in the direction you wish to go, and click on it. Use the double arrows to move all devices.

E. This is the Scanlist tab for a 1756-DNB, which is in a ControlLogix chassis.

Now that we have the devices we wish to scan on the network in the Scanlist we can proceed to mapping the devices.

Input Data Mapping

Mapping sets up how each network device interfaces with the PLC data table. In this section we will look at mapping inputs to a ControlLogix PLC. Figure 11-35 is an input map view from RSNetWorx. Note the following pieces of the mapping view:

A. This is a ControlLogix system, as we are using a 1756-DNB DeviceNet scanner.

B. All input devices in the scan list are listed here. Notice the arrow is pointing at node 6, which is a Bulletin 160 variable-frequency drive. Refer back to the Scanlist (Figure 11-34) and RSNetWorx view (Figure 11-33). Also notice that under the Size column is the number 4. Since DeviceNet is a byte-based network, the size of the input data we need to map for the drive is 4 bytes.

C. This is a list of tag addresses within the ControlLogix processor used for this application. The arrow is pointing at 1:I.Data[2]; the full tag for this is Local:1:I.Data[2]. Even though the word "Local" is not displayed, it is understood. As a review of the ControlLogix tag or address configuration, Local identifies this input module in the local chassis, as compared to a remote chassis. The: 1 signifies this module is in chassis slot one. The: I identifies this an input module, while .Data[2] tells us that we are looking at (input) data at double integer 2.

D. This is the data map. This identifies where each device is mapped into the ControlLogix input tags in the ControlLogix processor. Keep in mind that ControlLogix is a 32-bit processor, so by default tags will be 32 bits wide. Refer to Figure 11-36 below for a closer look at the data mapping.

1. Notice in Figure 11-36 that our Bulletin 160 drive at node 6 is mapped to the upper 16 bits of Local:I:1.Data[2]. Bits 16 through 31 include this drive input information. The first word of the drive input structure contains status bits coming into the PLC from the drive, so each will be addressed to the bit level. We are not going to go into the specific drive input structure in this chapter, but as an example, the first bit of the structure will be addressed as Local:1:I.Data[2].16.

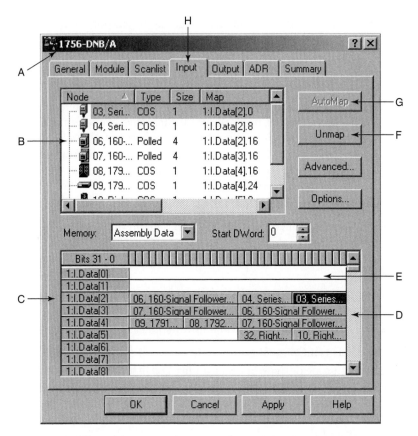

Figure 11-35 Input Data mapping for a ControlLogix DNB using RSNetworx for DeviceNet.

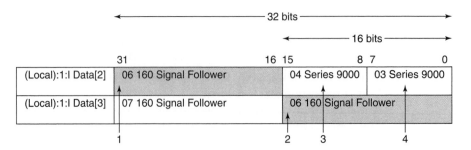

Figure 11-36 ControlLogix input data mapping.

2. The next double integer down is addressed as Local:1:I.Data[3] and the lower 16 bits also contain input data coming to the PLC processor from the drive. This 16-bit word contains the speed reference information from the drive. Sixteen

bits is an integer, so the input tag or address for the drive speed reference will be Local:1:I.Data[3], and its data type will be an INT (integer). This completes the drive input structure.

3. Node 4 is an Allen-Bradley Series 9000 photo switch. One byte has been designated for the photo switch input data. Typically a photo switch has only a couple of bits of information; however, DeviceNet will use the minimum of one byte to represent that data. Series 9000 photo switches have two bits of information to transfer. Bit 0 is the photo switch's output status, on or off. The tag address for bit zero is Local:1:I.Data[2].8. Bit 1 is the margin; a 0 indicates margin is OK, while a 1 indicates the margin is low. The tag address for bit one is Local:1:I.Data[2].9.

4. Node 3 is also an Allen-Bradley Series 9000 photo switch. One byte has been designated for the photo switch input data. Bit 0 is the photo switch's output status, on or off. The tag address for bit zero is Local:1:I.Data[2].0 Bit 1 is the margin; a 0 indicates the margin is OK, while a 1 indicates the margin is low. The tag address for bit one is Local:1:I.Data[2].1.

E. The areas (Figure 11-35) that show no devices mapped are unused memory and can be mapped manually at a later time.

F. Click here (Figure 11-35) to unmap a device.

G. The auto map button (Figure 11-35) can be clicked to have the software select the area of memory and map the tag address for the device currently. The downside to auto mapping is that you have no control over the tag address that your newly mapped device will have. Remember this tag address will have to correlate directly to the ladder logic tag addresses for this device. Manual mapping is always an option.

Output Data Mapping

Mapping sets up how each network device interfaces with the PLC data table. In this section we will look at mapping outputs to an SLC 500 PLC. Remember, the SLC 500 is a 16-bit PLC, so the output status table is 16 bits wide. Figure 11-37 is an output map view from RSNetWorx. Note the following pieces of the mapping view:

A. Node 6 is the same Bulletin 160 drive we looked at above; however, we are looking at output data from the SLC 500 processor to the drive now. Note that we need to map four bytes.

B. This column identifies the PLC output status table address where the information is coming from for use by the network devices.

C. Currently the four bytes are mapped to SLC addresses O:1.2 and O:1.3 FOR NODE 6. Again the basic drive structure is two words or four bytes. The first output word is typically command bits to the drive from the PLC ladder program and processor's output status file. The command bits addresses start with O:1.2/0. The second word is typically the speed command from the PLC. The speed command is an integer, and its address is O:1.3.

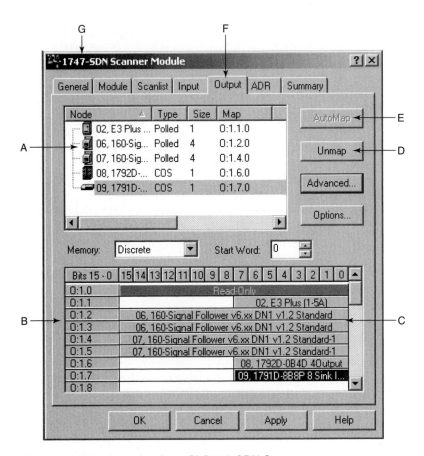

Figure 11-37 Output data mapping for a SLC 500 SDN Scanner.

D. To unmap a network device, select it, then click here.
E. Click here to auto-map a new device. Keep in mind the downside of auto mapping.
F. This is the output mapping tab from the SLC 500 DeviceNet scanner properties found in RSNetWorx.
G. This is a 1747-SDN SLC 500 DeviceNet scanner output properties page.

Configuring and Monitoring Device Parameters

Part of setting up the DeviceNet network is to program the parameters of each device on the network. Figure 11-38 illustrates the general properties tab in RSNetWorx for a Right-Sight retroreflective sensor. Note the following items:

A. The device's properties are displayed after right-clicking on the device and selecting Properties.

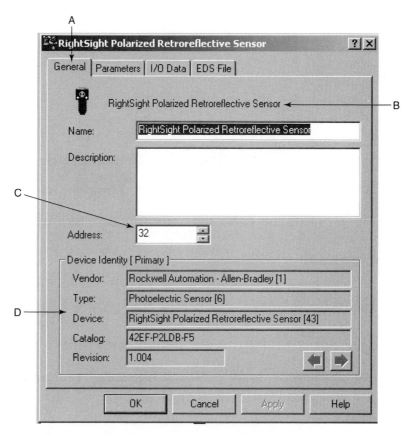

Figure 11-38 RSNetWorx screen showing general properties page for a RightSight polarized retroreflective photoelectric sensor.

B. This area identifies what this device is.
C. This is the node address of this device. The node address can also be changed here.
D. The device's identity is listed in this area. Notice the numbers in the square brackets; these are needed to determine the EDS file number.

The Parameters tab in Figure 11-39 contains all the device's parameters. Parameters can be read-only, or read/write (configurable by the user). When setting up each device, the device's properties will be set up on this screen. Note the following:

A. Select the Parameters tab to view or modify the device's parameters.
B. Parameter numbers are listed on the left.
C. The lock identifies read-only parameters.
D. When monitoring, select if you want to monitor a single parameter or all parameters.

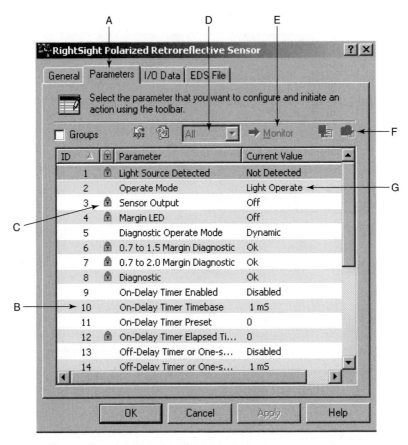

Figure 11-39 RSNetWorx screen showing parameters for a polarized photoelectric sensor.

E. Click here to monitor parameters.

F. Note the icons for uploading or downloading to the device.

G. Current value of a parameter. To modify a parameter, left click and select from the drop-down box as illustrated in Figure 11-40.

1. Select the parameter to edit.

2. Click here to access the Options drop-down box.

3. Select the parameter you want.

ELECTRONIC DATA SHEETS

The electronic data sheets, or EDS files, as they are called, contain information regarding the personality of the device. The correct EDS file must reside within each device before it can become a working part of the DeviceNet network. The EDS file must match the firmware level of the network device. If the EDS file is not current, you may have to access

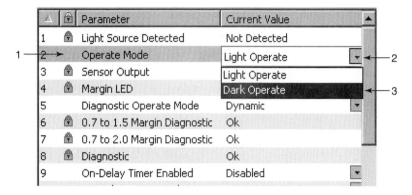

Figure 11-40 Options to modify parameter after left-click on current value.

either the manufacturer's or the odva.org Web site to download the correct file to your personal computer. Once the file is on your personal computer, use the EDS wizard to update or register the network device. Figure 11-41 displays one of the EDS Wizard screens. The screen illustrated is where the programmer selects the action to take. To register a new device or EDS file, select "Register an EDS file(s)" radio button.

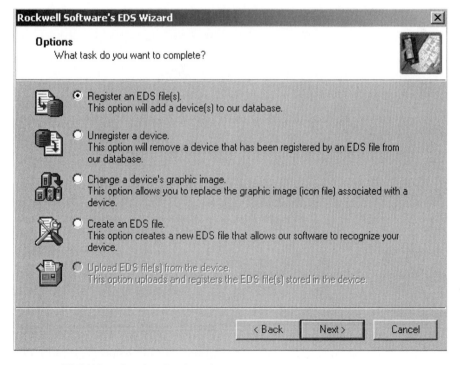

Figure 11-41 EDS Wizard task selection view.

Once the EDS file has been downloaded from either the manufacturer's Web site or ODVA, browse for it on the Registration view as illustrated in Figure 11-42.

A. Select how many files to register.
B. Browse for the file on your personal computer.
C. Note the file displayed in hexadecimal format in its directory.
D. Click Next to continue the registration process.

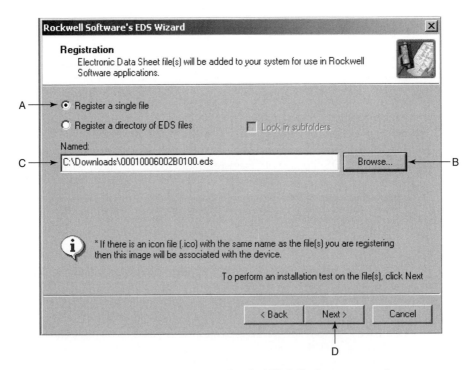

Figure 11-42   Options to browse for a downloaded EDS file from www.odva.org.

Determining an EDS File Value

When you go to either a device manufacturer or the ODVA Web site, the file that will be downloaded will be in hexadecimal. To determine what the hexadecimal value of the file will be, so you can browse for it when using the EDS wizard, you will have to know how to construct the file number from the general properties page. Figure 11-43 is a copy of a general properties page from the photo sensor we looked at above in the EDS registration view. The numbers in the brackets and the revision value need to be converted to hexadecimal. From your workings with hexadecimal, you should be able to construct an EDS

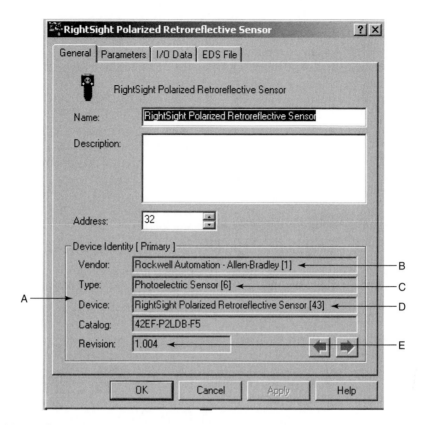

Figure 11-43   General properties page showing use of device identity information to construct EDS file number.

file number from the following information:

    A. Device Identity
    B. Vendor: Rockwell Automation—Allen-Bradley [1]
    C. Type: Photoelectric Sensor [6]
    D. Device: RightSight Polarized Retroreflective Sensor [43]
    E. Revision: 1.004

Hexadecimal file number conversion from the above decimal information:

        [1] = 0001

        [6] = 0006

     [43] = 002B

 [1.004] = 0100   (Since hex is four digits, drop the decimal point, add a leading zero, and discard the last digit 4 for this example)

Refer back to Figure 11-42 file number and you should see 00010006002B0100.eds.

SUMMARY

ControlNet is an open control network used for real-time data transfer of time-critical and non-time-critical data between processors or I/O on the same link. The ControlNet data transfer rate is five million bits per second. Up to 99 nodes can be connected on a ControlNet network, depending on how much data needs to be transferred and how efficiently the bandwidth of the network is set up. ControlNet is basically a combination of Data Highway Plus and Remote I/O. The network provides high-speed peer-to-peer information exchange and I/O networking on the same network.

Sharing data between a processor and I/O in a remote ControlLogix chassis can be done over a ControlNet network. ControlNet is typically used between a ControlLogix processor, SLC 500, or PLC 5 for processor-to-processor scheduled data exchange, or for local ControlLogix or PLC 5 chassis connection to a remote chassis for high-speed remote I/O connectivity, interlocking, or synchronization of two nodes, such as starting two variable frequency drives.

DeviceNet is also an open network intended to link low-level devices such as sensors, push-button stations, distributed I/O blocks, intelligent motor starter overload devices, and variable-frequency drives to higher-level devices such as PLCs. DeviceNet, similar to ControlNet, is managed by an independent group called the Open DeviceNet Vendors Association or ODVA. Their Web address is http://www.odva.org. Because it is an open network, devices or nodes from many different manufacturers can be nodes on the same DeviceNet network. A DeviceNet network is made of the following components: a trunk line, drop lines, nodes, a minimum of one power supply, two termination resistors, and a PLC scanner.

RSNetworx software will be used to set up the Scanlist, device properties including input and output mapping, node address, setting up the devices parameters, and registering a devices EDS file.

This chapter's intent was to introduce you only to the main concepts of ControlNet and DeviceNet. There was no intent to introduce or cover all aspects of either network, configuration, software setup, or interface to any PLC. There is much more information to consider before you would be able to select, design, configure, or install either network.

CONTROLNET REVIEW QUESTIONS

1. Define UMAX and explain why it is important.
2. Define SMAX and explain why it is important.
3. What is slot time?
4. How does slot time fit into our network configuration?
5. What is the difference between a CNB and a CNBR?
6. To configure the following ControlNet network, what is needed, and how will RSNetWorx parameters be set up?

   15 scheduled nodes

   6 unscheduled nodes

3 node addresses for maintenance personal computer connectivity

2 spare nodes for future scheduled service

2 spare nodes for future unscheduled service

Assume standard light industrial RG-6 coax cable

ControlLogix PLCs used with 1756-CNBs

    A. Maximum cable length is _____ meters.
    B. How many taps are required?
    C. How many termination resistors are required?
    D. UMAX will be set at _____ .
    E. SMAX will be set at _____ .

7. How many nodes can go on a segment?

8. Calculate the maximum trunk line cable length ControlNet network with 22 nodes, standard light industrial RG-6 coax.

9. A ControlNet tap length is:
    A. No longer than 20 feet
    B. As long as desired if using repeaters
    C. No longer than six feet
    D. Fixed at 39 inches
    E. No longer than 39 inches

10. Terminating resistors are:
    A. Used at each end of the network no matter how many repeaters are used
    B. Options and are used only in noisy environments
    C. Used with only one required
    D. Always required at the end of each segment
    E. Used at the end of the first and last segment
    F. Terminating resistors are optional

11. ControlNet networks can be configured as:
    A. Star
    B. Trunk line/drop line
    C. Ring
    D. Square
    E. All of the above
    F. Only A and B

12. A maintenance worker walks up to the network and connects a personal computer to a NAP on a CNB. The RSLinx driver for this computer has the computer as node 25. The worker wants to download a PLC project node 14. The network has the following parameters:
    a. SMAX set at 15
    b. UMAX set at 20
    c. Nodes 17 and 18 unused at this time

Will communication take place? Explain your answer.

13. What is a NAP, and where is it found?

14. Define the following pieces of the network update:
    A. NUT
    B. scheduled band
    C. unscheduled band
    D. maintenance or guard band

15. RSNetWorx for ControlNet is a software package that is:
    A. Required to set up scheduled and unscheduled service
    B. Required to set up scheduled service
    C. Used to set up network properties
    D. Used to schedule the network
    E. Needed to set up node addresses on nodes
    F. A and D
    G. C and D

16. An electrician needs to add another PLC on a ControlNet network. The PLC would send scheduled data every 5 ms, and would use two message instructions in the ladder project. RSNetWorx parameters are as follows:
    a. NUT 10 ms
    b. SMAX = 22
    c. UMAX = 30

    Current scheduled nodes are 1-10, 12-18, 20-22.

    Current unscheduled nodes are 23 to 28.

    Nodes for NAP are 29 and 30.

    Nodes 11 and 19 are for future expansion.

    Questions:
    1. What node can the PLC be connected to?
    2. Explain how slot time fits into the NUT.
    3. Any other considerations?

DEVICENET REVIEW QUESTIONS

1. A DeviceNet network can support up to _____ decimal nodes depending on the total amount of data transferred between the nodes.

2. A DeviceNet network was intended for a low-level device that exchanges small amounts of data, usually either a _____ or _____.

3. DeviceNet was not intended to support a number of operator interface terminals, each exchanging many words. Hardware such as this should be placed on what network?

4. If data needs to be shared between a DeviceNet node and the Ethernet operator interface, a pair of Ethernet and DeviceNet _____ can be used in a ControlLogix chassis.

5. When doing DeviceNet mapping, the minimum amount of data mapped, even though the device may only exchange one or two bits, is _____.

6. DeviceNet, similar to ControlNet, is managed by an independent group called the _____ or _____.

7. Being a(n) _____ network, devices or nodes from many different manufacturers can be nodes on the same DeviceNet network.

8. A DeviceNet network is made of the following components:
   A. Trunk line
   B. Drop lines
   C. Nodes
   D. Minimum of one power supply
   E. Two termination resistors
   F. Only A and C
   G. All the above

9. Nodes can be easily connected to the flat cable by _____ connectors.

10. Each end of the trunk line is required to have one _____ ohm _____ watt terminating resistor.

11. The module used to interface between the ControlLogix PLC and the DeviceNet network is a:
    A. 1747-SDN
    B. 1771-SDN
    C. 1784-DNS
    D. 1756-DBridge
    E. 1756-DNB

12. The communication module used interface between the PLC and DeviceNet is also called a _____.

13. DeviceNet network is popular because connecting devices together on a network can save on _____.

14. DeviceNet nodes attach to one cable called the _____.

15. DeviceNet cable has _____ wires, _____ signal and _____ for power.

16. Thick round and thin round can both be used for the _____ of the network.

17. While thin round is in many cases used for drop lines, it can also be used as the _____ in small applications with low current draw.

18. The maximum trunk line distance is defined as the maximum cable distance between _____.

19. The maximum length for any one drop is _____.
20. As with other network connections, either a _____, for installation into a desktop or industrial computer, or a _____-type card, for installation into a notebook personal computer, are needed to communicate to the network.
21. The safest way to commission a new device is to have a separate _____.
22. The maximum length of the cable is determined by the _____ type and the network _____.
23. If a device, or node, is not in the _____ as far as the scanner is concerned, it does not exist.
24. _____ sets up how each network device interfaces with the PLC data table.
25. The RSNetWorx _____ page contains device identity information used to construct an EDS file number.
26. Determine the EDS file name for the following device:
    A. Vendor: Rockwell Automation—Allen-Bradley [1]
    B. Type: General Purpose I/O [7]
    C. Device: 1792D-OB4D 4 output [1086]
    D. Catalog number: 1792D-OB4D
    E. Revision: 2.001

A partial list of the EDS files currently on your computer is listed in Figure 11-44 below. Select the correct file.

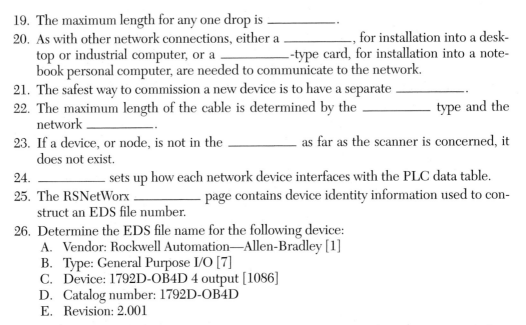

00010007044F0200.eds   00010007043B0200.eds   00010007043C0100.eds

00010007043D0100.eds   00010007043D0200.eds   00010007043E0100.eds

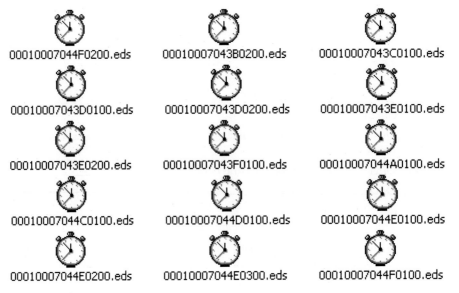

00010007043E0200.eds   00010007043F0100.eds   00010007044A0100.eds

00010007044C0100.eds   00010007044D0100.eds   00010007044E0100.eds

00010007044E0200.eds   00010007044E0300.eds   00010007044F0100.eds

Figure 11-44 Selected EDS files.

27. The _____, or EDS files, contain information regarding the personality of the device.

28. Explain why a separate node commission station is the best way to commission a new network node.

29. The correct _____ file must reside within each device before it can become a working part of the DeviceNet network.

30. A node commissioning station would consist of a personal computer with _____ and _____ software, the proper DeviceNet _____ and cables, a _____ 24 volt power supply, and the device to commission.

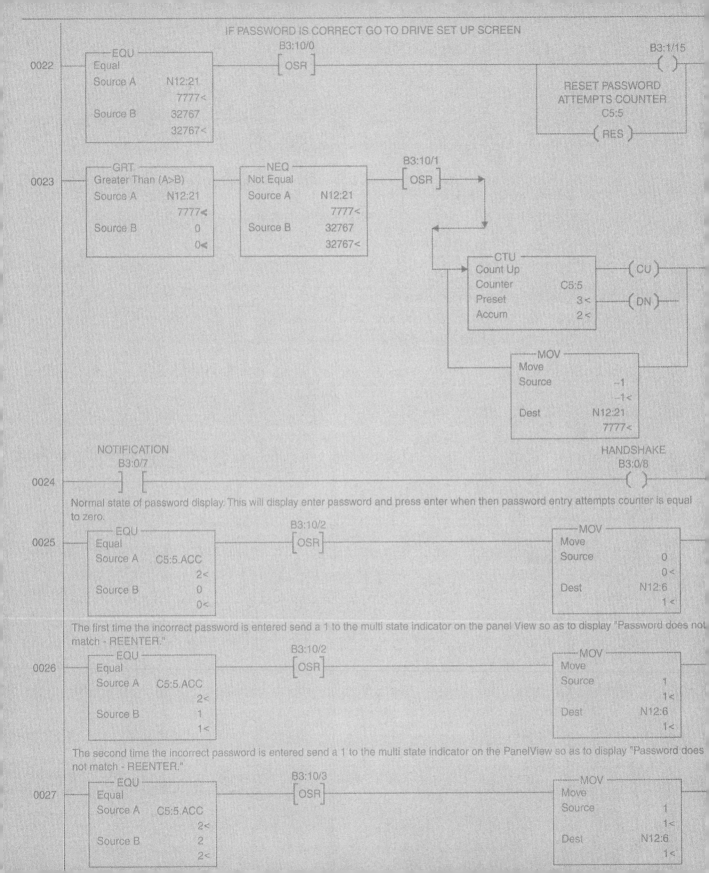

IF PASSWORD IS CORRECT GO TO DRIVE SET UP SCREEN

0022

```
┌─── EQU ────┐        B3:10/0
│ Equal      │       ─[ OSR ]─
│ Source A   N12:21 │
│            7777<   │
│ Source B   32767   │
│            32767<   │
└────────────┘
```

B3:1/15
─( )─

RESET PASSWORD
ATTEMPTS COUNTER
C5:5
─(RES)─

0023

```
┌─── GRT ───────────┐     ┌─── NEQ ───────────┐     B3:10/1
│ Greater Than (A>B) │     │ Not Equal          │    ─[ OSR ]─
│ Source A   N12:21  │     │ Source A   N12:21  │
│            7777<   │     │            7777<   │
│ Source B   0       │     │ Source B   32767   │
│            0<      │     │            32767<   │
└───────────────────┘     └───────────────────┘
```

```
┌─── CTU ──────────┐
│ Count Up         │   ─( CU )─
│ Counter   C5:5    │
│ Preset    3<      │   ─( DN )─
│ Accum     2<      │
└──────────────────┘
```

```
┌─── MOV ──────────┐
│ Move             │
│ Source      -1    │
│             -1<   │
│ Dest      N12:21  │
│             7777<  │
└──────────────────┘
```

NOTIFICATION
B3:0/7
0024 ─] [─

HANDSHAKE
B3:0/8
─( )─

Normal state of password display. This will display enter password and press enter when then password entry attempts counter is equal to zero.

0025

```
┌─── EQU ────┐        B3:10/2
│ Equal      │       ─[ OSR ]─
│ Source A   C5:5.ACC │
│            2<      │
│ Source B   0       │
│            0<      │
└────────────┘
```

```
┌─── MOV ──────────┐
│ Move             │
│ Source      0     │
│             0<    │
│ Dest      N12:6   │
│             1<    │
└──────────────────┘
```

The first time the incorrect password is entered send a 1 to the multi state indicator on the panel View so as to display "Password does not match - REENTER."

0026

```
┌─── EQU ────┐        B3:10/2
│ Equal      │       ─[ OSR ]─
│ Source A   C5:5.ACC │
│            2<      │
│ Source B   1       │
│            1<      │
└────────────┘
```

```
┌─── MOV ──────────┐
│ Move             │
│ Source      1     │
│             1<    │
│ Dest      N12:6   │
│             1<    │
└──────────────────┘
```

The second time the incorrect password is entered send a 1 to the multi state indicator on the PanelView so as to display "Password does not match - REENTER."

0027

```
┌─── EQU ────┐        B3:10/3
│ Equal      │       ─[ OSR ]─
│ Source A   C5:5.ACC │
│            2<      │
│ Source B   2       │
│            2<      │
└────────────┘
```

```
┌─── MOV ──────────┐
│ Move             │
│ Source      1     │
│             1<    │
│ Dest      N12:6   │
│             1<    │
└──────────────────┘
```

PART

PLC Instructions

Now that we have a good understanding of the hardware associated with the programmable controller, we need to explore the inner workings of how data is organized and stored. We will also look at the instructions that make the programmable controller able to control output devices. The repertory of instructions available for a specific PLC processor is called the **instruction set.**

There are many PLCs in the marketplace, and even though they all operate in basically the same way, there are many differences in features and instruction sets. All PLCs have the same basic instructions, but they may be programmed differently. Rather than attempt to introduce all the instruction sets and their differences, the remaining chapters in the book will focus on the Allen-Bradley SLC 500 PLC instruction set. Chapter 12 will deal with how data is organized inside the SLC 500 processor. Chapter 13 will introduce the basic relay-type instructions. Chapter 14 will look closely at programming relay instructions. Chapter 15 will introduce documentation and explain its importance to your ladder diagram and the associated instructions. Chapter 16 will look into timers and counters, and Chapter 17 will delve into the data-handling instructions. Chapter 18 will show you how the sequencer instruction is programmed. Chapter 19 will introduce program flow instructions. Other instructions will be introduced in a future second-term text and *Lab Manual*.

The accompanying *Lab Manual* will provide the opportunity to develop programs using each of the basic SLC 500 instructions.

This text will introduce the basic SLC 500 instructions. Lab exercises will provide hands-on programming experience using the RSLogix 500 Allen-Bradley Windows programming software and a personal computer to program basic relay instructions, timers and counters, copy and move instructions, and sequencers.

IF PASSWORD IS CORRECT GO TO DRIVE SET UP SCREEN

0022 — EQU —
      Equal
      Source A   N12:21
                  7777<

B3:10/0
—[OSR]—

B3:1/15
—( )—
RESET PASSWORD
ATTEMPTS COUNTER
C5:5
—(RES)—

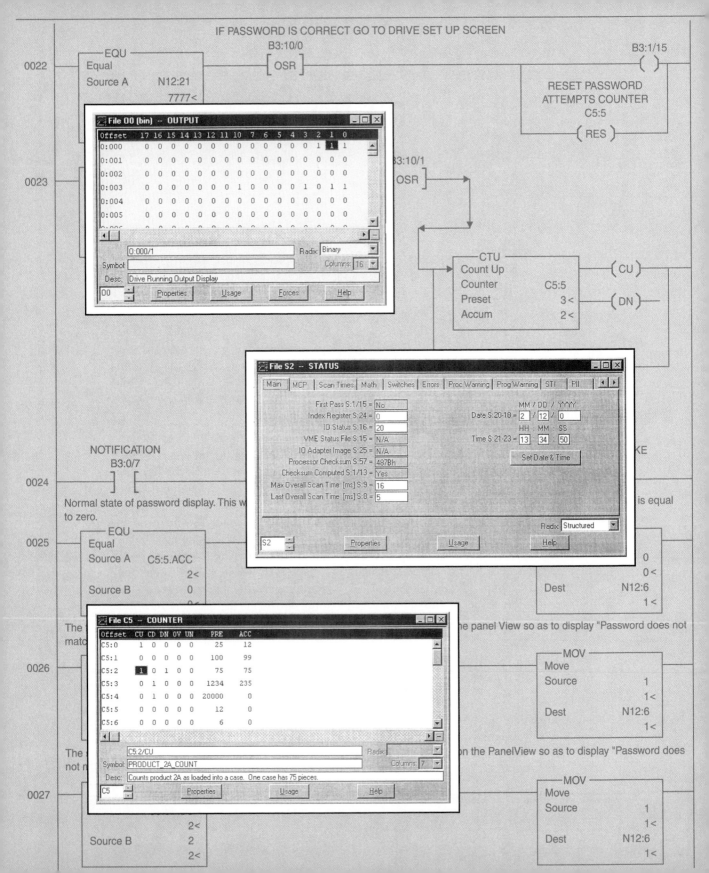

0023

B3:10/1
OSR

—CTU—
Count Up
Counter   C5:5
Preset   3<
Accum   2<
—(CU)—
—(DN)—

NOTIFICATION
B3:0/7
0024 —] [—

Normal state of password display. This w... is equal
to zero.

0025 — EQU —
      Equal
      Source A   C5:5.ACC
                  2<
      Source B   0

Dest   N12:6
        0
        0<
        1<

The s... ne panel View so as to display "Password does not
mat...

0026   —MOV—
                                             Move
                                             Source   1
                                                         1<
                                             Dest   N12:6
                                                     1<

The s... on the PanelView so as to display "Password does
not n...

0027   —MOV—
                                             Move
                                             Source   1
            2<   1<
      Source B   2   Dest   N12:6
                  2<   1<

File 00 (bin) -- OUTPUT

Offset	17	16	15	14	13	12	11	10	7	6	5	4	3	2	1	0
0:000	0	0	0	0	0	0	0	0	0	0	0	0	0	1	1	1
0:001	0	0	0	0	0	0	0	0	0	0	0	0	0	0	0	0
0:002	0	0	0	0	0	0	0	0	0	0	0	0	0	0	0	0
0:003	0	0	0	0	0	0	1	0	0	0	0	1	0	1	1	1
0:004	0	0	0	0	0	0	0	0	0	0	0	0	0	0	0	0
0:005	0	0	0	0	0	0	0	0	0	0	0	0	0	0	0	0

0:000/1   Radix: Binary
Symbol:
Desc: Drive Running Output Display
00   Properties   Usage   Forces   Help

File S2 -- STATUS

Main | MCP | Scan Times | Math | Switches | Errors | Proc Warning | Prog Warning | STI | PII

First Pass S:1/15 = No
Index Register S:24 = 0
IO Status S:16 = 20
VME Status File S:15 = N/A
IO Adapter Image S:25 = N/A
Processor Checksum S:57 = 487Bh
Checksum Computed S:1/13 = Yes
Max Overall Scan Time [ms] S:9 = 16
Last Overall Scan Time [ms] S:8 = 5

MM / DD / YYYY
Date S:20-18 = 2 / 12 / 0
HH : MM : SS
Time S:21-23 = 13 : 34 : 50
Set Date & Time

Radix: Structured

S2   Properties   Usage   Help

File C5 -- COUNTER

Offset	CU	CD	DN	OV	UN	PRE	ACC
C5:0	1	0	0	0	0	25	12
C5:1	0	0	0	0	0	100	99
C5:2	1	0	1	0	0	75	75
C5:3	0	1	0	0	0	1234	235
C5:4	0	1	0	0	0	20000	0
C5:5	0	0	0	0	0	12	0
C5:6	0	0	0	0	0	6	0

C5:2/CU   Radix:
Symbol: PRODUCT_2A_COUNT   Columns: 7
Desc: Counts product 2A as loaded into a case. One case has 75 pieces.
C5   Properties   Usage   Help

CHAPTER

12

Processor Data Organization

OBJECTIVES

After completing this chapter, you should be able to:

- describe a processor file
- explain which two files make up a processor file
- explain the function of a program file
- explain the function of a data file
- list the contents of a program file
- explain the contents of a bit file
- explain how a bit file is made up
- explain the contents and makeup of an integer file
- explain what an element and a subelement are
- list which instructions have multiple word elements
- describe what status bits are and where they are used

INTRODUCTION

In Chapter 10 we introduced the PLC processor, or CPU, hardware. The instructions we use to develop a user program and the data associated with the program must be stored inside the processor in an orderly manner. Data is usually grouped together and stored in areas called data files.

SLC 500 FILES

There is a lot of information that is, or can be, stored in an SLC 500 CPU. Information stored in the CPU's memory includes the output status file, input status file, processor status information, user ladder program, timer and counter data, 16-bit-word data storage, single-bit storage, storage for floating-point data, and storage of ASCII data (for message displays). Data needs to be stored in an orderly manner so that the processor and the human operator can access and monitor the desired data easily.

SLC 500 program development is started by creating a **project file.** The project file is displayed in the RSLogix 500 software as the project tree. A project file is the file stored on your personal computer's hard drive. The project file consists of the **processor file** and **database files.** The processor file contains all information relating to a user program. Information contained in a processor file is divided into a number of different files, two of which are **program files** and **data files.** A processor file can be transferred between the personal computer and the SLC 500 processor's memory.

Database files contain text information that is displayed on a ladder program as documentation. Documentation such as symbols, page titles, rung comments, instruction comments, and address descriptions are stored in their respective database files. Database files are not transferred to the PLC processor when the project is downloaded.

SLC 500 processor memory is also divided into files. Think of the program and data files as a two-drawer file cabinet. Program files are in one drawer and data files are in the other (refer to Figure 12-1).

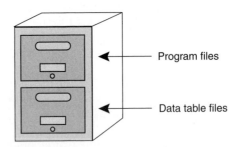

Figure 12-1   Processor memory is like a two-drawer file cabinet.

Inside each file drawer, data is organized into folders called files, which are used to store like information. The processor stores system and configuration information and the user program in one group of files called program files. The processor stores data used by the processor in conjunction with solving the user program in the other group of files, called data files. Figure 12-2 shows processor and data files as separate groups of memory. This is represented by the two file drawers. Think of the entire file cabinet as the processor file. Figure 12-2 illustrates a file cabinet as a representative grouping of program files. Program files consist of file 0, system files, file 1, reserved files, and file 2, the main ladder program. Files 3 through 255 are available as separate subroutine programs.

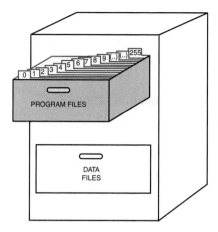

Figure 12-2   Program files in their separate file drawer, or section of processor memory.

The project tree is similar to a table of contents listing all processor files and the database files. Figure 12-3 illustrates the project tree from RSLogix 500 software. The figure identifies the program files folder and system files 0 and 1. The main ladder file, LAD 2, and one subroutine file, LAD 3 - SUBROUTINE, are shown.

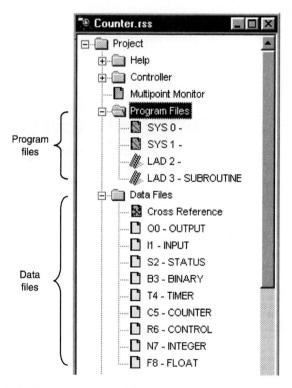

Figure 12-3   RSLogix 500 project tree for the project named Counter.rss.

DATA FILES

Data files contain information, or data, used in conjunction with your user ladder program. As the user program is running, it will fetch needed information from the appropriate data file, which consists of 256 separate files. Think of the data file as a separate file cabinet drawer with 256 file folders, as illustrated in Figure 12-4.

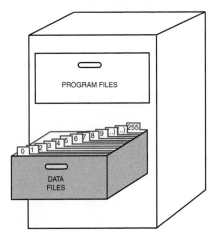

Figure 12-4   A data file is made of 256 separate files. Each file folder will contain like information.

Default files 0 through 8 are created when a processor file is named and created. Each file is automatically created by the processor and assigned a file number and alphabetical file identifier, as listed in Figure 12-5.

DEFAULT DATA FILES		
File Type	**File Identifier**	**File Number**
Output Status File	O	0
Input Status File	I	1
Processor Status File	S	2
Bit File	B	3
Timer File	T	4
Counter File	C	5
Control File	R	6
Integer File	N	7
Floating-Point File	F	8

Figure 12-5   SLC 500 default data file.

Each of these files contains specific information needed by the processor to complete its job of running the user program. The default data files contain the following data.

File 0, the Output Status File

File 0 is the default output status file. There can be only one output status file per processor file. The output status file is made up of single bits grouped into 16-bit words. Each bit represents the ON or OFF state of one output point. There is one bit in the output status file for each output module point in your PLC system. A 16-bit output file word is reserved for each slot in an SLC 500 modular chassis. Figure 12-6 lists the bits 0 through 15 across the top row from right to left. These bits represent outputs 0 through 15 for the module identified in the slot addressed in the right-most column of the table. The column on the right in Figure 12-6 lists the output module address.

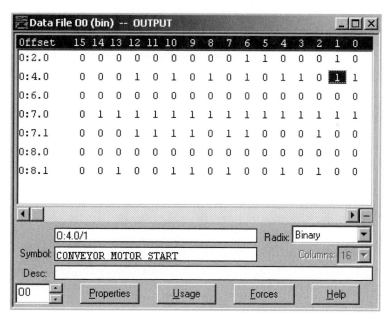

Offset	15	14	13	12	11	10	9	8	7	6	5	4	3	2	1	0
O:2.0	0	0	0	0	0	0	0	0	0	1	1	0	0	0	1	0
O:4.0	0	0	0	1	0	1	0	1	0	1	0	1	1	0	1	1
O:6.0	0	0	0	0	0	0	0	0	0	0	0	0	0	0	0	0
O:7.0	0	1	1	1	1	1	1	1	1	1	1	1	1	1	1	1
O:7.1	0	0	0	1	1	1	1	0	1	1	0	0	0	1	0	0
O:8.0	0	0	0	0	0	0	0	0	0	0	0	0	0	0	0	0
O:8.1	0	0	1	0	0	1	1	0	1	0	0	1	0	1	0	0

O:4.0/1   Radix: Binary

Symbol: CONVEYOR MOTOR START   Columns: 16

Desc:

00   Properties   Usage   Forces   Help

Figure 12-6   RSLogix 500 software output status file.

There will be a word created in the table only if the processor finds an output module residing in a particular slot when the I/O configuration is performed. If there is no module in a particular slot, no word will be created.

File 1, the Input Status File

File 1 is the default input status file. Only one input status file is allowed per processor file. The input status file is made up of single bits grouped into 16-bit words. Each bit represents the ON or OFF state of one input point. There is one bit in the input status file for each input module point in your PLC system. A 16-bit input file word is reserved for each slot in a modular chassis. Figure 12-7 lists the bits 0 through 15 across the top row from right to left. These bits represent outputs 0 through 15 for the module identified in the slot addressed in the right-most column of Figure 12-7. The column on the right in the table lists the input module address. Anywhere that there is a 1 in the table, the processor has seen a valid input signal from the input module.

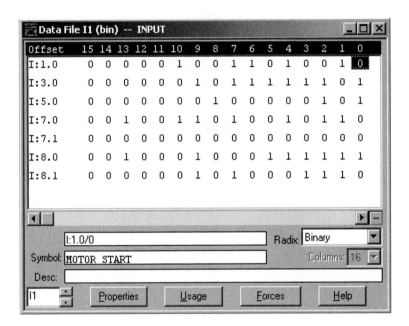

Figure 12-7 RSLogix 500 software input status file.

The table in Figure 12-7 has the words I:1.0, I:3.0, I:5.0, I:7.0, I:7.1, I:8.0, and I:8.1 in the input status file. Looking at this table, there are input modules in slots one, three, five, seven, and eight. The slots with no input status words—two, four, and six in this example—either have output modules or have no module installed. There will be a word created in the table only if the processor finds an input module residing in a particular slot. If there is no module in a particular slot, no status word will be created. There will not be an I:0 word for a modular PLC, as slot zero is always reserved for the processor.

File 2, the Processor Status File

The default processor status file is file 2. There can be only one processor status file per processor file. The processor status file contains an extensive amount of data regarding the processor and its operation. Information contained in the processor status file includes a wealth of information on how your PLC processor's operating system works. Some processor status information can only be monitored, while other processor options can be modified. Status file information falls into three classifications. First, status information, whether in words, bytes, or bits, is seldom monitored or modified. Second, dynamic configuration status words, bytes, or bits are used to select processor options while in run mode. Third, static configuration status words, bytes, or bits are used to select processor options before entering run mode. A general overview of what is found in an SLC 500 status file is listed below:

    operating system information

    monitoring of hardware and software faults

    clearing of hardware and software faults

    setup of memory module programs

    monitoring of arithmetic flags

    setting of the watchdog timer value

    battery low bit

    I/O chassis slot enable bits

    memory error bits

    going to run errors

    runtime errors

    I/O errors

    average scan time information

    communication bits

Figure 12-8 shows the main tab for an SLC 500 status file using RSLogix 500, software.

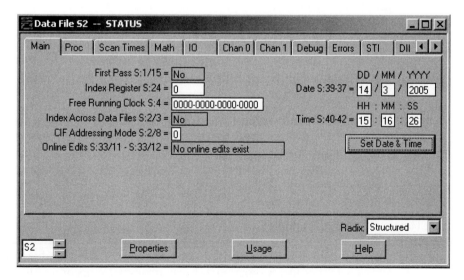

Figure 12-8 RSLogix 500 software processor status file S2.

As the functionality of the PLC increases, starting with the fixed SLC 500 and progressing through the MicroLogix 1000, 1200, 1500, 5/01, 5/02, 5/03, 5/04, and 5/05, additional status information and processor options will be contained within the status file. (See Appendix B for an overview of the status file.)

File 3, the Bit File

The default bit file is file 3. A bit file is used to store single bits in a 16-bit-word format. There can be many bit files for a single processor file. Any data file greater than file 8 can be assigned as an additional bit file. Each bit file will have up to 256 sixteen-bit words.

A bit file is made up of single bits grouped into a 16-bit word. One 16-bit file word is one element. Figure 12-9 lists the bits 0 through 15 across the top row from right to left. The column on the right of the table lists the bit file element addresses as full words. Each 1 or 0 in the file is a single bit.

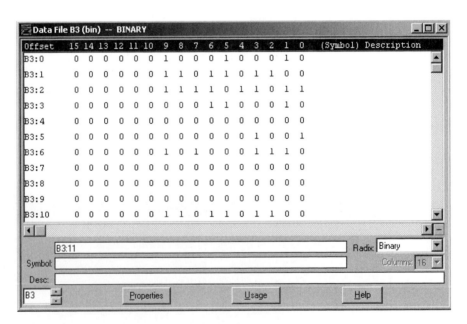

Figure 12-9 RSLogix 500 software binary, orbit file B3.

File 4, the Timer File

The default timer file is file 4. The timer file is used to store timer data. Each timer consists of three 16-bit words, collectively called a "timer element." There can be up to 256 timer elements for a single processor file. Any data file greater than file 8 can be assigned as an additional timer file. A programmer may separate timers into different files in cases where a large program has many timers. Separating timers into separate files according to their function or the process they control simplifies tracking and troubleshooting. As an example, if all timers associated with manufacturing process A are grouped in timer file T12, all timers associated with process B are grouped in timer file T13, and all timers associated with process C are grouped in timer file T14, troubleshooting problems with process B will lead to investigating only the timers in file T13. If all timers were in the default file 4, troubleshooting process B's timer problems could be quite a challenge as it might mean that 75 timers for process B were mixed in with all the other 256 timers contained in file 4.

Timer File Elements A timer instruction is one element. A timer element is made up of three 16-bit words. Each element consists of three parts (refer to Figure 12-10).

Word zero is for status bits.

Word one is for the preset value.

Word two is for the accumulated value.

EN	TT	DN	Reserved Bits
Preset Value			
Accumulated Value			

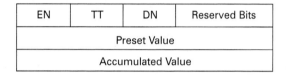

Figure 12-10 One timer element is made of three 16-bit words.

Figure 12-11 shows the default timer file for an SLC 500.

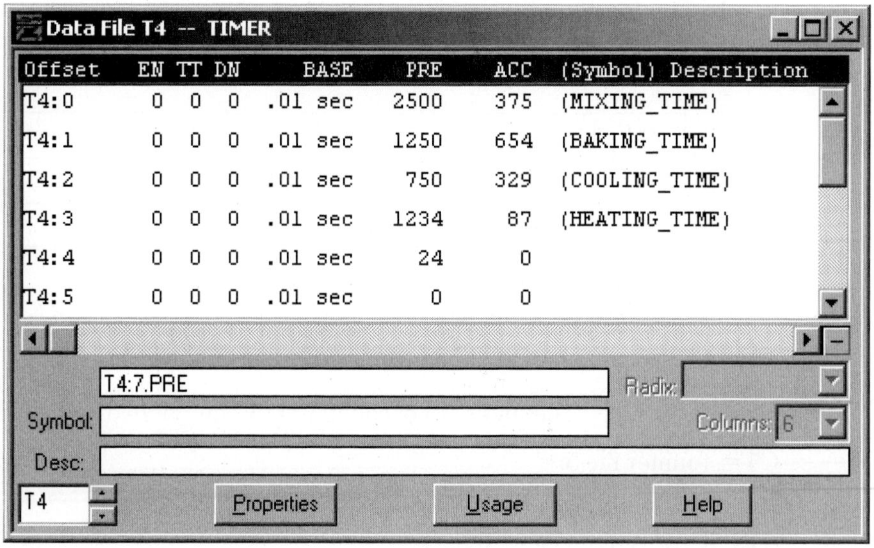

Data File T4 -- TIMER
Offset EN TT DN BASE PRE ACC (Symbol) Description
T4:0 0 0 0 .01 sec 2500 375 (MIXING_TIME)
T4:1 0 0 0 .01 sec 1250 654 (BAKING_TIME)
T4:2 0 0 0 .01 sec 750 329 (COOLING_TIME)
T4:3 0 0 0 .01 sec 1234 87 (HEATING_TIME)
T4:4 0 0 0 .01 sec 24 0
T4:5 0 0 0 .01 sec 0 0
T4:7.PRE
Radix:
Symbol:
Columns: 6
Desc:
T4
Properties
Usage
Help

Figure 12-11 RSLogix 500 software timer file T4.

Timer File Status Bits Status bits for timer file 4, timer element number two (T4:2), are as listed below. Refer back to Figure 12-11. Status bits are addressed as follows:

T4 = timer file 4.

:2 = timer element #2 (0–255 timer elements per file).

/DN = status bit in element.

T4:2/DN is the address for the done bit of timer file 4, timer two.

T4:2/TT is the address for the timer-timing bit of timer file 4, timer two.

T4:2/EN is the address for the enable bit of timer file 4, timer two.

Symbol or description is "Cooling Time."

Each status bit will be covered in depth in chapter 16.

File 5, the Counter File

The default counter file is file 5. The counter file is used to store counter data. Each counter consists of three 16-bit words, called "counter elements." There can be many counter files for a single processor file. Any data file greater than file 8 can be assigned as an additional counter file. Each counter file will have 256 counter elements. Separating counters into separate files and grouping them by their function or the process they control follows the same reasoning as for timers.

Counter File Elements A counter instruction is one element. A counter element is made up of three 16-bit words (refer to Figure 12-12).

Word zero is for status bits.

Word one is for the preset value.

Word two is for the accumulated value.

CU	CD	DN	OV	UN	UA	Reserved Bits
Preset Value						
Accumulated Value						

Figure 12-12 One counter element is made of three 16-bit words.

Figure 12-13 shows an SLC 500 counter file.

Counter Status Bits The status bits of counter file 5, counter two, are addressed as follows:

C5 = counter file 5.

:2 = counter element #2 (0–255 counter bits per file).

/DN = status bit in element.

C5:2/UA is the update accumulator bit (for fixed PLCs with a high-speed counter).

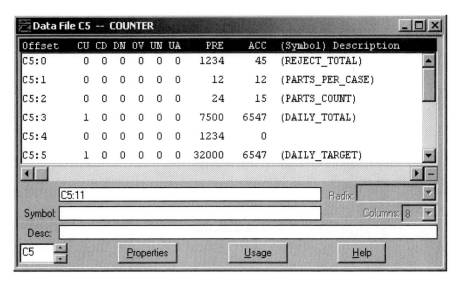

Figure 12-13   RSLogix 500 software counter file C5.

C5:2/UN is the underflow bit.

C5:2/OV is the counter's overflow status bit.

C5:2/DN is the address for the done bit of counter file 5, counter two.

C5:2/CD is the address for the count-down enable bit of counter file 5, counter two.

C5:2/CU is the address for the count-up enable bit of counter file 5, counter two.

Symbol or description is "Parts Count."

Each of these status bits will be covered in depth in chapter 16.

File 6, the Control File

The control file, file 6, is used to store status information for bit shift, first in and first out stacks (FIFO), last in and first out stacks (LIFO), sequencer instructions, and certain ASCII instructions. Each control file consists of a three-word element. The control file contains status and control information for these instructions. Word zero contains status bits. Word one contains the length of the bit array, or file. Word two is the current position in the bit array, or file. Figure 12-14 illustrates one control element.

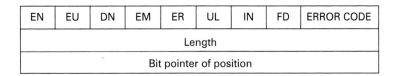

Figure 12-14   One control element is made of three 16-bit words.

Control File Status Bits The control file identifier is R. The status bits for control file 6, element 2 (R6:2), are addressed as follows:

R6 = control file six.

:2 = control element #2 (0–255 control elements per file).

/FD = status bit in element.

R6:2/FD is the found bit.

R6:2/IN is the inhibit bit.

R6:2/UL is the unload bit.

R6:2/ER is the files error bit.

R6:2/EM is the empty stack bit.

R6:2/DN is the done bit.

R6:2/EU is the unload enable bit.

R6:2/EN is the control files enable bit.

Figure 12-15 shows an SLC 500 control file, R6.

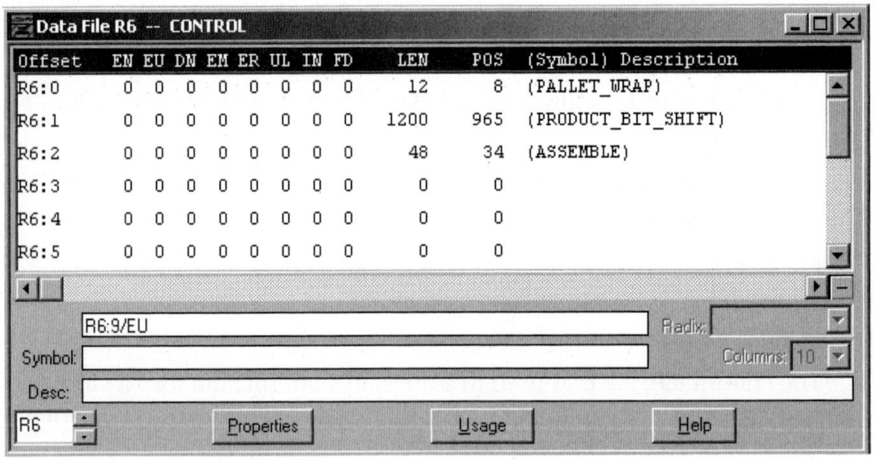
Data File R6 -- CONTROL
Offset EN EU DN EM ER UL IN FD LEN POS (Symbol) Description
R6:0 0 0 0 0 0 0 0 0 12 8 (PALLET_WRAP)
R6:1 0 0 0 0 0 0 0 0 1200 965 (PRODUCT_BIT_SHIFT)
R6:2 0 0 0 0 0 0 0 0 48 34 (ASSEMBLE)
R6:3 0 0 0 0 0 0 0 0 0 0
R6:4 0 0 0 0 0 0 0 0 0 0
R6:5 0 0 0 0 0 0 0 0 0 0
R6:9/EU
Radix:
Symbol:
Columns: 10
Desc:
R6
Properties
Usage
Help

Figure 12-15 RSLogix 500 software counter file R6.

File 7, the Integer File

An integer file element is a 16-bit word representing one whole number. The integer file, default file 7, is used to store integers, that is, whole numbers. Any whole number (such as 1; 2; 3; 4; 5; 34; 56; 234; 2,345; or 32,767) can be stored in the integer file. Each integer file has 256 one-word elements (0–255). Each element can store the binary equivalent of one 16-bit word. Data stored in the integer file can be addressed as an entire word or at the bit level. There can be many integer files for a single processor file. Any unused data file greater than file 8 can be assigned as an additional integer file. Figure 12-16 illustrates the default integer file N7 with integer data displayed.

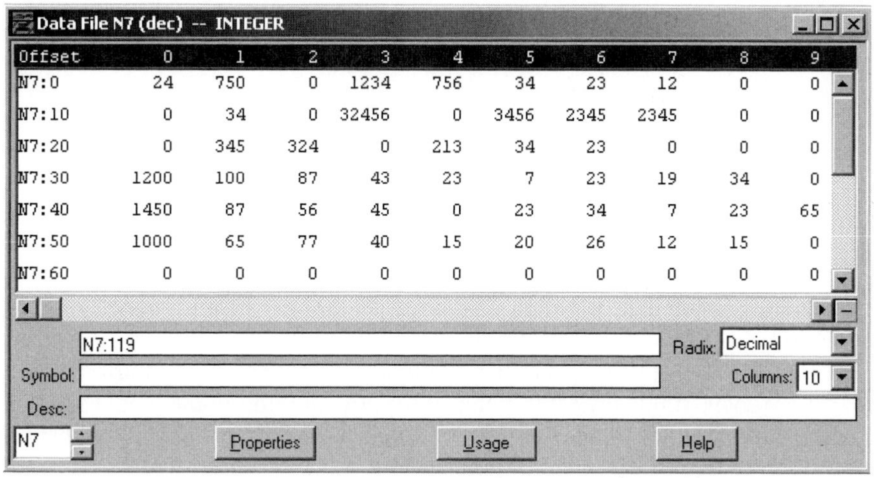

Data File N7 (dec) -- INTEGER									
Offset	0	1	2	3	4	5	6	7	8	9
N7:0	24	750	0	1234	756	34	23	12	0	0
N7:10	0	34	0	32456	0	3456	2345	2345	0	0
N7:20	0	345	324	0	213	34	23	0	0	0
N7:30	1200	100	87	43	23	7	23	19	34	0
N7:40	1450	87	56	45	0	23	34	7	23	65
N7:50	1000	65	77	40	15	20	26	12	15	0
N7:60	0	0	0	0	0	0	0	0	0	0

N7:119   Radix: Decimal

Symbol: Columns: 10

Desc:

N7   Properties   Usage   Help

Figure 12-16   RSLogix 500 software integer file N7.

The integer file illustrated in Figure 12-16 is very similar to what you would see if you were monitoring integer file 7 using Allen-Bradley RSLogix 500 programming software. The integer file consists of 10 elements across the top of the file. Each element can hold an integer up to 32,767. Looking at Figure 12-16, element starting addresses are in the left column. The N7:0 address breaks down into:

N7 = integer File 7

:0 = integer element within integer file 7

The numbers across the top row, 0 through 9, identify the element within file N7. As an example, the first column to the right of the N7:0 address is labeled 0 and has a value of 24 where the first row and that column intersect. This tells you that element address N7:0 contains the integer 24. The next column in the top row is a 1. It intersects with the row marked N7:0 at the number 750. Integer address N7:1 equals the integer 750. The second row has the element address N7:10. If you go across to where this row intersects with the address marked 1, this is element N7:11. Address N7:11 contains the integer 34. In our example, integer file N7:0 contains the number 24, while N7:1 contains 750, N7:2 contains 0, N7:3 contains 1,234, N7:30 contains 1,200, and N7:31 contains 100.

File 8, the Floating-Point File

When fractional numerical data or numerical data values greater than 32,767 are needed in a PLC application, the floating-point data file can be used. Floating-point data consists of two parts: an integer and an exponent. Floating-point data is stored in two-word elements. One word is used to store the integer. The second word is used to store the exponent. As an example, the number 500,000,000 would be stored as 5e+08. This number is the same as 5.00×10^8.

200,000 would be displayed as 200000

2,000,000 would be displayed as 2e+07

.123456 would be displayed as .123456

.12345678987 would be displayed as .123457

.000023 would be displayed as 2.3e−05

Floating-point data file data can store a number within a range of 1.1754944e−38 to 3.40282347e+38. Floating-point files are available on SLC 500 5/03 processors with operating system OS301 and above, plus the SLC 500, 5/04, and 5/05 processors. Newer MicroLogix 1200 and 1500 will also support floating-point files. Figure 12-17 illustrates the floating-point values listed above, plus other data as it would be displayed by RSLogix 500 software.

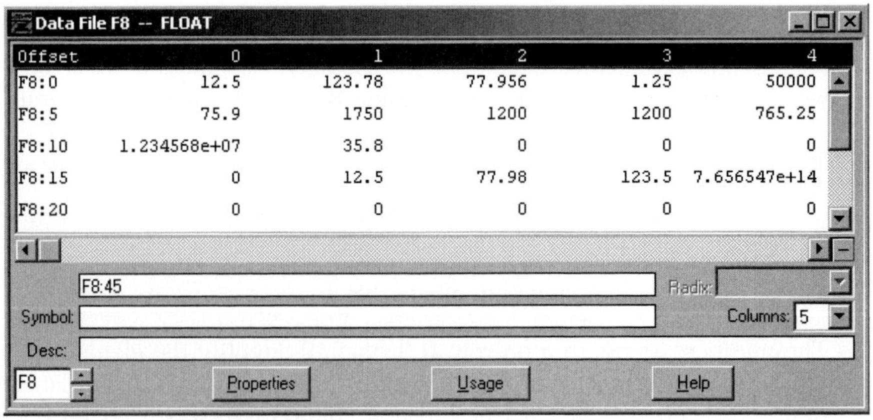

Data File F8 -- FLOAT					
Offset	0	1	2	3	4
F8:0	12.5	123.78	77.956	1.25	50000
F8:5	75.9	1750	1200	1200	765.25
F8:10	1.234568e+07	35.8	0	0	0
F8:15	0	12.5	77.98	123.5	7.656547e+14
F8:20	0	0	0	0	0

F8:45   Radix:

Symbol: Columns: 5

Desc:

F8   Properties   Usage   Help

Figure 12-17   RSLogix 500 software floating-point file F8.

File 9, the Common Interface File

The common interface file (CIF) is used as the target file in an SLC 500 processor when transferring data from one SLC 500 to another on the Allen-Bradley Data Highway 485 network. Each SLC 500 on the network must have a unique identifier, or address, called a **node** address. The CIF file is always file 9. Data going from one processor to another over the network is stored in file 9 before it is sent out over the network. In addition, data received by a processor is sent to file 9. After the data is received in file 9, it can be moved to any desired file for use.

A "message instruction" is used to send data between SLC 500 processors. The message instruction is programmed on the user ladder program as either a read or a write instruction. A message write instruction sends, or writes, data out of the current processor. A message read instruction reads, or brings information into, the resident CIF file from another processor. The CIF is called the target file. The device we want to send data to is called the target device. The target device does not always have to be another SLC 500 processor. Other devices, including data display devices, bar code readers, and other types of processors, can accept data written from, and write data to, a resident CIF.

ASCII Data Files

ASCII stands for American Standards Code for Information Interchange. Each time you press a key on a computer keyboard, a 7- or 8-bit word is stored in computer memory to represent the alphanumeric, function, or control data represented by the specific keyboard key that was depressed.

ASCII data is not stored in a default data file in the SLC 500. Instead, the programmer can create an ASCII file in any unused data file numbered from 9 to 255. ASCII is a valid file type only for SLC 5/03, OS 301 and above, and SLC 5/04 and 5/05 processors. The ASCII data file information contains one-word elements. Two hex characters are typically used to represent an ASCII character. As an example, the following alphabetical letters represent the hex characters in Figure 12-18.

ASCII	HEX	ASCII	HEX	ASCII	HEX
A	41	J	4A	S	53
B	42	K	4B	T	54
C	43	L	4C	U	55
D	44	M	4D	V	56
E	45	N	4E	W	57
F	46	O	4F	X	58
G	47	P	50	Y	59
H	48	Q	51	Z	5A
I	49	R	52		

Figure 12-18 ASCII to HEX conversion table.

ASCII Application Example:

A bar code reader can read bar codes on boxes passing on a conveyor line. Bar code data is sent into the PLC, which uses the bar code data read from the boxes passing on the conveyor to determine which diverters are to be energized in a sorting application. When a good bar code is read, alphanumeric data is sent to the PLC data table. If the bar code reader cannot read the bar code on a particular box, a no-read message is sent to the PLC in place of the 10-digit bar code. The PLC data table will look like Figure 12-19, which illustrates ASCII file A12. Words are numbered across the top from zero through nine. The far left column of the table identifies the file and starting word numbers.

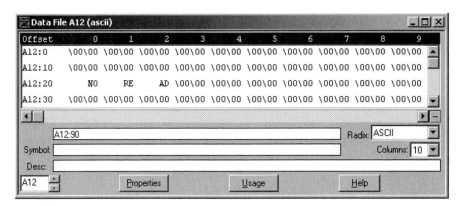

Figure 12-19   RSLogix 500 software ASCII data file A12.

String Data

String data consists of a group, or string, of ASCII characters that make up a word or group of words. The difference between an ASCII file and a string file is that the string file strings together a number of ASCII characters rather than treating them separately.

ASCII string data is not stored in a default data file in the SLC 500. Instead, the programmer can create an ASCII string file in any unused data file numbered from 9 to 255. ASCII is a valid file type only for SLC 5/03, OS 301 and above, 5/04, and SLC 5/05 processors. Each ASCII string data file element is comprised of 42 words.

A string data file located in file 12 (ST12) will look similar to Figure 12-20. The figure illustrates string file 12. ST identifies the file as a string file, and the file number is 12. The number after the colon is the word number. In this example, word number zero of string file 12 contains the elements "NO READ." The length tells us how many ASCII characters are contained in the string. Notice that the space between "NO" and "READ" counts as one character.

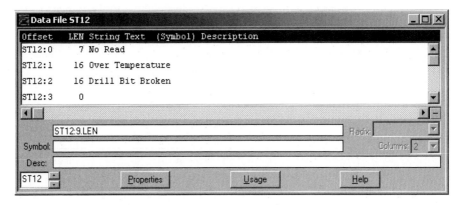

Figure 12-20   RSLogix 500 software string file ST12.

A string file can contain up to 255 elements, with each element made up of 42 words. Any unused data file or files other than the default data files can be designated as a string file.

USER-DEFINED FILES

There are a total of 256 available data files in the SLC 500 processor. Files beyond the automatically created default files are available as "user-defined files." User-defined files are files that the user can designate as needed for a specific application. Only one output status file, input status file, and processor status file are allowed per processor. Any other data file may be assigned a file number between 9 and 255. No two files may share the same file number. If additional timer or counter files are desired, they may be assigned any available file number between 9 and 255. As an example, as a convenient method to separate timers and counters used in different parts of the process, timer files numbered 10, 11, and 12 could be created. Also, separate counter files numbered 13, 14, and 15 could be created. To keep integer data in logical groups, such as numerical data for differing recipes, integer files 16, 17, 18, 19, and 20 could be created. Each different integer file could contain recipe data for one of, say, five different products being manufactured. Figure 12-21 lists some of the common user-defined data files. There are additional files. Refer to the RSLogix 500 software help screens for additional information on user-defined files. Remember, a file number can be used only once. Once a user-defined file is created, its file number cannot be reused for another user-defined file.

USER-DEFINED DATA FILES		
File Type	File Identifier	File Number
Bit File	B	9-255
Timer File	T	9-255
Counter File	C	9-255
Control File	R	9-255
Integer File	N	9-255
Floating-Point File	F	9-255
String File	St	9-255
ASCII	A	9-255

Figure 12-21   SLC 500 user-defined files.

The default file alphabetical identification and file numbers are fixed and unchangeable. User-defined data files can be assigned to unused file numbers from 9 to 255; remember that the alphabetical file identifier identifies the file type, not only for your reference, but also for the processor. The particular processor you have and the operating system firmware level will dictate which data files are available for your use. Data files 0 through 7 are default files for all processors, including the SLC 500 fixed PLCs.

Figure 12-22 lists each file type and the number of words per element.

File Type	Data Contained	Words per Element
Output Status	Output status table	Varies
Input Status	Input status table	Varies
Processor Status	Processor status data	Varies
Bit	Single bits of data	1
Timer	Timer instruction parameters	3
Counter	Counter instruction parameters	3
Control	Control parameters and data	3
Integer	Whole numbers	1
Floating Point	Floating-point math	2
String	ASCII string data	42
ASCII	ASCII data	3

Figure 12-22   Data files are made of elements, words, and bits.

SUMMARY

The Allen-Bradley SLC 500 system refers to the group of files that make up the total PLC program as the processor file. A processor file can stand on its own and be transferred between computers or PLCs. It contains files associated with the user program, called program files. Data associated with a program and its program files are called data files. Data files contain input status; output status; processor status; bit, timer, and counter information; control status; integer, floating-point, and common interface files; and ASCII data. Each time a processor file is created, default data files are automatically created, too. Files 0 through 7 are default files for all SLC 500 processors. Which particular SLC 500 family member you have will dictate the default data files. Starting with the SLC 500 5/03 processor with operating system OS302, the floating-point file F8 became a default file for any newer processor, such as 5/04 (OS401), and any 5/05 processor.

File numbers 9 through 255 are user-configurable. These files can be made into whatever additional files are needed for a particular application. There are a few exceptions, however. Only one of the following file types is allowed per system: output status files, input status files, and processor status files. As an example, if additional timers are needed or it is wise to separate timers into two or more separate files to more easily keep track of the various timers, additional timer files can be created. Each timer file will contain 256 timers. Any of the user-defined files can be used for extra data storage or for ease of keeping track of data. As an example, it might be easier to keep track of recipe data if the data corresponding to each recipe is stored in its own separate file. Since most recipe data are usually whole numbers or integers, additional integer files could be created to keep the recipes data separate. For example, integer files N:14, N:15, N:16, and N:17, each containing 256 16-bit words, could be easily created for this purpose.

REVIEW QUESTIONS

1. SLC 500 program development is started by creating a _____.
2. Information contained in a processor file is divided into two types of files: _____ files and _____ files.
3. Default data files are assigned a number and alphabetical letter for easy identification. Fill in the file identifiers for the files listed in Figure 12-23.
4. File _____ is the default output status file.
5. There can be only one _____ status file per processor file.
6. The output status file is made up of single bits grouped into a _____-bit word.
7. A 16-bit output file word is reserved for each _____ in a modular chassis.
8. Figure 12-24 represents an output status file. In the column at the far right, list the address assuming one output module in the following slots of a seven-slot modular chassis: slots: two, five, and six.

DEFAULT DATA FILES		
File Type	**File Identifier**	**File Number**
Output Status File		
Input Status File		
Processor Status File		
Bit File		
Timer File		
Counter File		
Control File		
Integer File		
Floating-Point File		

Figure 12-23   SLC 500 default data files.

Address	15	14	13	12	11	10	9	8	7	6	5	4	3	2	1	0
	1	1	1	0	0	1	0	1	0	0	1	1	0	1	1	0
	0	0	1	1	1	0	1	0	1	1	1	0	1	1	1	1
	1	1	1	1	1	0	1	1	0	0	0	0	1	1	0	1

Figure 12-24   Output status file.

9. True or false? The output point in slot five, terminal 5, is a 1.
10. True or false? The output point in slot six, terminal 12, is a 1.
11. File _____ is the default input status file.
12. The input status file is made up of _____ bits grouped into a 16-bit word.
13. There is one bit in the input status file for each _____ in your PLC system.
14. Figure 12-25 represents an input status file. In the column at the far right, list the address assuming one input module in the following slots of a seven-slot modular chassis: slots one, three, four, and five.

Address	15	14	13	12	11	10	9	8	7	6	5	4	3	2	1	0
	1	1	1	0	0	1	0	1	0	0	1	1	0	1	1	0
	0	0	1	1	1	0	1	0	1	1	1	0	1	1	1	1
	1	1	0	0	0	0	1	0	0	0	0	1	1	1	1	1
	1	1	1	1	1	0	1	1	0	0	0	0	1	1	0	1

Figure 12-25   Input status file.

15. Looking at the status tables from questions 9 and 15, determine what our PLC I/O module order would look like. What type of modules are in the following slots?

    Slot zero (slot reserved for processor only):

    Slot one:

    Slot two:

    Slot three:

    Slot four:

    Slot five:

    Slot six:

16. True or false?: The I/O configuration for question 16 is acceptable. If false, explain what the problem is.

17. The default bit file is file _____.

18. One 16-bit bit-file word is one bit-file _____.

19. Determine the bit contained in the address in Figure 12-26.

Element	15	14	13	12	11	10	9	8	7	6	5	4	3	2	1	0
B3:156	1	1	1	0	0	1	0	1	0	0	1	1	0	1	1	0
B3:157	0	0	1	1	1	0	1	0	1	1	1	0	1	1	1	1
B3:158	1	1	0	0	0	0	1	0	0	0	0	1	1	1	1	1
B3:159	0	0	1	0	1	1	1	1	0	0	1	1	1	0	0	1
B3:160	1	1	1	1	1	0	1	1	0	0	0	0	1	1	0	1

Figure 12-26   Each bit-file element consists of one 16-bit word.

    A. B3:156/12 contains a _____.
    B. B3:156/9 contains a _____.
    C. B3:158/15 contains a _____.
    D. B3:159/0 contains a _____.
    E. B3:160/2 contains a _____.
    F. Bit 14, element 157, contains a _____.
    G. Bit 3, element 156, contains a _____.

20. The default timer file is file _____.

21. Each timer consists of three 16-bit words, called a _____.

22. Each timer file will have _____ elements.

23. Why would a programmer separate timers into different files in cases where a large program has many timers?

24. A timer instruction consists of three parts:

> Word zero is for _____.
>
> Word one is for the _____.
>
> Word two is for the _____.

25. The default counter file is file _____.

26. Each counter file will have _____ counter elements.

27. The counter instruction consists of three parts:

> Word zero is for _____.
>
> Word one is for the _____ value.
>
> Word two is for the _____ value.

28. The control file, file _____, is used to store status information for bit shift, first in and first out stacks, last in and first out stacks, and sequencer and certain ASCII instructions.

29. Each control file consists of a _____ word element.

30. The integer file, default file _____, is used to store _____.

31. Each integer file has _____ elements.

32. Data stored in the integer file can be addressed as an _____ or to the _____ level.

33. Integer file N78 is illustrated in Figure 12-27. Identify the address in which the following data is stored.

Element	0	1	2	3	4	5	6	7	8	9
N78:0	2	3	4	11	11	3220	2321	31	40	41
N78:10	7	3218	9	5	2	8	2	9	6	7
N78:20	13	3338	9	390	11	6621	31	21	21	22
N78:30	60	4471	2380	2	21	7767	389	99	1212	1213
N78:40	55	66	16	4324	2224	1	321	4	1333	1334

Figure 12-27   Integer file N78.

> _____ = 390
>
> _____ = 1,334
>
> _____ = 2
>
> _____ = 3,220
>
> _____ = 9
>
> _____ = 4,471

34. Floating-point data is stored in _____ elements.
35. ASCII stands for _____ .
36. Each time you press a key on a computer keyboard, a _____ or _____-bit word is stored in computer memory.
37. The ASCII data file information is comprised of _____ elements.
38. String data consists of a group, or _____ , of _____ characters that make up a word or group of words.
39. True or false?: The difference between an ASCII file and a string file is that the string file strings together a number of ASCII characters to make groups of words that humans can understand.

IF PASSWORD IS CORRECT GO TO DRIVE SET UP SCREEN

0022

```
┌── EQU ──────────────┐        B3:10/0                              B3:1/15
│ Equal               │        [ OSR ]                               ( )
│ Source A    N12:21  │
│             7777<   │                                    RESET PASSWORD
│ Source B    32767   │                                    ATTEMPTS COUNTER
│             32767<  │                                         C5:5
└─────────────────────┘
                                                              ( RES )
```

0023

```
┌── GRT ──────────────┐   ┌── NEQ ──────────────┐     B3:10/1
│ Greater Than (A>B)  │   │ Not Equal           │     [ OSR ]
│ Source A    N12:21  │   │ Source A    N12:21  │
│             7777<   │   │             7777<   │
│ Source B    0       │   │ Source B    32767   │
│             0<      │   │             32767<  │
└─────────────────────┘   └─────────────────────┘
                                                    ┌── CTU ──────────┐
                                                    │ Count Up        │── ( CU )
                                                    │ Counter   C5:5  │
                                                    │           3<    │── ( DN )
                                                    │           2<    │
                                                    └─────────────────┘

                                                            -1
                                                            -1<
                                                          N12:21
                                                          7777<
```

NOTIFICATION
B3:0/7

0024
```
─] [─
```
Normal state of passw...
to zero.

HANDSHAKE
B3:0/8
( )

...try attempts counter is equal

0025

```
┌── EQU ──────────────┐                                   ┌── MOV ──────────┐
│ Equal               │                                   │ Move            │
│ Source A    C5:     │                                   │ Source      0   │
│                     │                                   │             0<  │
│ Source B            │                                   │ Dest    N12:6   │
│             0<      │                                   │             1<  │
└─────────────────────┘                                   └─────────────────┘
```

The first time the incorrect password is entered send a 1 to the multi state indicator on the panel View so as to display "Password does not match - REENTER."

0026

```
┌── EQU ──────────────┐        B3:10/2                    ┌── MOV ──────────┐
│ Equal               │        [ OSR ]                    │ Move            │
│ Source A  C5:5.ACC  │                                   │ Source      1   │
│             2<      │                                   │             1<  │
│ Source B    1       │                                   │ Dest    N12:6   │
│             1<      │                                   │             1<  │
└─────────────────────┘                                   └─────────────────┘
```

The second time the incorrect password is entered send a 1 to the multi state indicator on the PanelView so as to display "Password does not match - REENTER."

0027

```
┌── EQU ──────────────┐        B3:10/3                    ┌── MOV ──────────┐
│ Equal               │        [ OSR ]                    │ Move            │
│ Source A  C5:5.ACC  │                                   │ Source      1   │
│             2<      │                                   │             1<  │
│ Source B    2       │                                   │ Dest    N12:6   │
│             2<      │                                   │             1<  │
└─────────────────────┘                                   └─────────────────┘
```

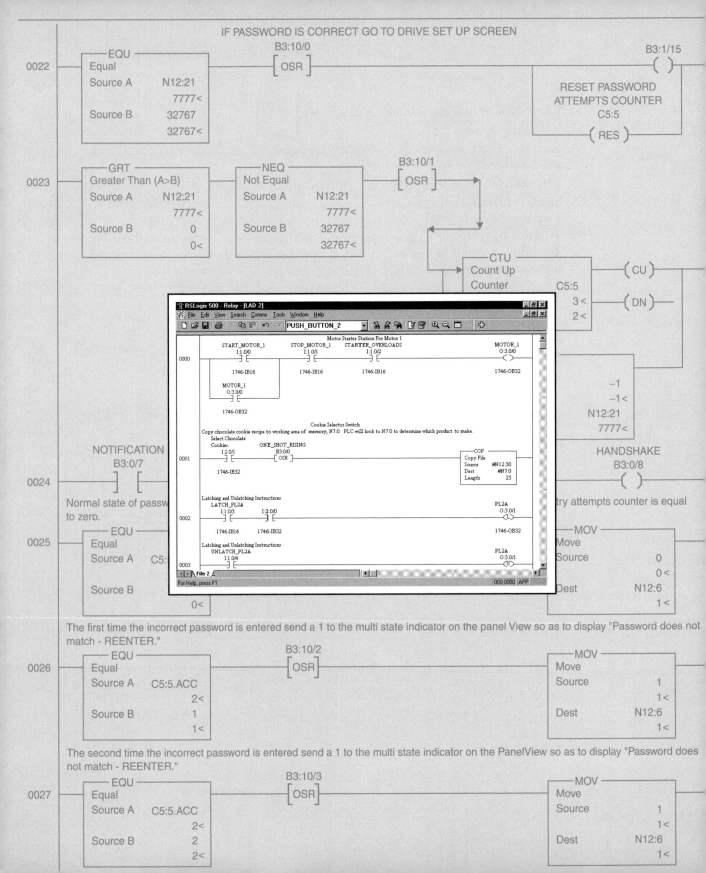

RSLogix 500 - Relay - [LAD 2]

File Edit View Search Comms Tools Window Help

PUSH_BUTTON_2

Motor Starter Station For Motor 1

```
      START_MOTOR_1    STOP_MOTOR_1   STARTER_OVERLOADS              MOTOR_1
         I:1.0/0          I:1.0/1          I:1.0/2                   O:3.0/0
0000     ─] [─            ─]/[─            ─] [─                      ─( )─
        1746-IB16        1746-IB16        1746-IB16                 1746-OB32
        MOTOR_1
        O:3.0/0
        ─] [─
        1746-OB32
```

Cookie Selector Switch
Copy chocolate cookie recipe to working area of memory, N7:0. PLC will look to N7:0 to determine which product to make.

```
      Select Chocolate
      Cookies.        ONE_SHOT_RISING
         I:2.0/1          B3:0/0                              ┌── COP ────────────┐
0001     ─] [─            [ OSR ]                             │ Copy File         │
        1746-IB32                                            │ Source    #N12:30 │
                                                             │ Dest      #N7:0   │
                                                             │ Length    25      │
                                                             └───────────────────┘
```

Latching and Unlatching Instructions

```
      LATCH_PL2A                                                    PL2A
         I:1.0/3          I:2.0/0                                  O:3.0/1
0002     ─] [─            ─] [─                                    ─( )─
        1746-IB16        1746-IB32                                1746-OB32
```

Latching and Unlatching Instructions

```
      UNLATCH_PL2A                                                  PL2A
         I:1.0/4                                                   O:3.0/1
0003     ─] [─                                                     ─(U)─
```

File 2

For Help, press F1   000:0000 APP READ

CHAPTER

13

The Basic Relay Instructions

OBJECTIVES

After completing this chapter, you should be able to:

- describe the function of the normally open, or examine if closed, instruction
- describe the function of the normally closed, or examine if open, instruction
- explain the function of a one-shot instruction
- explain the function and programming of the latch and unlatch instructions
- explain input and output instruction formatting for the SLC 500 and MicroLogix PLCs
- given an address, identify the input or output point on an SLC 500 fixed or modular PLC and on MicroLogix PLCs

INTRODUCTION

Each manufacturer's PLC processors have their own vocabulary of instructions. A PLC processor's repertoire of instructions is called its instruction set. While different processors have different instruction sets, there are basic instructions shared by all PLC processors. This chapter will introduce the basic relay instructions. Each instruction will be illustrated by looking at an actual rung of logic printed from a PLC program development software package. The software package we will use to print our sample rungs will be Rockwell Automation's RSLogix 500 software. If you are using other SLC 500 software packages, your rungs will look very similar. After looking at the sample rung, we will explain instruction addressing and show how each instruction will function in a program.

BIT, OR RELAY, INSTRUCTIONS

Contacts and coils are the basic symbols found on a ladder diagram. Normally open or normally closed contact symbols are programmed on a given rung to represent input conditions that are to be evaluated by the processor as it solves the user ladder program. Rung contacts are evaluated to determine how output instructions are to be controlled by the PLC. Each output is represented by a coil symbol. Contacts and coils are also referred to as bit or relay instructions. Each input or output is represented by a separate bit in the input or output status file. This information is used to represent actual inputs coming in from, and going out to, the outside world. Figure 13-1 presents a generalized overview of the basic instructions available in most PLC processors.

BIT INSTRUCTIONS		
Instruction	**Symbol**	**Use This Instruction**
Normally Open or Examine ON	—| |—	As a normally open, or examine if ON, input instruction on your ladder rung
Normally Closed or Examine OFF	—|/|—	As a normally closed, or examine if OFF, input instruction on your ladder rung
One-Shot	—(OSR)—	To input a single digital pulse from a maintained input signal
Latch Output Coil	—(L)—	To latch an output ON. Output stays ON until the unlatch instruction becomes true
Unlatch Output Coil	—(U)—	To unlatch a latched ON instruction with the same address
Output Coil	—( )—	As an output instruction that becomes true when all inputs on the rung are true
Negated Output	—(/)—	As an output instruction that passes power at all times except when all rung inputs are true

Figure 13-1   Overview of bit instructions.

Instructions direct the PLC as to how to respond to bits found in its memory. Bits in PLC memory are typically input status file bits representing ON or OFF signals input into the status file from an input module. Figure 13-2 provides an overview of how the normally open instruction is evaluated by the PLC processor's CPU.

THE NORMALLY OPEN INSTRUCTION

The normally open instruction is used by all PLCs; however, each manufacturer may have a different name for the instruction. The SLC 500 and MicroLogix use the term "examine if closed" (XIC) to represent the normally open instruction. Other PLC manufacturers might use "examine if ON" to identify the normally open instruction. Since this section of this text focuses on the Allen-Bradley SLC 500 family of PLCs, we will use that terminology.

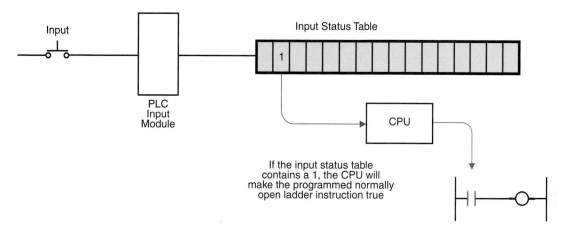

Figure 13-2   Normally open instruction interaction between field device, input module, input status table, and the CPU.

The Examine if Closed Instruction

The examine if closed instruction tells the processor to test for an ON condition from the reference address bit. The reference address bit could be an input device in the input status file, an output bit from the output status file, or an internal bit used as either an output or a status bit from other instructions.

THE OUTPUT INSTRUCTION

The output instruction is typically represented as an output coil. The SLC 500 family and MicroLogix refer to the output coil as an output energize instruction (OTE). Every rung must have a minimum of one output instruction, and it may have more than one output. Multiple outputs are almost always programmed in parallel. The output instruction is always the last instruction before the right power rail. An output instruction represents the action that is to be taken when the solved input logic results in a logically true rung. Figure 13-3 illustrates a rung with an XIC input instruction and one OTE output instruction. The OTE instruction is controlling a field hardware device connected to terminal one of a 1746-OB16 module.

Figure 13-3   Ladder rung containing an examine if closed input instruction and an output enable instruction.

Explanation of Rung Operation

If input conditions on a rung that precedes an OTE instruction are true, the processor, when solving the ladder program, will cause the OTE instruction to either become true or maintain its true status. OTE status will be updated as part of the update output's portion of the processor scan.

If a rung's input conditions that precede an OTE instruction direct the rung to go false, the processor will cause the OTE instruction to either become false or maintain its false status. Figure 13-4 provides a table listing input conditions and the resulting processor evaluation of a normally open, or examine if closed, instruction.

If Input Status Table Is	And the Programmed Instruction Is	The Processor Will Evaluate the Instruction as
1	—| |—	True, or pass logical continuity
0	—| |—	False, or will not pass logical continuity

Figure 13-4   XIC input instruction interaction with input status table.

MODULAR SLC 500 INSTRUCTION ADDRESSING

Figure 13-3 illustrates an examine if closed instruction with a reference of I:1/0 and an OTE instruction with a reference of O:2/1. Each reference is called the instruction's address. An SLC 500 input address consists of three pieces of important information:

- identification as input address
- identification of the location of the module in the SLC 500 modular chassis
- identification of the screw terminal on the input module with which this particular input is associated

Figure 13-5 illustrates the basic input or output addressing format.

Data File Type	:	Module Slot	/	Module Screw Terminal

Figure 13-5   Basic SLC 500 input or output addressing format.

The address I:1/0 breaks down into the following parts:

I = identifies this as an input instruction. Input address data is stored in the input file one, identified as I file type.

: = the element delimiter. It serves simply to separate the file type and file number from the other parts of the address.

1 = identifies the slot in which the addressed module resides in the SLC chassis.

/ = the bit delimiter. It separates the input bit reference from the input module slot reference.

1 = screw terminal number of this input reference on the input module residing in slot one.

Directly below the input address and instruction symbol is the part number of the input module on which this input will be found. The input module in slot one of our SLC 500 chassis is a 1746-IA8, which is an eight-point, 120 VAC input module.

The address O:2/1 breaks down into the following parts:

O = identifies this as an output instruction. Output address data is stored in the output file 0, identified as O file type.

: = the element delimiter. It serves simply to separate the file type and file number, when used, from the other parts of the address.

2 = identifies the slot in which the addressed module resides in the SLC chassis.

/ = the bit delimiter. It separates the output bit reference from the output module slot reference.

1 = screw terminal number of this output reference on the output module residing in slot two.

Directly below the output address and instruction symbol is the part number of the output module on which this output will be found. The output module in slot two of our SLC 500 chassis is a 1746-OB16, which is a 16-point, DC output module for transistor sourcing.

Slot Identification in an SLC 500 Chassis

To accurately identify a specific input point among multiple input modules, each slot in a modular chassis is assigned a slot number. Figure 13-6 illustrates a seven-slot SLC 500 modular chassis.

	Slot Zero	Slot One	Slot Two	Slot Three	Slot Four	Slot Five	Slot Six
Power Supply	Reserved for the Processor	Input or Output Module	Input or Output Module	Input or Output Module	Input or Output Module	Input or Output Module	Input or Output Module

Figure 13-6   Seven-slot SLC 500 chassis slot identification.

On the far left of Figure 13-6 is the power supply. The slot next to the power supply is reserved for the processor; this is slot zero. Slots are numbered in decimal numbers from left to right: slot zero, one, two, three, four, five, six, and so on. There are four chassis sizes available for a modular SLC 500 PLC: four-, seven-, ten-, and thirteen-slot chassis. Slot numbering for all chassis follows the numbering convention illustrated in Figure 13-6.

Fixed SLC 500 Family Slot Numbering

Slot zero on a modular SLC 500 is always reserved for the processor. Modular I/O modules are inserted in the chassis beginning with slot one. To maintain consistency, the processor for a fixed SLC 500 or a MicroLogix PLC will also be identified as slot zero. A fixed PLC's I/O is built into the same physical unit that houses the processor. Since both the processor and the I/O are built into the same assembly, fixed inputs will be addressed as slot zero inputs, I:0. Likewise, fixed outputs are addressed as slot zero, O:0. Figure 13-7 illustrates the basic terminal format for the fixed 1747-L20A SLC 500. The MicroLogix is very similar.

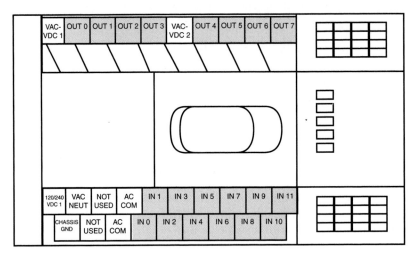

Figure 13-7 SLC 500 fixed PLC, part number 1747-L20A.

Address Format

All fixed SLC 500 and MicroLogix PLCs are addressed similar to modular units, except that all I/O points on the fixed controller are slot zero. Input addresses are represented as:

I:0/12 = input, slot zero, screw terminal 12

I:0/3 = input, slot zero, screw terminal 3

Output addresses are represented as:

O:0/5 = output, slot zero, screw terminal 5

O:0/12 = output, slot zero, screw terminal 12

Notice that there is an input 12 and an output 12, both of which are addressed as slot zero. A fixed PLC will have both inputs and outputs starting with address zero. The only difference in their address script is the I or O designation. There will be a word zero in the input status file and a word zero in the output status file to store these bits.

THE NORMALLY CLOSED INSTRUCTION

The normally closed instruction is also called the examine off, or examine if open, instruction, depending on the particular PLC manufacturer. The SLC 500 and MicroLogix use the term "examine if open" (XIO) to identify this instruction.

The normally closed instruction is programmed whenever the processor is to test the instruction for an OFF, or open, condition. Since the XIO instruction is normally closed, the instruction will normally be represented as a 0 in the input status file. Finding a 0 in the status file referencing the instruction's address means that the device controlling the bit address is in the OFF condition. With the field input device OFF, the normally closed instruction will remain closed. A normally closed instruction that is evaluated as closed is a true instruction. Finding the normally closed instruction true, the instruction will continue to provide continuity through the instruction on the rung. If the status table bit is found to be a 0, the device controlling this bit is in its ON state. Finding the instruction's associated status bit in the 0 condition will direct the processor to open the normally closed instruction and make it false. With the normally open contacts open, continuity will be disrupted to any remaining contacts or instructions on the rung. Figure 13-8 is a ladder rung with a normally closed, or examine if open, instruction.

Figure 13-8   Normally closed, or examine if open, instruction controlling an OTE instruction.

Explanation of Rung Operation

If the field hardware input device connected to the SLC 500 input module residing in slot one, screw terminal one, is not energized, the input status bit will remain a zero. With the input status bit a zero, the XIO instruction will remain closed and pass logical continuity. As long as XIO instruction I:1/1 is passing logical continuity, the OTE instruction with the address of O:2/0 will be true. Being true, the field hardware output device attached to the output module screw terminal zero in SLC 500 chassis slot two will either be turned ON or remain ON.

Figure 13-9 summarizes physical input device condition, input status table bit condition, and how the processor will evaluate the normally closed instruction.

If the Input Device's Physical Condition Is	And the Input Status Table Bit Is	And the Programmed Instruction Is	The Processor Will Evaluate the Instruction as
OFF	0	—|/|—	True
ON	1	—|/|—	False

Figure 13-9   Physical input conditions and the normally closed XIO instruction.

Addressing

The normally closed, or examine if open, instruction is addressed in the same manner as the normally open, or examine if closed, instruction for the SLC 500 or the MicroLogix.

Next we will investigate the one-shot instruction, which is an input instruction used when you want one pulse, or one shot, as an input signal on your ladder rung.

THE ONE-SHOT INSTRUCTION

The one-shot rising instruction, OSR, is an input instruction that allows an event to occur only once. Figure 13-10 is a rung with input I:2/1 as input logic to the OSR instruction, B3:0/0. The OSR instruction controls the one-shot output for output address O:2/2. The internal bit is not the actual output—O:2/2 is. The bit shown in Figure 13-10, which depicts the SLC 500 family one-shot ladder rung, is used only as a reference to remind the instruction of the previous rung state. As with any other input instruction, always use a unique bit address for the OSR instruction.

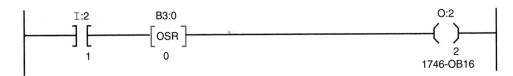

Figure 13-10   SLC 500 family one-shot ladder rung.

The "rising" portion of the one-shot rising instruction means that the instruction is looking for an OFF-to-ON, or false-to-true, transition in the leading-edge pulse input. Figure 13-11 illustrates a leading-edge one-shot timing diagram in comparison to a trailing-edge one-shot instruction. Input I:2/1 is illustrated as the top line. Notice that I:2.1 is ON for multiple scans. Line two is the leading-edge one-shot timing line. Notice that the leading-edge one-shot instruction is energized as the input's leading edge goes from OFF to ON. The bottom timing line, entitled the trailing edge, changes state on the trailing edge of the input pulse, at the end of scan three.

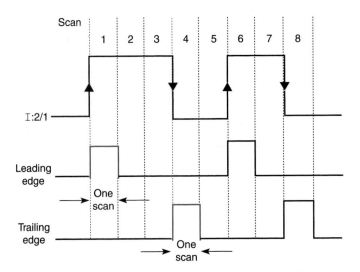

Figure 13-11   Leading-edge versus trailing-edge one-shot timing diagram.

The SLC 500 family instruction is the one-shot rising instruction, OSR. A one-shot instruction operates in a manner similar to an output instruction. If a continuous path of input instructions on a ladder rung preceding the one-shot instruction is true, thus providing logical continuity, the one-shot instruction will be energized. When input instruction I:2/1 goes from false to true, the one-shot (OSR) referenced output instruction goes true for one scan. Following the scan in which the OSR instruction is true, the referenced output instruction goes false. The OSR instruction holds the referenced output false even though the input logic to the OSR instruction may remain true. Refer to the leading-edge timing trace in Figure 13-11. The OSR instruction will only allow the referenced output to become true again after input logic goes false and then transitions from false to true again.

When to Use the One-Shot Instruction

You will use this instruction to start an event triggered by a push-button input into your PLC where you want the event to happen only once per actuation of the push button, no matter how long it is held in. As an example:

1. A one-shot can be used to reset desired conditions in a single scan.
2. You can use a push button to read the current value of a thumbwheel switch one time.
3. A momentary push-button actuation could be used to increment speed on a motor. Speed would increment one step for each push of the button.
4. A single push button could be used to start and stop a motor. The first closure of the button would start the motor and the second closure would stop it.
5. Use the one-shot instruction with a math instruction to perform a calculation once per scan.
6. A one-shot instruction can be used to bring in changing analog input data, which can be sampled at a predetermined rate.
7. Use a push button or an internal bit and the one-shot instruction to send data to output display devices. A typical output device could be an LED numerical display. The one-shot instruction will allow rapidly changing data to be "frozen" and output to an LED display in a timely manner. Timely updates ensure a readable, stable display. If data were allowed to be output continuously and were changing rapidly, the display could be hard to read.

   An example of an LED display where data changes rapidly is a filling station gasoline pump displaying total gallons as they are pumped. Another example is the display on the pump showing the total dollars spent on gasoline. As you fill your tank, both displays flash numbers rapidly. These displays are instantaneously updated. This is a situation in which receiving instantaneous updates is not a problem. As the individual pumping the gasoline, you want to know exactly how much gasoline has been dispensed and how much money has been spent.

   Figure 13-12 illustrates a rung where the OSR instruction will allow the conversion of integer data to BCD data. The PLC conversion instruction, TOD (to BCD), will be true for one scan. The TOD instruction will convert decimal integer data from the accumulated value of the timer file 4, timer zero

Figure 13-12   A rung of logic in which a one-shot rising instruction controls and converts an integer value to BCD instruction (TOD instruction).

(T4:0.ACC) and send it as an output only once to the destination O:5.0. If the timer was running and displayed continuously, the display might change rapidly and be unreadable. The one-shot instruction allows the TOD instruction to send converted data out only when I:2/1 triggers the OSR instruction. A timer used in conjunction with a one-shot instruction could provide a stable display that could be updated every few seconds rather than appear as a blur of numbers.

8. A one-shot instruction could be used to trigger a copy instruction, as illustrated in Figure 13-13. The one-shot instruction will allow the copy instruction to copy data from the source to the destination only once per trigger.

Figure 13-13   Internal bit B3:0/3 controlling a one-shot instruction to enable data to be copied only once per trigger of bit B3:0/3.

OSR Instruction Addressing

Use either a bit file address or an integer file address as the address assigned to the OSR instruction. The address assigned to each OSR instruction is used only as a reference to the previous ON or OFF state of its rung. The output(s) associated with the one-shot instructions are the actual instructions that are energized for one shot. The address used for the OSR must be unique and not used elsewhere in your program.

The OSR instruction we just looked at is used only on the SLC 500 modular processors and the MicroLogix 1000. The MicroLogix 1200, 1500, and the ControlLogix use the PLC 5 one shot instructions. The instructions are listed below.

ONS   This is a one-shot rising input instruction similar to the OSR for the SLC 500 processors and MicroLogix 1000.

OSR   The OSR instruction for the MicroLogix 1200, 1500, PLC 5, and ControlLogix is a one-shot rising instruction, but it is an output instruction for these PLCs.

OSF   The OSF is a one-shot falling output instruction, also for the MicroLogix 1200, 1500, PLC 5, and ControlLogix.

For additional information on these instructions, refer to RSLogix instruction set help or your accompanying RSLogix 500 programming lab manual.

THE OUTPUT-LATCHING INSTRUCTION

An output-latching instruction is an output instruction used to maintain, or latch, an output ON even if the status of the input logic that caused the output to energize changes.

When any logical path on the ladder rung containing the latching instruction has continuity, the output referenced to the latching instruction is turned ON and remains ON, even if the rung's logical continuity or PLC system power is lost. Since the latch instruction retains its state through a system power loss, the latching instruction is called a *retentive instruction*. Remember, the processor's battery must be in good condition for the latching status to be remembered (or retained by the processor) in case of a power failure.

The latched instruction will remain in a latched ON condition until an unlatch instruction with the same reference address is energized. Latch and unlatch instructions are always used in pairs. Each instruction is typically located on a separate rung.

Figure 13-14 illustrates two ladder rungs. The first rung contains the latching instruction, while the second rung contains the unlatching instruction.

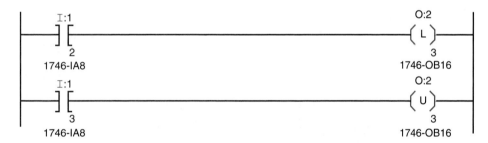

Figure 13-14 Latching and unlatching ladder logic.

Rung Operation

When input I:1/2 on the first rung is energized, the output-latch instruction, address O:2/3, is energized. The output-latch instruction will remain latched ON and will be unaffected, no matter how input I:1/2 changes. You must use the output-unlatch instruction to turn OFF, or unlatch, the output that was turned on by the latching instruction. The unlatch instruction with the same address must be energized to unlatch the output address that was latched ON. When input I:1/3 is energized, the unlatch instruction will turn off output O:2/3. These particular example rungs have only one input each. Any valid input logic instructions may be used as inputs to the latch and unlatch instructions. The following rules pertain to most latch and unlatch instructions:

1. Latch and unlatch instructions are used in pairs.
2. Latch and unlatch pairs of instructions must have the same reference address.

3. The latch and unlatch ladder rungs do not need to be grouped together in the ladder program.

4. Latching and unlatching instructions are retentive, provided your PLC system battery is installed and in good condition.

5. Not all PLCs allow internal bits to be latched. SLC 500 family PLCs will allow internal bits to be latched from either a bit file or an integer file.

6. Use an unlatch instruction to unlatch, or clear, status bits. Figure 13-15 illustrates an unlatch instruction unlatching an overflow trap status bit. This status bit is unlatched in applications where a math overflow or divide by zero operation occurs and could otherwise fault the processor.

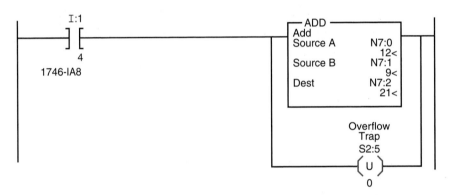

Figure 13-15 Unlatch instruction used alone to unlatch an addition instruction's overflow status bit.

7. If an unlatch instruction is left energized, the associated latching output cannot be latched.

8. Output latch instructions are retentive. This means that if the processor loses power, is switched to program or test mode, or detects a major fault, the output-latch instruction will retain the state of the latched bit in processor memory. Even though all outputs will be turned OFF during these processor conditions, retentive outputs will return to their previous states when the processor returns to run mode.

Programming Considerations

The placement of the latch and unlatch instruction rungs within your ladder program can affect the behavior of these instructions. Figure 13-16 illustrates a latching instruction programmed before the unlatch instruction. If both instructions are true, the last instruction programmed on the rung will take precedence over the other instruction. In this example, the output instruction will always be unlatched.

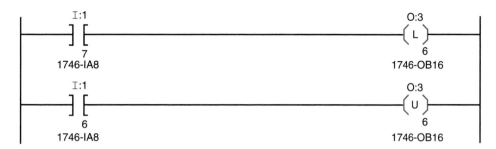

Figure 13-16   Latching instruction programmed before the unlatch instruction.

Figure 13-17, on the other hand, has the unlatch instruction programmed after the latch instruction. In this case, the last instruction—the latch instruction—will take precedence and keep the output latched, provided both the latch and unlatch rungs are true.

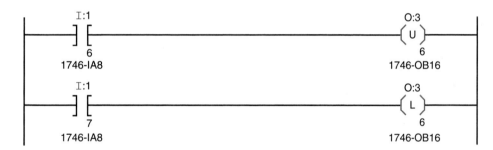

Figure 13-17   Latching instruction programmed after the unlatch instruction.

THE NEGATED OUTPUT INSTRUCTION

The negated output instruction is the opposite of a normal output instruction. Even though the normal output instruction is false when the ladder rung is logically false, the negated output instruction is true when the rung's input logic is false. The negated output instruction is also referred to as a not output coil or instruction. Not all PLCs have a negated or not output instruction. Rockwell Automation PLCs do not have this instruction.

INTERNAL BIT-TYPE INSTRUCTIONS

There are instances when you will need to control instructions other than an output on a rung or rungs other than the current rung. An instruction is needed that is easily programmed on the current rung and that will not represent a real field device. An internal bit-type instruction that could be programmed as either normally open or normally closed would really help in program development.

Most PLCs have some method of incorporating internal bits into the user program when other than real-world field devices are needed as input or output reference instructions. Different manufacturers name these internal bit-type instructions differently. Some use the terms internal bits, internal coils, or internal relays to identify internal bits programmed as non-real-world field devices.

An internal bit used as an output is sometimes referred to as an internal relay, internal coil, or internal output. An internal bit would be used as a rung output when a real output is not desired. An internal bit as an output (used like a control relay, as an example) would be used when the logical resultant of a rung is used to control other internal logic. Figure 13-18 illustrates a simplified usage of an internal bit as an output. Input I:1/2 controls output bit B3:0/0. Output bit B3:0/0 is not a real output such as you can see on an output module. B3:0/0 is a bit found in bit file 3. In this example, B3:0/0 or B3:0/1, from another portion of the ladder program, controls output O:2/1.

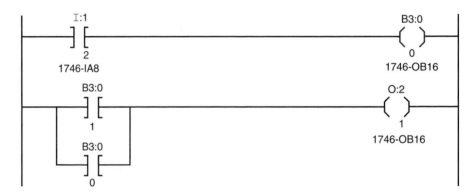

Figure 13-18 Internal bit B3:0/0 used as an output.

Internal bit-type instructions are usually stored in their own files. The SLC 500, PLC 5 and MicroLogix have a separate storage location for internal bits, called bit file 3.

File 3, the Bit File

The default bit file is file 3. A bit file is used to store single bits in a 16-bit-word format. There can be many bit files for a single processor file. Any data file greater than file 10 can be assigned as an additional bit file. Each bit file will have 256 16-bit words, or 4,096 individual bits. Figure 13-19 illustrates a portion of bit file 3 from RSLogix 500 software.

Bit File Addressing   Figure 13-19 lists the bits 0 through 15 across the top row from right to left. The column on the left of the table, entitled "Offset," lists the bit file addresses as full words. The address B3:0 is broken down as follows:

B = the designator for a bit file

3 = identifies this as bit file 3 (the default bit file; additional bit files may be designated as any unused file between 9 and 255)

:0 = element number 0 (the colon is the file separator)

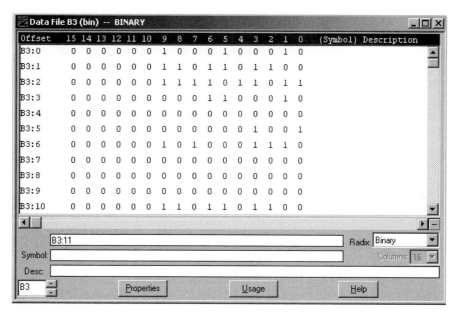

Figure 13-19   Each bit file element consists of one 16-bit word.

Addressing to the Bit Level   If bit number 5 in bit file 3, element three, is the single bit needed to be addressed for use in the user program, the addressing format is as follows:

B3:3/5

B = the designator for a bit file

3 = identifies this as bit file 3

:3 = element number three (the colon is the file separator)

/5 = bit 5 (the slash is called the bit separator)

SUMMARY

This chapter introduced the basic relay-type instructions that are used when programming PLC ladder programs. No matter if you are using a handheld programmer or a personal computer and whether you wish to develop, edit, or troubleshoot a user program, you will find many of the instructions on the ladder program are the familiar normally open and normally closed instructions.

Figure 13-20 summarizes XIC and XIO instruction use.

Keep in mind that the PLC ladder program is only a logical representation of how real-world input devices are to interact with real-world outputs. The ladder program is only a graphical representation of what the processor is to do in response to programmed input conditions. As you look at a ladder rung, remember that the rung will be true if there is logical continuity from the left power rail, through a minimum of one continuous path of instructions, and onto the output instruction. The actual logical status of all input

PROGRAMMING XIC AND XIO INSTRUCTIONS		
Instruction	Input's Physical State	To Make Instruction TRUE
XIC	Open	Close input device or set bit
XIC	Closed	Instruction is true
XIO	Open	Instruction is true
XIO	Closed	Open input device or reset bit

Figure 13-20   RSLogix 500 XIC and XIO instructions.

instructions is found in each instruction's associated status table and the address specifi-cally referencing the particular instruction. When the processor determines if the rung is logically true or false, it will send a status bit to the address assigned to the output in-struction in the associated status table. Remember that not all input and output instruc-tions represent real-world field devices. Internal bits are used in many situations where only internal program control is desired.

REVIEW QUESTIONS

1. If there is a 1, or ON bit, in the particular normally open instruction's corresponding status table address, the instruction will be evaluated as _____.
2. The SLC 500 and MicroLogix use the term _____ to represent the normally open instruction.
3. When using an XIC instruction, the reference address could be an:
   A. input device in the input status file
   B. output bit from the input status file
   C. internal bit used as an output
   D. internal bit used as a status bit from other instructions
   E. B, C, and D
   F. A, C, and D
4. The output instruction is always the last instruction before the _____.
5. The output instruction, sometimes referred to as an output coil:
   A. may be a real output connected to an output module
   B. may be a real input connected to an input module
   C. may be an internal bit such as a bit-file bit
   D. may be an external bit such as an external bit-file bit
   E. A and D
   F. A and C
   G. A, B, and C
6. An SLC 500 input address consists of which pieces of important information?
   A. identification that this is an input address
   B. identification of the location of the module in the SLC 500 chassis

   C. identification of the module for ladder diagram placement
   D. identification of the screw terminal on the input module
   E. A, B, and C
   F. A, B, and D

7. The address I:4/7 breaks down into the following parts:

   I =

   : =

   4 =

   / =

   7 =

8. The address I:1/12 identifies a module that is:
   A. an input module residing in slot zero of the PLC chassis
   B. an input module residing in slot one of the PLC chassis
   C. an input module residing in slot two of the PLC chassis
   D. an input module residing in slot three of the PLC chassis
   E. an input module residing in slot four of the PLC chassis
   F. an input module residing in slot five of the PLC chassis

9. The address I:3/15 would be found on:
   A. an input module residing in a 4-slot chassis
   B. an input module residing in a 7-slot chassis
   C. an input module residing in a 10-slot chassis
   D. an input module residing in a 13-slot chassis
   E. an input module residing in any of the above-mentioned chassis

10. Directly below the input address and instruction symbol on an RS Logix ladder print-out is the _____ of the input module on which this input will be found.

11. The normally closed instruction is also called the _____, or _____, instruction.

12. The normally closed instruction is programmed whenever the processor is to test for an _____, or open, condition.

13. A normally closed instruction that is evaluated as closed is a _____ instruction.

14. If the status table bit is found to be a _____, the device controlling this bit is in its ON state.

15. The one-shot rising instruction, OSR, is an input instruction that allows an event to occur _____.

16. After the scan in which the OSR instruction is true, the instruction goes false. The OSR instruction remains _____ even though the input logic may remain true.

17. A one-shot instruction being energized on the leading edge of an input pulse means:
   A. The one-shot instruction is energized when the PLC processor is cycled from OFF to ON.
   B. The one-shot instruction is energized when the PLC processor is switched from program to run mode.

   C. The one-shot instruction is energized when it registers the output changing from false to true.
   D. The one-shot instruction is energized when it registers previous input logic on its rung changing from logically false to logically true.
   E. The one-shot instruction is energized when it registers previous input logic on its rung changing from ON to OFF.
18. An output _____ instruction is an output instruction used to maintain, or latch, an output ON even though the status of the input logic that caused the output to energize changes.
19. Since the latch instruction retains its state through a system power loss, the latching instruction is called a _____ instruction.
20. The latched instruction will remain latched ON until an unlatch instruction with the same _____ is energized.
21. When using latching and unlatching instructions, each instruction is typically located on a _____.
22. If both a latching and an unlatching instruction are true, the _____ instruction programmed on the rung will take precedence over the other instruction.
23. The _____ instruction is also referred to as a not output coil or instruction.
24. Using an internal bit as an output is sometimes referred to as an internal relay, _____, or _____.
25. _____ bit-type instructions are usually stored in their own status files.
26. The default bit file for an SLC 500 family or a MicroLogix PLC is file _____.
27. Any data file greater than file _____ can be assigned as an additional bit file.
28. A bit file is made up of single bits grouped into a _____-bit word.
29. The address B3:0 is broken down as follows:

   B =
   3 =
   :0 =

30. Fill in the requested bits in the bit file in Figure 13-21.

Element	15	14	13	12	11	10	9	8	7	6	5	4	3	2	1	0
B3:0										1						
B3:1																1
B3:2		0			1	1										
B3:3	0															
B3:4			1										1			

Figure 13-21   Bit file for Question 30.

    A. B3: 2/14 = 0
    B. B3: 1/0 = 1
    C. B3: 4/13 = 1
    D. B3: 4/2 = 1
    E. B3: 3/15 = 0
    F. B3: 2/10 = 1
    G. B3: 0/6 = 1
    H. B3: 2/11 = 1

31. Fill in the missing information in Figure 13-22.

BIT INSTRUCTIONS		
Instruction	Symbol	Use This Instruction
Normally Open or Examine if ON	‑| |‑	
Normally Closed or Examine if OFF	‑|/|‑	
One-Shot	‑(OSR)‑	
Latch Output Coil	‑(L)‑	
Unlatch Output Coil	‑(U)‑	
Output Coil	‑( )‑	
Negated Output	‑(/)‑	

Figure 13-22   Bit instruction table for Question 31.

IF PASSWORD IS CORRECT GO TO DRIVE SET UP SCREEN

0022

```
     ┌─── EQU ──────────┐            B3:10/0                                              B3:1/15
     │ Equal            │           ┌─ OSR ─┐                                             ─( )─
     │ Source A   N12:21│           [       ]
     │            7777<  │                                              RESET PASSWORD
     │ Source B   32767 │                                              ATTEMPTS COUNTER
     │            32767< │                                                   C5:5
     └──────────────────┘
                                                                         ─( RES )─
```

0023

```
     ┌─── GRT ──────────────┐       ┌─── NEQ ──────────┐     B3:10/1
     │ Greater Than (A>B)   │       │ Not Equal        │    ┌─ OSR ─┐──────┐
     │ Source A   N12:21    │       │ Source A   N12:21│    [       ]      │
     │            7777<      │       │            7777<  │                  │
     │ Source B   0         │       │ Source B   32767 │            ┌──────┘
     │            0<         │       │            32767< │            │
     └──────────────────────┘       └──────────────────┘            │
                                                            ┌────────┴────┐
                                                            │  ┌─ CTU ────────┐   ─( CU )─
                                                            │  │ Count Up     │
                                                            └──│ Counter  C5:5│
                                                               │ Preset    3< │   ─( DN )─
                                                               │            2< │
                                                               └──────────────┘
```

```
                                                                          −1
                                                                          −1<
                                                                        N12:21
                                                                        7777<
```

NOTIFICATION
B3:0/7

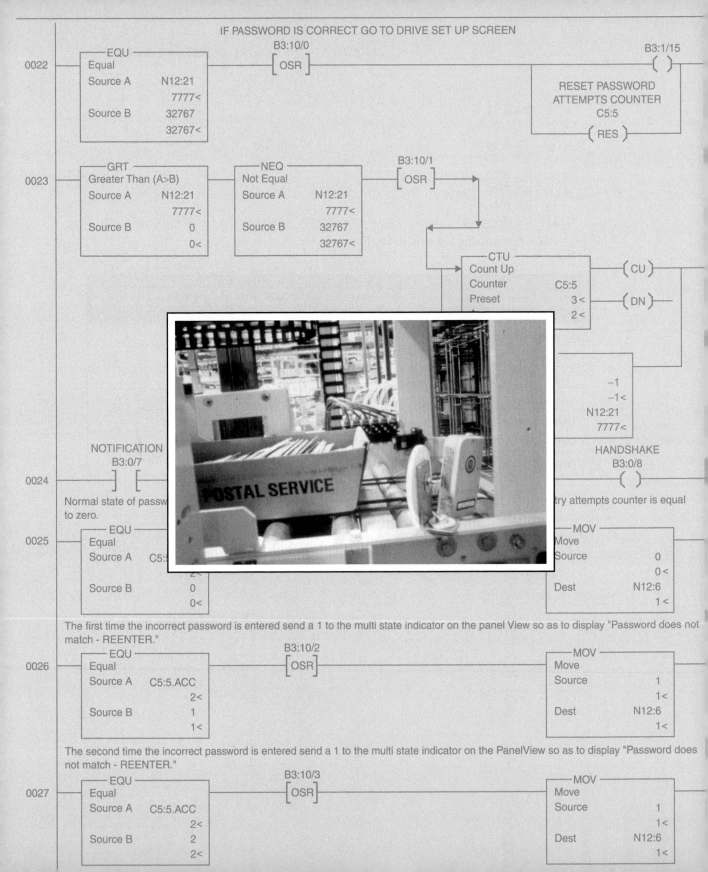

HANDSHAKE
B3:0/8

0024

```
     ─] [─                                                                  ─( )─
```

Normal state of passw...   ...try attempts counter is equal
to zero.

0025

```
     ┌─── EQU ──────┐                                              ┌─── MOV ──────┐
     │ Equal        │                                              │ Move         │
     │ Source A  C5:5│                                             │ Source     0 │
     │           2<  │                                             │            0< │
     │ Source B  0   │                                             │ Dest  N12:6  │
     │           0<  │                                             │            1< │
     └──────────────┘                                              └──────────────┘
```

The first time the incorrect password is entered send a 1 to the multi state indicator on the panel View so as to display "Password does not match - REENTER."

0026

```
     ┌─── EQU ──────────┐           B3:10/2                         ┌─── MOV ──────┐
     │ Equal            │           ┌─ OSR ─┐                       │ Move         │
     │ Source A  C5:5.ACC│          [       ]                       │ Source     1 │
     │           2<      │                                          │            1< │
     │ Source B  1       │                                          │ Dest  N12:6  │
     │           1<      │                                          │            1< │
     └──────────────────┘                                          └──────────────┘
```

The second time the incorrect password is entered send a 1 to the multi state indicator on the PanelView so as to display "Password does not match - REENTER."

0027

```
     ┌─── EQU ──────────┐           B3:10/3                         ┌─── MOV ──────┐
     │ Equal            │           ┌─ OSR ─┐                       │ Move         │
     │ Source A  C5:5.ACC│          [       ]                       │ Source     1 │
     │           2<      │                                          │            1< │
     │ Source B  2       │                                          │ Dest  N12:6  │
     │           2<      │                                          │            1< │
     └──────────────────┘                                          └──────────────┘
```

CHAPTER

14

Understanding Relay Instructions and the Programmable Controller Input Modules

OBJECTIVES

After completing this chapter, you should be able to:

- explain proper programming of normally closed and normally open PLC input signals
- describe hardware relay operation and its correlation to proper PLC interface and programming
- convert a conventional start-stop schematic to a PLC ladder rung diagram
- explain why PLC ladder logic for a motor starter interface has an additional contact in comparison to a conventional schematic
- hook up a start-stop station to a PLC input module
- hook up and develop a PLC program to correctly control a motor starter

INTRODUCTION

One major difference in managing a programmable controller system, as compared to relay control systems, is the necessity to think of the programmable controller's operation as a function of input signals read by the CPU rather than the physical wiring configuration of field input devices. As an example, if a normally open push button is connected to a PLC as an input device, the PLC has no way of knowing that a normally open push button is actually connected. The PLC cannot determine if the push button is normally open or normally closed. The processor can only see the ON or OFF incoming signals as a bit in the input status file. Current ON or OFF input status information from the status file is used to solve the user program.

A programmer must develop a user program based on the status of input signals received and stored in the input status file in addition to the physical status of the field input devices. An effective troubleshooter must understand how the user program interacts with the sensors, push buttons, and limit switch input signals in addition to their physical ON or OFF status. Figure 14-1 illustrates a common start-stop latching circuit such as we would expect to see on a conventional schematic diagram.

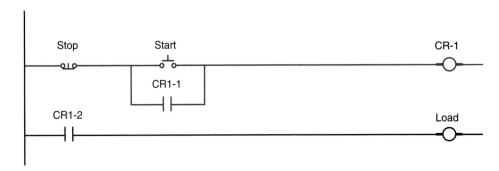

Figure 14-1 Typical hard-wired start-stop latching circuit.

A PLC program could be developed directly from this ladder rung, as illustrated in Figure 14-2. Unfortunately, this PLC rung will not function properly in systems managed by programmable controllers.

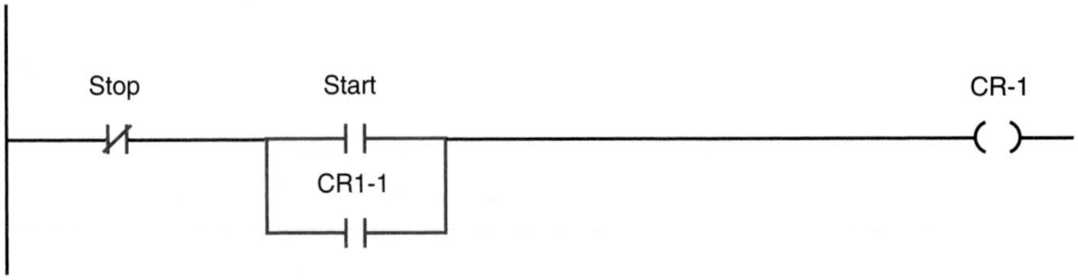

Figure 14-2 Incorrect conversion of conventional start-stop schematic to PLC control.

In this chapter we will investigate issues to consider when connecting real-world hardware devices as discrete inputs to PLC input modules. Then we will see how these input signals affect the programming instructions that are chosen as the user program is developed. Because the programmable controller was designed to simulate relay ladder logic, relay operation is a key building block in understanding how input signals from hardware input devices must interact with programming instructions.

First, let's review basic relay logic and how a hard-wired relay operates. When the relay is at rest, the normally open push-button input is not actuated and the relay coil is not energized. The relay's normally closed contacts will remain closed and Light A will be lit in Figure 14-3.

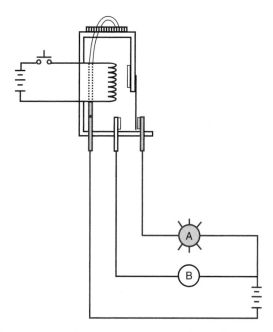

Figure 14-3 Normally open push button and nonenergized relay.

When the push button is closed, the relay coil will be energized and its associated contacts will switch. The normally open contacts, now closed, will pass power. Light A will turn off as its contacts are open, and Light B will energize through its normally open contacts, which are closed (see Figure 14-4).

Figure 14-5 illustrates a normally closed push button, which allows power to flow continuously to the relay coil. With the relay coil energized, Pilot Light B will remain on as long as the normally closed push button allows power to flow to the relay coil. In this example, the relay's normally open contacts are being held closed because the relay coil is being energized.

When the normally closed push button is depressed, as illustrated in Figure 14-6, power will be interrupted to the relay's coil. As the coil loses power, the normally open contacts, which have been held closed, will open. Normally closed contacts will pass power as they return to their normal state. Pilot Light A will then energize.

Figure 14-7 summarizes the relay's status in conjunction with the input push buttons.

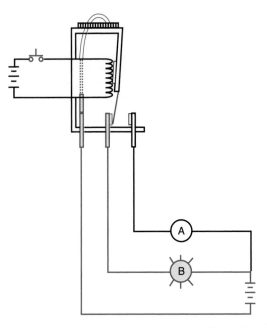

Figure 14-4 Normally open push button energizing relay coil and Pilot Light B.

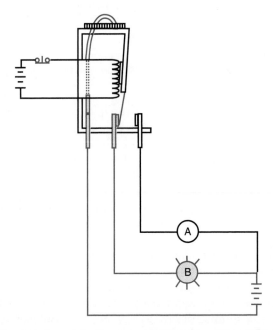

Figure 14-5 Normally closed push button energizing relay coil and Pilot Light B.

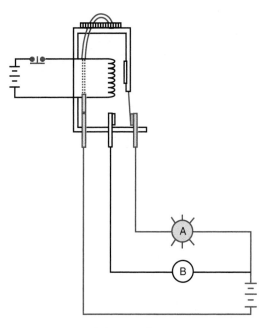

Figure 14-6   Normally closed stop push button depressed, causing an open input circuit.

Input Switch	Relay Coil	No Relay Contacts	NC Relay Contacts	Pilot A	Pilot B
Normally Open (No)					
Pressed	ON	Closed	Open	OFF	ON
Not Pressed	OFF	Open	Closed	ON	OFF
Normally Closed (NC)					
Pressed	OFF	Open	Closed	ON	OFF
Not Pressed	ON	Closed	Open	OFF	ON

Figure 14-7   Relay status in conjunction with input push buttons.

INTERFACING A START-STOP PUSH-BUTTON STATION TO A PROGRAMMABLE CONTROLLER

Start-stop push buttons are sometimes built into a small enclosure called a start-stop station. The typical start-stop push button has one momentary, normally open, start push button and one momentary, normally closed, stop push button.

The U.S. Occupational Safety and Health Administration (OSHA) states that stop push buttons must fail safely. To fail safely, the stop push button must be normally closed. As a normally closed device, the stop push button will continually pass power until depressed.

When depressed, the stop push button will open the circuit and shut off the controlled device. If a wire was broken in the control circuit, the output device would be de-energized, thus avoiding a potentially dangerous situation. If a normally open stop push button was used for a stop circuit, the stop circuit could not fail safely if there was a bad connection or broken wire because manual intervention would have to take place to kill power to the circuit.

Circuit Operation

In a conventional circuit, when you actuate the start button, power flows through the normally closed stop push button to the control relay coil (refer to Figure 14-1). The coil contacts, CR 1-1, close around the normally open start push button to maintain current flow after the push button is released. The coil contacts, CR1-1, latch the circuit ON. A second set of relay contacts, CR1-2, provides power to the load. Depressing the normally closed stop push button interrupts current flow, and the CR-1 coil de-energizes.

To interface a start-stop push button station to a PLC, we need to evaluate the conventional schematic and convert it to ladder logic for PLC control. The first step in schematic conversion to PLC management is to determine inputs and outputs from the conventional ladder symbols.

From the conventional schematic illustrated in Figure 14-1, it can be seen that the normally open start push button and normally closed stop push button are input devices controlling output.

In a PLC control application, the PLC and the user program take the place of the relay coil as the decision maker. The PLC ladder rung does not show the decision-making part of the system; as a result, only inputs and outputs are represented on a PLC ladder rung. Our conventional control circuit from Figure 14-1 is being converted to a PLC ladder rung in Figure 14-8.

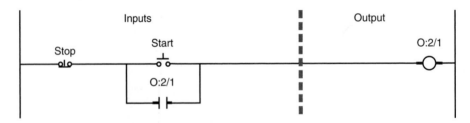

Figure 14-8   Input and output separation.

When converting to PLC ladder logic, the CR-1 coil and its associated contacts become internal to the PLC and are not represented on the PLC ladder program. Because CR1-1 and its associated contacts no longer exist, the PLC output instruction will be used to latch the ladder logic.

Generally, to receive maximum flexibility from a PLC, you should connect each input to the PLC input module separately so that the PLC ladder program can be used to combine each input as required to satisfy any control problem. Figure 14-9 illustrates

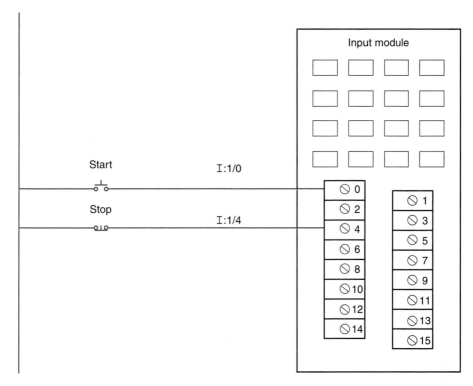

Figure 14-9 Separated inputs connected to PLC input screw terminals.

separated inputs wired to a PLC input module. The PLC ladder program's job is to logically recombine inputs into the control logic necessary to operate hardware output devices. The input module screw terminals to which the inputs are connected in this example were randomly selected.

Typical PLC Input Module Wiring

Let us look more closely at wiring this input module and the signals being input to the CPU. Now that we have separated the inputs and outputs, the addresses need to be determined. For this example, our start push button will be input zero on the input module in slot one, address I:1/0. The stop push button will be input four on the input module in slot one, address I:1/4. Our latching instruction is referenced to O:2/1. The normally open start push button is open; no ON signal will be input to the CPU when it scans this input address. A binary zero signal level will be placed in the input status file (illustrated in Figure 14-10).

The stop button provides a constant ON signal to the input module (input address I:1/4), just as it would in a conventional circuit. A binary one will be placed in the input status file reflecting this ON condition (see Figure 14-11).

We have an understanding of the signals that each input places in the input status file, but how are proper instructions selected as the user program is developed?

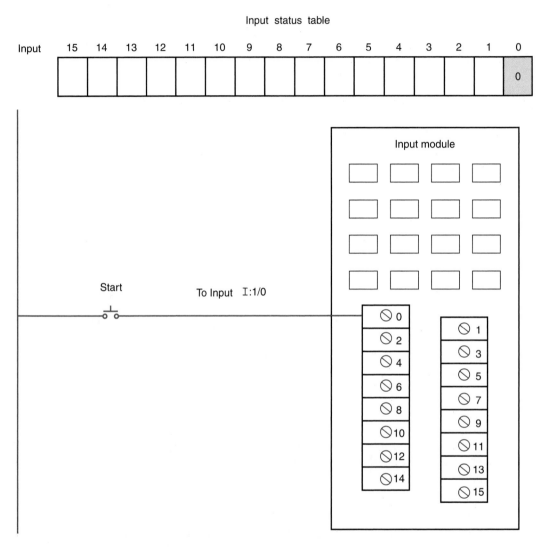

Figure 14-10   Start push-button status reflected in the input status table. This OFF signal will be reflected as a 0 in this input status table position.

USER PROGRAM DEVELOPMENT FOR A START-STOP PUSH-BUTTON INTERFACE

In this section we will use what we have learned regarding physical input signals and the PLC's input status file to develop a user program incorporating a start-stop rung of PLC ladder logic.

Two programming instructions are available, the normally open instruction (represented as —| |—) and the normally closed instruction (represented as —| / |—). By using a PLC, we are using software rather than hardware relays to control our application. Although

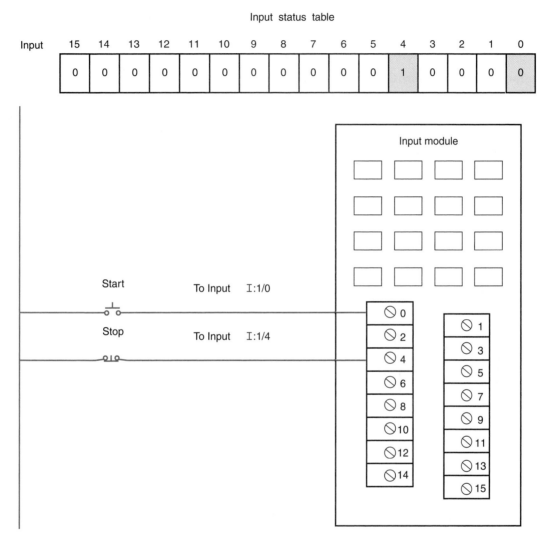

Figure 14-11   Start-stop input bit status from input hardware in a nonenergized state.

there is no actual relay coil in the PLC's input section, the relay coil's ON or OFF state could be represented by the input status file bit. The hardware relay's normally open and normally closed contacts are represented by the normally open and normally closed instructions, programmed on the ladder rung. The SLC 500 family of PLCs and the MicroLogix use the term **examine if closed (XIC)** instruction to represent a normally open contact and an **examine if open (XIO)** instruction to represent a normally closed instruction.

When programming the XIC instruction, we ask the processor to test to see if the instruction is closed. The XIC instruction examines the PLC input status file address for an ON, or binary 1 condition. If the processor finds an ON condition, the XIC instruction is true.

When programming the XIO instruction, the processor is testing to see if the instruction is open. The XIO instruction examines the PLC input status file for an OFF, or binary 0, condition. If the processor finds an OFF condition, the XIO instruction is true.

Programming a Normally Open Push Button as an Input to a PLC

Figure 14-12 portrays a normally open push button used as a PLC input to address I:1/2. The ladder program area illustrates a normally open instruction and a normally closed instruction, each on a separate rung. The top rung has a normally open instruction as the input controlling output O:2/1. The bottom rung has I:1/2 as a normally closed instruction controlling O:2/2. Let us investigate how the two programming instructions react in relation to the input signal.

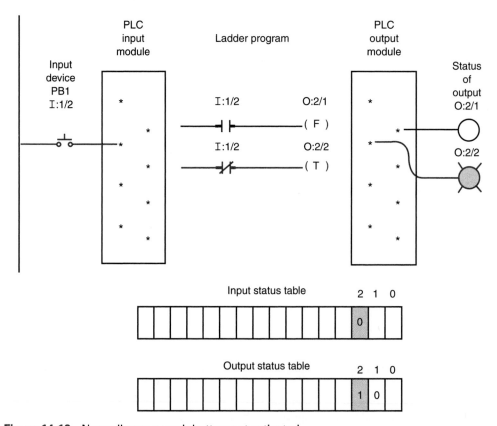

Figure 14-12   Normally open push button not activated.

First we will look at programming a normally open (XIC) instruction with a normally open field input device.

Programming a Normally Open Instruction   A normally open field input device will input a binary 0 to input status file bit position I:1/2 (refer to Figure 14-12). The normally open, or XIC, instruction examines input status file location I:1/2 for an ON condition. In this example the processor detects a binary 0 in the status file address and the instruction is false. Since this instruction is false, a 0 will be placed in the output status file bit position for this rung's output instruction. As illustrated in Figure 14-12, when the PLC updates outputs, output O:2/1 does not energize.

Programming a Normally Closed Instruction   Next we will look at programming a normally closed (XIO) instruction with a normally open field input device. The normally open input device will input a 0 into input status file bit position I:1/2. Remember, when programming the XIO instruction, the processor is testing to see if the instruction is open. The XIO instruction examines the PLC input status file for an OFF, or binary 0, condition. If the processor finds an OFF condition, the XIO instruction is true. When the PLC updates outputs, output O:2/2 will energize (as illustrated in Figure 14-12).

The normally open instruction will be false until someone pushes the normally open push button. When the normally open push button is pressed, an ON signal will be seen at the PLC input module. A 1 will be stored in the input status file address I:1/2. With a 1 in the input status file, the associated normally open instruction will close. Finding a 1 in the input status file address I:1.2, the processor will make the XIO instruction false. As illustrated in Figure 14-13, output O:2/1 will be true and output O:2/2 will be false.

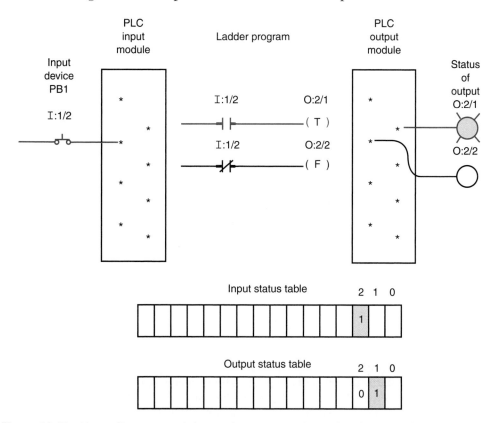

Figure 14-13 Normally open push-button input pressed, causing the normally open programming instruction to pass power, or become true.

The next example will look at how a normally closed field input device interacts with the XIC and XIO instructions:

Normally Closed Stop Push-Button Interface to the PLC

Figure 14-14 illustrates the normally closed push-button input device. This push button will send a continuous signal into the PLC input module. The input status file will have a 1 in the bit position associated with this input address (I:1/2), as illustrated in Figure 14-14.

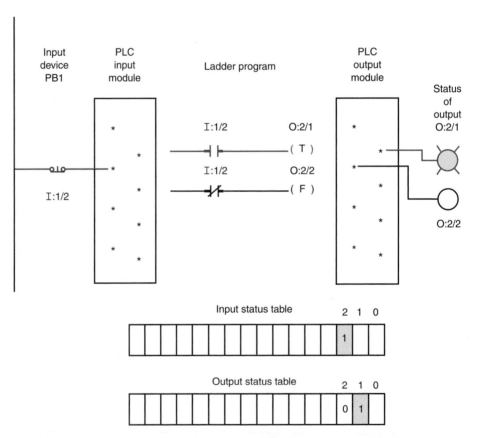

Figure 14-14 Normally closed input push button providing a constant ON signal to the PLC input module.

Programming a Normally Open Instruction   A normally closed input device sends a 1 to input status file bit position I:1/2. The normally open contact instruction on this rung will be energized. If you program a normally open instruction for the normally closed input stop push button, the XIC instruction will be true. With the XIC instruction true (as in Figure 14-14), output O:2/1 will be true.

Programming a Normally Closed Instruction   Programming a normally closed, XIO, instruction with a normally closed field input device will result in the following.

A normally closed field input device sends a continuous binary 1 to input status file bit position I:1/2. Remember, when programming the XIO instruction, the processor is testing to see if the instruction is open. If the XIO instruction examines the PLC input status file I:1/2 and the processor finds a binary 1 condition, the XIO instruction is false. When the PLC updates outputs, output O:2/2 will not energize (as illustrated in Figure 14-14).

If you program a normally open instruction for a normally closed input such as a stop push button, the XIC instruction will be true. With the XIC instruction true, O:2/1 will be true (as in Figure 14-14).

Programming a normally closed instruction for the normally closed input stop push button will result in the instruction being false. If the normally closed instruction was programmed, the operator would have to hold the stop button in its open position continuously to energize the start-stop circuit. Refer to the status of the normally closed instruction in Figure 14-14.

An operator will press the normally closed stop push button to stop the controlled hardware device (typically a motor). When pushing the normally closed push button, the electrical signal to the PLC input module will be interrupted. With the input signal interrupted, the input module will send a 0 input status signal to the associated bit address in the input status file. Figure 14-15 shows the normally closed push button being pressed.

With a 0 in the input status file, the normally open instruction will be open and the normally closed instruction will close. Power will be interrupted on the ladder rung if a normally open instruction was programmed. With power interrupted to the normally open instruction, this instruction will go false, shutting down the motor. This was the operator's intention upon pressing the normally closed stop push button.

Latching the Rung after Release of a Momentary Start Input

The start momentary push button supplies an input signal only if the button is held in. It would be unreasonable to expect an operator to hold the start button in for the time the motor is to run. To allow operators to go on with their tasks and allow the motor to run, we must seal a motor starter circuit or rung in the ON condition.

When the operator pushes the start button, the output instruction O:2/1 becomes true. A motor starter circuit has a normally open hardware auxiliary contact on the motor starter to seal the circuit in the ON state after the normally open start push button is pressed and released. This normally open auxiliary contact is a physical field input device that will be wired as a separate input to an unused input address. For this example we will choose I:1/3 as our auxiliary contact.

When I:1/3 is true, its associated normally open contacts will close; this is the seal-in contact. With I:1/3 true, a parallel path of logical continuity will be provided around the momentary start push button. Figure 14-16 illustrates that the parallel I:1/3 instruction will hold, or seal in its output coil, after the momentary normally open start push button has been pressed and released.

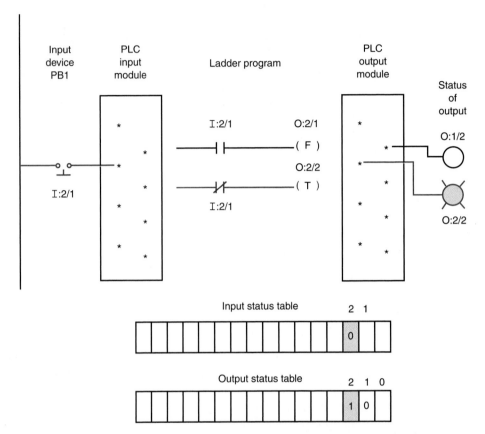

Figure 14-15 Normally closed push-button input to PLC being depressed by the operator. While being held open, no input signal is sent to the PLC input module.

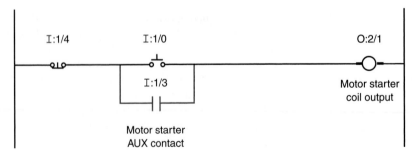

Figure 14-16 After start push button is released, the energized output will be latched through CR1-1.

When an operator presses the stop push button, the O:2/1 output instruction will lose power (become false). The output module's output point will be turned off as the PLC rung is no longer true. As it becomes false, the motor starter coil will return to its unenergized condition. Now de-energized, the motor starter contacts return to their normally open state. Normally open auxiliary contacts (addressed as I:1/3) will return to their open state, which will input an OFF condition into the input module. The I:1/3 instruction will also return to its normally open state.

From these examples we are now prepared to develop a ladder rung representing our start-stop station's input and properly control our output hardware device. First, program the stop input. The normally closed stop push button will be programmed as a normally open instruction (as illustrated in Figure 14-17).

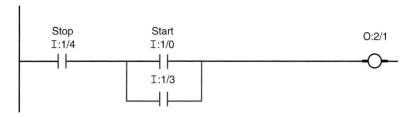

Figure 14-17   Conventional schematic start-stop logic from Figure 14-16 converted to a PLC-managed ladder program rung.

Program the normally open push button as a normally open instruction in series, or ANDed, with the normally open stop programming instruction. To program the hold-in, or sealing, contact, program a normally open instruction in a parallel branch around the start instruction. After all input instructions are programmed, program the output instruction.

MOTOR STARTER CIRCUITS AND THE OVERLOAD CONTACTS

The typical motor is controlled by a circuit much like the start-stop circuit we have been working with in Figure 14-17. The typical motor starter circuit also has the addition of a set of normally closed hardware overload contacts that are part of the starter's overload relay. Figure 14-18 illustrates the overload contacts as they are represented in a typical start-stop rung.

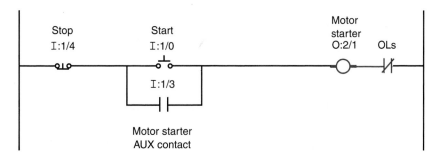

Figure 14-18   A conventional motor starter schematic diagram.

Many conventional schematics place the normally closed overload contact after the starter coil on the rung. Because most PLCs require that all input instructions be programmed before the output instruction, the conventional schematic must be converted.

- The first step in converting a conventional schematic to PLC management is to determine inputs and outputs. The overload relay's normally closed contacts will be a physical hardwired input.
- After they have been determined, separate inputs from outputs.
- While separating inputs and outputs, move the overload contacts to the input side of the output.
- Assign inputs their addresses and enter the proper instructions on the ladder.

The overload contacts in a motor starter are closed during normal operation. Being a normally closed set of contacts like the normally closed stop push button, the proper instruction on a PLC ladder rung will be normally open. This is illustrated in Figure 14-19.

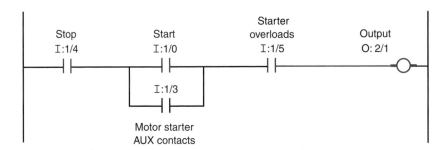

Figure 14-19 Conventional motor starter circuit converted for PLC system management.

Overload relay contacts are programmed as normally open for the same reason as the normally closed stop push button. This input will always pass power except when the motor starter is in an overload condition. In an overload condition, the normally closed overload contacts will open and de-energize the start-stop control circuit.

There are additional considerations when incorporating PLC management into a motor starter application.

SPECIAL CONSIDERATIONS FOR PLC CONTROL OF MOTOR STARTERS

When a PLC manages a motor starter, the logic represented in Figure 14-20 is usually recommended. This ladder rung differs by the inclusion of an additional input on the parallel branch. The added input instruction and its address, O:2/1, refer back to the output address instruction.

As starters become physically larger, they become slower. As size increases, a starter's pickup and dropout times also increase. Mechanical sluggishness in larger motor starters could cause the auxiliary input (I:1/3 from Figure 14-19) to switch much more slowly than

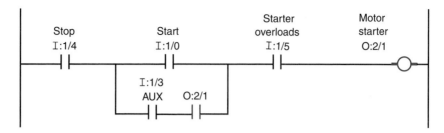

Figure 14-20 Recommended PLC ladder rung for PLC management of motor starters.

a smaller starter. As a result, the fast-scanning PLC may not unlatch the motor starter output even though the stop push button is depressed. To unlatch the rung, the latching auxiliary contact must open before the stop push button is released. The PLC sensed the stop push button as depressed yet returned to its normal undepressed, "true" state before the auxiliary contacts could physically open. Sensing the stop push button as true, along with the parallel seal-in contacts still remaining physically unlatched, the logically true rung will re-energize the output. This is because the PLC still sees the rung as true, the result of slow-acting auxiliary contacts unable to respond quickly enough.

To solve this problem, you can program a normally open input instruction on the latching parallel branch and refer it back to the output address. Program the parallel branch of I:1/3 along with an input bit representing O:2/1 so that they are logically ANDed. This is illustrated in Figure 14-20. The normally open instruction, O:2/1, is the same bit in the output status file as the output instruction, O:2/1.

At the end of the first program scan, after the stop push button is depressed, the PLC will change input instruction O:2/1 to false, along with output O:2/1. Since the O:2/1 instruction on the parallel branch is the same bit as the output instruction, it is also made false. The AND logic between I:1/3 and O:2/1 on the parallel branch will now become false. With the rung now false, the slowness of the physical auxiliary contact will become irrelevant. The rung no longer has logical continuity because of the O:2/1 input instruction opening.

SUMMARY

A hardware input device can be in any of the following states: normally open; normally open, held closed; normally closed; or normally closed, held open. Each input is seen as either an ON or OFF signal to the input module. A status bit corresponding to each input module address and the ON or OFF signal received from the input device is stored in the input status file. Because the CPU has no way of knowing the normal state of each hardware device, users must communicate to the CPU through the proper use of programming instructions. Choosing the proper ladder diagram instruction is the key to proper PLC interaction with input signals.

Remember, the CPU can see only as far as the input status file. As the CPU scans the user logic, it looks to see if the ladder diagram instruction is programmed as normally open or normally closed. The CPU examines the input status bit referenced by each instruction for its ON or OFF status.

If the CPU finds instructions programmed as normally open but not logically true, it will not pass power. Normally closed instructions referenced to that input are permitted to become true.

Instructions programmed as normally open that are found to be logically true will be allowed to pass power, while their corresponding normally closed instructions will be false.

Instructions programmed as normally closed but logically true will be allowed to pass power, while their related normally open instructions will be evaluated as false.

Instructions programmed as normally closed that are found to be logically false will not be allowed to pass power, while their corresponding normally open instructions will be true.

This chapter has introduced the proper programming of a start-stop station interface to a PLC.

REVIEW QUESTIONS

1. True or false? If a stop push button is physically normally closed, you should use a normally closed instruction on your PLC ladder diagram.
2. The ON or OFF signal that a field input device sends into a PLC input module is stored in what portion of the processor's data file?
3. If a normally closed limit switch is held open, what rung instruction would you program to become true when the object holding the limit switch open moves away?
4. When programming a typical motor starter ladder rung into a PLC managed system (select one or more of the following):
   A. Auxiliary (AUX) contacts are moved to the input side of the rung.
   B. The overload contact is moved to the input side of the rung.
   C. The overload instruction is programmed as normally open.
   D. The overload instruction is placed in the parallel branch around the start push button.
   E. AUX contacts are programmed as normally closed.
   F. The stop push button and the AUX contacts are programmed as normally closed.
5. As a PLC system troubleshooter, why is it necessary to understand how the PLC user program actually interacts with the sensors, push buttons, and limit switch signals reflected in the input status file, rather than their physical ON or OFF status?
6. How does the input status file bit status relate to the physical ON or OFF status of the field device?
7. If input bit I:1/2 is a one and the rung is to be true when the input status file bit is a one, what instruction should you program on your PLC ladder rung?
8. If input bit I:1/3 is a zero and your rung needs to be true when the input status file bit is a one, what instruction should you program on your PLC ladder rung?

9. Figure 14-2 in the text refers to the incorrect conversion of a conventional start-stop schematic to PLC control. Explain in detail why programming your PLC ladder in this manner will cause it to not operate correctly.

10. Explain what is meant when we say that OSHA requires that a stop push button be fail-safe.

11. When converting a conventional schematic to PLC control, what is the first task you should do?

12. Why do we separate inputs so that each input provides a separate input signal to the input module?

IF PASSWORD IS CORRECT GO TO DRIVE SET UP SCREEN

0022
```
┌─ EQU ──────────┐         B3:10/0                                            B3:1/15
│ Equal          │        ┌─ OSR ─┐                                            ─( )─
│ Source A   N12:21│       └───────┘
│              7777<│                                              RESET PASSWORD
│ Source B   32767 │                                              ATTEMPTS COUNTER
│             32767<│                                                   C5:5
└────────────────┘                                                   ─( RES )─
```

0023
```
┌─ GRT ──────────┐      ┌─ NEQ ──────────┐       B3:10/1
│ Greater Than (A>B)│   │ Not Equal      │      ┌─ OSR ─┐
│ Source A   N12:21 │   │ Source A   N12:21│     └───────┘
│              7777<│   │             7777<│
│ Source B      0  │   │ Source B   32767 │
│                  │   │            32767<│
└────────────────┘     └────────────────┘
```

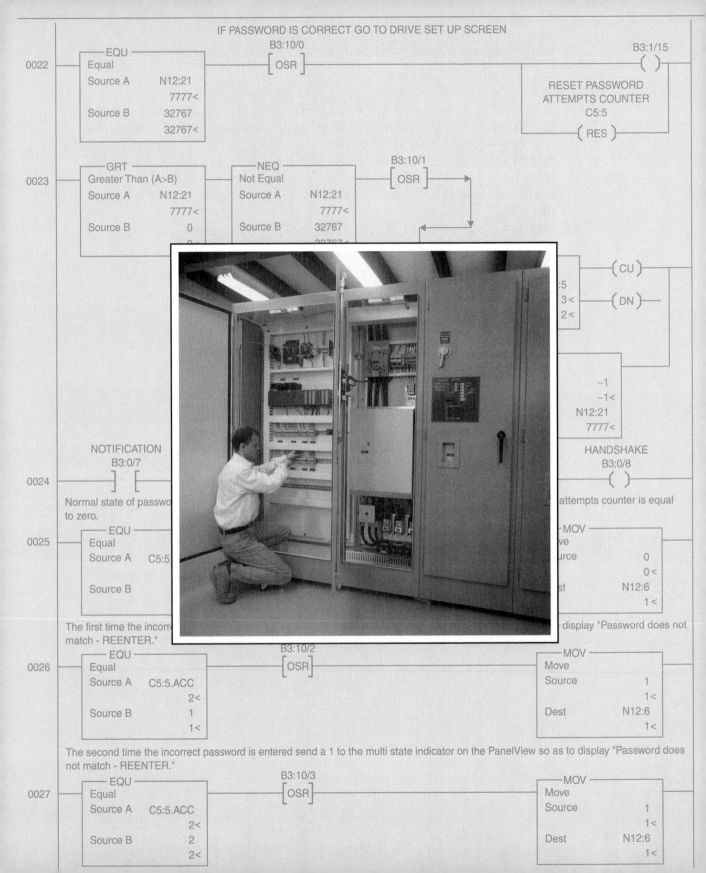

```
                                                              ─( CU )─
                                                         :5   ─( DN )─
                                                        3<
                                                        2<

                                                        −1
                                                        −1<
                                                        N12:21
                                                        7777<
```

0024
```
      NOTIFICATION                                           HANDSHAKE
        B3:0/7                                                 B3:0/8
      ─┤   ├─                                                 ─( )─
```
Normal state of passwo...   ...attempts counter is equal
to zero.

0025
```
┌─ EQU ──────────┐                                      ┌─ MOV ──────┐
│ Equal          │                                      │ ...ove     │
│ Source A   C5:5.│                                      │ ...urce    0 │
│                 │                                      │           0< │
│ Source B        │                                      │ ...st   N12:6 │
│                 │                                      │           1< │
└────────────────┘                                      └────────────┘
```

The first time the incorr...   ...display "Password does not
match - REENTER."

0026
```
┌─ EQU ──────────┐         B3:10/2                       ┌─ MOV ──────┐
│ Equal          │        ┌─ OSR ─┐                      │ Move       │
│ Source A  C5:5.ACC│      └───────┘                      │ Source    1 │
│              2<  │                                      │           1< │
│ Source B      1  │                                      │ Dest   N12:6 │
│               1< │                                      │           1< │
└────────────────┘                                      └────────────┘
```

The second time the incorrect password is entered send a 1 to the multi state indicator on the PanelView so as to display "Password does not match - REENTER."

0027
```
┌─ EQU ──────────┐         B3:10/3                       ┌─ MOV ──────┐
│ Equal          │        ┌─ OSR ─┐                      │ Move       │
│ Source A  C5:5.ACC│      └───────┘                      │ Source    1 │
│              2<  │                                      │           1< │
│ Source B      2  │                                      │ Dest   N12:6 │
│               2< │                                      │           1< │
└────────────────┘                                      └────────────┘
```

CHAPTER

15

Documenting Your PLC System

OBJECTIVES

After completing this chapter, you should be able to:

- explain what is contained in a system documentation package
- describe why a system documentation package should be developed
- state when the documentation package is usually developed
- list what documentation features the typical PLC software development package offers
- explain what value documentation has during troubleshooting
- describe the value of documentation if you have to expand the present system
- explain what a rung description is
- describe a rung symbol and the advantage of using symbols
- state the value of documentation during system start-up
- explain the value of documentation during an outside service person's service call

INTRODUCTION

A PLC system documentation package is an orderly collection of information concerning the configuration and operation of your system. These records must contain up-to-date hardware and software information. Typically the documentation package is put together as the system is developed. Information contained in a system's documentation package is

467

used as a system is designed and also during installation and debugging, start-up, maintenance, modification, and system duplication.

WHY ADD DOCUMENTATION TO YOUR USER PROGRAM?

There are many times during the life of manufacturing equipment when the need for information on how the system is put together or wired arises. A few of the instances when
system information is necessary are listed below:

- Assist system developers to assign hardware and software addresses as the system is
  developed.
- Assist installation personnel as they install and start up the system.
- Assist maintenance personnel to troubleshoot and maintain the system.
- Provide system information to operators who must understand how the system
  operates.
- List available system hardware capabilities when system expansion is needed.
- Assist in system upgrade decisions (e.g., can I add a modem for remote monitoring?
  Can I add or expand a communication network?).
- Enable other people, such as unfamiliar maintenance or outside service personnel,
  to answer questions, diagnose possible problems, or make system modifications if
  system requirements change.
- Allow easy reproduction of the system if an additional manufacturing line is
  needed.

INFORMATION TO INCLUDE IN PLC DOCUMENTATION

Your documentation package should contain the following information.

System Overview

Simply put, a system overview should contain a complete and clear statement of the control problem.

Diagram of System Configuration

This is a drawing that shows the hardware elements defined in the system overview. This
illustration of the system is a simplified layout of major hardware components as defined
in the system overview. The system configuration is a simplified pictorial drawing containing minimum details on major hardware components. Included in this pictorial representation of the system is hardware location, a simplified representation of connections, I/O
rack address assignments, I/O module type identification, and I/O expansion rack identification information. Figure 15-1 is a sample of a diagram of systems configurations.

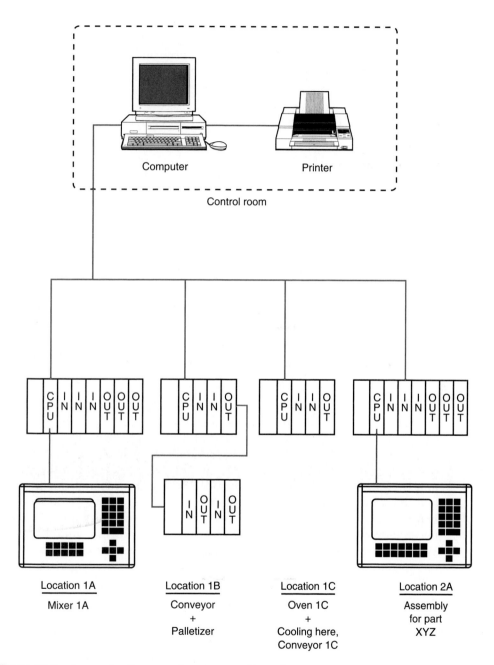

Figure 15-1   A sample diagram of system configuration.

I/O Address Assignments

There should be a sheet made up to identify each and every input and output point. Included should also be a reference about what each specific I/O address is connected to in the real world.

Internal Coil Address Assignments

Internal bit instruction references are analogous to control system relays that do not drive real-world devices. Most PLCs allocate a portion of their data memory for the storage of internal references. If your particular PLC does not have internal bit references, you can assign unused real-world output references for internal use.

Register or Data Table Assignments

Each data table needs an assignment sheet for each data table address, whether used or not. Typical register storage or data tables contain registers for the timer, counter, whole numbers (integers), ASCII data for message displays, floating-point data, and control. Making changes to these data tables is just as important as tracking I/O points.

Program Printout

There should be a current hard-copy printout of the control logic program.

Control Program Backup

It is important that there be a current backup copy of the control program. Your control program should be saved on a personal computer's hard drive and backed up on tape or floppy disk on a PROM chip in addition to the hard copy. A hard-copy program backup is particularly important if the program was developed using a handheld programming terminal. Some PLCs allow a tape backup program to be copied directly from the PLC's CPU to a cassette tape. Whatever method of program backup your particular PLC supports, make sure you use it and keep your backup program current. You never know when you may need to reload a control program after a processor failure. With no backup copy, you will be out of business. In fact, if there is no backup copy, the entire control program may have to be rewritten. This rewriting could put your manufacturing plant down for days. Think of your current backup copy of every control program as job insurance.

Wiring Diagrams

There should be a complete set of wiring diagrams. Included should be wiring diagrams for each I/O module, address identification for each I/O point, and rack identification for the specific module represented.

Figure 15-2 shows typical wiring representation for an input module, and Figure 15-3 shows typical output module wiring.

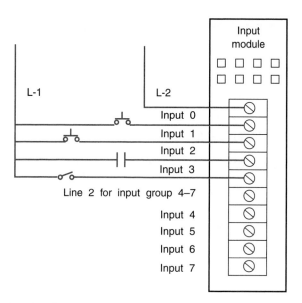

Figure 15-2 Typical wiring for a 120 VAC input module.

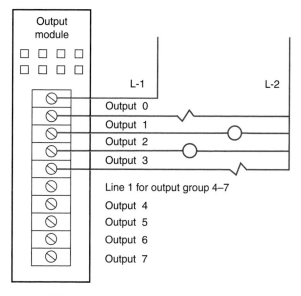

Figure 15-3 Typical wiring for a 120 VAC output module.

DOCUMENTATION AND SOFTWARE PACKAGES

One advantage of using a personal computer and PLC program development software is the readily available documentation features, automatically generated cross-references, and system and hardware data tables that are available by simply printing out a hard copy of the user program.

TYPES OF DOCUMENTATION

Though each manufacturer and software package differ, most software development packages offer a documentation feature such as hard copy of the user ladder diagram; text comments such as rung descriptions, instruction descriptions, and instruction symbols associated with ladder rungs, instructions, or addresses; cross-references sorted by I/O addresses; a listing of all data tables incorporated into your PLC program; a database showing I/O addresses and their associated text comments; a processor configuration report; and a hardware report listing hardware components in the system. Figure 15-4 is a sample table of contents printout from Rockwell Software's PLC 500 DOS software. Rockwell Software is the business unit of the Rockwell Automation Company that develops programming software for Allen-Bradley PLCs, including the SLC 500 family. Currently there are two software packages for programming the SLC 500 family of PLCs. RSLogix 500 is the Windows-based programming software, while PLC 500 A.I. software is the DOS programming version. Figure 15-5 is a sample of four program rungs created with Rockwell Software's PLC 500 A.I. Programming Software for programming the Allen-Bradley SLC 500 PLC.

```
PLC-500 LADDER LOGISTICS Report header (c) RSI. 1989-1995
              PLC-500 Report Table of Contents
                  ABC MANUFACTURING COMPANY

Table of Contents
_____

   Page   Rung    Description
   ----   ----    -----------
    1              Program File List
    2              Ladder Diagram Dump
    2      0       File #2  Proj:BARCODE
    3      4       THIS IS A PAGE TITLE. USE 80 CHARACTERS TO DESCRIBE A GROUP OF RUNGS OPERATION
    4              Cross Reference Report
    5              Data Table File List
    6              Data Table Dump Report
    7              Data Table Usage Report
    8              Data Base Form
   12              Unused Address Report
   15              I/O Part List
   18              Rack Des Report
   20              Processor Config Report
```

Figure 15-4   Rockwell Software PLC 500 DOS software table of contents printout.

In Figure 15-5, notice the project or program name in the upper left-hand corner. To the left of the left power rail are rung numbers 0 through 3. These reference numbers help identify which rung you are currently looking at. Also identified is the file number. File number 2 identifies the ladder as the main program file. This helps to identify the program being worked with. Next, let us look at rung descriptions.

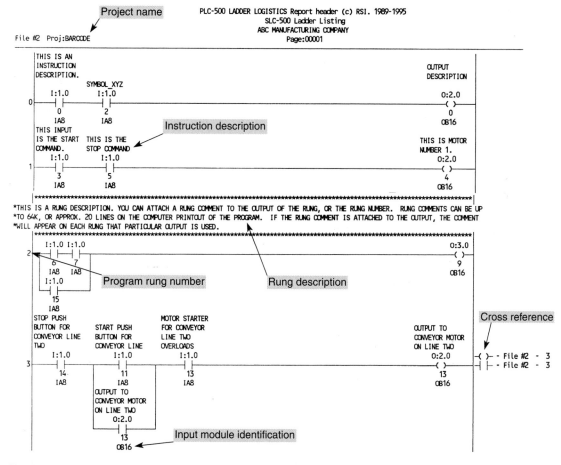

Figure 15-5   Rockwell Software PLC 500 DOS software ladder printout with rung descriptions.

Page Title

RSLogix software has a feature called page titles as part of the documentation. A page title is one line of up to 85 characters. A page title can be used to break your ladder program into groups of rungs called pages. By segmenting your rungs into pages, after printing a hard copy of your ladder program you will find a new printed piece of paper for each page title. This way rungs can be placed in a loose-leaf binder with page dividers identifying each section of the program. Examples of using page titles to separate logic into separate segments would include:

Mixing section

Filling section

Capping section

Labeling section

Quality control section

Packaging section

A second feature is that page titles can be searched for specifically. If you know exactly what you are searching for, you can search for a specific page title. Once the page title is displayed, you can drill down one level to a list of all outputs on that "page." Clicking on a specific output will take you directly to it on your ladder. See Figure 15-6 to view a page title.

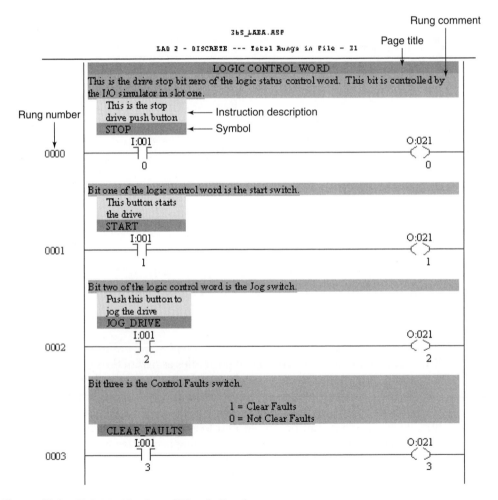

Figure 15-6   PLC 5 ladder from RSLogix 5 software.

Rung Comments

Rung comments give the reader an explanation of an individual rung or group of rungs (refer to Figure 15-6). A typical rung comment can be up to six lines and a total of 80 ASCII characters long. To help keep track of rung comments, they must be associated with some part of the rung for which they are providing a description. Since the output is the resulting action of the rung, the rung comment is associated with the output. If this output is used in another rung of your user program, the rung comment will appear above that rung, also.

Instruction Descriptions

An instruction description is associated with an instruction type and its address. Figures 15-5 and 15-6 illustrate a rung of ladder logic with an instruction description. An instruction description can be up to five lines long with 10 ASCII characters per line. How much text an instruction description can contain is decided by the software manufacturer. Typically, all programming instructions in the user program with the same address and instruction type will have the same instruction description. The PLC software will automatically insert the instruction description each time the same instruction and address are programmed. Notice that on rung zero, Figure 15-5, the upper left-hand description simply says, "This is an instruction description." The output instruction on this rung says, "Output description." The second rung has instruction descriptions that are more realistic for real applications. As an example, if there were a normally open instruction with the address I:1.0/5 with the comment, "This is the Stop Command," any time this instruction and address were used in the program, the instruction comment would automatically be inserted by the programming software. Information below each input and output point identifies the type of I/O module with which this instruction is associated. The first instruction on rung zero has IA8 below the normally open contact symbol. This signifies that this input is on a 1746-IA8 input module. A 1746-IA8 input module is an eight-point, 120 VAC module.

Instruction Symbols

Instruction symbols are identifying labels that serve as substitutes for addresses. Once a symbol has been associated with an address, either one may be entered when programming, as any instruction with the same address will have the same symbol. Figure 15-6 is a printout of a PLC 5 ladder from RSLogix 5 software. Notice the rung numbers, page titles, rung comments, instruction symbols, and address descriptions.

As you enter your program, you can simply enter the symbol, because the symbol is associated with the address. The software symbol will automatically insert the address for you. Address symbols are unique. A symbol cannot be applied to two different addresses in the same program. Notice that to the right of the right power rail in Figure 15-5 is cross-reference information. This alerts you to other instances where this I/O address and its associated symbol are used in the ladder program.

Data Table Usage Report

Figure 15-7 is a RSLogix 500 report listing the usage of the output status table. This report identifies which outputs are currently programmed. The W signifies that the entire module is used, while individual Xs identify which output points have been used on that specific output module. If there is a period displayed, that output point is available. Figure 15-7 only shows the data file usage for the output status file. Reports are available for each data file used.

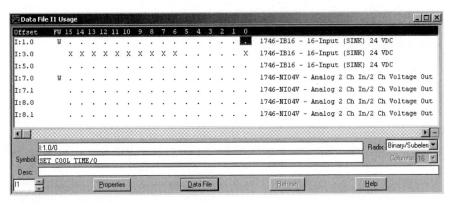

Figure 15-7 RSLogix 500 report listing the usage of the output status table.

This report is an excellent way to determine what references have been used. Whenever a PLC program is to be updated and additional I/O points or instructions added, these tables will show you where to find the available addresses.

I/O Card Usage Report

Some PLCs have an I/O card usage report, which lists the I/O cards that are currently installed in the PLC system represented by this report. The report can be valuable in determining the spare parts inventory. You can determine which I/O cards are in the system and how many there are of each by simply looking at this report.

Figure 15-8 illustrates an I/O configuration report from RSLogix 500 software for an SLC 500 PLC consisting of two chassis configured in a local I/O configuration. Local I/O configuration means that the two chassis are connected by a cable. When using an SLC 500 PLC, local I/O is a direct connection between one chassis and another. Remote I/O would require a special communication module in the chassis, and the I/O configuration would look different. The dotted line between I/O modules 9 and 10 illustrates the break between the two chassis.

The Processor Status File Report

The processor status file report is a report from Status file S2. This report provides information on status bits, processor information, scan time data, processor channel

PARTS.RSS

I/O Configuration

0	1747-L542C	5/04 CPU - 32K Mem. OS401 Series C
1	1746-IB16	16-Input (SINK) 24 VDC
2	1746-IB16	16-Input (SINK) 24 VDC
3	1746-IB16	16-Input (SINK) 24 VDC
4	1746-OB16	16-Output (TRANS-SRC) 10/50 VDC
5	1746-OB16	16-Output (TRANS-SRC) 10/50 VDC
6	1746-OB16	16-Output (TRANS-SRC) 10/50 VDC
7	1746-OB16	16-Output (TRANS-SRC) 10/50 VDC
8	1746-NI8	Analog 8 Channel Input - Class 1
9	1746-NO4I	Analog 4 Ch. Current Output
--	---------------	------------------------------------
10	1746-IA16	16-Input 100/120 VAC
11	1746-IA16	16-Input 100/120 VAC
12	1746-IO8	4-Input 100/120 VAC, 4-Output (RLY)
13	1746-OA16	16-Output (TRIAC) 100/240 VAC
14	1746-OA16	16-Output (TRIAC) 100/240 VAC
15	1746-OA16	16-Output (TRIAC) 100/240 VAC
16	1747-SN	RIO Scanner
17	1747-SDN	DeviceNet Scanner Module
18	1746-NT8	Analog 8 Ch Thermocouple Input
19	1746-NO4I	Analog 4 Ch. Current Output
20	1746-NI4	Analog 4 Channel Input Module
21	1746-NI4	Analog 4 Channel Input Module
22	1746-NI4	Analog 4 Channel Input Module

Figure 15-8   I/O configuration for an SLC 500 PLC from RSLogix 500 software consisting of one 10-slot and a second 13-slot chassis.

configuration, processor fault status bits, and which I/O slots are enabled. This report consists of many pages. Only a small part is displayed in Figure 15-9.

Cross Reference Report

Figure 15-10 is a cross-reference report from RSLogix 5 software for the PLC 5. The report is a good cross-reference source when looking for a cross-reference as to where else a specific address is used within your project. Notice that the report lists the I/O address, the instruction type, ladder file where the instruction is used, type of information, and the rung number. This is an example of a simple ladder program. If there were multiple additional references for a specific entry they would all be listed.

Data File List

Figure 15-11 illustrates a data file list printout for an SLC 500. The report lists the data files, usage scope as local or global, number of words used in the file, number of elements in each file, and the last address.

```
                                 PRINTING.RSS

                      Data File S2 (hex)  --  STATUS
```

Main

```
First Pass S:1/15 = No                                       DD / MM / YYYY
Index Register S:24 = 0                        Date S:39-37 = 0 / 0 / 0
Free Running Clock S:4 = 0000-0000-0000-0000
Index Across Data Files S:2/3 = No                             HH : MM : SS
CIF Addressing Mode S:2/8 = 0                  Time S:40-42 = 0 : 0 : 0
Online Edits S:33/11 - S:33/12 = No online edits exist
```

Proc

```
OS Catalog Number S:57 = 0          User Program Type S:63 = 1025
OS Series S:58 = A                  User Program Functionality Index S:64 = 95
OS FRS S:59 = 0                     User RAM Size S:66 = 0
Processor Catalog Number S:60 = 0  OS Memory Size S:66 = 0
Processor Series S:61 = A
Processor FRN S:62 = 0
```

Scan Times

```
Maximum (x10 ms) S:22 = 0
Average (x10 ms) S:23 = 0
Current (x10 ms) S:3 (low byte) = 0
Watchdog (x10 ms) S:3 (high byte) = 10
Last 1ms Scan Time S:35 = 0
Scan Toggle Bit S:33/9 = 0
Time Base Selection S:33/13 = 0
```

Math

```
Math Overflow Selected S:2/14 = 0   Math Register (lo word) S:13 = 0
Overflow Trap S:5/0 = 0             Math Register (high word) S:14-S:13 = 0
Carry S:0/0 = 0                     Math Register (32 Bit) S:14-S:13 = 0
Overflow S:0/1 = 0
Zero Bit S:0/2 = 0
Sign Bit S:0/3 = 0
Floating Point Flag Disable S:34/2 = 0
```

IO

```
I/O Interrupt Executing S:32 = 0        Interrrupt Latency Control S:33/8 = 0
                                        Event Interrupt 10 uS Time Stamp S:44 = 0


I/O Slot Enables: S:11 _S:12
0              10            20           30
11111111       11111111      11111111     11111111

I/O Slot Interrupt Enables: S:27 _S:28
0              10            20           30
11111111       11111111      11111111     11111111

I/O Slot Interrupt Pending: S:25 _S:26
0              10            20           30
00000000       00000000      00000000     00000000
```

Chan 0

```
Processor Mode S:1/0- S:1/4 = Remote Program Mode
Channel Mode S:33/3 = 0              DTR Control Bit S:33/14 = 0
Comms Active S:33/4 = 0              DTR Force Bit S:33/15 = 0
Incoming Cmd Pending S:33/0 = 0      Outgoing Msg Cmd Pending S:33/2 = 0
Msg Reply Pending S:33/1 = 0         Comms Servicing Sel S:33/5 = 0
DH485 Pass-Thru Disabled Bit S:34/0 = 0   Msg Servicing Sel S:33/6 = 0
DF1 Pass-Thru Enable Bit S:34/5 = 0   Modem Lost S:5/14 = 0
```

Figure 15-9 Status file report from RSLogix 500 software.

36S_LABA.RSP

Cross Reference Report - Sorted by Address

```
O:000          - BTD - File #2 DISCRETE - 18
O:001          - MVM - File #2 DISCRETE - 17
O:021/0        - OTE - File #2 DISCRETE - 0
O:021/1        - OTE - File #2 DISCRETE - 1
O:021/2        - OTE - File #2 DISCRETE - 2
O:021/3        - OTE - File #2 DISCRETE - 3
O:021/4        - OTE - File #2 DISCRETE - 4
O:021/5        - OTE - File #2 DISCRETE - 5
O:021/6        - OTE - File #2 DISCRETE - 6
O:021/7        - OTE - File #2 DISCRETE - 7
O:021/10       - OTE - File #2 DISCRETE - 8
O:021/11       - OTE - File #2 DISCRETE - 9
O:021/12       - OTE - File #2 DISCRETE - 10
O:021/13       - OTE - File #2 DISCRETE - 11
O:021/14       - OTE - File #2 DISCRETE - 12
O:021/15       - OTE - File #2 DISCRETE - 13
O:021/16       - OTE - File #2 DISCRETE - 14
O:021/17       - OTE - File #2 DISCRETE - 15
O:022          - MOV - File #2 DISCRETE - 16
O:023          - MOV - File #2 DISCRETE - 20
O:024          - MOV - File #2 DISCRETE - 21
O:025          - MOV - File #2 DISCRETE - 22
O:026          - MOV - File #2 DISCRETE - 23
O:027          - MOV - File #2 DISCRETE - 24
I:000/0        - XIC - File #2 DISCRETE - 8
I:000/1        - XIC - File #2 DISCRETE - 9
I:000/2        - XIC - File #2 DISCRETE - 10
I:000/3        - XIC - File #2 DISCRETE - 11
I:000/4        - XIC - File #2 DISCRETE - 12
I:000/5        - XIC - File #2 DISCRETE - 13
I:000/6        - XIC - File #2 DISCRETE - 14
I:000/7        - XIC - File #2 DISCRETE - 15
I:001/0        - {STOP} This is the stop drive push button
                 XIC - File #2 DISCRETE - 0
I:001/1        - {START} This button starts the drive
                 XIC - File #2 DISCRETE - 1
I:001/2        - {JOG_DRIVE} Push this button to jog the drive
                 XIC - File #2 DISCRETE - 2
I:001/3        - {CLEAR_FAULTS}
                 XIC - File #2 DISCRETE - 3
I:001/4        - XIC - File #2 DISCRETE - 4
I:001/5        - XIC - File #2 DISCRETE - 5
I:001/6        - XIC - File #2 DISCRETE - 6
I:001/7        - XIC - File #2 DISCRETE - 7
I:021          - MVM - File #2 DISCRETE - 17, 18
I:022          - MOV - File #2 DISCRETE - 19
I:023          - MOV - File #2 DISCRETE - 25
I:024          - MOV - File #2 DISCRETE - 26
I:025          - MOV - File #2 DISCRETE - 27
I:026          - MOV - File #2 DISCRETE - 28
I:027          - MOV - File #2 DISCRETE - 29
N9:0           -       - Data File - S:16
FILE N9:0 LEN:48 -     - Data File - S:16
N20:0          -       - Channel Configuration - Channel 1B:Diagnostic File
FILE N20:0 LEN:40 -    - Channel Configuration - Channel 1B:Diagnostic File
N107:10        - MOV - File #2 DISCRETE - 16
N107:11        - MVM - File #2 DISCRETE - 18
                 BTD - File #2 DISCRETE - 18
N107:20        - MOV - File #2 DISCRETE - 19
N107:31        - MOV - File #2 DISCRETE - 20
N107:32        - MOV - File #2 DISCRETE - 21
N107:33        - MOV - File #2 DISCRETE - 22
N107:34        - MOV - File #2 DISCRETE - 23
N107:35        - MOV - File #2 DISCRETE - 24
N107:51        - MOV - File #2 DISCRETE - 25
N107:52        - MOV - File #2 DISCRETE - 26
N107:53        - MOV - File #2 DISCRETE - 27
N107:54        - MOV - File #2 DISCRETE - 28
```

Figure 15-10   Cross-reference report for a PLC 5 project using RSLogix 5 software.

PARTS.RSS

Data File List

Name	Number	Type	Scope	Debug	Words	Elements	Last
OUTPUT	0	O	Global	No	288	96	O:95
INPUT	1	I	Global	No	294	98	I:97
STATUS	2	S	Global	No	0	164	S:163
BINARY	3	B	Global	No	50	50	B3:49
TIMER	4	T	Global	No	300	100	T4:99
COUNTER	5	C	Global	No	525	175	C5:174
CONTROL	6	R	Global	No	168	56	R6:55
INTEGER	7	N	Global	No	234	234	N7:233
FLOAT	8	F	Global	No	152	76	F8:75
LINE ONE	9	C	Global	No	75	25	C9:24
LINE TWO	10	C	Global	No	75	25	C10:24
LINE THREE	11	C	Global	No	75	25	C11:24
LINE 1	12	T	Global	No	300	100	T12:99
LINE 2	13	T	Global	No	84	28	T13:27
LINE 3	14	T	Global	No	33	11	T14:10
RECIPES	15	N	Global	No	167	167	N15:166
PV INTFACE	16	B	Global	No	100	100	B16:99
PROD TOTAL	17	N	Global	No	200	200	N17:199

Figure 15-11   RSLogix 500 Data file list report.

Address/Symbol Database Report

The address/symbol database report provides a list of all addresses and their respective symbols. The report also identifies the scope of the symbol, a description, and the group a particular symbol is in, assuming the programmer grouped the symbols into logical groups. Figure 15-12 illustrates a small portion of an RSLogix 500 address symbol database.

Address/Symbol Database

Address	Symbol	Scope	Description	Sym Group
I:1.0	PAINT_CODE_THUMBWHL	Global	Selects paint color	START_GROUP
I:3.0/1	TEST_BUTTON_1	Global	Push Button	TEST_BUTTONS
I:3.0/2	TEST_BUTTON_2	Global	Push button	TEST_BUTTONS
I:3.0/3	STOP_PB	Global	Stops part paint conveyor	START_GROUP
I:3.0/4	START	Global	Starts part painting conveyor	START_GROUP
I:3.0/5	PART_ENTER_POSITION	Global		
I:3.0/6	PART_EXIT_POSITION	Global		
I:3.0/14	PART_IN_OVEN	Global	Part in paint dry oven	
I:7.0	WATER_TEMP_DIAL	Global		ANALOG
O:7.0	WATER_TEMP_GUAGE	Global		ANALOG

Figure 15-12   RSLogix 500 Address/Symbol Database report.

Here is a last note on documentation and reports. The previous figures have provided you with an overview of some of the reports available for SLC 500, MicroLogix, and PLC 5s. There are many other reports available. The desired reports are configured in the RSLogix Report options window. Figure 15-13 shows a report options window for a

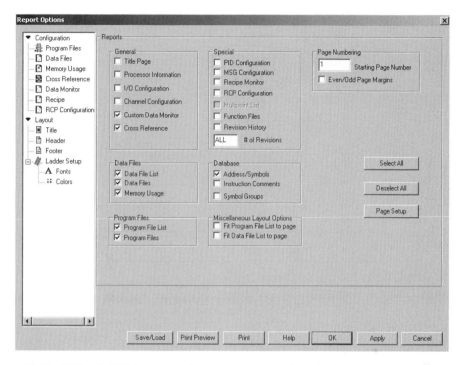

Figure 15-13   RSLogix 500 Report options setup screen for printing reports.

MicroLogix 1500 PLC. You simply check the boxes on the screen for the reports you desire. After these selections have been made, click on each selection in the far left column to refine your selections. Notice along the bottom of the screen Save, Print Preview, and Print buttons.

Clearly, by cross-referencing information between reports, you can get a good look at the PLC system without leaving your desk. The most important point is to make sure the documentation or printout is current. Documentation can be an invaluable tool to the troubleshooter or anyone interested in system configuration or program information. Unfortunately, often the documentation package is printed out and filed when the system is installed, only to be forgotten as changes are made.

ADDING DOCUMENTATION WHEN DEVELOPING A NEW APPLICATION

When developing a new PLC application, documentation should be created through each step of the development process. To illustrate the procedure, we will go through a sample process (Figure 15-14), which is part of a drum-filling application. We will step through

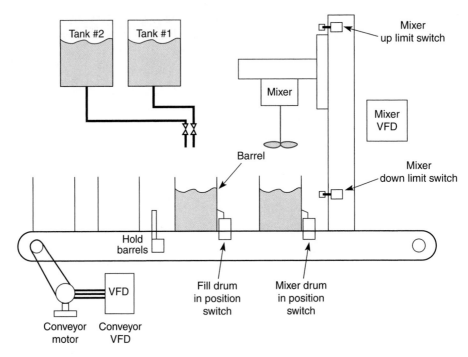

Figure 15-14   Diagram of system configuration.

the documentation process. Our documentation package will consist of the following information:

1. *System overview:* This is section 2B of the process to manufacture and package "Super Cleaner" #1, #2, #3, and #4. Secret ingredient in tank #1 and secret ingredient in tank #2 are to be mixed as specified in the recipe stored in the integer file. After filling the drum, move the drum to the mixing station. When at the mixing station, stop the conveyor, lower the mixer station, and mix for the time and speed specified in the recipe. When mixing is completed, raise mixer to the top position and move the drum to the next position. In the process, the next drum, which has been filled, is now moved into position for mixing.

2. *Diagram of system configuration:* A diagram of system configuration is an overview drawing such as illustrated in Figure 15-14.

3. *Real-world I/O determination:* Determine real-world hardware input and output devices that will be required for the application. All hardware devices are listed in Figure 15-15. An X has been placed in the input or output column to reflect the type of hardware device.

Ask yourself these questions: Do any of these devices have multiple inputs and outputs? Both VFDs have multiple inputs and outputs. Are any of these devices controlled by another device and thus not a PLC input or output? Both motors are output devices from their respective VFDs and not a direct output of the PLC.

Input	Output	Hardware
	X	Mixer motor
X	X	Mixer variable-frequency drive (VFD)
X		Mixer UP limit switch
X		Mixer DOWN limit switch
X		Solenoid to hold empty drums before fill position
X		Fill drum in position photo switch
X		Mix drum in position
	X	Tank #1 valve
	X	Tank #2 valve
	X	Conveyor motor
X	X	Conveyor variable-speed drive
		Control Panel Mixing Section
	X	Pilot light "Mixing"
	X	Pilot light "Mixer AUTO"
	X	Pilot light "Mixer MANUAL"
	X	Pilot light "Mixer STOPPED"
X		VFD START
X		VFD STOP
X		VFD JOG
		Control Panel Conveyor Section
	X	Pilot light "Conveyor AUTO"
	X	Pilot light "Conveyor MANUAL"
	X	Pilot light "Conveyor STOPPED"
X		VFD START
X		VFD STOP
X		VFD JOG
X		VFD FORWARD/REVERSE
		Control Panel Product Selection
X		Recipe four position selector switch

Figure 15-15   Chart for determining I/O.

4. *Determine and allocate I/O reference numbers to real-world hardware devices.* Fill in a table like Figure 15-16 for your outputs and one for your inputs. This table will not only start your documentation of I/O process, it will help organize your I/O points as they are allocated and programmed.

OUTPUT ADDRESS ALLOCATION TABLE			
Address	I/O Type	Device	Function
O:2/0	110 VAC	Solenoid	Tank #1 valve
O:2/1	110 VAC	Solenoid	Tank #2 valve
O:2/3	110 VAC	Pilot light	Mixing
O:2/4	110 VAC	Pilot light	Mixer AUTO
O:2/5	110 VAC	Pilot light	Mixer MANUAL
O:2/6	110 VAC	Pilot light	Mixer STOPPED
O:3/0	Contact closure	Drive	Conveyor drive start
O:3/1	Contact closure	Drive	Conveyor stop
O:3/2	Contact closure	Drive	Conveyor speed select switch 1
O:3/3	Contact closure	Drive	Conveyor speed select switch 2
O:3/4	Contact closure	Drive	Conveyor speed select switch 3
O:3/5	Contact closure	Drive	Conveyor reverse
O:3/6	Contact closure	Drive	Mixer start
O:3/7	Contact closure	Drive	Mixer stop
O:4/0	Contact closure	Drive	Mixer jog

Figure 15-16   Output address allocation table.

There are six 120 VAC output points and nine contact closure, or relay output, points. We will select one 120 VAC output and two eight-point relay output modules. Fill out a separate table for all input addresses (see Figure 15-17). There are 15 120 VAC input points, so we will choose a 120 VAC 16-point input module.

5. After selecting I/O modules, fill in module positions in your system configuration drawings (Figure 15-18). Once real-world I/O references have been determined, internal bits need to be allocated.

6. Internal bit instruction references are analogous to control system relays that do not drive real-world devices. Most PLCs allocate part of their data memory for the storage of internal references. If your particular PLC does not have internal bit references, you can assign unused real-world output references for internal use. An internal bit allocation table is illustrated in Figure 15-19.

Develop a table for each data table in your system. Figure 15-20 illustrates a sample timer address allocation table. Figure 15-21 is a sample counter address allocation table. Figure 15-22 illustrates control file documentation. Integer data is documented in an integer file allocation table, as illustrated in Figure 15-23.

INPUT ADDRESS ALLOCATION TABLE			
Address	I/O Type	Device	Function
I:1/0	120 VAC	Photo switch	Fill drum in position
I:1/1	120 VAC	Photo switch	Mix drum in position
I:1/2	120 VAC	Selector switch	Recipe select #1
I:1/3	120 VAC	Selector switch	Recipe select #2
I:1/4	120 VAC	Selector switch	Recipe select #3
I:1/5	120 VAC	Selector switch	Recipe select #4
I:1/6	120 VAC	Momentary push button (PB)	Conveyor start
I:1/7	120 VAC	Momentary PB	Conveyor stop
I:1/8	120 VAC	Momentary PB	Conveyor reverse
I:1/9	120 VAC	Momentary PB	Mixing
I:1/10	120 VAC	Selector switch	Mixer AUTO
I:1/11	120 VAC	Selector switch	Mixer manual
I:1/12	120 VAC	Selector switch	Mixer stopped
I:1/13	120 VAC	Momentary PB	Mixer start
I:1/14	120 VAC	Momentary PB	Mixer stop
I:1/15	120 VAC	Momentary PB	Mixer jog

Figure 15-17   Input address allocation table.

	Slot Zero	Slot One	Slot Two	Slot Three	Slot Four	Slot Five	Slot Six
POWER SUPPLY	CPU	120 VAC Input	120 VAC Out	Relay Out	Relay Out	Empty	Empty

Figure 15-18   Seven-slot chassis with I/O modules assigned.

INTERNAL BIT ADDRESS ALLOCATION TABLE			
Address	Bit Type	Device	Function
B:3/0			
B:3/1			
B:3/1			

Figure 15-19   Internal bit allocation table.

TIMER ADDRESS ALLOCATION TABLE			
Address	Timer Type	Device	Function
T4:0	TON		Fill valve #1
T4:1	TON		Fill valve #2
T4:2	TON		Mixing time
T4:3			
T4:4			

Figure 15-20   Timer address allocation table.

COUNTER ADDRESS ALLOCATION TABLE			
Address	Counter Type	Device	Function
C5:0	UP		Count total barrels produced
C5:1			
C5:2			

Figure 15-21   Counter address allocation table.

CONTROL FILE ALLOCATION TABLE		
Address	Device	Function
R6:0		
R6:1		
R6:2		

Figure 15-22   Control file allocation table.

If you have message displays that accept the message they are to display in ASCII, you will need an ASCII file to store the messages in. Figure 15-24 illustrates an ASCII file address allocation table.

7. Now that you have determined and organized all of your data files, development of the user program can begin. This is the process of either converting a conventional ladder diagram and its conventional symbols to the proper symbology for programmable controller system management or developing a new user program.

8. *PLC user program printout:* Print all the reports necessary from your PLC software package.

INTEGER FILE ALLOCATION TABLE			
Address	**Value**	**For**	**Function**
N7:0	66	Recipe #1	Secret ingredient #1
N7:1	33	Recipe #1	Secret ingredient #2
N7:2	45	Recipe #1	Mixing time
N7:3	1	Recipe #1	Mixing speed
N7:4	50	Recipe #2	Secret ingredient #1
N7:5	25	Recipe #2	Secret ingredient #2
N7:6	30	Recipe #2	Mixing time
N7:7	2	Recipe #2	Mixing speed

Figure 15-23   Integer file allocation table.

ASCII FILE ADDRESS ALLOCATION TABLE			
Address	**Target Device**		**Message**

Figure 15-24   ASCII address allocation table.

9. *Control program backup:* Although you have a copy of your user program on the hard drive of your personal computer, make sure you also have a CD-ROM or DVD backup or a backup on floppy disks.

10. *Wiring diagrams:* Develop a complete set of wiring diagrams. Included should be a wiring diagram for each I/O module, address identification for each I/O point, and rack identification for each module represented. Unused I/O points should be identified, not only for maintenance and troubleshooting, but also to reference when I/O points are desired for future use.

SUMMARY

A documentation package is an orderly collection of information concerning the configuration and operation of your system. These records must contain up-to-date hardware and software information. Typically, the documentation package is put together as the system is developed. Information contained in a system's documentation package is used as a system is designed and during installation and debugging, start-up, maintenance, modification, and system duplication. Proper documentation should define each input, each output, and each internal coil or bit address and function. Tables should be included with every I/O point available for use, even if the point is currently being used. Unused I/O points marked for future use will make system expansion easier. Tables for internal data functions such as timers, counters, data storage locations for numerical or ASCII data, and text-type data need to be listed and identified. Drawings showing system configuration, wiring diagrams, locations of remote racks, networked PLCs, and their network addresses will help to provide a system layout. One of the most important, and frequently overlooked, pieces of documentation is a minimum of one backup copy of the control program. To have a backup copy of your control program on your personal computer's hard drive is not enough. What happens if your personal computer breaks or your hard drive crashes? A backup copy of your hard drive's control programs should be either on backup tapes or floppy disks. Remember that you can be caught short without a control program backup.

REVIEW QUESTIONS

1. A PLC system _____ is an orderly collection of information concerning the configuration and operation of your system.
2. _____ are identifying labels that serve as substitutes for addresses.
3. Documentation records must contain up-to-date _____ and also _____ information.
4. Typically, the documentation package is put together as the system is _____.
5. Once a symbol has been associated with an address, either may be entered when programming, as any instruction with the same address will have the same _____.
6. Tables for internal data functions such as _____, _____, and _____ locations for numerical, ASCII, or text-type data need to be listed and identified.
7. The _____ is a simplified pictorial drawing containing minimum details on major hardware components contained in the PLC system.
8. A backup copy of your hard drive's control programs should be either on _____ or _____.
9. Comment on the following statement: A backup copy of your PLC's control program on your personal computer's hard drive is really all that is necessary for a secure backup system.
10. The system configuration pictorial representation of your PLC system contains which of the following:
    A. hardware location
    B. simplified representation of connections
    C. I/O rack address assignments
    D. I/O module type identification
    E. I/O expansion rack identification information
    F. C and D only
    G. all of the above

11. Proper documentation should define each _____, _____, and each internal coil or bit address and function.

12. Typically the documentation package is put together as the system is developed. Information contained in a system's documentation package is used for:
    A. system duplication and modification
    B. installation and debugging
    C. start-up and maintenance
    D. system design
    E. only A, B, and C
    F. all of the above

IF PASSWORD IS CORRECT GO TO DRIVE SET UP SCREEN

0022

EQU
Equal
Source A   N12:21
            7777<
Source B   32767
            32767<

B3:10/0
[OSR]

B3:1/15
( )

RESET PASSWORD
ATTEMPTS COUNTER
C5:5
(RES)

0023

GRT
Greater Than (A>B)
Source A   N12:21

Source B

NEQ
Not Equal
Source A   N12:21

B3:10/1
[OSR]

(CU)

(DN)

-1
-1<
21
77<

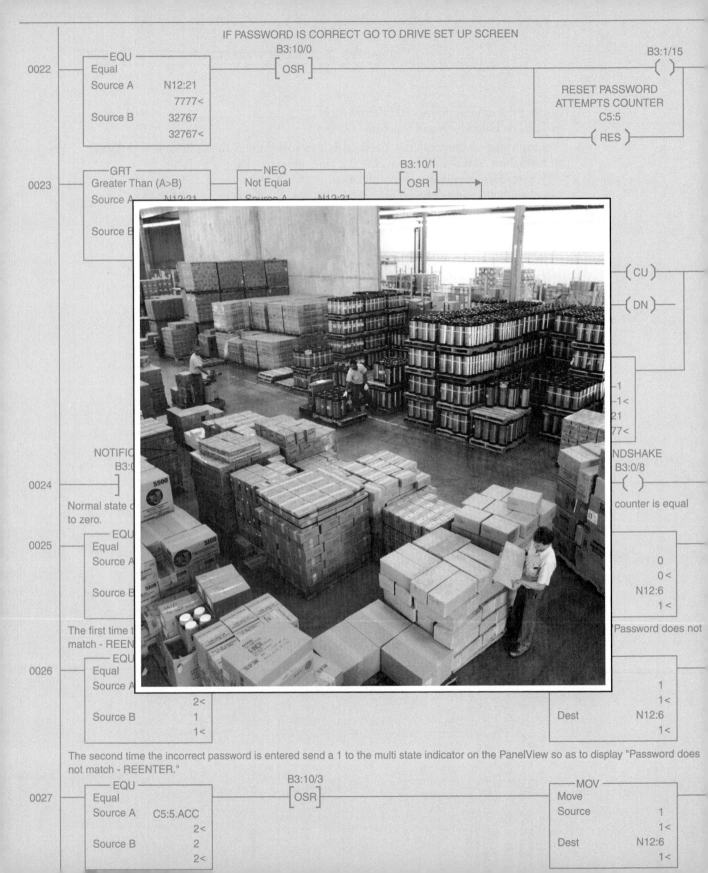

NOTIFIC
B3:0

0024

Normal state o
to zero.

0025

EQU
Equal
Source A

Source B

The first time t
match - REEN

0026

EQU
Equal
Source A

            2<
Source B   1
            1<

NDSHAKE
B3:0/8
( )

counter is equal

0
0<
N12:6
1<

Password does not

1
1<
Dest   N12:6
            1<

The second time the incorrect password is entered send a 1 to the multi state indicator on the PanelView so as to display "Password does not match - REENTER."

0027

EQU
Equal
Source A   C5:5.ACC
            2<
Source B   2
            2<

B3:10/3
[OSR]

MOV
Move
Source   1
            1<
Dest   N12:6
            1<

CHAPTER

16

Timer and Counter Instructions

OBJECTIVES

After completing this chapter, you should be able to:

- describe the function of an on-delay timer
- describe the function of an off-delay timer
- describe in what instances one would use a retentive timer
- describe the function of an up-counter
- describe the function of a down-counter
- describe in what instances one would use an up- versus a down-counter
- define preset, accumulative value, and the timer or counter address
- explain how the various timers and counters are reset

INTRODUCTION

The programmable controller was conceptualized as an easy-to-use, easily changeable, relay-replacing device. One of the major enhancements to the original programmable controller was to add timing and counting abilities. Early PLCs had optional timing circuit cards that the user could slide into the CPU to add timer or counter functionality to the PLC. Timing cards had physical solid-state timing chips installed on slide-in timer boards.

Today's PLCs use modern microprocessor technology and have timers and counters included in the instruction set. As an example, each SLC 500 can have up to 256 timers or counters in each of multiple timer or counter files. Software timers or counters are easily included in your ladder programs by simply programming the desired instruction on the ladder rung. This chapter will introduce the Allen-Bradley SLC 500 and MicroLogix 1000 timers and counters. Allen-Bradley's PLC 5 timers and counters are almost identical to

SLC 500 timers and counters with the exception of the number of timers or counters allowed per file and the specific data files that can be designated as timer or counter files. When using a PLC 5, a short review of the PLC 5 *Instruction Set Reference Manual* will acquaint you with the differences between the two PLCs.

TIMER INSTRUCTIONS

All PLCs have timer instructions. Even though each manufacturer may represent timers differently on the ladder diagram, most timers operate in the same manner. A timer consists of the following parts: timer address, preset value, time base, and accumulated value. Figure 16-1 illustrates an SLC 500 timer ladder rung from Rockwell Software's RSLogix 500 Programming Software for programming the SLC 500 and MicroLogix 1000. Notice the timer instruction name, timer on delay (TON), timer address (T4:0), time base (1.0 seconds), a preset value of 100, and the accumulated value of 0.

Figure 16-1   SLC 500 on-delay timer.

There are bits associated with the current state of the timer called status bits. The timer address is the timer's unique identifier in PLC memory. A timer's preset value is the length of time for which the timer is to run, while the time base specifies at what rate the timer will increment. A typical time base could be in seconds, tenths of a second, or hundredths of a second. The time base is also referred to as the timer's accuracy. The accumulated value is the current elapsed time.

The three types of timers are on-delay, off-delay, and retentive. Figure 16-2 describes the common timer instructions.

SLC 500 TIMERS

SLC 500, MicroLogix, and PLC 5 timers are stored in the timer file, file 4, in the data file section of the PLC's memory. The default timer file is file 4. The timer file is used to store timer data. Each timer consists of three 16-bit words, called "timer elements." There can be many timer files for a single processor file. Any data file greater than file 8 can be assigned as an additional timer file. Figure 16-3 illustrates an RSLogix 500 or RSLogix 5 timer file. The only difference between the SLC 500 and PLC 5s is that the SLC can have 256 timers per timer file where the PLC 5 can have 1,000 timers per file.

TIMER INSTRUCTIONS		
Instruction	Use This Instruction to	Functional Description
On Delay	Program a time delay before instruction becomes true	Use an on-delay timer when an action is to begin a specified time after the input becomes true. As an example, a certain step in the manufacturing process is to begin 30 seconds after a signal is received from a limit switch. The 30-second delay is the on-delay timer's preset value.
Off Delay	Program a time delay to begin after rung inputs go false	For example, for an external cooling fan on a motor, the fan is to run all the time the motor is running and for five minutes after the motor is turned off. This is a five-minute off-delay timer. The five-minute timing cycle begins when the motor is turned off.
Retentive	Retain accumulated value through power loss, processor mode change, or rung state going from true to false	Use a retentive timer to track the running time of a motor for maintenance purposes. Each time the motor is turned off, the timer will remember the motor's elapsed running time. The next time the motor is turned on, the time will increase from there. To reset this timer, use a reset instruction.
Reset	Reset the accumulated value of a timer or counter	Typically used to reset a retentive timer's accumulated value to zero.

Figure 16-2 Timer instructions common to most PLCs.

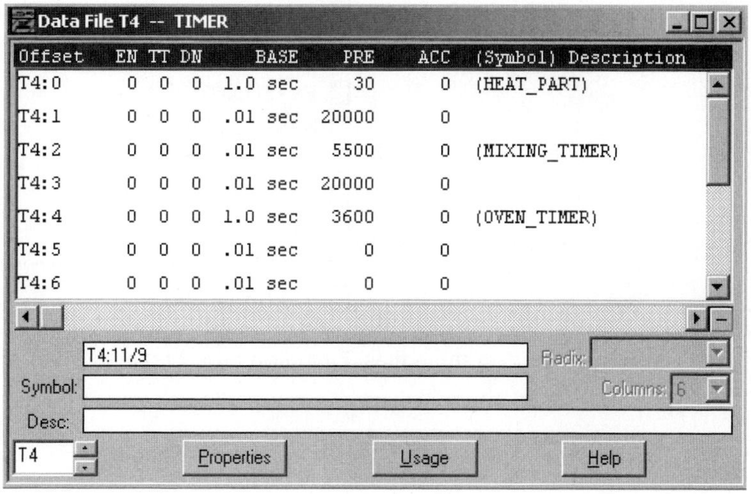

Figure 16-3 SLC 500 or PLC 5 timer file from RSLogix software.

Separating timers into separate files, and thus grouping them in terms of the function they play in the process being controlled, simplifies tracking and troubleshooting. As an example, if all timers associated with manufacturing process A are grouped in timer file T12, all timers associated with process B are grouped in timer file T13, and all timers associated with process C are grouped into timer file T14, troubleshooting problems with process B will lead to an examination of the timers in file 13. If all timers were in the default file, file 4, troubleshooting process B's timer problems could be quite a challenge. Figure 16-4 shows data files from either RSLogix 5 or RSLogix 500. Notice the default data files, and the newly created data files including T12, T13, and T14.

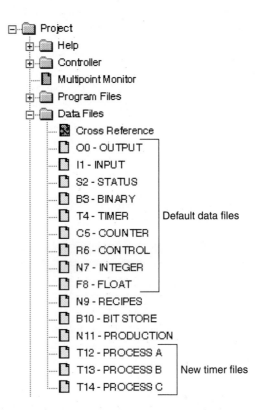

Figure 16-4   RSLogix data table files, including newly created timer files T12, T13, and T14.

Timer Element

A timer instruction is one element. A timer element is made up of three 16-bit words:

- Word zero contains the three status bits, EN, TT, and DN (status bits will be covered as we introduce each timer instruction).
- Word one is for the preset value.
- Word two is for the accumulated value.

Figure 16-5 illustrates the format of a single timer element.

EN	TT	DN	Reserved Bits
Preset Value			
Accumulated Value			

Figure 16-5   One timer element is made of three 16-bit words.

Timer Addressing

Timer addressing is done as follows:

1. To address the timer element, the address format is T4:2.

    T = T identifies this as a timer file.

    4 = This is timer file 4, the default. SLC timers can be assigned any unused file from 9 to 255.

    :2 = This is timer two in file 4. There are 256 timers available in each file. Timers 0 through 255 are available.

2. To address timer two's accumulated value, the address used is T4:2.ACC.

    T = T identifies this as a timer file.

    4 = This is timer file 4, the default.

    :2 = This is timer two in file 4.

    . = The point is called the word delimiter. The word delimiter separates the timer number, called the structure, from the subelement. The subelement is ACC, for accumulated value.

3. To address timer two's preset value, the address used is T4:2.PRE.

    T = T identifies this as a timer file.

    4 = This is timer file 4.

    :2 = This is timer two in file 4.

    . = The point is called the word delimiter. The word delimiter separates the timer number, called the structure, from the subelement. The subelement is PRE for preset value.

4. Timers have three status bits. Word zero, bit 13, is the done bit, identified as DN. The done bit is set when the timer's accumulated value is equal to the preset value. Word zero, bit 14, is the timer-timing bit. The timer-timing bit is identified as TT. This bit is set when the timer is timing.

    Word zero, bit 15, is the **enable** bit, represented as EN. Bit 15 is set whenever the timer is enabled. To address T4:2's status bits, simply place their identifier or bit number after the slash identified as the bit delimiter. The proper bit addressing format is listed below for timer status bits.

T4:2/DN is the address for timer file 4, timer two's done bit.

T4:2/TT is the address for timer file 4, timer two's timer-timing bit.

T4:2/EN is the address for timer file 4, timer two's enable bit.

SLC 500 Time Base

SLC 500 timer time bases are listed for each processor. MicroLogix 1000 has a selectable time base of either .01 second or 1.0 second. For fixed and 5/01 processors, the time base is .01 second. SLC 5/02, 5/03, and 5/04 processors have selectable time bases of .01, or 1.0 second. MicroLogix 1200 and 1500 have time bases of .001, .01, or 1.0 second.

Timer Accuracy

Timer accuracy is related to the timer time base and the processor scan length. When looking at a timing cycle, accuracy is the degree of error between the actual time and the accumulated value in the timer instruction representing the timing cycle. An excessive program scan time will disrupt the normal timer update from the processor. SLC 500 timer accuracy is −0.01 to 0 second, with a program scan of up to 2.5 seconds. As an example, a 1-second timer will maintain accuracy with a program scan of 1.5 seconds or less. For long program scan times, a possible solution to avoid timer inaccuracy is to repeat the timer rung in your program. By repeating the timer instruction in another rung within your ladder program, the timer instruction can be scanned within the limits of the 2.5-second maximum program scan for an SLC 500 PLC. When using other than an SLC 500 PLC, consult your particular specifications to decide the accuracy of timers associated with that particular PLC.

Now that we have looked at what makes up an SLC 500 timer element, we will investigate each timer instruction.

THE ON-DELAY TIMER INSTRUCTION

Use the on-delay timer instruction if you want to program a time delay before an instruction becomes true.

Timer Addressing

Timer addressing is as follows: T (Timer file number): (Timer element number). The timer address T4:0 is addressing timer file 4, timer element 0.

- The default timer file number is four.
- Any unused data file greater than 10 and up to and including 255 may be a timer file.
- Additional timer files may be created in the memory map screen.
- Each timer file may have up to 256 timer elements.

Example: T4:1 = timer file 4, timer element 1 (remember, a timer element is simply another name for timer number 1 in timer file 4)

T4:12 = timer file 4, timer element 12

T4:112 = timer file 4, timer element 112

T16:34 = timer file 16, timer element 34

T16:2 = timer file 16, timer element 2

Figure 16-6 illustrates an on-delay timer, called a timer on delay (TON) for Allen-Bradley RSLogix software along with the associated ladder rungs.

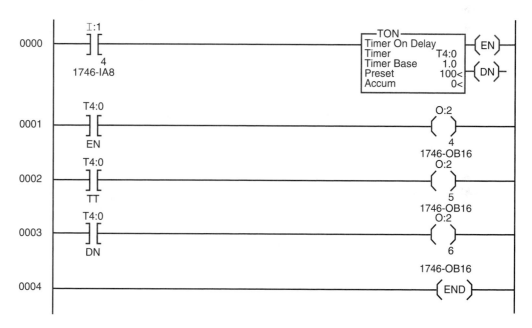

Figure 16-6   SLC 500 on-delay timer and its associated ladder rungs.

Ladder Explanation

As long as I:1/4 is true, the timer on delay T4:0 will increment every 1 second toward the preset value of 100 seconds. The current number of seconds that have passed will be displayed in the accumulated value portion of the instruction. When the accumulated value is equal to the preset, the timer's done bit will be energized, or set. The timer's done bit is on rung 3. As long as the accumulated value is equal to the preset value, the done bit will remain set. As long as the done bit is set, the T4:0/DN, examine if closed, instruction on rung 3 will be true. With this instruction true, the output instruction O:2/6 will also be true. An on-delay timer is not retentive. Any loss of continuity to the timer instruction on rung 0 will cause the timer to reset itself to an accumulated value of 0.

Associated Status Bits

The timer done bit, bit 13, is set when the accumulated value is equal to the preset value. This bit is commonly used to control other logic when the elapsed time (the accumulated value) has transpired.

The timer-timing bit, bit 14, is illustrated on rung 2. This bit is set any time the rung conditions are true and the timer is timing. The timer is timing whenever the rung is true and the accumulated value is less than the preset value. When the done bit is set, the timer-timing bit is reset. In Figure 16-6, the timer-timing bit will be true, or set, whenever input I:1/4 is true and as long as the accumulated value is less than the preset value of 100 seconds. Output O:2/5 will be on when the timer is timing between 0 seconds and 100 seconds. As the done bit is set and O:2/6 turns on, the timer-timing bit goes false and O:2/5 turns off.

The timer enable bit, bit 15, is set, or true, anytime the timer instruction's rung 1 is true. As long as I:1/4 is true, the timer instruction is considered enabled. The enable bit will be true when the timer-timing bit is true. The timer enable bit will stay set through the transition from the timer-timing bit to the timer done bit. As long as there is continuity through all input instructions to the timer instruction, no matter the relationship between the preset and accumulated values, the timer enable bit will be set. The enable bit is reset when the rung goes false.

THE OFF-DELAY TIMER INSTRUCTION

Use the off-delay timer instruction if you want to program a time delay to begin after rung inputs go false. As an example, an external cooling fan on a motor is to run all the time the motor is running and for 100 seconds after the motor is turned off. This involves a 100-second off-delay timer. The 100-second timing cycle begins when the motor is turned off. Figure 16-7 illustrates an off-delay timer, which Allen-Bradley calls a timer off delay (TOF), and its associated ladder rungs.

Ladder Explanation

When I:1/4 becomes true, the TOF T4:1 instruction becomes true. The done bit is set as long as the rung is true. As a result, O:2/6, the external cooling fan output, energizes as soon as the rung goes from false to true.

When the motor is turned off, rung 0 transitions from true to false, and the TOF instruction begins timing. The done bit and the external cooling fan's output (O:2/6) will remain on, or true, for the preset value of 100 seconds. The time period between the point when the rung becomes false and the point when the 100-second preset time expires for T4:1 is called the delay after the input goes false, or the off delay. The done bit and its associated output stay true until the off delay of 100 seconds expires. The time expires when the accumulated value reaches the preset value.

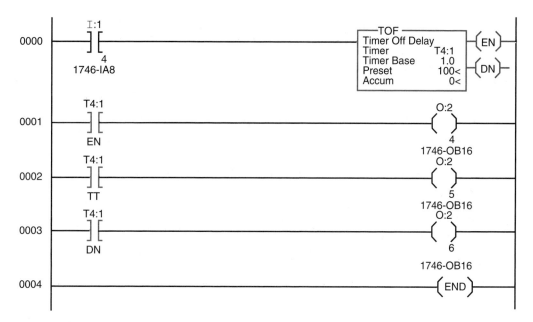

Figure 16-7 SLC 500 off-delay timer and its associated ladder rungs.

Associated Status Bits

The timer done bit, bit 13, is set when the rung becomes true. The done bit remains set through the true-to-false rung transition and until the accumulated value is equal to the preset value. This bit is commonly used to control other logic when an output needs to be turned on or off after its rung has been off for the preset time interval.

The timer-timing bit, bit 14, is illustrated on rung 2. This bit is set any time the rung conditions are false and the timer is timing. The timer times whenever the rung transitions from true to false and the accumulated value is less than the preset value. When the done bit is reset or the rung goes true, the timer-timing bit is reset. In Figure 16-7, the timer-timing bit will be true, or set, after input I:1/4 goes true and then false. The timer will start timing at this point.

The timer enable bit, bit 15, is set, or true, anytime the timer instruction's rung 1 is true. As long as I:1/4 is true, the timer instruction is considered enabled. The enabled bit will be true when the timer-timing bit is true. The enable bit is reset when the rung goes false.

The Reset and TOF Instructions

A reset instruction should *not* be used with timer off delay (TOF) instructions. The reset instruction resets the accumulated value, done bit, timing bit, and enable bit. Resetting these bits on a TOF timer causes the timer to reset itself, thus eliminating the off delay portion of the timing cycle.

Many applications of a TOF-type timer involve cooling. If the timer is reset, the cooling cycle could be disrupted.

THE RETENTIVE TIMER INSTRUCTION

Use this instruction if you want to retain accumulated value through power loss, processor mode change, or change in the rung state from true to false. The retentive timer will measure the cumulative time period for which its rung is true. As an example, a retentive timer can be used to track the running time of a motor for maintenance purposes. The timer is used to track the accumulated time the motor has run. For our sample application, our motor needs maintenance after 28,800 seconds or 8 hours of running time. Each time the motor is turned off, the timer needs to remember the motor's total elapsed running time. The next time the motor is turned on, the timer will increase the accumulated running time from where it left off. When the total accumulated motor running time has been reached, a maintenance reminder pilot light will be lit. A retentive timer is used in this application. Figure 16-8 illustrates a retentive timer, called a retentive timer on (RTO).

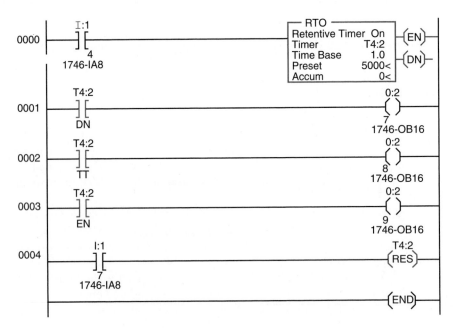

Figure 16-8 SLC 500 retentive timer and its associated ladder rungs.

Ladder Explanation

The retentive timer on RTO instruction behaves similar to the timer on delay with the exception that when the RTO instruction goes false, it will retain, or remember, its accumulated value.

The retentive timer will retain its accumulated value during any of the following:

- Its rung goes false.
- The processor loses power. The battery for memory backup must be in good condition.
- If the processor operating mode is changed from remote run or remote test to remote program mode.
- The processor faults.

Associated Status Bits

The timer done bit, bit 13, is set when the accumulated value is equal to the preset value. In this application, the done bit and output O:2/7 could control the maintenance reminder pilot light.

The timer-timing bit, bit 14, is illustrated on rung two. This bit is set any time the rung conditions are true and the timer times. The timer times whenever the rung is true and the accumulated value is less than the preset value. When the done bit is set, the timer-timing bit is reset as well. In Figure 16-8, the timer-timing bit will be true, or set, whenever input I:1/4 is true and as long as the accumulated value is less than the preset value of 5,000 seconds. Output O:2/8 will be on when the timer is timing between 0 seconds and 5,000 seconds. As the done bit is set and O:2/7 turns on, the timer-timing bit goes false and O:2/8 turns off.

The timer enable bit, bit 15, is set, or true, anytime the timer instruction's rung 0 is true. As long as I:1/4 is true, the timer instruction is considered enabled. The enabled bit will be true when the timer-timing bit is true. The timer enable bit will stay set through the transition from the timer-timing bit to the timer done bit. As long as there is continuity through all input instructions to the timer instruction, no matter the relationship between the preset and accumulated values, the enable bit will be reset when the rung goes false.

THE RESET INSTRUCTION

Use this instruction if you want to reset the accumulated value of a retentive timer. When a maintenance reminder pilot light comes on and a maintenance individual completes the required preventive maintenance, he or she will depress the push button connected to input I:1/7. Input I:1/7 is the reset maintenance timer signal. This will be a momentary normally open push-button field device. Pressing this button will reset the RTO's accumulated value back to 0. The motor is then ready to run for another cycle. Figure 16-9 illustrates a sample RTO timer and its associated reset instruction (RES) on rung 0001.

Now that you have a good understanding of the three types of timers, we will introduce PLC counters.

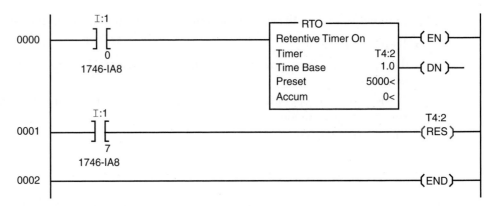

Figure 16-9   Retentive timer and its reset instruction.

PLC COUNTER INSTRUCTIONS

Every PLC has counter instructions. Although most PLC counters work the same, the instruction symbols used and method of programming will change for different manufacturers. Consult your specific PLC user manual for particulars on the counters.

The typical counter counts from 0 up to a predetermined value, called the "preset" value. As an example, if you wanted to count from 0 to 100, you would be counting up and would use a count-up or up-counter. The predetermined value of 100 would be the preset value. The "accumulated value" is the current or accumulated count. If our counter had counted 45 pieces that had passed on a conveyor, the accumulated count, or value, would be 45. When all 100 pieces had passed on the conveyor, the counter accumulated value and preset value would be equal. At this point the counter would signal other logic within the PLC program that the batch of 100 was completed and it should now take some action. The PLC might move the box containing 100 parts on to the next station for carton sealing. To start counting the next batch, a reset instruction would be used to reset the counter's accumulated value back to 0. Figure 16-10 illustrates a ladder rung and a count-up counter (CTU) for an Allen-Bradley SLC 500, MicroLogix, or PLC 5.

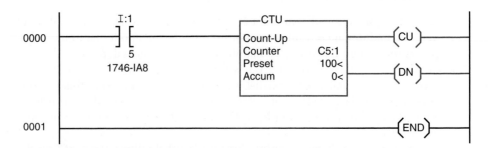

Figure 16-10   An RSLogix 5 or 500 count-up counter.

To count from 100 down to 0, you would choose a down-counter. A down-counter counts from the accumulated value of 100 down to the preset value of 0. Figure 16-11 illustrates a count-down counter.

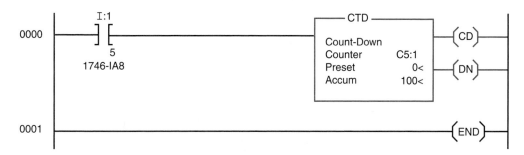

Figure 16-11   An SLC 500 count-down counter.

Every PLC has counter instructions. Although they may program differently, they operate very similarly. Figure 16-12 lists the generally available counters.

COUNTER INSTRUCTIONS		
Instruction	**Use This Instruction to**	**Functional Description**
Count Up	Count from zero up to a desired value	Counting the number of parts produced during a specific work shift or batch. Also counting the number of rejects from a batch.
Count Down	Count down from a desired value to zero	An operator interface display shows the operator the number of parts remaining to be made for a lot of 100 parts ordered.
High-Speed Counter	Count input pulses that are too fast separately from normal input points and modules	Most fixed PLCs will have a high-speed set of input points that allow interface to high-speed inputs. Signals from an incremental encoder would be a typical high-speed input. Check your specific PLC for the maximum pulse rate.
Counter Reset	Reset a timer or counter	Used to reset a counter to zero so another counting sequence can begin.

Figure 16-12   Common counters found in programmable logic controllers.

ALLEN-BRADLEY COUNTERS

This section will introduce the counters available for the Allen-Bradley SLC 500, PLC 5, and MicroLogix PLCs.

The default counter file is file 5. The counter file is used to store counter data. Each counter consists of three 16-bit words and is called a "counter element." There can be many counter files for a single processor file. Any data file greater than file 8 can be assigned as an additional counter file. Each RSLogix 500 counter file can have 256 counter elements. Separating counters into separate files and grouping them by their function or the process they control is done for the same reasoning as for timers.

A counter instruction is one element. A counter element is made up of three 16-bit words. Thus, the counter instruction contains three parts:

- Word zero is for status bits. Status bits CU, CD, DN, OV, UN, and UA will be covered, with their associated instructions.
- Word one is for the preset value.
- Word two is for the accumulated value.

Figure 16-13 illustrates a single counter element.

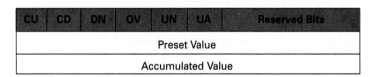

CU	CD	DN	OV	UN	UA	Reserved Bits
Preset Value						
Accumulated Value						

Figure 16-13   One timer element is made of three 16-bit words.

Counter Addressing

Allen-Bradley counter addressing is outlined as follows.

1. To address the counter as a unit, the address format is C5:3.

   C = C identifies this as a counter file.

   5 = This is counter file 5, the default. Counters can be assigned any unused RSLogix 500 file from 10 to 255.

   :3 = The colon is called the file separator. The colon separates the file, file 5, from the specific counter, in this case, counter 3 in counter file 5. There are 256 counters in each RSLogix 500 counter file. This happens to be counter 3. Figure 16-14 illustrates counter file 5 and individual counters within that file.

Figure 16-14 illustrates the relationship between file 5 and the individual counters within the file. Remember that each counter is an element and each counter element is three words.

2. To address counter 12's accumulated value, the address used is C5:12.ACC.

   C = C identifies this as a counter file.

   5 = This is counter file 5.

   :12 = The colon is called the file separator. The colon separates the file, file 5, from the specific counter, in this case counter 12 in counter file 5.

   . = The point is called the word delimiter. The word delimiter separates the counter number, called the structure, from the subelement. The subelement is ACC for the accumulated value.

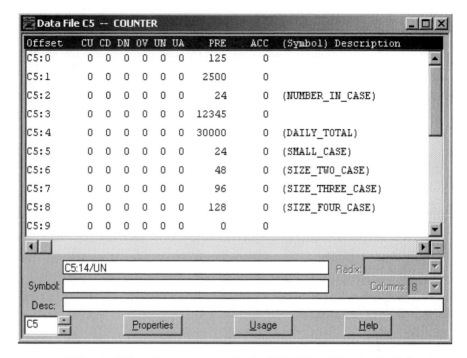

Figure 16-14   RSLogix 500 software counter file for SLC 500 or MicroLogix PLCs.

3. To address counter 12's preset value, the address used is C5:12.PRE.

C = C identifies this as a counter file

5 = This is counter file 5. File 5 is the default.

:12 = The colon is called the file separator. The colon separates the file, file 5, from the specific counter, in this case counter 12 in counter file 5.

. = The point is called the word delimiter. The word delimiter separates the counter number, called the structure, from the subelement. The subelement is PRE for preset value.

4. Counters have six status bits. Word zero, bits 0 through 9, are reserved for the processor.

Word zero, bit 10, is the update accumulator bit, identified as UA. The UA bit is only available on a fixed-style SLC 500 PLC. This bit is used in conjunction with the high-speed counter built into the fixed SLCs.

Bit 11 is the underflow bit, identified as UN. The underflow bit is set when the accumulated value of a count-down counter has reached the lowest possible accumulated value, that is −32,768. The counter will automatically wrap around and start counting down from the maximum positive value, +32,767.

Bit 12 is the overflow bit, identified as OV. The overflow bit is set when the accumulated value of a count-up counter has reached the highest possible accumulated value, which is +32,767. The counter will automatically wrap around and start counting up from the maximum negative value, −32,768.

Bit 13 is the done bit, identified as DN. The done bit is set when the counter's accumulated value is equal to or greater than the preset value. Much like with timers, a counter's done bit is used to signal the processor that it is time to take action.

Bit 14 is the count-down–enabled bit, identified as CD. This bit is set when counting down and the rung conditions are true. Use this bit to monitor when the counter is counting down.

Bit 15 is the count-up–enabled bit. The count-up–enabled bit is identified as CU. This bit is set when counting up and the rung conditions are true. Use this bit to monitor when the counter is counting up.

To address C5:12's status bits, simply place their identifier after the slash is identitied as the bit delimiter. The proper addressing format is listed below for counter status bits:

C5:12/DN is the address for counter file 5, counter 12's done bit.

C5:12/CU is the address for counter file 5, counter 12's count-up–enabled bit.

C5:12/EN is the address for counter file 5, counter 12's enable bit.

HOW COUNTERS WORK

A counter instruction is an output instruction. Most PLC counters count the false-to-true transitions of the rung's input logic. The counter instruction counts each time the input logic changes the rung from false to true. Input logic can be a signal coming from an external device, such as a limit switch or sensor, or a signal from internal logic. Each time the counter instruction sees a false-to-true rung transition, a count-up counter's accumulated value is incremented by one.

The down-counter works a little differently. Each time a count-down counter sees a false-to-true rung transition, its accumulated value is decremented by 1. Since the accumulated value is decremented each time the input logic changes the rung from false to true, the accumulated value must be the starting point of the count. In our example, we are counting a batch of 100 parts. Each time a part is made, the remaining total is displayed on an operator interface display device so the operator can see how many more parts need to be manufactured. The accumulated value will be programmed with the value of 100, whereas the preset value will be 0. As each part is made, the accumulated value will be decremented by 1. When all 100 parts have been made, the accumulated value and the preset value will be 0.

One important consideration when working with counters is the speed or frequency with which an input device is sending the false-to-true signals to the processor. The speed of the processor scan must be taken into consideration. As an example, if parts are passing an inductive proximity sensor on a conveyor faster than the processor scan can see them, some parts will not be counted. A high speed counter module could be used when counts come in faster than the processor can see them.

Counting range is the numerical range within which a counter can count. The counters can count within the range of −32,768 to +32,767. This is the range of a 16-bit signed integer. (Refer to chapter 4 for a review of signed integers.)

If a counter counts above +32,767, an overflow is detected and the overflow bit is set. Conversely, if a down-counter counts below −32,768, an underflow is detected and the underflow bit is set. In either an overflow or underflow, the counter will set the appropriate bit, wrap around, and begin counting from the other end of the counting range. As an example, suppose a counter overflows, sets the overflow bit, and then begins counting up from −32,768. A down-counter counts below −32,768, sets the underflow bit, and then continues counting down from +32,767. To avoid overflowing or underflowing a counter, a reset instruction can be used to reset the counter back to 0.

Counters are retentive. Assuming that the processor's battery is in good condition, a counter will retain its accumulated value and the on or off status of the done, overflow, and underflow bits through a power loss.

THE COUNT-UP INSTRUCTION

Use the count-up instruction if you want a counter to increment one decimal value each time it registers a rung transition from false to true. Again, the rung's transition from false to true must be seen by the processor during the processor scan before the input transitional signal will be recorded by the counter. Use the count-up instruction if you want to keep an accumulated tally of the number of times some predetermined event transpires.

Figure 16-15 illustrates a sample count-up (CTU) counter and its associated status bits.

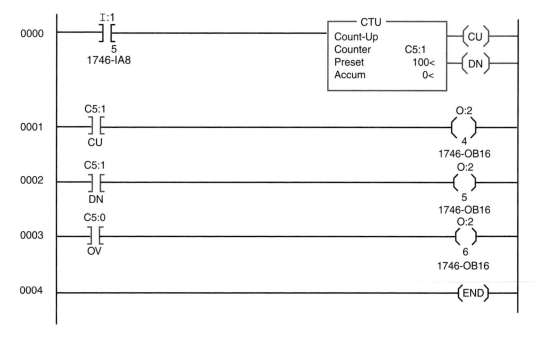

Figure 16-15   RSLogix 5 or 500 count-up instruction and its associated status bit rungs.

Ladder Explanation

Each time input I:1/5 transitions from off to on, counter C5:1 will increment its accumulated value by one decimal value.

Associated Status Bits

The count-up overflow bit, OV, on rung 0003, is set whenever the count-up counter's accumulated value wraps from +32,767 around to −32,768. The counter will only alert you to the overflow condition through setting the overflow bit. Counting will continue in the up direction, beginning at −32,768 and progressing to +32,767.

The done bit (DN, on rung 0002) is set when the accumulated value is equal to or greater than the preset value. In the event of a wrap from +32,767 to −32,768, the accumulated value becomes less than the preset value and the done bit will not be reset.

The count-up–enabled bit, on rung 0001, is set when the rung conditions are true, or enabled.

THE COUNT-DOWN INSTRUCTION

Use this instruction if you want to count down over the range of +32,767 to −32,768. Each time the instruction sees a false-to-true transition, the accumulated value will be decremented by one count. Assume that you want to display the remaining number of parts to be built for a specific order of 100 parts. The remaining parts to be built will be displayed on an operator display device so workers can see how many parts are needed to complete the lot. For this example, the accumulated value will be set at 100 and the preset value will be 0. Each time a part is completed and passes the sensor, the accumulated value will be decremented one decimal value, as illustrated in Figure 16-16.

Ladder Explanation

The accumulated value for count-down counter C5:5 is set at 100. The preset value is 0. Each time input I:1/5 transitions from off to on, the accumulated value will decrement one decimal value. The done bit will be set, or on, during the entire count from 100 to 0. If the done bit is programmed as an examine if closed instruction (as on rung 0001 in Figure 16-16), the instruction will be true during the entire count-down sequence. If, on the other hand, the done bit was programmed as an examine if open instruction (as on rung 0002 in Figure 16-16), the output would be off, or false, during the count-down cycle. The done bit will change state when the counter's accumulated value transitions from 0 to −1.

Rung 0004 in Figure 16-16 contains the reset instruction for our count-down counter. An important note is that resetting the counter will cause the accumulated value to be reset to 0. Our application specifies that we count down from 100 to 0. After each complete cycle and counter reset, the accumulated value is reset to 0. To replace the accumulated value of 100 back into the counter's accumulator, the **move** instruction on rung 0003

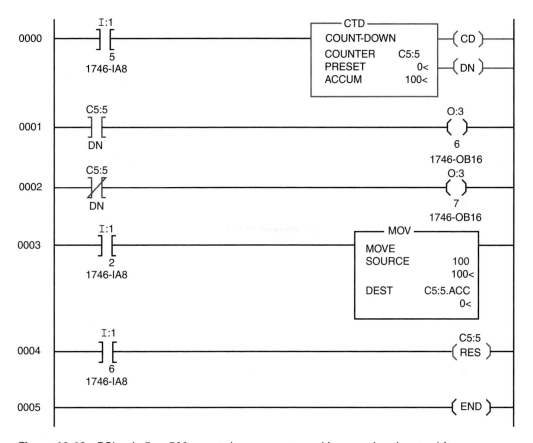

Figure 16-16   RSLogix 5 or 500 count-down counter and its associated status bits.

will be used. The move source is integer 100. The move instruction's destination for the integer 100 is C5:5 ACC. Now our counter is set to count down from 100 to 0. Chapter 17 will explore the move instruction and other data-handling instructions.

As with the up-counter, you must be careful as to the frequency of incoming counts. Counts that come in too fast for the processor and its program scan to register will be lost.

Associated Status Bits

The count-down underflow bit, bit 11 of counter element word zero, will be set when the accumulated value wraps from a $-32{,}767$ to a $+32{,}768$. The counter will only alert you to the underflow condition through setting the underflow bit. Counting will continue in the down direction, beginning at $+32{,}768$ and progressing on to 0.

The done bit, bit 13 of word zero, is set as long as the accumulated value is greater than or equal to the preset value. This means that since we are counting down from 100 to 0, the accumulated value will remain set until the last part is produced. When the accumulated value becomes less than 0, the done bit is reset. The ladder program can be programmed to initiate the desired action when it sees this transition.

The count-down–enabled bit, bit 14 of word zero, is set whenever rung conditions are true.

THE HIGH-SPEED COUNTER INSTRUCTION

The high-speed counter instruction is only used with the SLC 500 fixed PLCs. The MicroLogix 1000 has its own high-speed counter instruction. The MicroLogix 1000 high-speed counter has additional features. When using a modular SLC 500 chassis and a modular processor, a separate high-speed counter module is used.

Figure 16-17 illustrates a rung of logic containing the SLC MicroLogix 1000 high-speed counter instruction. One high-speed counter is allowed per MicroLogix program. The MicroLogix 1000's high-speed counter instruction address is fixed at C5:0. This counter can be programmed as either an up-counter or a bidirectional counter. The counter in Figure 16-17 is an up-counter. This counter is similar to the CTU; however, the high-speed counter instruction is only enabled by the rung on which it resides. The high-speed counter instruction does not count rung transitions. These transitions are counted from input I:0/1.

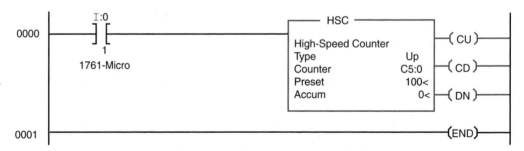

Figure 16-17   High-speed counter instruction for the MicroLogix 1000.

Associated Status Bits

Since the high-speed counter counts transitions seen at the input point rather than rung transitions, many counts may come in between program scans. Counts coming in between program scans could be lost, resulting in an inaccurate accumulator value. To solve this problem and provide an accurate update of the accumulated value, a separate hardware accumulator is used in conjunction with the high-speed counter. Status bit 10, word zero, is the update accumulator (UA) bit. Setting the UA bit will cause the instruction image accumulator (in the hardware accumulator used with the high-speed counter) to update the processor's accumulator.

THE COUNTER RESET INSTRUCTION

Use this instruction if you want to reset a timer-on delay, retentive timer, count-up counter, or count-down counter's accumulated value to zero. The reset instruction's address must match the address of the timer or counter that is to be reset. Only one address is allowed per reset instruction. Resetting multiple timers or counters will require multiple reset instructions. Figure 16-18 illustrates a reset instruction on rung 0001, which will reset counter C20:7 on rung 0000.

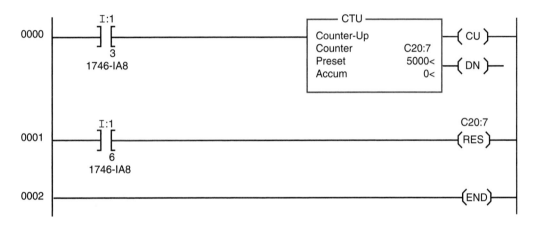

Figure 16-18   Reset instruction programmed for resetting count-up counter C20:7.

Ladder Explanation

When input I:1/13 is set, or true, the reset instruction will reset the counter with the same address as the reset instruction. When resetting a timer or counter instruction, accumulated values and status bits are zeroed.

THE CLEAR INSTRUCTION

The clear instruction (CLR) is an output instruction that can also be used to set the accumulated value of a timer or counter to zero. CLR can be used to set the destination of most data words, in addition to timers or counters, to zero. Figure 16-19 illustrates a rung of logic using parallel clear instructions to clear the accumulators of counters 5:0 and C5:1 to zero.

COMBINING TIMERS AND COUNTERS

Timers and counters can be programmed to work together. Timers or counters can be connected, or cascaded, together to increase the time or count. One counter can be used to count the number of cycles another counter has completed. There are many applications and ways in which timers and counters can be programmed to work together. Figure 16-20 illustrates two timers cascaded together so as to lengthen the time that can

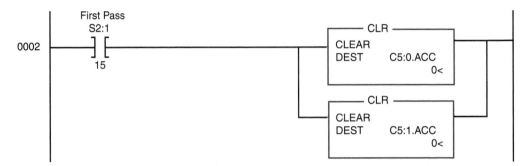

Figure 16-19   Using the clear instruction to set C5:0.ACC and C5:1.ACC to zero.

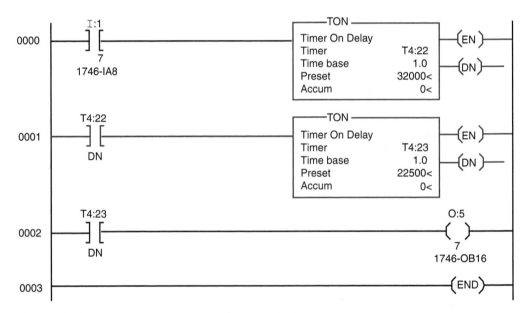

Figure 16-20   Two timers cascaded to lengthen timing.

be counted. The maximum preset of an SLC 500 timer is 32,767 seconds (546.11 minutes or 9.10 hours). Figure 16-20's timer on rung zero times for 32,000 seconds. Timer T4:23, on rung one, continues for an additional 22,500 seconds. When the total time has elapsed, T4:23's done bit on rung 0002 will energize output O:5/7. Thus, cascading two timers together is one method of lengthening a timing cycle.

Figure 16-21 illustrates a counter working in conjunction with a timer. The timer preset value is 10 seconds. Every 10 seconds, the T4:43 DN bit on rung 0001 will pulse at output O:5/9. Counter C5:43, a parallel output, will count the number of timing cycles.

Figure 16-22 illustrates two counters cascaded together to lengthen the total count. Counter C5:8 counts up to 1,250. When C5:8 reaches its preset value, 1,250, C5:8's done

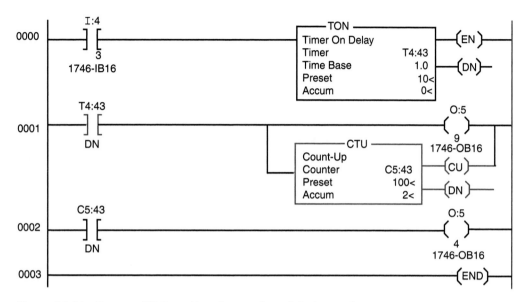

Figure 16-21   Counter C5:0 tracking the number of timing cycles.

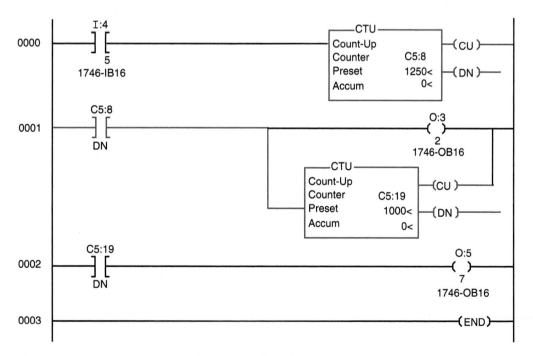

Figure 16-22   Counters cascaded together to lengthen count.

bit is set. Setting the done bit will energize output O:3/2 and increment counter C5:19. After C5:8 has gone through 1,000 cycles, done bit C5:19 will energize output O:5/7. In this example, O:5/7 is energized at a count of 1,250,000.

Combining Up- and Down-Counters to Get a Bidirectional Counter

Figure 16-23 illustrates a separate count-up instruction and a separate count-down instruction with the same address, C5:0. Input I:1/1 will increment CTU C5:0. Input I:1/2 will decrement CTD C5:0. Since each of these counters has the same address, the accumulator will reflect the state of the counter addresses as C5:0. Since there is an up-counter and a down-counter addressed as C5:0, the accumulator will be shared between the two counter instructions. Because the accumulator is shared, it will reflect any count seen by either counter instruction.

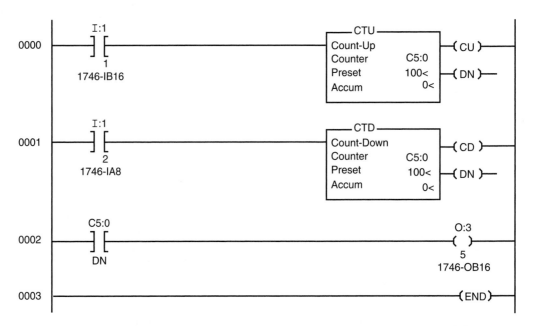

Figure 16-23 Using a CTU and a CTD with the same address to create an up- and down-counter.

SUMMARY

This chapter introduced the basics of timers and counters used with Allen-Bradley's SLC 500, PLC 5, and MicroLogix timers and counters.

Today's PLCs use modern microprocessor technology and have timers and counters included in the instruction set. As an example, each SLC 500 can have up to 256 timers or counters in each of its multiple timer or counter files. Timers and counters are easily included in your ladder programs.

All PLCs have timer instructions. Even though each manufacturer may represent timers differently on the ladder diagram, most similar timers operate in the same manner.

Summary of Timer Features

- A timer is an output instruction.
- A timer consists of the following parts: timer address, preset value, time base, and accumulated value.
- Status bits are associated with the current state of the timer.
- The timer address is the timer's unique identifier in PLC memory.
- A timer's preset value is the length of time the timer is to time for, while the time base specifies at what rate the timer will increment.
- A typical time base may be seconds, tenths of a second, or hundredths of a second. The time base is also referred to as the timer's accuracy.
- The accumulated value is the current elapsed time.

Summary of Counter Features

- A counter instruction is an output instruction.
- Almost all PLC counters count the false-to-true transitions of the rung's input logic.
- As an output, the counter instruction counts each time the input logic changes from a false to a true rung.
- Input logic can be a signal coming from an external device, such as a limit switch or sensor, or from internal logic.
- Each time the counter instruction registers a false-to-true rung transition, a count-up counter's accumulated value is incremented by one.
- Each time a count-down counter registers a false-to-true rung transition, its accumulated value is decremented by one.
- One important consideration when working with counters is the speed or frequency at which an input device is sending the false-to-true signals to the processor. The speed of the processor scan must be taken into consideration. As an example, if parts are passing an inductive proximity sensor on a conveyor faster than the processor scan can see them, some will not be counted.
- Counting range is the numerical range within which a counter can count. The RSLogix 5 and 500 counters can count within the range of $-32,768$ to $+32,767$. This is the range of a 16-bit signed integer.

Finally, we looked at a couple of ways in which timers and counters can be paired to track each other's operating cycles or to lengthen the total time or counting ranges. While we illustrated only a couple of possibilities, there are many other applications in which timers and counters can be teamed to achieve the desired application results.

The accompanying lab manual will provide the opportunity to develop a number of programs using each of these timer and counter applications.

REVIEW QUESTIONS

1. Even though each manufacturer may represent timers differently on the ladder diagram, all timers operate _____.
2. A timer instruction is composed of six parts. List the parts.
3. The _____ is the timer's unique identifier in PLC memory.
4. A timer's _____ is the length of time the timer is to time.
5. The _____ specifies at what rate the timer will increment.
6. A typical _____ could be in seconds, tenths of a second, or hundredths of a second.
7. The _____ value is the current elapsed time.
8. A counter instruction is an _____ instruction.
9. Almost all PLC counters count the _____ of the rung's input logic.
10. As an output, the counter instruction counts each time the _____ changes from a false to a true rung.
11. A counter's input signal can come from an external device, such as a limit switch or sensor, or from _____.
12. Each time the counter instruction sees a _____, a count-up counter's accumulated value is incremented by one.
13. Each time a count-down counter sees a _____, its accumulated value is decremented by one.
14. One important consideration when working with counters is the _____, or _____, in which an input device is sending the false-to-true signals to the processor.
15. _____ is the numerical range within which a counter can count.
16. The SLC 500 counters can count within the range of negative _____ to positive _____.
17. An SLC 500 counter's counting range is the range of a 16-bit _____.
18. If an SLC 500 counter counts above _____, an overflow is detected and the overflow bit is set.
19. If an SLC 500 down-counter counts below _____, an underflow is detected.
20. In the event of either an overflow or underflow, the counter will set the appropriate bit and _____ will begin counting from the other end of the counting range.
21. An SLC 500 counter that overflows will set the _____ bit and then begin counting up from _____.

22. A down-counter that counts below $-32,768$ sets the _____ bit and then continues counting down from _____.

23. To avoid overflowing or underflowing a counter, a _____ instruction will reset the counter back to zero.

24. SLC 500 counters are retentive. Assuming that the processor _____ is in good condition, a counter will retain its accumulated value, along with the on or off status of the _____, _____, and _____ bits through a power loss.

IF PASSWORD IS CORRECT GO TO DRIVE SET UP SCREEN

0022

| EQU |
| Equal |
Source A	N12:21
	7777<
Source B	32767
	32767<

B3:10/0

[OSR]

B3:1/15

( )

RESET PASSWORD
ATTEMPTS COUNTER
C5:5

(RES)

0023

| GRT |
| Greater Than (A>B) |
Source A	N12:21
	7777<
Source B	0
	0<

| NEQ |
| Not Equal |
Source A	N12:21
	7777<
Source B	32767
	32767<

B3:10/1

[OSR]

CTU
Count Up

(CU)

(DN)

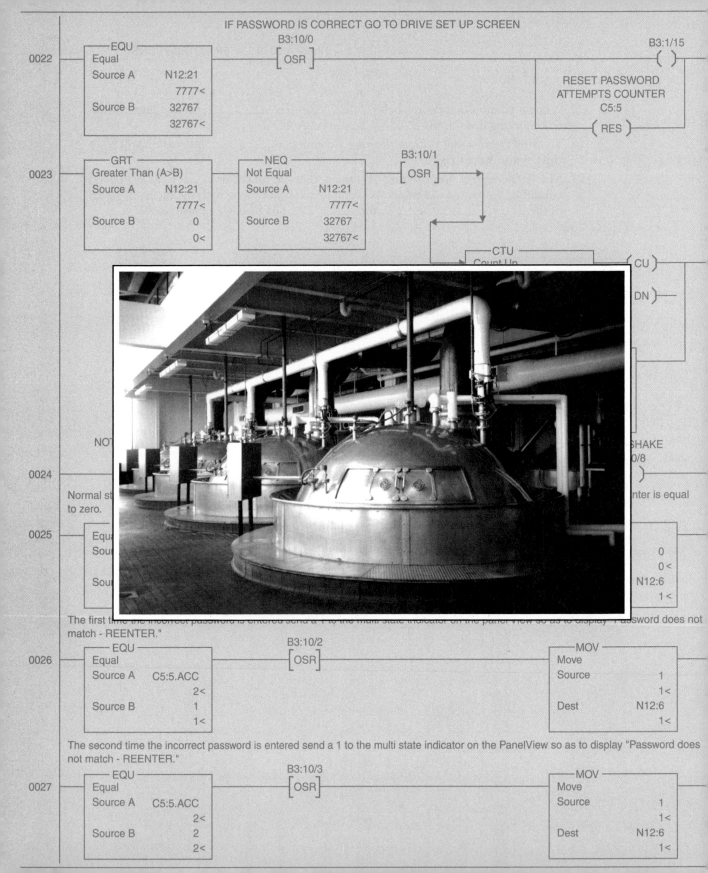

NOT

SHAKE
0/8

( )

0024

Normal st   nter is equal
to zero.

0025

| Equa |
| Sour |
| Sour |

0
0<
N12:6
1<

The first time the incorrect password is entered send a 1 to the multi-state indicator on the panel view so as to display "Password does not match - REENTER."

0026

| EQU |
| Equal |
Source A	C5:5.ACC
	2<
Source B	1
	1<

B3:10/2

[OSR]

| MOV |
| Move |
Source	1
	1<
Dest	N12:6
	1<

The second time the incorrect password is entered send a 1 to the multi state indicator on the PanelView so as to display "Password does not match - REENTER."

0027

| EQU |
| Equal |
Source A	C5:5.ACC
	2<
Source B	2
	2<

B3:10/3

[OSR]

| MOV |
| Move |
Source	1
	1<
Dest	N12:6
	1<

CHAPTER

17

Comparison and Data-Handling Instructions

OBJECTIVES

After completing this chapter, you should be able to:

- describe the function of the common comparison instructions
- explain the difference between a move instruction and a copy instruction
- determine the data resulting after execution of a masked move instruction
- explain how a copy instruction can be used to copy recipe data from one source to another
- explain which instruction to use to have the PLC convert from an integer to a BCD value or from a BCD value to an integer

INTRODUCTION

The main advantage of programmable controllers over relay control systems is the ability of the PLC to be easily programmed to meet the current control needs. As microprocessor power was incorporated into the PLC, a whole new horizon of functionality was opened by incorporating instructions that allowed data-handling capabilities. Major advances in PLC flexibility and control were now possible compared to earlier PLC relay replacers. Today's PLCs have vast instruction sets that include instructions that allow the PLC to store huge amounts of information pertaining to the manufacturing process. As an example, stored data could include numerous recipes, one for each of the different products produced. Data-handling instructions allow stored data to be moved or copied from processor memory into the PLC ladder program currently being executed. Input or output data can be compared for different levels of equality, logical functions can be performed, and input or output data can be converted into different data formats.

This chapter will introduce selected basic SLC 500 and MicroLogix 1000 data-handling instructions.

COMPARISON INSTRUCTIONS

The first group of instructions we will cover are instructions that test for equality.

The Equal Instruction (EQU)

The equal instruction is an input instruction used to test when two values are equal.

Instruction Usage. Programming the equal instruction consists of two steps. First, the equal (EQU) instruction is placed on the ladder rung as an input instruction. Second, the two parameters, source A data and source B data, are entered.

When source A and source B are equal, the instruction is true; otherwise, the instruction is false.

Source A: This is the address of the data to test for equality.

Source B: Source B can be either a constant or a word address.

If you use a constant as source B, the constant will be tested for equality with the data residing in the address specified in source A. If you use an address as source B, the data contained in that address will be tested for equality with the data residing in the address specified in source A.

Sample Ladder Rung

Figure 17-1 illustrates an equal instruction controlling an output instruction.

Figure 17-1   An equal instruction controlling OTE instruction address O:2/6.

Ladder Explanation

The ladder rung in Figure 17-1 illustrates an equal instruction where source A is the address integer file N7:0 and source B is a constant, the number 12. Directly under the address N7:0 is the value 12. This is the integer stored in integer file 7, word zero (N7:0). The value 12 is also stored in source B as a constant. Since the value in source A is equal to the value in source B, the instruction is true. As it is, output instruction **OTE,** addressed as O:2/6, is also true. Directly below the output instruction is noted that this output is output screw terminal six on a 1746-OB16 module.

The Not Equal Instruction (NEQ)

The not equal (NEQ) instruction is an input instruction used to test two values for inequality. Use this instruction to determine if two specified sources of data are not equal. The NEQ instruction is true when the data stored in the address specified as source A is not equal to either the data stored in the address specified as source B or a constant entered in source B.

Source A: This is the address of the data to test for inequality.

Source B: Source B can be either a constant or a word address.

If you use a constant as source B, the constant will be tested for equality with the data residing in the address specified in source A. If you use an address as source B, the data contained in that address will be tested for equality with the data residing in the address specified in source A.

Ladder Explanation

Figure 17-2 illustrates a not equal instruction where source A is the address of integer file N7:0, which contains the number 12. Source B has a constant, the number 12, as its parameter. Looking directly under the address N7:0, you will find the value of the integer stored in integer file 7, word zero (N7:0). Since the value in source A is equal to the value in source B, the instruction is false. The not equal instruction is true only when source A and source B are not equal. When source A is anything other than the constant 12, as in source B, the instruction is true. In this example, as a result of the instruction being false, output instruction OTE, addressed as O:2/7, is also false. The output is addressed as output screw terminal seven on a 1746-OB16 module.

Figure 17-2 Not equal input instruction controlling OTE instruction address O:2/7.

The Less Than Instruction (LES)

The less than (LES) instruction is an input instruction used when you want to test if one value (one source of data) is less than another. The LES instruction is true when the data stored in the address specified as source A is less than either the data stored in the address specified as source B or a constant entered in source B.

Source A: This is the address of the data to test to see if it is less than source B.

Source B: Source B can be either a constant or a word address.

If you use a constant as source B, source A will be tested to see if it is less than the data residing in the constant specified in source B. If you use an address as source B, the data contained in source A will be tested to see if it is less than the data residing in the address specified in source B. Figure 17-3 illustrates a less than instruction ladder rung.

Figure 17-3 The less than instruction tests to see if the value in source A is less than the value stored in or addressed as source B.

The LES instruction in Figure 17-3 tests to see if the value in source N7:1 is less than the value represented in source B. Integer file N7, word one, contains the integer value 7. Source B contains the constant integer 12. Since A < B, the not equal instruction is true. As the input instruction is true, the output instruction will also be true.

The Less Than or Equal Instruction (LEQ)

The less than or equal (LEQ) instruction is an input instruction much like the less than instruction, with the exception that if source A is equal to or less than source B, the instruction will be true.

Use this instruction to determine if one source of data is less than or equal to another. This instruction is true when the data stored in the address specified as source A is less than or equal to either the data stored in the address specified as source B or a constant entered when entering this instruction.

Source A: This is the address of the data to test to see if it is less than or equal to source B.

Source B: Source B can be either a constant or a word address.

If you use a constant as source B, source A will be tested to see if it is less than or equal to the data residing in the address specified in source A. If you use an address as source B, source A will be tested to see if it is equal to the data residing in the address specified in source B.

The PLC ladder rung in Figure 17-4 illustrates a less than or equal input instruction. If source A is less than or equal to source B, the instruction will be true. This rung has a value of 12 stored in source A, N7:0. Since the constant entered into source B is 12, which is equal to the value in source A, the instruction will be true (refer to Figure 17-5).

Figure 17-4   Less than or equal instruction tests to see if A ⩽ B.

If	Then the Instruction Will Be	Example
A is equal to B (A = B)	True	If A = 12 and B = 12
A is greater than B (A > B)	False	If A is 13 or greater, this rung will be false.
A is less than B (A < B)	True	If A is 11 or less, this rung will be true.

Figure 17-5   Table showing A and B relationship for A ⩽ B logic.

The Greater Than Instruction (GRT)

The greater than (GRT) instruction is an input instruction that tests to see if one source of data is greater than another source of data. The GRT instruction is true when the data stored in the address specified as source A is greater than either the data stored in the address specified as source B or a constant entered as source B when programming this instruction.

Source A: This is the address of the data to test to see if it is greater than source B.

Source B: Source B can be either a constant or a word address.

If you use a constant as source B, source A will be tested to see if it is greater than the data residing in source B. If you use an address as source B, the data contained in that address will be tested to see if it is greater than the data residing in the address specified in source A. Figure 17-6 illustrates a greater than instruction on a PLC rung.

Figure 17-6   The greater than instruction tests to see if A > B.

Source A, integer file N7:2 (containing the integer 12), is greater than the constant integer 10 (entered as source B). Because source A is greater than source B, the instruction is true and the OTE instruction will be true.

Figure 17-7 illustrates greater than logic.

If	Then the Instruction Will Be	Example
A is equal to B (A = B)	False	If A = 10 and B = 10
A is greater than B (A > B)	True	If A is 11 or greater, this rung will be true.
A is less than B (A < B)	False	If A is 10 or less, this rung will be false.

Figure 17-7   Table showing A and B relationship for A > B logic.

The Greater Than or Equal Instruction (GEQ)

The greater than or equal (GEQ) instruction determines if one source of data is greater than or equal to another. Use this instruction if you want to determine if one source of data is greater than or equal to another.

The instruction is true when the data stored in the address specified as source A is greater than or equal to either the data stored in the address specified as source B or a constant entered when entering this instruction. Figure 17-8 illustrates a greater than or equal instruction on a PLC rung.

Figure 17-8   Greater than or equal instruction tests to see if A ⩾ B.

The greater than or equal instruction has the integer 12 stored in source A. Source B has the integer 12 stored, also. Since these two values are equal, the GEQ instruction will be true. Figure 17-9 illustrates greater than or equal logic.

If	Then the Instruction Will Be	Example
A is equal to B (A = B)	True	If A = 12 and B = 12
A is greater than B (A > B)	True	If A is 12 or greater, this rung will be true.
A is less than B (A < B)	False	If A is 11 or less, this rung will be false.

Figure 17-9   Table showing A and B relationship for A ⩾ B logic.

DATA-HANDLING INSTRUCTIONS

Data-handling instructions are used when data needs to be moved, or copied, from one data file source to a desired destination. There are also instances when data must be converted from one format to another before it is sent on to a desired destination. The next section will address the SLC 500 and MicroLogix data-handling instructions.

The Move Instruction

The move instruction is an output instruction that moves a copy of one data file word to a specified destination. Use this instruction if you want to move a copy of information stored in one data file location to another.

When the instruction is true, a copy of information specified as source A is moved to a memory location specified as source B. The move instruction moves a copy of one word of data. The term "move" is a bit deceiving as it implies that data is physically moved from one location to another, which would mean that the original data would not remain in the original location. However, this is not the case with the move instruction. This instruction moves a copy of the data in the original location, source A, to the destination. It is important to understand that only a copy of source A's data is moved to the destination and the original data still remains in source A. A copy of the original data also is stored in the specified destination.

Source A: Source A contains the address of the data file, a copy of which is to be moved. Source A can also contain a constant value, of which a copy will be moved to the destination.

Destination: The destination is the address where the move instruction sends a copy of the information specified in source A.

Figure 17-10 illustrates a ladder rung containing the move instruction.

Figure 17-10   A move instruction, when true, will move a copy of the information stored in the specified source to the specified destination.

When input B3:0/5 is true, the one-shot instruction will make the move instruction execute one time. The move instruction will move a copy of the data contained in source N7:1 to the destination, integer file N12:15. Integer file N7:1 contains the integer seven. After execution of the move instruction, both integer file 7, word one, and integer file 12, word 15, will contain the integer seven.

Associated Status Bits.   The move instruction has status bits similar to math instructions. Move instruction status bits include: a carry bit, an overflow bit, a zero bit, and a sign bit.

Upon completion of the execution of math, logical, and masked move instructions, status bits are used to keep track of the result by alerting the processor if there is an overflow condition, whether the result is negative, positive, or zero, and whether it generated a carry or borrow. The four common status bits, sometimes called "arithmetic flags," are defined as follows:

Carry Bit
This bit is set by the processor when a carry condition is encountered as the result of executing an addition instruction. Similarly, the carry bit is set if a borrow is generated as a result of executing a subtraction instruction. If the destination resultant does not result in a carry or borrow, the carry bit will be reset.

Overflow Bit
The overflow bit is set by the processor whenever the result of a math instruction execution produces a destination result too large to fit in the destination. If the destination contains a value that does not overflow the destination, the bit will be reset.

Zero Bit
The processor sets the zero bit when the result of a math, logical, or move instruction execution produces a zero value in the destination. If the destination contains a value other than zero, the bit will be reset.

Sign Bit
If the result of a math, logical, or masked move instruction produces a negative value in the destination, the sign bit will be set. The sign bit will be reset if the result in the destination is a positive value.

Status Bit Addressing Format.   S:0/2 is an example of a status file bit address. Status bit addressing consists of the following components:

   S = status file identification
   : = element delimiter
   0 = element number
   / = bit delimiter
   2 = bit number

Status bits are found in status data file 2. See Appendix B for a look at the status file. The appendix will assist you in identifying the location of the status bits.

Move instruction status bits are as follows:

- The carry bit (S:0/0) is not affected by the move instruction and is always reset.
- The overflow bit (S:0/1) is not affected by the move instruction and is always reset.
- The zero bit (S:0/2) is set if the result is zero; otherwise, it is reset.
- The sign bit (S:0/3) is set if the result is negative; otherwise, it is reset.

Example Using the Move Instruction. Each position of a three-position selector switch is used to select the current length specification of a product to be cut for a manufacturing process (see the ladder program illustrated in Figure 17-11). The length parameter is stored as a constant in the source parameter of each move instruction. Switch position one equals 6 inches, position two equals 12 inches, and position three equals 24 inches. Contacts for each selector switch position are separate inputs to a PLC input module. One ladder rung is programmed for each switch position. When a particular rung is true, the appropriate move instruction on each rung sends a copy of the length information to N7:10 for use in the program.

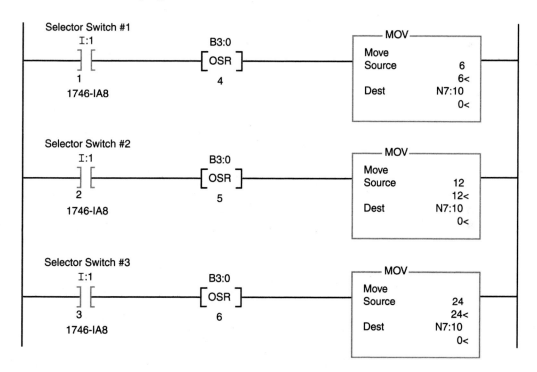

Figure 17-11 Selector switches used to select length specifications using a move instruction.

The Masked Move Instruction (MVM)

The masked move instruction is an output instruction that moves a copy of one data file word through a mask to a specified destination. A mask is used to filter out source bits that are not to be transferred to the destination.

Use this instruction to move a copy of one data file word from one memory location (called the source), through a mask, and on to a specified destination. When the MVM instruction is true, a copy of the data specified as source A is moved through a hexadecimal mask to the destination memory location specified as source B. The move instruction moves one word of data. Figure 17-12 illustrates a masked move instruction where data in

Figure 17-12   RSLogix 500 software masked move instruction.

source B3:12 is moved through a hexadecimal mask labeled 00FF. Refer to Appendix C for the mechanics of hex masking.

Ladder Explanation.   When input B3:0/5 is true, the one-shot instruction allows the move instruction to execute once. The masked move instruction will move a copy of the data contained in source B3:12 through the hexadecimal mask 00FF and store the filtered data into the destination B12:25, illustrated in Figure 17-13.

Source B3:12	0101 1011 1010 0111
Mask = 00FF	0000 0000 1111 1111
Destination B12:25	???? ???? ???? ????

Figure 17-13   Source data moved through a mask to the destination.

For this example, the hexadecimal mask is 00FF. A 00FF mask will allow only the lower eight bits (byte) to pass on through to the destination.

Figure 17-14 is a masked move instruction from RSLogix 5000 software for the ControlLogix PLC. This is an example of performing the same operation as Figure 17-12, except with a ControlLogix PLC. Since ControlLogix is a 32-bit PLC, data is worked with in blocks of 32 bits. As a result, the source, hexadecimal mask, and destination for the instruction will be 32 bits. Even though this instruction looks a bit different than Figure 17-12, it works the same. The source is a memory location called Bits[3].

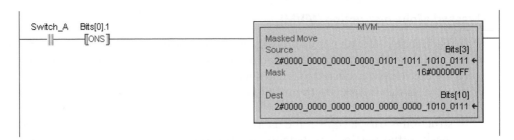

Figure 17-14   Masked move instruction from RSLogix 500 software for the ControlLogix PLC.

ControlLogix is a name-based PLC regarding identification of data and I/O points. These names are called tags. The name of this information is simply Bits[3]. The source contains 0000 0000 0000 0000 01011 1010 0111. The lower 16 bits are the same as the source bits in Figure 17-12. The mask is the same, with the exception that it is 32 bits wide and the upper word is all zeros. Notice the 2# designator before the source and destination and the 16# before the mask. ControlLogix uses this format to identify radix. This information is identified by 2# as binary, or radix two, while 16# signifies the mask as hex. The destination will hold the same information as Figure 17-12, with the exception that it will be 32 bits wide, too. See the table below for an overview of the instruction's execution.

Source tag is Bits[3]	0000 0000 0000 0000 0101 1011 1010 0111
Hexadecimal mask = 16#0000 00FF	0000 0000 0000 0000 0000 0000 1111 1111
Destination tag is Bits[10]	0000 0000 0000 0000 0000 0000 1010 0111

Mask Rules
1. The mask is a hexadecimal value.
2. Data is passed through the mask bit by bit. The mask bit in the same position as the source bit determines if the data is to pass or not. To pass data through the mask, set the appropriate bit (setting a bit means making it equal 1). To mask data from passing from the source to the destination, reset the appropriate bit (resetting a bit means making it equal 0).
3. Destination bits that correspond to zeros in the mask are not changed.
4. Mask bits can be either a constant or the address where that mask will be found.
5. Review Appendix C to refresh your memory on the mechanics of masks.

In another example, if the source data is 1010 1010 1010 1010 and the hexadecimal mask is 00FF (0000 0000 1111 1111), what will the destination contain?

Source:	1010 1010 1010 1010
Mask 00FF:	0000 0000 1111 1111
Destination:	0000 0000 1010 1010

Since the mask contained 0000 0000 1111 1111 (00FF), only the lower byte of the 16-bit word was allowed to pass through the mask from the source to the destination. (For more on masks, see chapter 18.)

Associated Status Bits.   The masked move instruction has status bits similar to other instructions. Masked move instruction status bits include a carry bit, an overflow bit, a zero bit, and a sign bit. These status bits are found in status file two. Move instruction status bits are listed below:

- The carry bit is not affected by the masked move instruction and is always reset.
- The overflow bit is not affected by the masked move instruction and is always reset.
- The zero bit is set if the result is zero; otherwise, it is reset.
- The sign bit is set if the result is negative; otherwise, it is reset.

Many machines use thumbwheels for numerical data input into the PLC. As an example, an operator can use a thumbwheel to dial in a mixing time in seconds for a particular manufacturing process. Data from the BCD thumbwheel will be input into the PLC and stored in the input status file in BCD format. This BCD data needs to be converted to integer format before the PLC can work with it.

The most common data conversion found in almost all PLCs is the instruction to convert to or from BCD data. Data coming in from BCD thumbwheels needs to be converted to integer data before the PLC can work with it. On the other hand, output data going to BCD formatted data displays needs to be converted from integer to BCD data before being output. The next section will look at BCD data conversion instructions.

Converting from BCD to an Integer (FRD)

The conversion from BCD (FRD) instruction is an output instruction that converts BCD values into integer data. Use this instruction if you want to convert BCD input data, such as thumbwheel data from the input status file data, into an integer with an FRD instruction. The resultant is stored in the selected destination. Figure 17-15 illustrates a possible application for an FRD instruction using an SLC 5/02 processor and above.

Figure 17-15   FRD instruction converting BCD input data for use in a timer preset value.

Figure 17-15 is a rung of ladder logic that converts BCD thumbwheel input data from address I:1/0, the source, and moves it into the preset value (T4:0.PRE) of timer file 4, timer zero, the destination.

Associated Data Bits.   The FRD instruction has status bits similar to other instructions. FRD status bits include a carry bit, an overflow bit, a zero bit, and a sign bit. These status bits are found in status file two. FRD instruction status bits are listed below:

- The carry bit is not affected by the FRD instruction and is always reset.
- The overflow bit is set at 1 if the value contained in the source is not a BCD value and at 2 if the value to be converted is greater than 32,767; otherwise, it is reset.
- The zero bit is set if the result is zero; otherwise, it is reset.
- The sign bit is not affected by the FRD instruction and is always reset.

Instruction Note
If you are using a fixed or SLC 500 modular 5/01 processor, the instruction source can only be the math register, S:13. The 5/02, 5/03, and 5/04 processors and the MicroLogix 1000 allow the source parameter to be a word in any data file, including the math register, S:13.

Converting an Integer to BCD (TOD)

Output data bound for a formatted BCD data display needs to be converted to BCD data before it is sent to the output module. The conversion from an integer to BCD (TOD) instruction is an output instruction that converts integer values into BCD data.

Use this instruction if you want to convert an integer value to BCD. The converted value is stored in a selected destination or sent to the output status file for output to an operator display. Figure 17-16 illustrates a possible application for a TOD instruction using an SLC 5/02 processor and above.

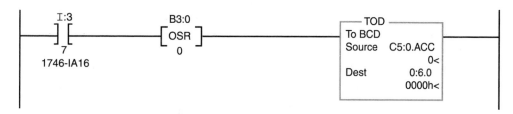

Figure 17-16 TOD instruction converting an integer value, counter file five, counter zero's accumulated value to BCD.

When input I:3/7 is true, the integer to BCD (TOD) instruction will take the source value, which is the accumulated value (C5:0.ACC) of counter file 5, counter zero, and convert it to BCD. After the conversion, the processor will send the converted data to the destination, O:6.0. Data sent to output O:6.0 is a 16-bit word in BCD format. The 16-bit word formatted for BCD will contain four BCD values. These BCD values will be sent out to a BCD operator display device.

Associated Data Bits.   The TOD instruction has status bits similar to other instructions. TOD instruction status bits include a carry bit, an overflow bit, a zero bit, and a sign bit. These status bits are found in status file two. The status bits are listed below:

- The carry bit is not affected by the TOD instruction and is always reset.
- The overflow bit is set at 1 if the result value is larger than 9999. An overflow results in a minor fault 0020, an error code represented in hexadecimal. If this fault occurs when running, the program will have to be corrected and re-downloaded. Otherwise, the overflow bit is reset.
- The zero bit is set if the result is zero; otherwise, it is reset.
- The sign bit is set if the source word is negative; otherwise, it is reset.

Instruction Note
If you are using a fixed or SLC 500 modular 5/01 processor, the instruction destination can only be the math register, S:13. The 5/02, 5/03, and 5/04 processors and the MicroLogix 1000 allow the destination parameter to be a word in any data file, including the math register, S:13.

When more than one word of data needs to copied from a source data file location to a working location, a copy instruction is used. As an example, in a beverage-mixing application, different amounts of each flavor will be mixed with water to arrive at the desired flavored drink. Integer data for flavor and water volumes and mixing times can be stored in integer file locations called recipes. To manufacture orange drink, the correct recipe can be copied to the integer file location used by the processor when mixing a batch of orange drink.

The Copy Instruction

Copy is an output instruction that copies a user-defined group of 16-bit data file words, called a "user-defined source file," to a destination file. Use this instruction to copy a pre-determined, consecutive group of words from one data file to another data file location. The difference between a move instruction and a copy instruction is that, while the move instruction moves a copy of one 16-bit data word to a new location, the copy instruction will copy up to 128 one-word elements or 42 three-word elements to another location.

Using the copy instruction, source and destination addresses may be different file types. As an example, timer, counter, or control file data may be copied to an integer file. Even though timers, counters, and control files each contain three words per element, their data can be copied to the integer file, which contains only one word per element. Copy instruction parameters are as follows:

Source
The source is the address of the file. The # symbol is used in front of the source address to specify that this is the beginning of a user-defined source file.

Destination
Destination is the starting address, from which the elements specified in the length parameter will be copied. A # symbol must be used in front of the destination address to specify that this is a user-defined file. Due to the fact that different file types may be copied between each other, the destination file type and the length parameter specify how many elements will be transferred.

Length
The length parameter is the number of elements that are to be copied. The designation file type determines the number of words that will be copied from the source. Figure 17-17 illustrates different file copy examples and the resulting number of words or elements transferred.

Analysis of Figure 17-17
The first example in Figure 17-17 is a copy between two integer files. The destination file type determines the number of words that will be copied. Since the destination file is formatted as 1 word per element, 10 words will be copied. An integer file element is 1 word, so 10 words or elements are copied.

If the Source Is	And the Destination Is	And the Length Is	What Will Be Transferred?
#N7:2	#N7:200	10	10 integer file elements or 10 words
#N7:0	#T4:12	6	The equivalent of 6 timer elements or 18 integer words or elements
#T4:0	#C5:12	6	Copies 6 timer elements or 18 words
#C5:2	#N7:0	6	Copies 6 words or 2 counter elements
#B3:0	#B3:150	75	The equivalent of 75 bit file elements or 75 words

Figure 17-17   Elements transferred for mixed file copy examples.

Example two in Figure 17-17 contains dissimilar files. The source file type is an integer file. Integer files have 1-word elements. A timer file is the destination file. Timer files have 3-word elements. The length is 6, and the destination file type determines how many words will be copied. A timer element is 3 words each multiplied by a length parameter of 6 (equals a total of 18 words). Eighteen words or integer file elements (1 word each) will be copied into 6 timer elements. Thus, 6 timer elements amount to 18 words.

Example three in Figure 17-17 also contains two dissimilar file types. The source file is a timer file and the destination file is a counter file. Each file type has 3-word elements. With a length parameter of 6, 6 timer elements will be copied to a destination of 6 counters. A total of 18 words will be copied.

Example four in the figure contains a counter file as a source and an integer file as the destination. Counter elements are 3 words, while integer file elements are 1 word each. With a length of 6 and the destination file type dictating the number of words to be copied, 6 words will be copied. If 6 words are copied from the source, which is a counter file, 2 counter elements will be copied from counter file C5:2 and C5:3. Since counter elements are 3 words each, C5:2 words will be copied to N7:0 through N7:2 (3 words), and C5:3 words will be copied to N7:3 through N7:5. Figure 17-18 illustrates a new copy instruction ladder rung.

Figure 17-18   Ladder rung containing copy instruction to copy 25 words, or elements, from one integer file to another.

As long as input B3:5 is true, 25 words will be copied from user-defined file N7:2 to user-defined file N12:0. A one-shot instruction could be used if the copy instruction is to be executed only once.

Within file N7, the programmer has defined a group of words 25 words long as a file. Figure 17-19 illustrates the user-defined files within files N7 and N12 and graphically illustrates how the data is copied from one to the other. The file within file N7 is called a user-defined file because the programmer has grouped together a number of words (25 in this instance) to be used together as a storage area for like data.

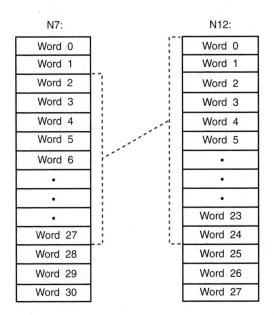

Figure 17-19 Words two through twenty-seven from file N7 copies to file N12, words zero through twenty-four.

Associated Status Bits. There are no status bits associated with the copy instruction.

Using the Copy Instruction. One popular use of the copy instruction is for setting up recipe data for different products just before they are manufactured. Let's assume we manufacture four different types of soft drinks: fruit punch, tropical punch, citrus punch, and orange. Each product has a different recipe and each recipe contains data regarding the amount of each ingredient that is needed to produce the product. Figure 17-20 lists the ingredients, in gallons, in each drink recipe.

Each list of ingredients for each drink product is its recipe. Since each amount is an integer value, we will store these values in integer file N12. We have arbitrarily selected and created integer file N12 for two reasons. First, we want to keep all the recipes together. Second, we want to keep the user-defined recipe files separate from the standard or working integer file, N7. This was done only to help avoid confusion and accidents of data being written over when mixing single integer storage with multiple integer storage. We will reserve 10 words to store data for each recipe.

Ingredients	Fruit Punch	Tropical Punch	Citrus Punch	Orange
Water	100	120	130	150
Sweetener	25	25	25	20
Grape Flavor	8	9	0	0
Orange Flavor	0	25	75	90
Pineapple Flavor	10	25	30	0
Apple Flavor	2	8	0	0
Pear Flavor	1	0	1	0
Strawberry Flavor	8	0	0	0
Passion Fruit Flavor	0	10	0	0

Figure 17-20   Recipes for our drink manufacturing process.

The operator will select which recipe will be manufactured by positioning a four-position selector switch. Figure 17-21 contains data regarding our input address allocation and data file allocation. Starting at N12:0, we will assign integer file addresses as follows:

Drink	Selector Switch Input Address	Recipe Storage Addresses	Total Words Reserved for Recipe
Fruit Punch	I:3/4	N12:0–N12:9	10
Tropical Punch	I:3/5	N12:10–N12:19	10
Citrus Punch	I:3/6	N12:20–N12:29	10
Orange	I:3/7	N12:30–N12:39	10

Figure 17-21   Data allocation for our drink manufacturing process.

Figure 17-22 is an example of the selection portion of our drink-manufacturing PLC ladder program. Notice that each selector switch position energizes a copy instruction. Each copy instruction copies the correct recipe to the working data files (N7:50 in our program).

Selector Switch Operation.   In many instances there is misunderstanding about how a selector switch operates in conjunction with the physical switch position and how the input signal gets to the PLC input module. Our application has four selections, one selection for each product to be manufactured. Referring back to Figure 17-22 and our PLC ladder diagram, we see that there are four input signals, one for each product. This means that each selector switch position has a separate circuit wired to the PLC input module. On the back of the selector switch you will find four contact blocks, one for each switch

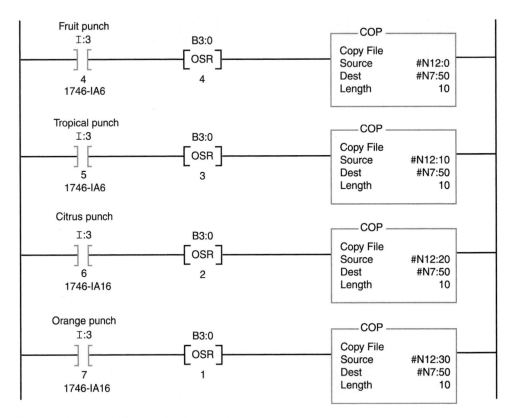

Figure 17-22 Drink flavor selection portion of our program.

position. Each contact block has one circuit. The proper contact block is to close only when its associated selector switch position is selected. All other positions must leave the contact circuit in an open condition. A closed signal must only be present when the switch is in position for selecting that specific product. A selector switch contact block "target table" is used to map out the behavior of each contact block circuit. Figure 17-23 illustrates the target table for our selector switch application. The left-most column lists the PLC input addresses. Each product is listed across the top of the table with its associated switch position. Notice an X placed in the position where the circuit is to be closed. The closed circuit will input an on signal to the PLC input module when the appropriate switch position is selected. Looking at the target table, only one input circuit can be energized at any one time.

Each input circuit from the selector switch is wired to its corresponding input module address. Each input address is reflected on the ladder rungs in Figure 17-22. Recipe data will be copied as determined by which selector switch is inputting an on signal. After the selector switch selects which recipe data is to be copied, the recipe is copied to a common working register, #N7:50. The OSR instruction is used to ensure that the copy

	SELECTOR SWITCH POSITIONS			
Input Address	Position 1, Fruit Punch	Position 2, Tropical Punch	Position 3, Citrus Punch	Position 4, Orange
I:3/4	X	0	0	0
I:3/5	0	X	0	0
I:3/6	0	0	X	0
I:3/7	0	0	0	X

Figure 17-23   Selector switch circuit correlation to drink selection process.

instruction is executed only once as the result of the selector switch going from false to true. If making fruit punch for 12 hours, the recipe needs to be copied only one time, not at every scan.

LOGICAL INSTRUCTIONS

Logical instructions are used to perform logical operations such as AND, OR, exclusive-OR, and NOT logic on two 16-bit data file words or one data file word and a constant. When programming logical instructions, the two words to be operated on are each called a "source." Since we are working with two words, we have two sources, A and B. The result of the operation is placed in the destination. One source and the destination need to be data file word addresses. The second source is usually a constant; however, it, too, can be a data file word address.

Logical instructions and their operations are listed in Figure 17-24.

LOGICAL INSTRUCTIONS		
Instruction	Use This Instruction to	Functional Description
AND	AND the bits in one word with the bits in another word.	When true, the word specified as source A is ANDed bit by bit with the word specified as source B. The result is stored in the user-specified destination.
OR	OR the bits in one word with the bits in another word.	When true, the word specified as source A is ORed bit by bit with the word specified as source B. The result is stored in the user-specified destination.
Exclusive-OR	Exclusive-OR the bits in one word with the bits in another word.	When true, the word specified as source A is exclusive-ORed bit by bit with the word specified as source B. The result is stored in the user-specified destination.
NOT	NOT the bits in one word with the bits in another word.	When true, the word specified as source A is complemented bit by bit with the word specified as source B. The result is stored in the user-specified destination.

Figure 17-24   Overview of logical instructions.

The AND Instruction

The AND instruction performs a logical AND operation on two 16-bit words. Use this instruction if you want to logically AND a 16-bit word in a data memory file with either a constant or another 16-bit word in a data file. Single-bit AND logic is illustrated in Figure 17-25.

Source A	Source B	Destination
0	0	0
0	1	0
1	0	0
1	1	1

Figure 17-25   AND truth table.

Figure 17-26 illustrates ANDing the contents of input word I:1.0 with the hexadecimal constant 006D.

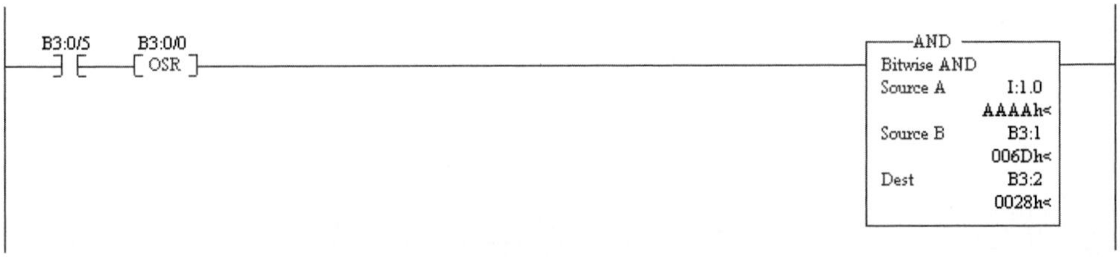

Figure 17-26   RSLogix 500 software AND instruction.

When bit address B3:0/5 is true, the one-shot instruction will cause the AND instruction to execute one time. The AND instruction will take the contents of source A, input I:1 (all 16 bits as a group), and then AND them with the contents of bit file word B3:1 bit by bit. The result will be stored in B3:2. A sample operation is illustrated in Figure 17-27.

Source A:	I:1	10101010	10101010
Source B:	B3:1	0000 0000	0110 1101
Destination:	B3:2	0000 0000	0010 1000

Figure 17-27   ANDing source A and source B and the resulting 16-bit word.

Various status bit conditions are related to the execution of the AND instruction (see Figure 17-28).

Address	Description	Explanation	Status of Bit
S:0/0	Carry Bit	Result of instruction execution is a carry generated in the destination.	Not used
S:0/1	Overflow Bit	Result of instruction execution is an overflow detected in the destination.	Not used
S:0/2	Zero Bit	If result of instruction execution in the destination equals 0.	Set
S:0/3	Negative Bit	If result of instruction execution in the destination's MSB equals 1.	Set

Figure 17-28   Status bit conditions from execution of the AND instruction.

Using Status Bits.   Only the zero and negative status bits are pertinent to AND instruction execution. If an operator needs to know when the destination result is either a negative value or a zero value, the status bit can be programmed in the ladder program. The S:0/2 bit can be used as an input instruction controlling an output bit instruction. As an example, Figure 17-29 illustrates a ladder rung using the zero status bit to turn on OTE instruction O:2/5.

Figure 17-29   Zero status bit used as an input controlling an OTE instruction.

Example of ANDing Instruction Usage.   Assume that your application required that all 16 inputs on a particular input module must be in specific on and off conditions before an output could be energized. The 16 inputs could represent 16 interlocks that must be made before the process can proceed. There are a couple of ways to accomplish this. First, you could program a ladder rung with all 16 input contacts as inputs controlling the desired output. Alternately, you could use the AND instruction to test all 16 inputs with one instruction.

The OR Instruction

The OR instruction performs a logical OR operation on two 16-bit words. Use this instruction if you want to logically OR a 16-bit word in a data memory file with either a constant or another 16-bit word in another data memory file. Single-bit OR logic is illustrated in Figure 17-30.

Source A	Source B	Destination
0	0	0
0	1	1
1	0	1
1	1	1

Figure 17-30 OR truth table.

Figure 17-31 illustrates an OR instruction. Allen-Bradley SLC 500 and the Micro-Logix 1000 formally call this a Bitwise Inclusive-OR instruction. Simply put, source A and source B bits are ORed bit by bit.

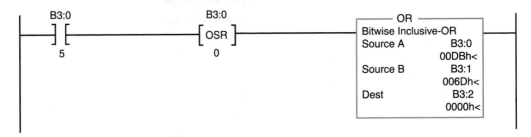

Figure 17-31 OR instruction on a ladder rung.

When bit address B3:0/5 is true, the one-shot instruction will cause the OR instruction to execute one time. The OR instruction will take the contents of source A, bit file B3:0 (all 16 bits as a group), and then OR them with the contents of bit file word B3:1 bit by bit. The result will be stored in B3:2. The operation is illustrated in Figure 17-32.

	HEX	Binary
Source A: B3:0	00DBh	0000 0000 1101 1011
Source B: B3:1	006Dh	0000 0000 0110 1101
Destination: B3:2	00FFh	0000 0000 1111 1111

Figure 17-32 ORing source A and source B and the resulting 16-bit word.

As the result of being executed, the OR instruction is stored in the designated destination. There are status bits that will provide the processor with important information regarding the condition of the resultant.

Associated Status Bits. The OR instruction execution updates the arithmetic status bits in the same manner as the AND instruction. Refer to Figure 17-28 for status bit behavior after executing an OR instruction.

The Exclusive-OR Instruction (XOR)

The exclusive-OR instruction is an output instruction that performs an exclusive-OR logical operation on two 16-bit words. Use this instruction if you want to perform exclusive-OR logic on one 16-bit word in data memory with either a constant or another 16-bit word in another data memory file. Single-bit exclusive-OR logic is illustrated in Figure 17-33.

Source A	Source B	Destination
0	0	0
0	1	1
1	0	1
1	1	0

Figure 17-33   Exclusive-OR truth table.

The difference between OR logic and exclusive-OR logic is as follows: OR logic is true when any or all input conditions are true. Two-input exclusive-OR logic, as an example, is true only when one or the other input condition is true. When both input conditions are true, two-input exclusive-OR logic is false. Figure 17-34 illustrates a rung of exclusive-OR logic.

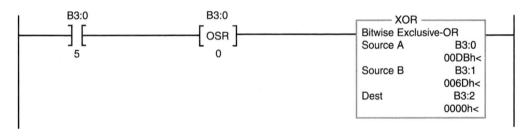

Figure 17-34   Exclusive-OR instruction on a ladder rung.

When bit address B3:0/5 is true, the one-shot instruction will cause the XOR instruction to execute one time. The XOR instruction will take the contents of source A, bit file B3:0 (all 16 bits as a group), and then exclusive-OR them with the contents of bit file word B3:1 bit by bit. The result will be stored in B3:2. The operation is illustrated in Figure 17-35.

		HEX	Binary
Source A:	B3:0	00DBh	0000 0000 1101 1011
Source B:	B3:1	006Dh	0000 0000 0110 1101
Destination:	B3:2	00FFh	0000 0000 1011 0110

Figure 17-35   Exclusive ORing source A and source B and the resulting 16-bit word.

As the result of being executed, the XOR instruction is stored in the designated destination. Status bits will provide the processor with important information regarding the condition of the resultant.

Associated Status Bits. The XOR instruction execution updates the arithmetic status bits in the same manner as the AND instruction and the OR instruction. Refer to Figure 17-28 for status bit behavior after executing an XOR instruction.

The NOT Instruction

The NOT instruction is an output instruction in which a logical NOT is performed bit by bit. Use this instruction if you want to invert the bit values of a 16-bit word address. The NOT instruction requires only one source word address. When rung input conditions are true, the source word is inverted, or "NOTed." The result is placed in the destination. Other terms associated with NOT logic are "complementing a 16-bit word" and "ones complementing a binary number." In the binary system, the complement of a 1 is a 0 and the complement of a 0 is a 1. Figure 17-36 illustrates a NOT truth table.

Source	Destination
0	1
1	0

Figure 17-36   Truth table for NOT logic.

Figure 17-37 is a PLC rung containing NOT logic.

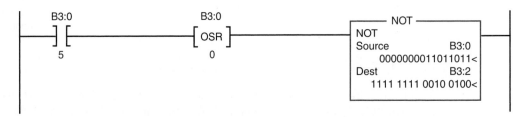

Figure 17-37   NOT logic as an output instruction on a PLC ladder rung. After execution, destination will contain 1111 1111 0010 0100.

Ladder Explanation.   When bit address B3:0/5 is true, the one-shot instruction will cause the NOT instruction to execute one time. The NOT instruction will take the contents of source bit file B3:0 (all 16 bits as a group) and then complement them bit by bit. The result will be stored in B3:2. The operation is illustrated in Figure 17-38.

As the result of being executed, the NOT instruction is stored in the designated destination. Status bits will provide the processor with important information regarding the condition of the resultant.

	Binary
Source A: B3:0	0000 0000 1101 1011
Destination: B3:2	1111 1111 0010 0100

Figure 17-38   NOTing, or complementing source A, and the resulting 16-bit word stored in destination B3:2.

Associated Status Bits.   The NOT instruction execution updates the arithmetic status bits in the same manner as the AND, OR, and exclusive-OR instructions.

The Limit Test Instruction

The limit test (LIM) instruction is available for the SLC 500 family of modular processors starting with the 5/02 processor and the MicroLogix family. The limit test instruction is used to test for values whether they are within or outside of the specified range. Programming the LIM instruction consists of entering three parameters: low limit, test, and high limit. The test parameter provides the information the high and low limits will be evaluated against. Instruction parameters can be programmed with the data types shown in Figure 17-39.

LIM Parameters	Parameter Data Formats	
Low Limit	Word address	Constant or word address
Test	Constant	Word address
High Limit	Word address	Constant or word address

Figure 17-39   Limit test instruction parameter data formats.

Figure 17-40 illustrates a possible application for an LIM instruction. Here the instruction has a low limit of 0 while the high limit is 1750. This illustrates an application where an operator is allowed to enter motor speed into an operator interface terminal. The motor speed reference is sent to an SLC 500 processor via a network connection. The SLC processor will use the limit test instruction to verify the operator input motor speed is within acceptable range. If the test input value from the operator interface is within the

Figure 17-40   The limit test instruction testing for values between 0 and 1750.

acceptable range, the instruction will be true and the new speed reference will be passed on to the variable-frequency drive. If the speed reference input is outside of the 0 to 1750 rpm range, the instruction will be false and no speed reference will be passed on to the drive. As a result of the instruction being false, the drive will run at the last commanded speed. This instruction can be used, as in this example, to limit information input into the PLC from an operator.

On the other hand, by programming the LIM instruction differently we can look at a temperature value and determine if it is outside a desired range. Figure 17-41 illustrates an LIM instruction where the low limit is 100 and the high limit is 0. In this case, if the temperature is anything outside the range of 0 to 100 degrees, the instruction will be true. A temperature value within the 0 to 100 degree range will make the instruction false.

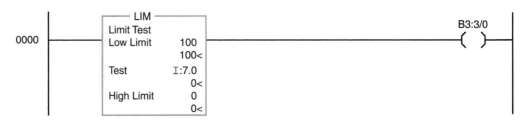

Figure 17-41   The limit test instruction testing for values outside the range of 0 to 100.

The Scale with Parameters Instruction

The scale with parameters (SCP) instruction is included in the instruction set for the SLC 500 modular processors with operating systems OS302, OS401, and OS500. The SCP instruction is also in the MicroLogix 1200 and 1500 controllers instruction set. The SCP instruction is used when:

1. Analog input data needs to be scaled (converted) before it can be used in the PLC program.
2. Input data from a field device needs to be scaled (converted) before it can be used in the PLC program.
3. Analog output data needs to be scaled (converted) before it can be sent from an analog module channel to field devices.
4. Output data needs to be scaled (converted) before it is sent to an operator interface device or variable-frequency drive.

Not all devices format data in the same manner. As a result, the data must be scaled, or converted, before they are worked within the PLC or output to a field device in an understandable format. As an example, a variable-frequency drive may represent motor speed as 0 to 32767 equaling 0 to 1750 rpm, or 0 to 60 Hz representing 0 to 1750 rpm. An operator would have problems entering 0 to 32767 as input data to an operator interface device representing the desired motor speed range within 0 to 1750 rpm. Or an operator would have difficulty viewing a value between 0 to 32767 on an operator interface terminal as motor speed feedback of 0 to 1750 rpm from a variable frequency drive. Figure 17-42 is a block diagram illustrating data flow from operator input (as 0 to 1750 rpm) into a PLC input status table, which is scaled and output to the variable-frequency drive.

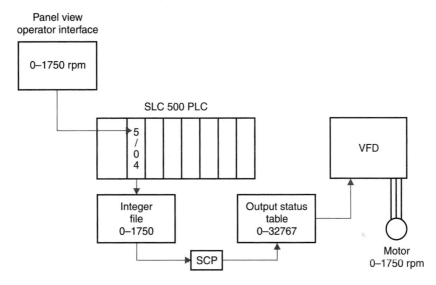

Figure 17-42   Data flow from a panel view operator interface terminal into the PLC and out to a VFD.

Motor speed reference data could be input from a variable-frequency drive as 0 to 32767 to represent 0 to 1750 rpm motor speed scaled and output to a Panel view operator interface terminal to display motor speed for the operator. Figure 17-43 is a block diagram illustrating data flow from a variable-frequency drive that is input into the SLC 500 input status table. The SCP instruction scales the data from 0 to 32767 to 0 to 1750 rpm and then outputs it to the Panel view operator interface terminal.

As another example, analog input data can be input from or output to a field device via the analog input module channel as represented in Figure 17-44. This input or output data must be scaled so the data is correctly formatted for the device using it. As an example, Figure 17-45 illustrates a possible application for an SCP instruction. The instruction has an input minimum of 0 while the input maximum is 1750. Programming this instruction in this manner can illustrate an application where an operator is allowed to enter motor speed sent to a variable-frequency drive from an operator interface device (similar to Figure 17-42), such as a Panel view terminal connected to an SLC 500 processor. The processor output parameter is the address that will send the speed reference out to a variable-frequency drive to control motor speed.

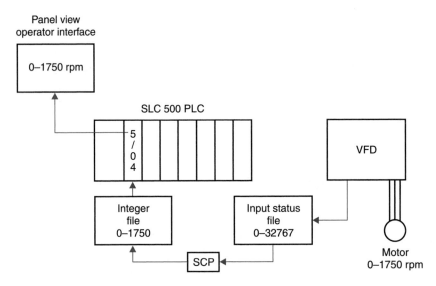

Figure 17-43   Data flow from a VFD into the PLC and out to a panel view operator interface terminal.

Analog Input or Output Levels	PLC Input/Output Values Represented As
+/−10 volts	−10250 to +10250
0 to 10 volts	−500 to +10250
1 to +10 volts	0 to 32767
0 to 5 volts	−500 to +5500
1 to 5 volts	+500 to +5500
0 to 20 mA	0 to +20500
4 to 20 mA	+3500 to +20500

Figure 17-44   Sample of possible analog input or output signal levels and their corresponding data ranges.

Instruction Parameters.   The instruction parameters shown in Figure 17-45 on the next page are identified in the following list:

   Input: This is the input value to be scaled. It can be an address, integer, or floating point address.

   Input Min.: The input minimum is the minimum input value. For our example, this value is 0.

   Input Max.: Input maximum is the maximum input value. For our example, this value is 1750.

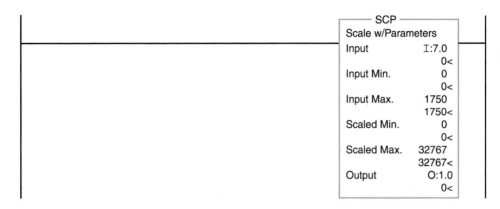

Figure 17-45 The scale with parameters instruction scaling the input value of 0 to 1750 to 0 to 32767.

Scaled Min.: This represents the input scaled minimum value. In our example, the minimum scaled value is 0.

Scaled Max.: Scaled maximum is the scaled or converted maximum value. This value is 32767 in our application example.

Output: Enter the address for the scaled input data to be output after instruction execution. This can be an address, integer or floating-point address. If any parameter is programmed as floating-point, then the entire instruction is treated as floating-point.

SUMMARY

Today's PLCs have vast instruction sets that include instructions that allow the PLC to store vast amounts of information pertaining to the manufacturing process. As an example, stored data could include numerous recipes, one for each different product produced. The flavored drink example illustrates data recipe storage.

Data-handling instructions allow stored data to be moved or copied from processor memory into the currently executing PLC ladder program. Input or output data can be compared for different levels of equality, logical functions can be performed, and input or output data can be converted into different data formats.

This chapter introduced selected data-handling instructions available specifically in the SLC 500 and MicroLogix 1000. Even though this chapter was specific to some brands of PLC, most PLCs have similar instructions that can be used to accomplish the manipulation and conversion of data.

Data-handling instructions can be grouped into the following types:

- an instruction to move (copy) a single word
- an instruction to copy groups of words

- an instruction to convert data from BCD to integer
- an instruction to convert data from integer to BCD
- instructions to test for equality
- instructions that will perform logical operations
- an instruction that will move data through a mask

REVIEW QUESTIONS

1. The _____ instruction is an input instruction used to test when two values are equal.
2. When executing the equal instruction, when source A and source B are equal, the instruction is _____; otherwise, the instruction is false.
3. When programming the equal instruction, if you use a constant as source B, the constant will be tested for equality with the data residing in the address specified in source _____.
4. The equal instruction is an _____ instruction.
5. The _____ instruction is used to test two values for inequality.
6. The _____ instruction is true when the data stored in the address specified as source A is not equal to either the data stored in the address specified as source B or a constant entered when entering this instruction.
7. The _____ instruction is used when you want to test if one value is less than another value.
8. The parameters listed below are for which instruction—Source A, the address of the data to test to see if it is less than source B; or source B, which can be either a constant or a word address?
   A. equal
   B. not equal
   C. greater than
   D. less than
   E. all of the above
9. If you use an address as source B, this instruction will cause the data contained in that address to be tested to see if it is greater than the data residing in the address specified in source A. What instruction is used to accomplish this?
10. The _____ instruction is true when the data stored in the address specified as source A is less than or equal to either the data stored in the address specified as source B or a constant entered when entering this instruction.
11. If you use a constant as source B, this instruction will cause the constant to be tested to see if it is equal to or greater than the data residing in the address specified in source A. What instruction did you program?

12. The greater than instruction, _____, is used to determine if one source of data is greater than another.

13. The LES instruction is an input instruction that tests to see if one source of data is _____ than another source of data.

14. The greater than or equal instruction's programming abbreviation is _____.

15. _____ to instructions determine if one source of data is greater than or equal to another.

16. Suppose the greater than or equal instruction has the integer 19 stored in source A and the integer 23 stored in source B. Is this instruction true or false?

17. Data-handling instructions are used when data needs to be _____ or _____ from one data file source to a desired destination.

18. The move instruction:
    A. moves a copy of data from source A to source B
    B. moves the data in source A into source B
    C. moves a copy of data from source A to source A
    D. moves a copy of data from source B to source A
    E. moves the data in source B into source A
    F. any of the above depending on how the instruction is programmed

19. When the move instruction is true, _____ of information specified as source A is moved to a memory location specified as source B.

20. The _____ instruction moves one word of data.

21. Explain what is contained in source A and the destination after programming a move instruction.

22. The _____ instruction is an output instruction that moves a copy of one data file word to a specified destination. A mask is used to filter out bits that are not to be transferred from the source to the destination.

23. Use the _____ instruction to copy the contents of many consecutive words in one file to another location.

24. Complete the rules of masking:
    A. The mask is a _____ value.
    B. Data is passed through the mask _____. The mask bit in the same position as the source bit determines if the data is to pass or not. To _____ data through the mask involves setting the appropriate bit. Setting a bit means making the bit equal to _____. To _____ data from passing from the source to the destination, you reset the appropriate bit. _____ a bit means to make the bit a zero.
    C. Destination bits that correspond to _____ in the mask are not changed. They remain as they were.
    D. Mask bits can be either a _____ or the _____ where that mask will be found.

25. A mask of _____ will allow all data to pass on through to the destination.

26. The mask 00FF will allow which bits to pass?
    A. upper nibble
    B. lower nibble
    C. upper 8-bit word
    D. lower 8-bit word
    E. bits 1 through 8
    F. bits 8 through 15

27. The mask 000F will allow which bits to pass?
    A. upper nibble
    B. lower nibble
    C. upper 8-bit word
    D. lower 8-bit word
    E. bits 1 through 8
    F. bits 8 through 15

28. The mask 0FFF will allow which bits to pass?
    A. upper nibble
    B. lower nibble
    C. upper 8-bit word
    D. bits 4 through 11
    E. bits 0 through 11
    F. bits 1 through 11

29. When _____ instruction is true, a copy of information specified as source A is moved through a hexadecimal mask to the memory location specified as source B.

30. The conversion from BCD (FRD) instruction is an output instruction that converts _____ values into _____ data.

31. The conversion from an integer to BCD (TOD) instruction is an output instruction that converts _____ values into _____ data.

32. Copy is an output instruction that copies a user-defined group of 16-bit data file words called a _____ to a destination file.

33. The copy instruction will copy up to _____ one-word elements or _____ three-word elements to another location.

34. When using the copy instruction, source and destination addresses may be _____ file types.

35. The copy file destination parameter is the starting address where the elements specified in the _____ parameter will be copied.

36. When using the copy instruction, different file types may be copied between a source and a destination. How is the amount of information to be transferred determined?
    A. The destination file type specifies how many bits will be transferred.
    B. The destination file type specifies how many nibbles will be transferred.
    C. The destination file type specifies how many bytes will be transferred.
    D. The destination file type specifies how many elements will be transferred.
    E. The destination file type specifies how many files will be copied.

37. Explain how status bits are affected with the execution of the copy instruction.

38. The source file type is an integer file. Integer files have one-word elements. A timer file is the destination file. Timer files have _____ word elements. The copy instruction's length parameter is six. How many words will be copied? _____

39. Logical instructions are used to perform a logical operation on two 16-bit _____ words or one data file word and a _____.

40. One source of a logical instruction, along with the _____, needs to be a data file word address.

IF PASSWORD IS CORRECT GO TO DRIVE SET UP SCREEN

0022

```
----- EQU -----
Equal
Source A    N12:21
               7777<
Source B     32767
             32767<
```

B3:10/0
─[OSR]─

B3:1/15
─( )─
RESET PASSWORD
ATTEMPTS COUNTER
C5:5
─(RES)─

0023

```
----- GRT -----
Greater Than (A>B)
Source A    N12:21
               7777<
Source B        .0
                0<
```

```
----- NEQ -----
Not Equal
Source A    N12:21
               7777<
Source B     32767
             32767<
```

B3:10/1
─[OSR]─

```
----- CTU -----
Count Up
Counter        C5:5
Preset           3<
                 2<
```
─(CU)─
─(DN)─

```
               −1
              −1<
N12:21
    7777<
```

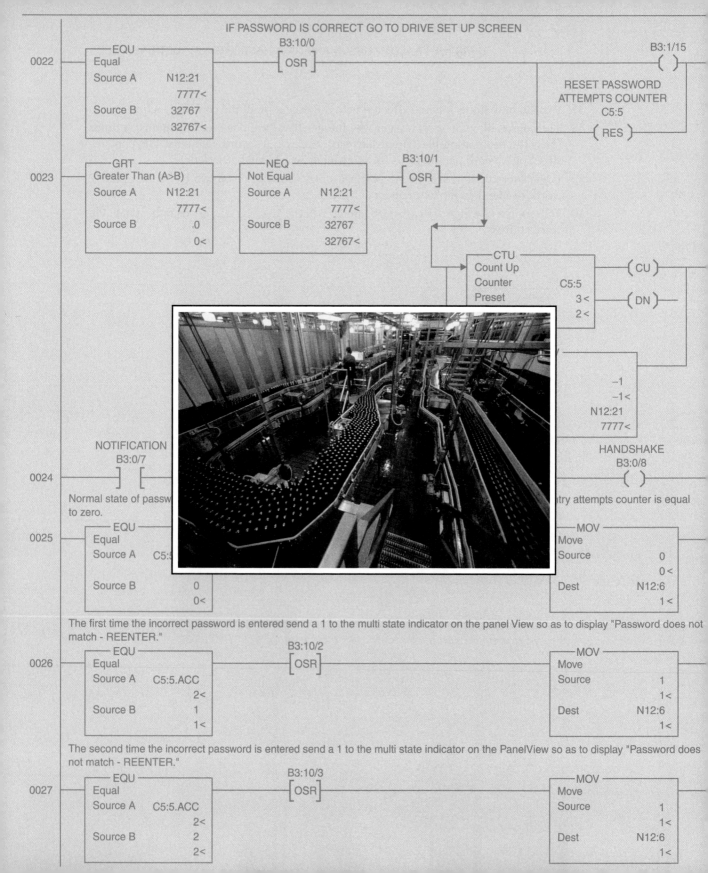

0024

NOTIFICATION
B3:0/7
─] [─

HANDSHAKE
B3:0/8
─( )─

Normal state of passw... ...try attempts counter is equal
to zero.

0025

```
----- EQU -----
Equal
Source A    C5:5...
Source B        0
                0<
```

```
----- MOV -----
Move
Source           0
                0<
Dest         N12:6
                1<
```

The first time the incorrect password is entered send a 1 to the multi state indicator on the panel View so as to display "Password does not
match - REENTER."

0026

```
----- EQU -----
Equal
Source A    C5:5.ACC
               2<
Source B        1
                1<
```

B3:10/2
─[OSR]─

```
----- MOV -----
Move
Source           1
                1<
Dest         N12:6
                1<
```

The second time the incorrect password is entered send a 1 to the multi state indicator on the PanelView so as to display "Password does not
match - REENTER."

0027

```
----- EQU -----
Equal
Source A    C5:5.ACC
               2<
Source B        2
                2<
```

B3:10/3
─[OSR]─

```
----- MOV -----
Move
Source           1
                1<
Dest         N12:6
                1<
```

CHAPTER

18

Sequencer Instructions

OBJECTIVES

After completing this chapter, you should be able to:

- describe the function of a sequencer
- describe the operation of the SLC 500 sequencer output (SQO) instruction.
- explain the function of each SQO status bit
- explain the role of the control (R6) file
- explain the role of the bit file while programming a sequencer instruction
- explain what a mask is and why it is used
- determine the appropriate hexadecimal mask for your application

INTRODUCTION

Early applications involving sequencing required the use of mechanical rotating-cam switches. Rotating-cam switches were used as control circuit devices on machinery with repetitive or sequential operating cycles. These cam switches were operated by manually adjustable cams mounted on a rotating shaft. A set of electrical contacts for each cam would close at the proper shaft position in relation to shaft rotation. The sequence of contact closures was determined by positioning the proper cams to open or close their contacts at the desired position between 0 and 360 degrees on a rotating shaft. Cams were selected by the degrees of rotation of the shaft and the cam's degrees of open contacts versus closed contacts.

Modern microprocessor technology has, in many instances, replaced the mechanical sequencer. As the microprocessor evolved and gained more functionality, sequencers were incorporated into the PLC instruction set. For example, the SLC 500 can have up to 256 sequencers per control data file, each with up to 256 steps, or sequences.

Programming an SLC 500 sequencer requires a number of steps, all of which need to be completed accurately for the sequencer to operate as desired. A knowledge of binary and hexadecimal numbers and of the SLC 500 data file structure, including understanding and creating a bit file, is necessary to program a sequencer successfully.

This chapter will introduce you to the Allen-Bradley sequencer output instruction.

THE SEQUENCER

The sequencer instruction has become one of the workhorse features of the PLC. The sequencer simply controls a predetermined sequence of events such as the control of a pallet stretch-wrap machine. Each step of the pallet-wrapping routine is controlled by preprogrammed sequencer steps entered into your PLC user program. Each step of the sequence performs a predetermined task such as:

1. determining if adequate stretch wrap material is available
2. sensing presence of a pallet to be wrapped
3. movement of the pallet to the proper position for wrapping
4. performing the wrapping process
5. The PLC is programmed to remember stretch-wrap patterns. Plastic wrap is staggered to improve transportability of the loaded pallet. Each staggered wrapping pass is a sequence in itself.
6. When the wrapping process is completed, the pallet is sequenced to exit the conveyor section to be loaded into a truck or possibly an automatic guided vehicle for transport to an indexed storage location.

SEQUENCER INSTRUCTIONS

Sequencer instructions are available on most PLCs. Even though the basic operating principle of the sequencer is the same regardless of manufacturer, each manufacturer will have its own specific programming procedures. The following sequencer instructions are specifically for the Allen-Bradley SLC 500; MicroLogix 1000, 1200, 1500; and the PLC 5 (refer to Figure 18-1).

SEQUENCER INSTRUCTIONS		
Instruction Use	**Use This Instruction to**	**Functional Description**
Sequencer Output	Control machine sequence	Each 16-bit word represents 16 outputs on an output module. Outputs are controlled by sequencing from one word to the next.
Sequencer Compare	Monitor inputs	Compare 16-bit internal data to a 16-bit input module's input points.
Sequencer Load	Load data into a data file sequentially	Load data into a data file from each step of a sequencer operation. Data can be from the input status file or another data file.

Figure 18-1 Sequencer instructions available for the Allen-Bradley SLC 500 family of PLCs.

The sequencer compare instruction for the SLCs is similar to the sequencer input instruction for a PLC 5.

THE SLC 500 SEQUENCER OUTPUT INSTRUCTION

The sequencer output (SQO) instruction is programmed as an output on a ladder rung. Each false-to-true rung transition causes the sequencer output instruction to increment to the next sequence step. Each step of the sequence is stored in a sequencer file, which is typically a user-defined bit file whose length represents the number of sequence steps. Stepping though the sequencer bit file one word at a time, the sequencer output instruction uses each bit file word to represent output status bits for the current sequencer step.

One bit in each sequencer file word is associated with the specified output module's output screw terminals. Upon execution of the instruction, the program will step through the predetermined sequencer file to control various output devices.

Elements of Sequencer Instruction Programming

Input logic is programmed first. Figure 18-2 illustrates an XIC instruction addressed as I:1/2 as input logic. Notice that below the input instruction is the module identification associated with that input address. Figure 18-2 illustrates the sequencer instruction.

Figure 18-2   SLC 500 PLC family sequencer output instruction on a ladder rung.

Even though this particular rung only has one input instruction, any input logic needed for your specific application is acceptable. This rung of logic was printed from Rockwell Software's RSLogix 500 software.

The sequencer output instruction consists of the following parameters. The file parameter is the address of the sequencer file. The sequencer file is typically a bit file and contains one bit file word representing the output action required for each step of the sequence.

To illustrate the purpose and operation of the sequencer file, let's look at a sample application. Assume we have a four-step process. Four outputs are to be energized on one 16-point output module for each of the four steps. Because of the way in which the sequencer instruction operates, all output points must be on a single output module. The sequence is as follows:

Step one: Outputs 0, 1, 2, and 3 will be energized.

Step two: Outputs 4, 5, 6, and 7 will be energized.

Step three: Outputs 8, 9, 10, and 11 will be energized.

Step four: Outputs 12, 13, 14, and 15 will be energized.

Figure 18-3 illustrates the bit file where we will store the output status for each of the four steps.

15	14	13	12	11	10	9	8	7	6	5	4	3	2	1	0	
0	0	0	0	0	0	0	0	0	0	0	0	0	0	0	0	Start
0	0	0	0	0	0	0	0	0	0	0	0	1	1	1	1	Step 1
0	0	0	0	0	0	0	0	1	1	1	1	0	0	0	0	Step 2
0	0	0	0	1	1	1	1	0	0	0	0	0	0	0	0	Step 3
1	1	1	1	0	0	0	0	0	0	0	0	0	0	0	0	Step 4

Figure 18-3   Bit file containing sequencer step data.

The start position and steps one through four are listed in the right column. The bit positions (also the output points) are listed across the top row.

The sequencer operates as follows:

1. When a valid true input starts the sequencer, it will be positioned at the start position. Before we start the sequence, we need a starting point where the sequencer is in a neutral position. The start position is all zeros; thus, all output points will be off.
2. When the SQO instruction registers a false-to-true transition from its input instructions, the sequencer will move to step one. When the sequencer is on step one, it is referred to as "pointing" to step one. Refer to Figure 18-4. It is also said that the pointer has moved to position one. Pointing at position one, the 16-bit word 0000 0000 0000 1111 will be sent through the mask. The mask is used to mask out, or block, selected bits from passing on to the output section of the PLC. (We will look at mask operation in detail later in this chapter.) After passing through the mask, the 16-bit output status word is sent to the output status file and then on to the output module. At this point, outputs 0, 1, 2, and 3 will be energized.
3. When the SQO instruction registers a second false-to-true transition from its input instructions, the sequencer will move to step two. When the sequencer is on step two, it is said to be pointing to step two. Pointing at position two, the 16-bit word 0000 0000 1111 0000 will be output through the mask and on to the output status file and the output module. Outputs 4, 5, 6, and 7 will be energized.
4. The SQO instruction registers a third false-to-true transition from its input instructions. The sequencer will move to step three. When the sequencer is pointing to step three, the 16-bit word 0000 1111 0000 0000 will be run through the mask, to the output status file, and then on to the output module. Outputs 8, 9, 10, and 11 will be energized.

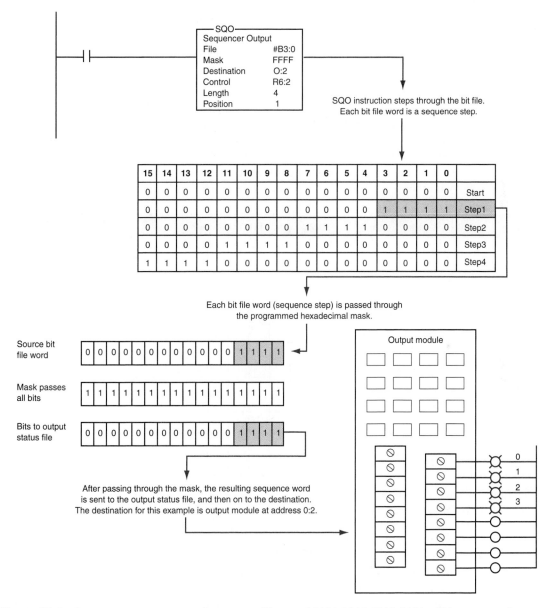

Figure 18-4 Sequencer at step one. Sequencer file word 0000 0000 0000 1111 will be sent to the output status file designated by the sequencer instruction, and then out to the output module.

5. When the SQO instruction registers a fourth false-to-true transition from its input instructions, the sequencer will move to step four. When the sequencer is pointing to step four, the 16-bit word 1111 0000 0000 0000 will be output to the output status file through the mask and then on to the output module. Outputs 12, 13, 14, and 15 will be energized.

6. When the next false-to-true rung transition is registered by the sequencer instruction, the pointer will return to step one. The sequencer will continuously repeat the cycle until no additional input signals are registered.

Sequencer File Addressing

The address contained in the file parameter in Figure 18-2 is #B3:0. This is the beginning address of the sequencer file. The # sign is necessary to designate this as a user-defined file and is called the file designator. Figure 18-5 illustrates bit file three, words 0 through 7. The first five words make up our sequencer file. These five words are grouped together to create a file within bit file three. The # file designation parameter defines the starting point of the file, while the length parameter defines how many words are contained in the file.

Status	Explanation	15	14	13	12	11	10	9	8	7	6	5	4	3	2	1	0
#B3:0 Sequencer File	B3:0, Step 0 Start-up position	0	0	0	0	0	0	0	0	0	0	0	0	0	0	0	0
	B3:1, Step 1 Turns on Output 0, 1, 2, 3	0	0	0	0	0	0	0	0	0	0	0	0	1	1	1	1
Length 5 words	B3:2, Step 2 Turns on Output 4, 5, 6, 7	0	0	0	0	0	0	0	0	1	1	1	1	0	0	0	0
	B3:3, Step 3 Turns on Output 8, 9, 10, 11	0	0	0	0	1	1	1	1	0	0	0	0	0	0	0	0
	B3:4, Step 4 Turns on Output 12, 13, 14, 15	1	1	1	1	0	0	0	0	0	0	0	0	0	0	0	0
	B3:5																
	B3:6																
	B3:7																

Figure 18-5 Bit file B3 words B3:0 through B3:7. The address #B3:0 with a length of 5 is illustrated along with words following the file.

For this example, our user-defined file within B3 started at word 0. This file could have been started at any point within the file, say at word 53. The file parameter would then have been #B3:53. With a length of 5 words, this file would start at B3:53 and end at B3:57. Multiple user-defined files may be created within any bit file. Be careful not to overlap file boundaries or attempt to use bit file words inside a user-defined file already designated for another purpose.

The Sequencer Mask

The mask parameter is the hexadecimal mask or the address where the hexadecimal mask will be found. Mask bits that are set, or 1, will allow data to pass. Mask bits that are reset, or 0, will mask data (restrict passage). Figure 18-2 shows 0FFFFh as the mask parameter.

The value FFFF is the hex mask, while the "h" designates the mask value as hex. Refer to Appendix C to review hexadecimal masks.

Sequencer Destination Parameter

The destination parameter of the sequencer instruction is the output module's address. Output devices controlled by the sequencer instruction will be interfaced to a single 16-point output module. In Figure 18-2, the sequencer destination, or output module address, is O:2.0. The output module residing in slot two of the SLC 500 chassis will be used as our output device for sequenced outputs. Simply, the destination is the output word or file to which the SQO will write output data from the sequencer bit file.

Sequencer Control File

A sequencer does not have a designated data file like a timer or counter. Since there is no designated data file to store information and status bits associated with the sequencer instruction, the control data file (file 6) will be used. Data file R6 was created for instructions that do not have specific data files assigned. The default control file is R6; however, any unused data file between 9 and 255 can also be designated as a control file. Figure 18-6 illustrates the three-word control element used for sequencer output instructions.

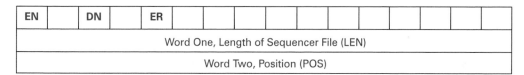

EN		DN		ER											
Word One, Length of Sequencer File (LEN)															
Word Two, Position (POS)															

Figure 18-6   Three-word control element for the SLC 500 sequencer output instruction.

The control parameter of the sequencer output instruction is the address of the control file. A control file element's address may be used only once. If there were additional sequencer instructions, another unused control file address would be selected. The next control file address is R6:1.

The control file element for a sequencer output instruction contains the instruction's status bits in word zero. Sequencer file length is stored in word one, while the current step of the sequence is stored in word two.

The instruction's length parameter designates how many steps are in the sequence. Each sequence step is represented by one word in the user-defined bit file. The total number of words that can be used is 256. (The MicroLogix 1000 length maximum is 104 words.) The address you entered in the file parameter is the address of step zero. Step zero is the start-up position of the sequencer. Data reflecting outputs is stored starting at step one. Step zero is not used for turning outputs on; it is only a neutral position from which to start the sequence. After the sequencer steps through all the steps, it wraps around to step one.

Sequencer Length Parameter

The length parameter is directly related to the file parameter. Let's assume that a sequencer application specified turning on each output of a 16-point output module in sequence. Step 1 turns on output 0, step 2 turns on output 1, and so on until step 16 turns on output 15. Since there are 16 steps and the starting step is 0, the length will be designated as 16 output steps plus 1 step for the 0. The length parameter will thus be 17. Figure 18-7 illustrates the 17 words we have reserved within B:3.

Status	Explanation	15	14	13	12	11	10	9	8	7	6	5	4	3	2	1	0
#B3:0 Sequencer File	B3:0, Step 0 Start-up position	0	0	0	0	0	0	0	0	0	0	0	0	0	0	0	0
Length 17 words	B3:1, Step 1 Turns on Output 0	0	0	0	0	0	0	0	0	0	0	0	0	0	0	0	1
	B3:2, Step 2 Turns on Output 1	0	0	0	0	0	0	0	0	0	0	0	0	0	0	1	0
	B3:3, Step 3 Turns on Output 2	0	0	0	0	0	0	0	0	0	0	0	0	0	1	0	0
	B3:4, Step 4 Turns on Output 3	0	0	0	0	0	0	0	0	0	0	0	0	1	0	0	0
	B3:5, Step 5 Turns on Output 4	0	0	0	0	0	0	0	0	0	0	0	1	0	0	0	0
	B3:6, Step 6 Turns on Output 5	0	0	0	0	0	0	0	0	0	0	1	0	0	0	0	0
	B3:7, Step 7 Turns on Output 6	0	0	0	0	0	0	0	0	0	1	0	0	0	0	0	0
	B3:8, Step 8 Turns on Output 7	0	0	0	0	0	0	0	0	1	0	0	0	0	0	0	0
	B3:9, Step 9 Turns on Output 8	0	0	0	0	0	0	0	1	0	0	0	0	0	0	0	0
	B3:10, Step 10 Turns on Output 9	0	0	0	0	0	0	1	0	0	0	0	0	0	0	0	0
	B3:11, Step 11 Turns on Output 10	0	0	0	0	0	1	0	0	0	0	0	0	0	0	0	0
	B3:12, Step 12 Turns on Output 11	0	0	0	0	1	0	0	0	0	0	0	0	0	0	0	0
	B3:13, Step 13 Turns on Output 12	0	0	0	1	0	0	0	0	0	0	0	0	0	0	0	0
	B3:14, Step 14 Turns on Output 13	0	0	1	0	0	0	0	0	0	0	0	0	0	0	0	0
	B3:15, Step 15 Turns on Output 14	0	1	0	0	0	0	0	0	0	0	0	0	0	0	0	0
	B3:16, Step 16 Turns on Output 15	1	0	0	0	0	0	0	0	0	0	0	0	0	0	0	0
Bit file 3, word 17	B3:17																
Bit file 3, word 18	B3:18																
Bit file 3, word 19	B3:19																
Bit file 3, word 20	B3:20																

Figure 18-7 Seventeen-word, sixteen-step sequencer file.

Important Note:

The programming software you are using will dictate how the length paramenters will be programmed. If you are using APS or AI programming software, the start-up position must be included in the length. As illustrated in the previous paragraph, the length would be 17.

When using RSLogix software, the length parameter would be 16. The start-up position would not be included.

The left column of Figure 18-7 shows the designated 17 words (words 0 through 16) as the sequencer file. Bit file 3, word 17, is the first word not included in the sequencer file. Words 17 and above are available for other usage.

PROGRAMMING THE SEQUENCER OUTPUT INSTRUCTION

Figure 18-8 is a printout of the sequencer instruction rungs for our current application. The sequencer instruction is on rung 0000, while the status bit rungs are rungs 0001 through 0003.

Sample Ladder Rung

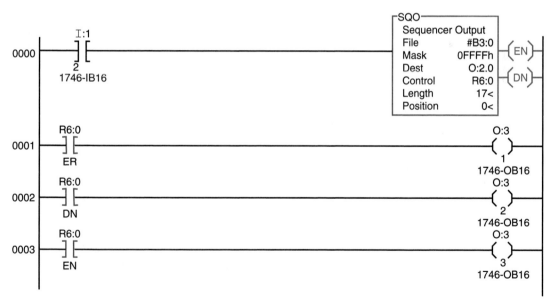

Figure 18-8   RSLogix 500 sequencer output instruction and associated status bit rungs.

Ladder Rung Explanation

Rung 0000 is the sequencer instruction. For each false-to-true transition of input instruction XIC, I:1/2, the sequencer output instruction will increment to the next step in the predetermined sequencer file #B3:0. Figure 18-7 is a representation of the sequencer file.

Associated Status Bits

The status bits and status words are contained in control data file R6. (Refer back to Figure 18-6 for the physical makeup of a control data file when used with the sequencer output instruction.) There are many status bits available with the control file. The sequencer output instruction only uses status bits 11, 13, and 15.

The sequencer error bit, R6:0/ER (bit 11) is set when the processor detects either a negative position or a negative length value. Bit 11 will also be set if the length value is a 0. If bit 11 is set when the processor evaluates the end program instruction, a major error will be generated and the processor will fault. Refer to status file word S:6 to view the hexadecimal fault code. The fault can also be cleared from the status file S:2.

The done bit, bit 13 (addressed as R6:0/DN or R6:0/13), is set by the SQO or SQC instruction after evaluation of the last word of the sequencer file. The done bit will be reset when the SQO instruction registers the next false-to-true transition.

The enable bit, bit 15 (addressed as R6:0/EN or R6:0/15), is set by each false-to-true transition from the rung's input logic that the sequencer instruction registers.

The Reset Instruction and Sequencers

The reset instruction can be used to reset a sequencer. Resetting a sequencer instruction will reset all status bits to zero except the FD (found) bit. The found bit is only used with the sequencer compare instruction. The position parameter will also be reset to zero. Remember to address the control file in your reset instruction. To reset our sequencer using R6:0 as its control file, address the reset instruction as RES R6:0.

MULTIPLE SEQUENCERS AND FILE DESIGNATION

If multiple sequencer output instructions were used in a particular application, there will be one file for each sequencer data file created within bit file 3. In many instances, for the sake of simplicity, sequencer data files may each be assigned their own bit file. As an example, bit file 3 could be left as the default file for single-bit data. Each sequencer file could be assigned its own separate bit file, such as bit file 15 and bit file 16.

Thirty-Two–Bit Sequencer

As we have seen, each sequencer instruction controls up to 16 output points. These output points are typically all located on the same output module. If we have an application where we have 32 steps in a sequence, we can put two sequencer output instructions in parallel.

Probably the easiest way to track separate sequencers is to create a separate bit file for each. Before creating additional files, always check a memory map to determine which are available. Figure 18-9 is a representation of an RSLogix 500 memory map illustrating all files, including our two sequencer files. Bit file 3 is used as a regular bit file. Bit file 15 (B15) is used for our first sequencer output instruction and bit file B16 is used for our second sequencer instruction.

Looking at our memory map, notice the makeup of the table and the data contained. Especially notice the elements and words columns, which list the number of elements and the number of words for each file, respectively. Remember, a control file element contains

Name	File	File Type	Elements	Words	Last
Output	0	O	2	2	O:2.2
Input	1	I	6	2	I:3.3
Proc. Status	2	S	83	83	S2:82
Bit File	3	B	32	32	B3:31
Timers	4	T	44	132	T4:43
Counters	5	C	44	132	C5:43
Control	6	R	2	6	R6:1
Integer	7	I	70	70	N7:69
Float	8	F	50	100	F8:49
ASCII	12	A	31	31	A12:30
Seq. One	15	B	50	50	B15:49
Seq. Two	16	B	50	50	B16:49

Figure 18-9 Memory map of our sequencer application.

three words. Since this application has two sequencer instructions, two control elements will be needed. R6:0 and R6:1 will be used. The elements column will list two elements (three words per element). The word column should list six words.

Each bit file was created with 50 words each. Since bit files B15 and B16 contain one word per element, the word and element columns will correspond on a one-to-one basis. Not all words need to be used at this time, nor do they need to all be used in conjunction with the sequencer.

Earlier we introduced the mask parameter for the sequencer instruction. Now we will explore the mask to understand how it is used with the sequencer output instruction. The mask is used to control the physical outputs that will be controlled by the sequencer output instruction. To accomplish this, a hexadecimal code (or the address of the mask) is entered as a parameter when programming the sequencer output instruction. As part of instruction execution, the processor moves the current sequencer file word through a mask before the word is sent to the output status file. The mask can be used to filter out, or mask, any bit or bits in the 16-bit sequencer file word from passing and going on to the output status file.

Since the mask is programmed using hexadecimal numerical values, you may wish to review hexadecimal numbers in chapter 4.

APPLYING HEXADECIMAL NUMBERS AND MASKS WITH SLC 500 PROGRAMMING INSTRUCTIONS

In some instances we may not desire that a particular sequencer instruction control all 16 outputs of the destination output module. We need a way to block or mask the bits we do not desire to use in a particular sequence operation.

Masking can be accomplished by passing the bits through a mechanism that controls the bits allowed to pass through to the destination output screw terminals.

For example, suppose an application requires sequencing the lower eight bits on a particular 16-point output module while the upper byte of the output status word (upper eight bits or screw terminals) of the 16-point module are not affected. To achieve this control, a mask will be used to mask out the upper eight bits so they do not fall under the control of the sequencer instruction.

Sequencer Mask

The SLC 500 mask parameter uses either a four-character hexadecimal numerical code as the mask word or the address where the mask word will be found. This hex word, when used in its binary equivalent form, is the mask through which the sequencer instruction will move the data. If the mask bits are set (a logical one), data will pass through the mask; if the bits are reset (a logical zero), the mask will hold data from passing.

Let's look at an example. If we had a 16-point output module and wished to control only certain output points with a sequencer instruction, we could control which bits passed through to the output status table by using a mask in the following way. The data table in bit file three has the following data:

B3:1 = 1101 1000 1010 1101

B3:2 = 0001 0110 1010 1011

B3:3 = 1111 1100 1000 1100

For our application, we wish to use the following bits:

For process A, we wish to use only bits or outputs 0–7.

For process B, we wish to use only bits or outputs 8–11.

For process C, we wish to use only bits or outputs 8–15.

For this example we will illustrate the use of a hex mask to control the bits for process A from the bit file to the output status table (refer to Figure 18-10).

1101	1000	1010	1101	Data from file 3
NO	NO	YES	YES	Data allowed to pass?
0000	0000	1010	1101	Desired data sent to output status table

Figure 18-10   Only bits 0 through 7 will be allowed to pass through the mask.

To set up the mask to pass the correct data, set a 1 in the mask bit position where each source data bit is to pass, as illustrated in Figure 18-11.

1101	1000	1010	1101	Data from step one
0000	0000	1111	1111	Bits set allow data to pass
0000	0000	1010	1101	Desired data sent to output status table

Figure 18-11   Only bits set to a one will allow data to pass.

To determine the hex mask value for the sequencer instruction parameter, convert the bit format to hex. See the table in Figure 18-12 for the correlation between binary and hex. The hex value will be entered into the mask parameter of the sequencer instruction. Figure 18-13 illustrates the binary bits changed into their hexadecimal equivalent.

HEXADECIMAL CORRESPONDING BIT PATTERNS			
Hex	**Binary**	**Hex**	**Binary**
0	0000	8	1000
1	0001	9	1001
2	0010	A	1010
3	0011	B	1011
4	0100	C	1100
5	0101	D	1101
6	0110	E	1110
7	0111	F	1111

Figure 18-12   Binary bit pattern and hex correlations.

The mask bit pattern	0000	0000	1111	1111
Hex value of	0	0	F	F

Figure 18-13   Bit pattern translated to hexadecimal.

It should be easy to determine the mask for process B as 0F00 from Figure 18-14. The mask for process C is FF00.

1101	1000	1010	1101	Data from step one
0000	1111	0000	0000	Bits set to allow data to pass
0000	1000	0000	0000	Desired data sent to output status table for process B

Figure 18-14   Only bits 8 through 11 will be allowed to pass through the mask.

In the next example, we will illustrate how to pass selected single bits in the lower byte of the sequencer data word to the destination. To set up the mask so as to pass the correct single data bits, you simply set a 1 in the mask bit position where source data is to pass. This is illustrated in Figure 18-15.

Initially the destination word is all zeros, but after the operation the destination is the result of the source word's interaction with the mask. Notice that in Figure 18-15, we mask

1101	1000	1010	1101	Data from step one
0000	0000	1100	0011	Bits set to allow data to pass
0000	0000	1000	0001	Desired data sent to output status table

Figure 18-15 Selected single mask bits set to a one.

out the upper two bits in the lower nibble along with the lower two bits in the upper nibble. Even though it may appear that bit one, a 0, in the upper nibble passed through the mask because there was a 0 in the source and in the destination, this is not the case. The bit in the destination word is filtered, or masked. The original 0 in the destination is still there.

Referring to Figure 18-16, the mask bit pattern equates to a hex C3.

The mask bit pattern	0000	0000	1100	0011
Hex value of	0	0	C	3

Figure 18-16 Bit pattern of Figure 18-14 translated to hexadecimal.

SUMMARY

Early machine-sequencing applications required the use of mechanical-rotating cam switches. Rotating-cam switches were used as control circuit devices on machinery that had repetitive or sequential operating cycles.

Modern microprocessor technology has, in many instances, replaced the mechanical sequencer. As the microprocessor evolved, gaining more functionality, sequencers were incorporated into the PLC instruction set.

The programmable controller was conceptualized as easy-to-use, easily changeable, and easily reprogrammable. Each manufacturer's PLC processors have their own vocabulary of instructions, called the instruction set. Most programmable controllers have some type of sequencing ability included in the instruction set. As an example, GE Fanuc Automation's Series 90-30 PLC has a bit sequencer function that shifts bits through a bit array. Even though the bit sequencer looks and programs differently from the SLC 500's sequencer output instruction, a similar application control can be achieved with either PLC. On looking at the majority of PLCs you will find some method to program a sequence of operations. Even though they will all look and program differently, the basics are still similar.

This chapter introduced the Allen-Bradley sequencer output instruction.

REVIEW QUESTIONS

1. Explain the main use for the SQO instruction.
2. In which file does the actual instruction reside?

3. Briefly explain each of the instruction's parameters:
   A. file
   B. mask
   C. destination
   D. control
   E. length
   F. position
4. Explain the addressing format for the file parameter. Why is the address represented as #B3:1?
5. The sequencer element contains how many words?
6. Explain the function of each status bit.
   A.
   B.
   C.
   D.
7. What is the mask used for?
8. What number system is used to represent the mask?
9. Why is this number system used?
10. Modern microprocessor technology has in many instances replaced the _____ sequencer and incorporated sequencers into the PLC instruction set.
11. The SLC 500 can have up to _____ sequencers per control data file.
12. The sequencer instruction sends a predetermined _____ word representing the desired output configuration for a specific output module.
13. The SLC 500 sequencer output instruction, SQO, steps through the sequencer file, usually a _____ file, whose bits have been previously set up to control the desired output devices.
14. Even though the basic operating principle of the sequencer is the same between manufacturers, each manufacturer will have _____ programming procedures.
15. Fill in the blank areas of our sequencer instruction table (Figure 18-17).

SEQUENCER INSTRUCTIONS		
Instruction	Use This Instruction to	Functional Description
Sequencer Output		
Sequencer Compare		
Sequencer Load		

Figure 18-17   SLC 500 sequencer instructions.

16. The sequencer output instruction (SQO) is programmed as an _____ on a ladder rung.

17. Each _____ rung transition from the rung's inputs registered by the SQO instruction causes the instruction to increment to the next sequence step.

18. Each step of the sequence is stored in a _____.

19. The sequencer file is typically a _____.

20. Stepping though the sequencer file, the sequencer output instruction steps through a predetermined sequencer file to control _____.

21. The file parameter is the _____ of the sequencer file.

22. Develop the sequencer data file for the following application: Assume we have a four-step process. Four outputs are to be energized on one 16-point output module for each of the four steps. Because of the way in which the sequencer instruction operates, all output points must be on a single output module. The sequence is as follows:

   Step one: Outputs 9, 10, 11, and 3 will be energized.

   Step two: Outputs 4, 6, and 7 will be energized.

   Step three: Outputs 3, 4, 5, and 8 will be energized.

   Step four: Outputs 14 and 15 will be energized.

   Figure 18-18 illustrates the bit file where we will store the output status for each of the four steps. The start position and steps one through four are listed in the right column. The bit position and the output points are listed across the top row.

16	14	13	12	11	10	9	8	7	6	5	4	3	2	1	0	
																Start
																Step 1
																Step 2
																Step 3
																Step 4

Figure 18-18 Question 22 sequencer data file.

23. When a valid true input starts the sequencer, the sequencer will be positioned at the _____ position.

24. Before we start the sequence, we need a starting point where the sequencer is in a neutral position. The start position is all zeros, representing our neutral position; thus, all _____ points will be off.

25. The address contained in the file parameter is #B3:0. This is the beginning address of the sequencer file. The # sign is necessary to designate:
   A. This is a bit file.
   B. This is a control data file.
   C. More than one bit will be used for this instruction.

   D.   More than one word of the bit file will be used with this instruction.
   E.   Both B and C are correct.
26. True or false?: The # is called the file designator.
27. #B3:0 with a length parameter of 25 means:
   A.   Bit file three has been designated as a sequencer file.
   B.   The first 25 bits of B3 have been designated as a sequencer file.
   C.   The first 25 words of B3 have been designated as a sequencer file.
   D.   The first 25 elements of B3 have been designated as a sequencer file.
   E.   The first 25 bytes of B3 have been designated as a sequencer file.
   F.   C and D are correct.
28. We need to sequence through 24 steps in our product assembly sequence. We will use 16 outputs on one module and 8 outputs on a second module. The length parameter for the first sequencer output instruction was entered as 17. What will you enter in the length parameter of the second sequencer output instruction?
   A.   8
   B.   9
   C.   12
   D.   16
   E.   24
   F.   does not matter
29. A sequencer file can have up to 256 steps. If we started our sequencer file at #B3:240, what would be a valid length parameter?
   A.   12
   B.   13
   C.   14
   D.   15
   E.   16
30. The mask parameter is the hexadecimal mask or the _____ where the hexadecimal mask will be found.

IF PASSWORD IS CORRECT GO TO DRIVE SET UP SCREEN

0022
```
   ─── EQU ───                          B3:10/0                              B3:1/15
   Equal                               ─[ OSR ]─                              ─( )─
   Source A    N12:21
                    7777<                                          RESET PASSWORD
   Source B    32767                                              ATTEMPTS COUNTER
                    32767<                                               C5:5
                                                                      ─( RES )─
```

0023
```
   ─── GRT ───                    ─── NEQ ───               B3:10/1
   Greater Than (A>B)             Not Equal                ─[ OSR ]─
   Source A    N12:21             Source A    N12:21
                    7777<                          7777<
   Source B    0                  Source B    32767
                    0<                             32767<
```

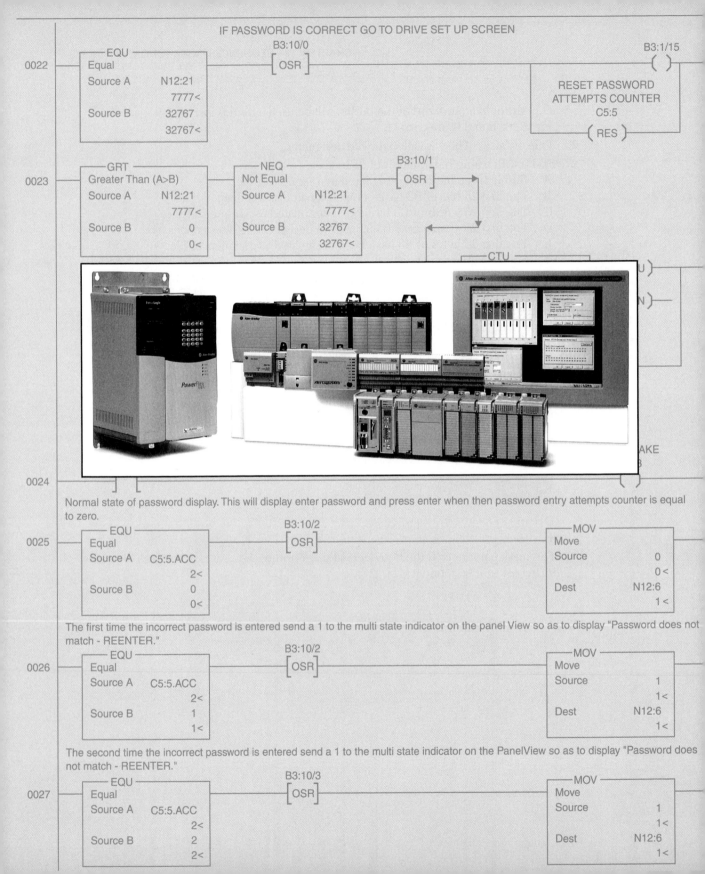

```
                                                                    ─── CTU ───
                                                                          U )
                                                                          N )
```

0024

Normal state of password display. This will display enter password and press enter when then password entry attempts counter is equal to zero.

0025
```
   ─── EQU ───                          B3:10/2                           ─── MOV ───
   Equal                               ─[ OSR ]─                          Move
   Source A    C5:5.ACC                                                   Source          0
                    2<                                                                    0<
   Source B    0                                                         Dest          N12:6
                    0<                                                                    1<
```

The first time the incorrect password is entered send a 1 to the multi state indicator on the panel View so as to display "Password does not match - REENTER."

0026
```
   ─── EQU ───                          B3:10/2                           ─── MOV ───
   Equal                               ─[ OSR ]─                          Move
   Source A    C5:5.ACC                                                   Source          1
                    2<                                                                    1<
   Source B    1                                                         Dest          N12:6
                    1<                                                                    1<
```

The second time the incorrect password is entered send a 1 to the multi state indicator on the PanelView so as to display "Password does not match - REENTER."

0027
```
   ─── EQU ───                          B3:10/3                           ─── MOV ───
   Equal                               ─[ OSR ]─                          Move
   Source A    C5:5.ACC                                                   Source          1
                    2<                                                                    1<
   Source B    2                                                         Dest          N12:6
                    2<                                                                    1<
```

CHAPTER

19

Program Flow Instructions

OBJECTIVES

After completing this lesson you should be able to:

- understand program flow using program control instructions
- create a subroutine
- access a subroutine
- create Jump to Label logic
- identify the program flow instructions and describe how they work
  - Jump to Label Instruction
  - Jump to Subroutine (JSR) Instruction
  - Subroutine (SBR) Instruction
  - Return (RET) Instruction
  - Master Control Reset (MCR)
  - Immediate Input with Mask (IIM)
  - Immediate Output with Mask (IOM)
  - Refresh (REF)

INTRODUCTION

To this point in our studies we have been working with ladder file two as our only ladder file. Ladder file two executes our ladder logic, starting with rung zero executing ladder logic from the left to right, starting with rung zero straight through to the last rung. The last rung contains the END Instruction. The end rung can not be deleted, and no logic can be programmed on this rung. The end rung directs the processor to proceed to the next step in the processor scan. The program scan consists of the following steps as illustrated in Figure 19-1.

1. Reads inputs and updates the input status file.
2. Executes ladder file two starting with rung zero, evaluating each rung left to right, straight through to the bottom of the ladder file and the END Instruction. As each rung is executed and the output instruction is updated, its status is immediately updated in its respective data file location. This is done so that if there are any other references to the output instruction in later rungs, there will be current status information in the data table.
3. Transfers output status table information one word at a time to turn on, off, or hold the current output status of each field device for each output module in the system.
4. Housekeeping and communication updates the personal computer screen if online and monitoring the project. For example, it updates message instruction data and the processor status file.
5. Resets the watchdog timer. If the scan is successful and the watchdog timer is reset, the processor will start another scan. The processor will continue this scan cycle continuously as long as the processor is in run mode.

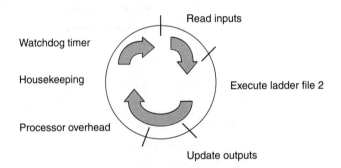

Figure 19-1   Basic PLC program scan using ladder file 2 only.

Program control instructions are input and output instructions that change how the processor scans. Program control instructions modify the sequence in which the program is executed, thereby interrupting the normal sequence of the processor operating cycle. Typically ladder file 2 in an SLC 500, MicroLogix, or PLC 5 is the main ladder program or file. Starting with ladder file 3, through 255 for an SLC 500 and up to 999 for a PLC 5 are designated for subroutines. Figure 19-2 illustrates the project tree from RSLogix 5 software for the PLC 5. The view would be very similar for the RSLogix 500 software that supports the MicroLogix and SLC 500. Under the Program Files folder, notice ladder file 2. It is identified as the main ladder program. For this example, ladder files 3 through 5 are subroutines.

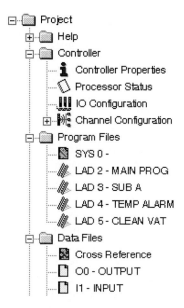

Figure 19-2   Project tree showing main ladder and subroutines for a PLC 5, MicroLogix, or SLC 500 PLC.

The ControlLogix family of PLCs has a main routine that is specified by the programmer. In Figure 19-3, notice the main routine and the ladder symbol identifying it as a ladder routine. Also notice that there is a paper-like icon with a 1 on it designating this as the main ladder routine. There are three subroutines in the example, Subroutines A, B, and C. They are ladder routines, because of the ladder icon to the left of the subroutine name. Since there is no other identifier on the ladder symbol, they are designated as subroutines. The number of routines when using ControlLogix is limited by processor memory.

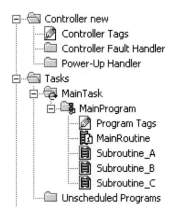

Figure 19-3   ControlLogix Controller Organizer showing main routine and subroutines.

TERMINOLOGY

Subroutine: A subroutine is a program file that performs a specific task.

JSR Instruction: The JSR Instruction redirects logic execution from the current ladder file to the specified subroutine file.

End: The last rung in any ladder file. Tells the processor that there is no additional logic here to execute and to move on to the next task.

SBR Instruction: The SBR is the first instruction in a subroutine file. This instruction identifies the type of subroutine file as either a plain subroutine or an interrupt file.

RET Instruction: Return Instruction directs the processor to leave the current routine and return to the previous file.

SUBROUTINES

As a programmer, there are two things to look at when the program has been completed. First, does the PLC do what it is supposed to do? Second, is the program efficient? Is efficiency important for this particular application? As an example, let's assume the PLC is controlling a carwash. One car can be washed at a time. Is efficiency important in this application? Possibly not. On the other hand, efficiency would be important for an application packaging 800 or more cans of fruit punch per minute. High-speed applications require additional consideration by the programmer to ensure that the PLC can react fast enough to control the application successfully. High-speed applications require the programmer to find ways to get the fastest throughput from the PLC. Throughput is defined as how fast the PLC can react to an input and update the respective output.

In the first application, in which efficiency is possibly not important, one option is to keep all the rungs in one ladder file, the main ladder, which is typically ladder file 2. Figure 19-4 shows an example of an application with all rungs in ladder file 2.

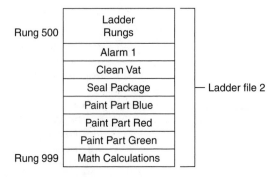

Figure 19-4   Ladder file 2 containing all program rungs.

In this example, assume that the figure represents 1000 rungs. The first 500 rungs are nothing special. The remaining rungs are groups of rungs with specific functions such as alarm 1, clean vat, seal package, paint part blue, paint part red, paint part green, and math calculations. One option is to leave all the rungs in ladder file 2 and execute starting with rung 0 on through to rung 999.

Let's reorganize the program to make it more efficient. Separating the rungs into separate files will improve organization for ease of understanding, interpretation, monitoring, and editing by electrical or maintenance personnel. Figure 19-5 illustrates reorganizing the groups of ladder rungs into separate subroutine files.

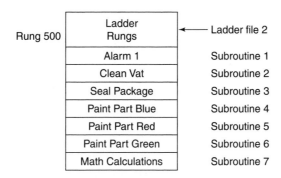

Figure 19-5   Groups of rungs starting with alarm 1 through math calculations are divided into separate subroutine files.

If someone wanted to monitor or edit the green painting rungs, all rungs are together in subroutine 6. Figure 19-6 shows part of an RSLogix 5 or RSLogix 500 project tree with our groups of ladder rungs assigned separate ladder files. The main ladder is in ladder file 2, whereas ladder files 3 through 9 contain our subroutines.

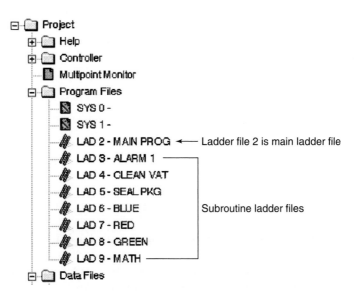

Figure 19-6   RSLogix 5 or 500 software showing subroutine ladder files.

Another consideration: If we were going to paint parts green for four hours, do we need to execute the red or blue painting rungs? Probably not. Scanning these rungs is not necessary during the time we are painting green parts. Hopefully we would not have any alarms for an hour, a shift, or possibly the entire day. Why would we need to execute these rungs every scan? If we were making chocolate chip cookies all day, why would we execute the vat-cleaning rungs during that time? Subroutines allow us to execute the rungs needed currently while we ignore the others until they are needed. We will jump to a subroutine file and execute the rungs only when we need them. If we are not jumping into a subroutine ladder file, as far as the processor is concerned the rungs are not there. By putting rungs in different ladder files or subroutines, we can execute only the rungs necessary for our current manufacturing situation.

Figure 19-7 illustrates how the program flow is changed as a result of executing a subroutine file such as the green painting subroutine, ladder file 8. Notice that the execution of ladder file 2 is interrupted and file 8 is scanned. When all rungs from ladder file 8 have been scanned, control is returned to ladder file 2. File 2 rungs will continue to be scanned until all rungs in ladder file 2 have been scanned or the file has been interrupted by another Jump to Subroutine Instruction. Assuming we were going to paint green parts for four hours, this sequence would be repeated every scan for the four hours. If after the four hours we were to paint blue parts for the next two hours, we would jump into ladder file 6, the blue painting rungs for the time required to complete the blue parts order.

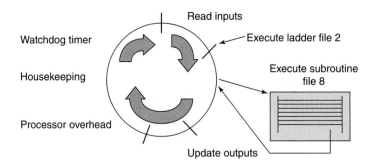

Figure 19-7 Processor ladder file scan sequence while jumping to ladder file 8 to execute the green painting rungs.

Jump to Subroutine Instruction (JSR)

The Jump to Subroutine Instruction is used to trigger the change in program flow. The Jump to Subroutine or JSR instruction is shown in Figure 19-8 for an SLC 500 using RSLogix 500 software. When input I:1.0/0 is true, the JSR Instruction is also true. The JSR now redirects ladder execution to subroutine ladder file 8.

Notice the U:8 on the JSR Instruction. The U is internal. The number eight identifies the ladder file that is to be scanned. Subroutine files for the SLC 500 are from file three

Figure 19-8   The Jump to Subroutine (JSR) Instruction on an RSLogix 500 ladder rung redirecting program flow to ladder file 8.

to 255. PLC 5 ladder files available for subroutines are from 3 to 999. Figure 19-9 provides a view from RSLogix 500 and a JSR Instruction redirecting program flow to subroutine ladder file 8.

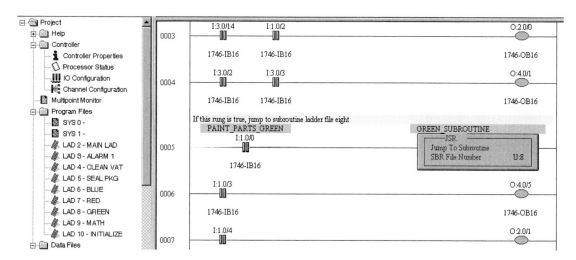

Figure 19-9   RSLogix 500 JSR Instruction redirecting program flow to subroutine ladder file 8.

Figure 19-10 illustrates a ControlLogix JSR Instruction. If the input is true the JSR Instruction will redirect program execution to the routine named Subroutine A. Since ControlLogix is a name (tag)-based controller there are no ladder files, thus only the name of the routine. Refer back to Figure 19-3 to see the subroutine A routine in the ControlLogix controller organizer. ControlLogix subroutines are limited only by processor memory.

Subroutine File

The SBR Instruction is optional. If used, the first instruction in a subroutine file is the SBR Instruction. The rung the SBR Instruction is on then must be programmed to be a verifiable rung. The rungs below the SBR rung are programmed normally. The instruction is used

Figure 19-10   Jump to Subroutine Instruction displayed for ControlLogix and RSLogix 5000 software.

with the SLC 500 to identify the type of subroutine. When using SLC 500 family PLC's there are two types of subroutine ladder files other than the main ladder, ladder file 2. There are plain subroutines, and there are interrupts. The SBR Instruction identifies this as a plain subroutine where an I/O Interrupt instruction identifies this as an interrupt routine.

The last rung of any ladder file contains the end instruction. The end rung directs the processor to return to the ladder file it came from, in this case ladder file 2. Notice a Return (RET) Instruction on rung 4. The Return Instruction directs the processor to return to where it came from. In Figure 19-11 the Return Instruction and the END Instruction do the same thing, so the Return Instruction is unnecessary and redundant.

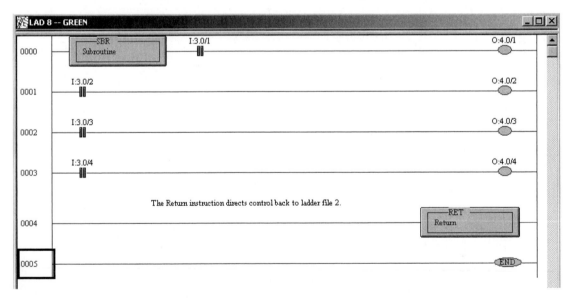

Figure 19-11   RSLogix 500 Subroutine file with redundant Return and END Instruction.

Conditioned Return in Subroutine

A Return Instruction can be conditioned, however. Figure 19-12 shows a subroutine with a conditioned Return. If I:3.0/8 is false, the Return is false and the remaining rungs are

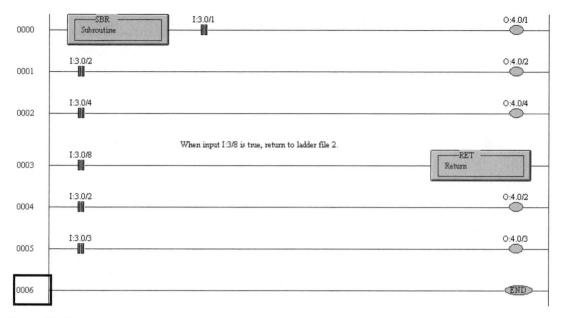

Figure 19-12   Conditioned Return used in a subroutine.

executed until the end rung is reached. The end rung directs the processor to return to the ladder file it came from. If I:3.0/8 is true, this makes the return true. The Return Instruction will direct the processor to return to the ladder file it came from at this point without completing the remaining rungs. This is called a conditioned return, as it has one or more inputs determining its true or false status. It is commonly used to get out of a subroutine early, as illustrated in Figure 19-12.

Nested Subroutines

One subroutine can use a JSR Instruction to jump to another subroutine. This is called nesting subroutines. In Figure 19-13 we have already jumped into ladder file 5, nested_sub. If rung 2 is true in ladder file 5, the JSR Instruction will jump to subroutine ladder file 6 (U6) and file 6's rungs will be executed. A Jump to Subroutine Instruction could be programmed in subroutine ladder file 6 to jump to another ladder file, or a conditioned Return could bring us back to ladder file 5, or the END Instruction in subroutine ladder file 6 could return us to subroutine ladder file 5. The SLC 500 or PLC 5s can nest up to eight levels, while ControlLogix nesting is limited only by processor memory.

Figure 19-14 illustrates three levels of nesting. We start with main ladder file 2. If rung 6 is true we will use the JSR to jump to subroutine ladder file 5. Rung 6 has a JSR Instruction. If this rung is true, the processor will jump to ladder file 7 and start executing its rungs. Rung 6 of ladder file 7 has a JSR that, if true, will jump to ladder file 8.

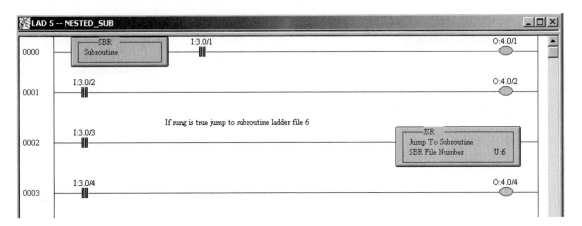

Figure 19-13 Nested subroutine where subroutine ladder file 5 uses JSR to jump to subroutine ladder 6.

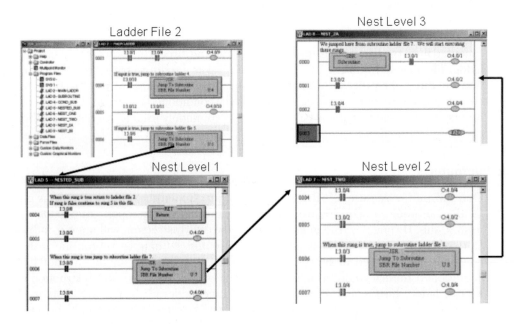

Figure 19-14 Nested subroutines to three levels with RSLogix 5 or 500 software.

Returning from nested subroutines must be in the reverse order that was used when going from one subroutine to the next. In Figure 19-14 we started with ladder file 2. The first JSR took us to ladder file 5. The JSR in ladder file 5 took us to ladder file 7. Ladder file 7's JSR took us to ladder file 8. You must go back from 8 to 7 to 5 and back to 2. If you attempt to jump back into ladder file 2, the processor will fault.

Input and Return Parameters for PLC 5 and ControlLogix Family PLCs

The PLC 5 and ControlLogix PLCs, when using RSLogix software, support input and return parameters in the JSR, SBR, and RET Instructions. Input parameters allow the option to bring information into the subroutine as they jump to the subroutine. The JSR Instruction will pass the input parameter(s) to the SBR Instruction, so the SBR Instruction is necessary. The input information is then current when used in the subroutine. The Return Instruction will then have a return parameter to return the resulting data back to the originating subroutine. If input parameters are used, the return instruction is required. If input parameters are programmed in the JSR and SBR Instructions, and a Return Instruction is not programmed, the processor will fault. Figure 19-15 shows a ControlLogix JSR Instruction with input parameters to program. Figure 19-16 illustrates a PLC 5 JSR with input parameters. The SLC 500 family of PLCs, which includes the MicroLogix, does not support input and return parameters in program flow instructions.

Figure 19-15   RSLogix 5000 and ControlLogix JSR Instruction with input and return parameters.

Figure 19-16   RSLogix 5 and PLC 5 JSR Instruction and return parameters.

JSR with Input and Return Parameters Example
Let's look at a very simple example for using input and return parameters with Subroutine Instructions. We have the following scenario: There are three variable-frequency drives on a network. One drive is at node 21, one is at node 22, and one is at node 23 on the

network. We want to input each drive's heat-sink temperature into our PLC, and then display each temperature on an operator-interface terminal. The temperatures from the drives come in degrees Centigrade. Operators want to see the temperature displayed in degrees Fahrenheit. Each temperature will need to be converted to the proper temperature format and stored in a memory location before being passed on to the operator interface. To do the conversion, we will use a subroutine with a compute instruction. By using the subroutine, we can jump to the subroutine each time we need to do the conversion, rather than program the same rung each time we need to do a temperature conversion. This will keep our ladder file shorter. Keep in mind that this is a very simple example; complex applications will contain multiple rungs within the subroutine. Figure 19-17 is that of PLC 5 ladder logic for our drive's temperature conversion scenario. Note the following:

A. Variable frequency drive on network node 21 temperature data is stored at address F8:0, the input parameter. The converted temperature data is stored at address F8:12, the return parameter.

B. Variable frequency drive on network node 22 temperature data is stored at address F8:1, the input parameter. The converted temperature data is stored at address F8:13, the return parameter.

C. Variable frequency drive on network node 23 temperature data is stored at address F8:2. The converted temperature data is stored at address F8:14.

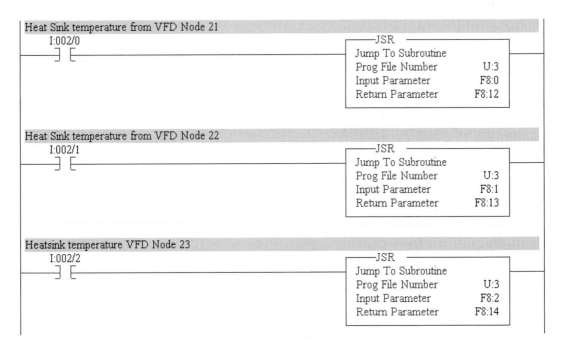

Figure 19-17 PLC 5 subroutine with a JSR Instruction with input and return parameters.

Let's assume the first rung, node 21 is true. The steps of operation are listed below and illustrated in Figure 19-18.

A. The heat sink temperature is stored in F8:0. This information will be passed to the SBR Instruction and F8:10.
B. The compute instruction contains the conversion formula and will use the information in F8:10 in the calculation. Notice F8:10 is the first part of the expression.
C. The converted temperature will be stored in the compute instructions destination, F8:11.
D. Notice the destination for the compute instruction is the same address as the return parameter for the Return Instruction.
E. The Return Instruction will pass the converted temperature stored in F8:11 back to the return parameter in the JSR Instruction on ladder file 2 and the address specified in the return parameter, which is F8:12.
F. The converted temperature is stored in F8:12 and can now be passed on to the operator interface device for display.

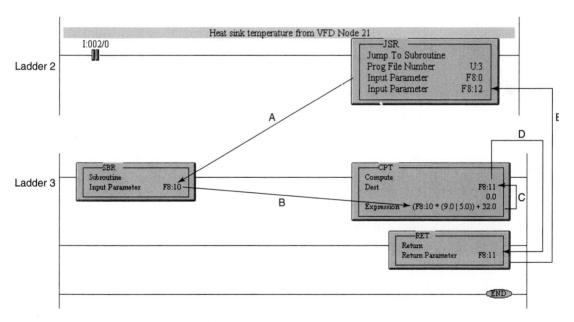

Figure 19-18   PLC 5 subroutine with input and return parameters data passing sequence.

Subroutines and System Initialization

An initialize subroutine is usually run only once, at machine startup. The subroutine can be used to return a jammed machine to a common startup position, delete data after being sent across an Ethernet connection to management, or start up a machine at the beginning of a work week. The subroutine is triggered by the status file first pass bit S:1/15. If this address is programmed with a normally open or examine if closed instruction, the instruction will only be true for the first scan through ladder file 2 when the PLC processor enters run mode. Typically there is no need to run the initialization subroutine except once at startup, or for the restart of a machine after a jam or malfunction. Putting the startup rungs in a subroutine removes them from the main ladder program, thus eliminating scanning of rungs that are only used at startup. Figure 19-19 shows a JSR triggered by the first pass bit for an SLC 500, MicroLogix, or PLC 5 using RSLogix software.

Figure 19-19   RSLogix 5 or RSLogix 500 first pass bit to trigger initialization subroutine.

We have been working with the Jump to Subroutine Instruction so far in this lesson. The next section will investigate the Jump to Label Instruction. The JSR Instruction redirects the processor to another ladder file, whereas the Jump to Label Instruction allows jumping over rings within the current ladder file.

JUMP TO LABEL INSTRUCTION

The SLC 500 and PLC 5 family, and the ControlLogix PLC instruction set, provide instructions that allow you to jump forward or backward in your ladder program to a corresponding target called a Label Instruction. To jump to a label, two instructions are necessary. The Jump (JMP) Instruction is an output, and the Label Instruction, an input, is the target. The JMP Instruction is assigned a unique identifier from 0 to 999. The number acts much like the address of the label or target, so the JMP Instruction knows where to direct the processor to jump when the rung is true. See rung 10 and 13 in Figure 19-20. Notice that the JMP Instruction has a Q2:1 above it. The Q2 is internal and provided by the software as you program the JMP Instruction. The Q2 simply identifies this as ladder file 2. A JMP Instruction in ladder file 3 would be Q3. The programmer will insert the target address, which is a 1 in this case. The label is the target that the processor jumps to when the Jump Instruction is true. The Label Instruction on rung 13 is programmed with the same unique identifier (Q2:1) as the Jump Instruction. As illustrated in Figure 19-20, the Label Instruction must be programmed as the first instruction on the rung where it resides. The Label Instruction is always true, and the remaining instructions on the rung must make up a verifiable rung. Since the instructions to the right of the LBL on the label rung are outside the jump zone, these instructions are not affected by the jump.

The logic operates as follows. If rung 10, which contains the JMP Instruction, is false, the Jump Instruction is false and the jump is not executed. Rungs 11 and 12 are executed

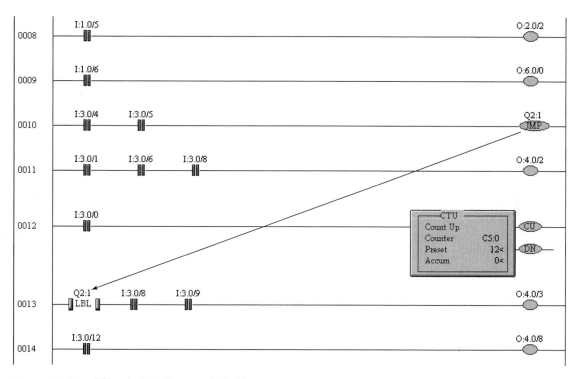

Figure 19-20   RSLogix 500 Jump to Label logic.

as normal and the LBL Instruction on rung 13 is transparent. If, on the other hand, rung 10 is true, the jump is executed and rungs 11 and 12 are jumped over. The Jump Instruction jumps to the target, the LBL Instruction. Instructions to the right of the LBL are out of the jump zone and will always be executed as a normal rung. This is called a forward jump, as we are jumping forward in our program. Sometimes jumping around in your program can be hard to interpret, so Jump to Label sequences should be used with care. Jumping backward to a label that is on an upper rung is possible, but it is in many cases considered bad programming practice. The problem in jumping backward is the watchdog timer. If the scan is not completed because of one or more backward jumps and the watchdog timer is not reset before it times out, the processor will fault. In some cases programmers might consider jumping backward to wait for an input to come in. They would want to know the exact instant an input arrived so they could take care of it immediately. The SLC 500 and PLC 5 family of PLCs have interrupt instructions that can be used to trap an input immediately. The immediate input instruction is introduced below.

Multiple Jumps to Same Label

Figure 19-21 shows how multiple Jump Instructions can be used to jump to the same label. The Jump Instructions on rungs 10 and 12 both jump to the same label, which is on rung 14. If rung 10 is true the jump to the label on rung 14 is executed. Likewise, if rung 12 is true the jump is also executed from rung 12 to rung 14.

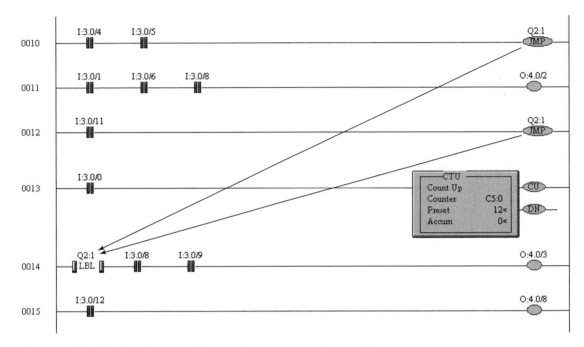

Figure 19-21   Multiple jumps to the same label using RSLogix 500 software.

Rung Operation Within a Jump Zone

Jumping over rungs does not afford the same scan time savings for efficiency as using subroutines. If the Jump Instruction is false, the rungs within the jump zone are executed normally. On the other hand, if the Jump Instruction is true the rungs are literally jumped over. This means that the rungs in the jump zone are not executed and their outputs are left in their last state. Referring to the figure above, if rung 10 is false, rungs 11, 12, and 13 are executed as normal rungs. The Label Instruction on rung 13 is transparent. However, if the jump rung, rung 10, is true, rungs 11, 12, and 13 are jumped over and left in their last state. If jumping over rung 11, if inputs I:3.0/1 and I:3.0/6 and I:3.0/8 go true, the output will not be energized, as the rung is not being updated when being jumped over. On the other hand, if we were currently not jumping over the rung and the three inputs all became true the output would become true. Then if on the next scan the Jump Instruction became true, and we started jumping over the rung and one or all of the inputs became false, thus making the rung false, since the rung is being jumped over, the output would be left in its last state and not be turned off. When jumping over rung 13, if the input to the counter I:3.0/0 goes true, the counter will not be updated, as it is left in a frozen state.

IMMEDIATE INPUT WITH MASK (IIM) INSTRUCTION

The Immediate Input with Mask Instruction is used, for example, when a high-speed application requires the processor to be interrupted and to immediately update an input or group of inputs. If you think about the scan of the processor and how it works (refer back to Figure 19-1), all inputs are read in one operation, and this information is used to update the input status file just before executing the ladder logic. This provides fresh information as we begin our scan of the ladder logic. As we scan the ladder rungs the input instructions use the input status file information to solve the ladder instructions. When the ladder rungs have all been evaluated, the outputs are updated. This is called the output scan. What if you had a high-speed application where because of the speed of the process, you could not wait until the processor went all the way around and updated the inputs on the next scan to get the information into the PLC and use it? For example, we are bottling fruit punch. The bottles are filled, capped, and labeled at about 800 to 1000 bottles per minute. There is a sensor pointing at the top of the bottle to verify that a cap was installed. If the cap is missing, or not on straight, the bottle needs to be identified, tracked, and removed from the line at the reject station. Now, thinking about the operation of the processor scan in relation to the speed of the manufacturing line, can the processor react quickly enough to identify, mark, track, and reject the bottle consistently? The longer the program, the less likely the processor would be able to react quickly enough. We need to be able to interrupt the processor and its scan of the ladder logic to catch and record this input.

The IIM Instruction has three parameters: slot, mask, and length. The slot is the chassis slot where the input module resides from which we wish to update input data. Notice in Figure 19-22 that the slot parameter is a whole word address. The mask is used to determine which input bit or bits we wish to update. Whereas the length specifies how many words of information are available from the module, a sixteen-point module would be a length of one.

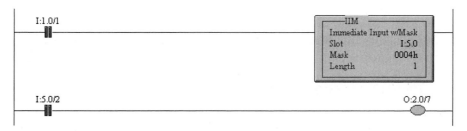

Figure 19-22   Immediate Input Instruction with Mask for RSLogix 500 and MicroLogix.

When the IIM Instruction is true, the processor stops the execution of ladder logic and does an immediate update of the input module specified in the slot parameter. Figure 19-22 is an SLC 500 family PLC IIM Instruction from RSLogix 500 software that will update input status file word I:5.0. The hexadecimal mask is used to determine which input or inputs will be updated in the specified input status table for use in the ladder. The mask 0004h specifies which bit or bits will be updated. As you remember from the masking exercises, 0004h is equal to 0000 0000 0000 0100 in binary. The mask signifies that bit 2 of the input word is to be updated. Remember, when counting bits, always start with the first bit as zero. The input address updated then is I:5.0/2. The length parameter specifies how much information is to be read from the module. More than one word can be read from an input module when using a 5/03, 5/04, or 5/05 modular processor, or a MicroLogix 1200 or 1500. For example, a 32-point input module requires two 16-bit words to store the input data. As a result, to immediately update one of the upper 16 bits, the length would have to be 2. If you have a 24-input MicroLogix 1200 or 1500, a length of 2 would have to be specified to be able to update bits 16 through 23, as they would be in the second word of the status table. Typically, the rung that needed fresh updated information is programmed on the next rung, as illustrated in the figure. Along with the Immediate Input with Mask (IIM) Instruction there is an Immediate Output with Mask (IOM) Instruction. The IOM Instruction works similar to the IIM Instruction, except that it does an immediate update of an output module. Figure 19-23 has the same immediate input rungs as illustrated in Figure 19-22; however, we have now added another rung with the Immediate Output Instruction. When executed, the IOM Instruction will perform an immediate update of the output module located in slot O:2.0. Because the mask is 0000 0000 1000 0000 (0080h) output bit seven will be updated (O:2.0/7). Let's assume a 16-point output module, thus the length of 1. After these rungs are executed, normal operation will return to the scan of the remaining ladder rungs as the interrupts are completed.

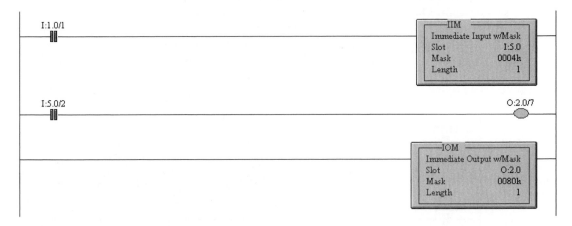

Figure 19-23   IIM and IOM Instructions used to do an immediate input update and immediate output update.

THE REFRESH INSTRUCTION

If a programmer wanted to do an immediate update of all inputs and outputs at a specific point in the ladder execution, the REF Instruction would be used. The REF Instruction interrupts normal ladder rung execution and:

- Updates all outputs
- Services communications
- Resets watchdog timer
- Reads all inputs and updates the inputs status file

When this is completed, the processor is returned to control for scanning and executing the remaining rungs with fresh input information. Figure 19-24 displays an SLC 500 family Refresh Instruction for a 5/03, 5/04, and 5/05 processor. Note that channel 0, which is always a serial port for a current Rockwell Automation PLC processor, can be selected or not selected for update. Also, channel 1 is selectable for update if desired. On a 5/03 processor, channel 1 is the DH-485 network connection on the processor. Channel 1 on a 5/04 processor is DH+, and channel 1 on a 5/05 processor is Ethernet. Because a 5/02 processor only has a DH-485 communications port, the instruction will look more like —(REF)- on a ladder rung.

Figure 19-24   SLC 500 Refresh Instruction for 5/03, 5/04, and 5/05 processors.

The PLC 5 has an instruction similar to the IIM. It is called the Immediate Input Instruction (IIN). The IIN Instruction does not have a mask associated with it. The entire input word is updated when the instruction is executed. Figure 19-25 shows an Immediate Input Instruction for the PLC 5 when using RSLogix software. When input I:002/5 is true, the IIN Instruction will interrupt ladder execution to update input word I:012 in the input status file. After the update, the ladder logic scan is resumed beginning with the next rung. PLC 5 addressing is a bit different than that of the SLC 500. The 012 address above the IIN identifies the input module to update as rack 01 group 2. The XIC Instruction on the next rung has the input address of I:012/3. This address identifies rack 01, group 2, and bit 3. This is bit 3 of the input word, the immediate update instruction updated. The PLC 5 and ControlLogix each have an Immediate Output Instruction (IOT) that works similar to the IIN, except that it is used to update the outputs.

Figure 19-25   PLC 5 using RSLogix 5 software Immediate Input Instruction.

ControlLogix does not have the IIM Instruction. When setting up the ControlLogix I/O configuration, the user can set up each individual digital input point to send data immediately into the controller solving the ladder logic whenever there is a change of state detected. Change of state from off to on, or on to off, can be used as a trigger to send immediate input information to the processor. See Figure 19-26.

Enable change of state   Local module in slot 2   Module part number
for each input point here

Module configuration tab

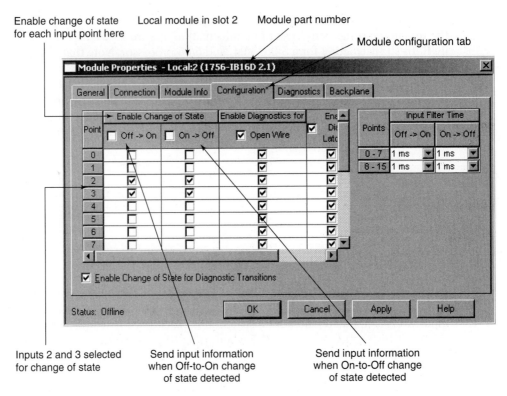

Inputs 2 and 3 selected   Send input information   Send input information
for change of state   when Off-to-On change   when On-to-Off change
                              of state detected   of state detected

Figure 19-26   ControlLogix I/O configuration screen showing change of state setup.

MASTER CONTROL RESET (MCR) INSTRUCTION

The MCR Instructions are used in pairs to create a program zone where all rungs with nonretentive outputs can be turned off, while outputs that are retentive will retain their state. The Master Control Reset Instruction, or MCR, is available in the PLC 5, SLC 500, and ControlLogix families of PLCs, and many other PLC brands. MCR zones typically are used to inhibit zones of the program while loading recipes, monitoring emergency stops, disabling logic, or establishing preconditions to sync up a machine on startup—or, after a machine jam or failure, or in zone control where valve states and set points are determined by which one of several MCR zones are active.

Figure 19-27 illustrates an MCR zone. Rungs outside the zone are not affected by the activation or deactivation of the zone. The first rung has a conditional MCR Instruction and the last rung is an unconditioned MCR rung. In this example, when input I:3.0/15 is true, all rungs within the zone act as if the zone does not exist and execute normally. When the input is false, all nonretentive outputs within the zone are turned off regardless of rung continuity. When I:3.0/15 goes false, OTE Instructions addressed as O:4.0/0, O:4.0/4, and

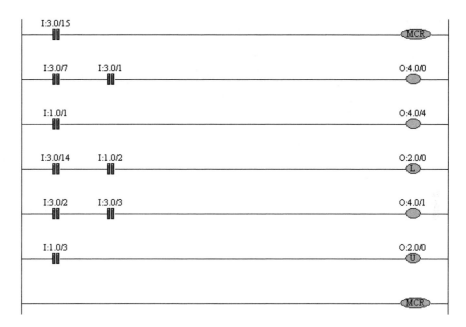

Figure 19-27   RSLogix software Master Control Reset Instruction zone.

O:4.0/1 will be turned off regardless of what rung input conditions dictate. Since the OTL and OTU Instructions are retentive instructions, they will retain their status.

MCR Usage Rules

When programming or interpreting an MCR zone remember that:

1. The MCR zone must end with an unconditional MCR Instruction.
2. If the end of zone MCR Instruction is omitted, the processor will use the file's end rung instruction as the end of the zone.
3. Nesting of MCR zones is not allowed.
4. Do not jump into an MCR zone. Jumping into the zone will activate the zone from the LBL Instruction to the end of the zone.
5. When an MCR zone goes false, TOF timers within the zone will automatically activate and begin their off delay cycle.
6. When an MCR zone goes false, TON timers within the zone will reset themselves.
7. Overlapping of MCR zones could result in unpredictable machine operation.

Always remember that the MCR Instruction is not a replacement for a hard-wired master control relay that provides emergency stop in the event someone pushed an emergency stop palm button. All emergency stop buttons must always be hard-wired. Refer to your PLC Hardware installation for more information on E-stop wiring and monitoring with the MCR Instruction.

SUMMARY

Program control instructions are input and output instructions that change how the processor scans. Program control instructions modify the sequence in which the program is executed, thereby interrupting the normal sequence of the processor operating cycle. Typically, ladder file 2 in an SLC 500, MicroLogix, or PLC 5 is the main ladder program or file. Starting with ladder file 3, through 255 for an SLC 500 and up to 999 for a PLC 5 are designated for subroutines. The Jump to Subroutine (JSR) Instruction redirects the program flow to another ladder file. Within the subroutine ladder file are the SBR and RET Instructions. These instructions are optional. Unless conditioned, the Return is optional. SBR and RET Instructions are required when input and return parameters are used with a PLC 5 or ControlLogix PLC. Subroutines can be nested to eight levels with the PLC 5s and SLC 500 family members, while ControlLogix nesting is limited by processor memory. The Jump to Label instruction and its associated Label Instruction are used to jump over rungs in the current ladder file. If the application requires an immediate update of input and output information due to high-speed processing applications, the Immediate Input and Immediate Output Instructions can be used to obtain fresh input information just before the rung is executed, and output fresh output information as the result of solving a specific rung or group of rungs. The Refresh Instruction can perform a complete update of all inputs and outputs. The Master Control Reset instructions are used in pairs to create a program zone where all rungs with nonretentive outputs can be turned off, while outputs that are retentive will retain their state.

REVIEW QUESTIONS

1. One subroutine can use a JSR Instruction to jump to another subroutine. This is called _____ subroutines.
2. The Jump to Label instruction jumps to the target, the _____ instruction.
3. You need to program an IIM Instruction where inputs I:12.0/2, I:12.0/3, and I:12.0/8 need to be updated when the instruction is executed.
   A. List the instructions parameters below:

   B. What PLC uses this instruction?

   C. Using the answer from A above, fill in the correct information for the parameters.

4. The PLC 5 has an instruction similar to the IIM; it is called the _____.
5. Returning from nested subroutines must be in the _____ order that was used when going from one subroutine to the next.
6. Typically, ladder file 2 in an SLC 500, MicroLogix, or PLC 5 is the main ladder program or file. Starting with ladder file 3, through _____ for an SLC 500 and 3 to _____ for a PLC 5 are designated for subroutines.

7. When the IIM Instruction is true, the processor stops the execution of ladder logic and does an _____ of the input module slot specified.

8. True or false?: When using multiple JMP Instructions to jump to the same label and the first JMP Instruction is identified as Q2:1, the next JMP will be identified as Q2:2.

9. Input parameters give programmers the option to bring information into the subroutine as they jump into the subroutine. The JSR Instruction will pass the input parameter(s) to the _____.

10. The _____ Instruction is not a replacement for a hard-wired master control relay that provides emergency stop.

11. True or False?: Instructions to the right of the LBL Instruction are out of the jump zone and will always be executed as a normal rung.

12. The mask for an IIM Instruction is ABC5h. What bits in the input word are we looking at?

13. The JMP Instruction is assigned a unique identifier from 0 to _____. The number acts much like the address of the label or target to jump to.

14. The _____ and _____ PLCs, when using RSLogix software, support input and return parameters in the JSR, SBR, and RET Instruction.

15. An IOM Instruction has the parameters listed below. What output(s) will be updated?

Slot	O:16.0
Mask	2A2Ch
Length	1

16. When programming MCR Instructions, the first rung has a _____ MCR Instruction and the last rung is an _____ MCR rung.

17. To jump to a label there are two instructions necessary. The _____ Instruction is an output, and the Label Instruction is the target.

18. When returning from nested subroutines:
    A. You must return the way you came.
    B. Jumping from a nested subroutine directly into ladder file 2 is an efficient way to get back to the main ladder.
    C. Jumping from a nested subroutine directly into ladder file 2 will fault the processor.
    D. The processor will automatically return when it executes the END Instruction of the last nested subroutine in the sequence.
    E. All of the above
    F. A and C

19. To stop the logic scan and do an immediate refresh of all inputs and outputs on an SLC 500 family PLC, which instruction would be used?
    A. IIM
    B. IIN

    C. IOM
    D. REF
    E. DO I/O
    F. None of the above

20. The MCR Instructions are used in pairs to create a program zone where all rungs with _____ outputs can be turned off, while outputs that are _____ will retain their state.

21. When the input on the first MCR Instruction at the beginning of a zone is _____, all nonretentive outputs within the zone are turned off regardless of rung continuity.

22. When an MCR zone goes false, _____ timers within the zone will automatically activate and begin their off delay cycle.

23. When an MCR zone goes false, _____ timers within the zone will reset themselves.

24. The MCR Instruction is not a replacement for a _____ master control relay that provides emergency stop.

APPENDIX

A

Instruction Set Reference

BIT INSTRUCTIONS		
Instruction	**Symbol**	**Use This Instruction**
Normally Open or Examine ON	—| |—	As a normally open, or examine if ON, input instruction on your ladder rung
Normally Closed or Examine OFF	—| / |—	As a normally closed, or examine if OFF, input instruction on your ladder rung
One Shot	—(OS)—	To input a single digital pulse from a maintained input signal
Latch Output Coil	—(L)—	To latch an output ON. Output stays ON until the unlatch instruction becomes true
Unlatch Output Coil	—(U)—	To unlatch a latched-ON instruction with the same address
Output Coil	—( )—	As an output instruction that becomes true when all inputs on the rung are true
Negated Output	—(/)—	As an output instruction that passes power at all times except when all rung inputs are true

TIMER INSTRUCTIONS		
Instruction	**Use This Instruction to**	**Functional Description**
On-Delay	Program a time delay before instruction becomes true	When you want an action to begin a specified time after the input becomes true.
Off-Delay	Program a time delay to begin after rung inputs go false	If an external cooling fan on a motor is to run all the time the motor is running and for five minutes after the motor is turned off, you have a five minute off-delay timer. The five-minute timing cycle begins when the motor is turned off.
Retentive	Retain accumulated value through power loss, processor mode change, or rung state going from true to false	Use a retentive timer to track the running time of a motor for maintenance purposes. Each time the motor is turned off, the timer will remember the elapsed running time. Next time the motor is turned on, the time will increase from that point. When you want to reset this timer, use a reset instruction.
Reset	Reset the accumulated value of a timer or counter	Typically used to reset a retentive timer's accumulated value to zero.

COMPARISON INSTRUCTIONS		
Instruction	**Use This Instruction to**	**Functional Description**
Equal	Test if two values are equal	If the value in source A is equal to the value in source B, this input instruction is true.
Not Equal	Test if one value is not equal to another	If the value in source A is not equal to the value in source B, this input instruction is true.
Greater Than	Test if one value is greater than another	If the value in source A is greater than the value in source B, this input instruction is true.
Less Than	Test if one value is less than another	If the value in source A is less than the value in source B, this input instruction is true.
Greater Than or Equal to	Test if one value is greater than or equal to another	If the value in source A is greater than or equal to the value in source B, this input instruction is true.
Less Than or Equal to	Test if one value is less than or equal to another	If the value in source A is less than or equal to the value in source B, this input instruction is true.
Masked Comparison for Equal	Test a portion of one value for equality with a portion of a second value	The 16-bit data in source A is compared to the 16-bit data in source B through a hexadecimal mask. If the bits that are not masked are true, this input instruction is true.
Limit Test	Test to see if test value is within the high and low limits established	If the test value is within or outside a specific range, this input instruction will be true or false, depending on how the limits are set.

COUNTER INSTRUCTIONS		
Instruction	**Use This Instruction to**	**Functional Description**
Count Up	Count from zero up to a desired value	Counting the number of parts produced during a specific work shift, or in the current batch. Also, counting the number of rejects from this batch.
Count Down	Count down from a desired value to zero	An operator interface display shows the operator the number of parts remaining to be made for a lot of 100 parts ordered.
High-Speed Counter	Count input pulses that are too fast to read for normal input points and modules	Most fixed PLCs will have a high-speed set of input points that will allow interface to high-speed inputs. Signals from an incremental encoder would be a typical high-speed input. Check your specific PLC for maximum pulse rate for your PLC.
Counter Reset	Reset a timer or counter	Used to reset a counter to zero so another counting sequence can begin.

SEQUENCER INSTRUCTIONS		
Instruction	**Use This Instruction to**	**Functional Description**
Sequencer Output	Control machine sequence	Each 16-bit word represents 16 outputs on an output module. Outputs are controlled by sequencing from one word to the next.
Sequencer Compare	Monitor inputs	Compare 16-bit internal data to a 16-bit input module's input points.
Sequencer Load	Load data into a data file sequentially	Load data into a data file from each step of a sequencer operation. Data can be from the input status file or another data file.

DATA-HANDLING INSTRUCTIONS		
Instruction	**Use This Instruction to**	**Functional Description**
Move	Move a copy of information from one memory location to another	When instruction is true, a copy of information specified as source A is moved to memory location specified as source B. The move instruction usually moves one word of data.
Masked Move	Move a copy of information from one memory location through a mask to another location	When instruction is true, a copy of information specified as source A is moved through a hexadecimal mask to memory location specified as source B. The move instruction usually moves one word of data.
Copy	Copy the contents of many consecutive words in one file to another location	When instruction is true, a copy of the information specified as source A is moved to memory location specified as source B. The copy instruction usually moves multiple words of data.
Convert a value to BCD	Convert to BCD from an integer	Convert an integer value to BCD. The converted value is stored in a selected destination or sent to the output status table for output to an operator display.
Convert a value from BCD	Convert from BCD to an integer	Convert a BCD source such as thumbwheel input from the input status table data into an integer. The resultant is stored in a selected destination.
Radian	Convert from degrees to radians	Convert the source in degrees into radians and store the result in the selected destination.
Degrees	Convert from radians to degrees	Convert the source in radians into degrees and store the result in the selected destination.

LOGICAL INSTRUCTIONS		
Instruction	**Use This Instruction to**	**Functional Description**
AND	AND the bits in one word with the bits in another word	When true, the word specified as source A is ANDed bit by bit with the word specified as source B. The result is stored in the user-specified destination.
OR	OR the bits in one word with the bits in another word	When true, the word specified as source A is ORed bit by bit with the word specified as source B. The result is stored in the user-specified destination.
Exclusive-OR	Exclusive-OR the bits in one word with the bits in another word	When true, the word specified as source A is exclusive-ORed bit by bit with the word specified as source B. The result is stored in the user-specified destination.
NOT	NOT the bits in one word with the bits in another word	When true, the word specified as source A is negated bit by bit with the word specified as source B. The result is stored in the user-specified destination.

BIT SHIFTING AND ROTATING INSTRUCTIONS		
Instruction	**Use This Instruction to**	**Functional Description**
Bit Shift Right	Shifts bits from left to right, usually to track parts	Bit shift right is an output instruction for loading bits into a bit array. Each time the inputs go from false to true the bits are shifted one position to the right. The LSB is shifted out of the array with each false-to-true input transition.

(continued)

	BIT SHIFTING AND ROTATING INSTRUCTIONS	
Instruction	**Use This Instruction to**	**Functional Description**
Bit Shift Left	Shift bits from right to left, usually to track parts	Bit shift left is an output instruction for loading bits into a bit array. Each time the inputs go from false to true the bits are shifted one position to the left. The MSB is shifted out of the array with each false-to-true input transition.
Rotate Right	Rotate bits from left to right through a bit array	Bit shift right is an output instruction for loading bits into a bit array. Each time the inputs go from false to true the bits are shifted one position to the right. The LSB is shifted to be the input bit at the MSB end of the array. Rather than being shifted out of the array, bits rotate through the array.
Rotate Left	Rotate bits from right to left through a bit array	Bit shift left is an output instruction for loading bits into a bit array. Each time the inputs go from false to true the bits are shifted one position to the left. The MSB is shifted to be the input bit at the LSB end of the array. Rather than being shifted out of the array, bits rotate through the array.

	DATA STORAGE INSTRUCTIONS	
Instruction	**Use This Instruction to**	**Functional Description**
LIFO	Last in first out stack	Load data words into a file and unload them in the opposite order as they were loaded.
FIFO	First in first out stack	Load data words into a file and unload them in the same order as they went in.

	MATH INSTRUCTIONS	
Instruction	**Use This Instruction to**	**Functional Description**
Add	Add two values together	Add the value in one source to another source and store the result in a desired destination file.
Subtract	Subtract one value from another	Subtract the value of one source from a second source and store the result in a destination file.
Multiply	Multiply two values together	Multiply the value of one source by a second source and store the result in a destination file.
Divide	Divide one value by another	Divide the value of one source by a second source and store the result in a destination file.
Square Root	Determine the square root of a value	Determine the square root of a value; the result is rounded and placed in a destination file.
Sine	Take the sine of a number	Take the sine of a value; the result is placed in a destination file.
Cosine	Take the cosine of a number	Take the cosine of a value; the result is placed in a destination file.
Tangent	Take the tangent of a number	Take the tangent of a value; the result is placed in a destination file.
Arc Sine	Take the arc sine of a number	Take the arc sine of a value; the result is placed in a destination file.

MATH INSTRUCTIONS		
Instruction	Use This Instruction to	Functional Description
Arc Cosine	Take the arc cosine of a number	Take the arc cosine of a value; the result is placed in a destination file.
Arc Tangent	Take the arc tangent of a number	Take the arc tangent of a value; the result is placed in a destination file.
Natural Log	Take the natural log of a number	Take the natural log of a value; the result is placed in a destination file.
Log to the base 10	Take the log base 10 of a number	Take the log base 10 of a value; the result is placed in a destination file.
X to the power Y	Take the value X and raise it to the power Y	Take the X value; raise it to the power Y and place the result in a destination file.
Scale	Scale analog input data into engineering units	To scale analog input data into linear data to be used internally in the processor. The formula to scale input data is $Y = mx + b$.
SCP	Scale with parameters	Use this instruction to scale incoming or outgoing data. As an example, speed reference data from a variable frequency drive could be seen as 0 to 32,767 (a 16-bit signed integer) by your SLC 500 processor. Use this instruction to easily scale the incoming data to 0 to 1,750 rpm.

APPENDIX

B

SLC 500 Status File Overview

STATUS FUNCTIONS APPLICABLE TO ALL PROCESSORS	
Word	**Status Function**
S:0	Arithmetic Flags
S:1	Processor Mode Status and Control
S:2	STI bits and DH-485 Communications
S:3L	Current and Last Scan Time
S:3H	Watchdog Scan Time Setting
S:4	Free-Running Clock
S:5	Minor Error Bits
S:6	Major Error Code
S:7 and S:8	Suspend Code/Suspend File
S:9 and S:10	Active DH-485 Nodes
S:11 and S:12	I/O Slot Enables
S:13 and S:14	Math Register
S:15L	Node Address
S:15H	Baud Rate

STATUS FUNCTIONS APPLICABLE TO 5/02, 5/03, AND 5/04 PROCESSORS	
Word	**Status Function**
S:16 and S:17	Test Single Step—Start Step On—Rung/File
S:18 and S:19	Test Single Step—Breakpoint—Rung/File
S:20 and S:21	Test—Fault/Power Down—Rung/File
S:22	Maximum Observed Scan Time
S:23	Average Scan Time
S:24	Index Register
S:25 and S:26	I/O Interrupt Pending
S:27 and S:28	I/O Interrupt Enabled

(continued)

STATUS FUNCTIONS APPLICABLE TO 5/02, 5/03, AND 5/04 PROCESSORS	
Word	**Status Function**
S:29	User Fault Routine File Number
S:30	Selectable Timed Interrupt Set Point
S:31	Selectable Timed Interrupt File Number
S:32	I/O Interrupt Executing

STATUS FUNCTIONS APPLICABLE TO 5/03 AND 5/04 PROCESSORS	
Word	**Status Function**
S:33	Extended Processor Status and Control Word
S:34	Pass-through Disabled (5/04 processor only)
S:35	Last 1 ms Scan Time
S:36	Extended Minor Error Bits
S:37	Clock/Calendar Year
S:38	Clock/Calendar Month
S:39	Clock/Calendar Day
S:40	Clock/Calendar Hours
S:41	Clock/Calendar Minutes
S:42	Clock/Calendar Seconds
S:43	STI Interrupt Time
S:44	I/O Event Interrupt Time
S:45	DII Interrupt Time
S:46	Discrete Input Interrupt File Number
S:47	Discrete Input Interrupt Slot Number
S:48	Discrete Input Interrupt Bit Mask
S:49	Discrete Input Interrupt Compare Value
S:50	Discrete Input Interrupt Preset
S:51	Discrete Input Interrupt Return Mask
S:52	Discrete Input Interrupt Accumulator
S:53 and S:54	Reserved
S:55	Last DII Scan Time
S:56	Maximum Observed DII Scan Time
S:57	Operating System Catalog Number
S:58	Operating System Series
S:59	Operating System Firmware Revision
S:60	Processor Catalog Number
S:61	Processor Series
S:62	Processor Revision
S:63	User Program Type

STATUS FUNCTIONS APPLICABLE TO 5/03 AND 5/04 PROCESSORS	
Word	**Status Function**
S:64	User Program Functionality Index
S:65	User RAM Size
S:66	Flash EEPROM Size
S:67 and S:68	Channel Zero Active Nodes

STATUS FUNCTIONS APPLICABLE TO 5/04 PROCESSORS	
Word	**Status Function**
S:69 to S:82	Reserved
S:83 to S:86	DH+ Active Nodes for Channel One
S:87 to S:98	Reserved
S:99	Global Status Word
S:100 to S:163	Global Status File

APPENDIX
C

Applying Hexadecimal Numbers in PLC Masking Applications

Most modern PLCs have instructions that allow the manipulation of data in different ways. Some PLC data manipulation instructions use masks, or filters, to block out unwanted data from passing from one location to another. A mask is simply a method to control data flow. Think of a mask as a line of 16 doors, with one door for each bit in a 16-bit word. If the door is open, the bit can pass; if the door is closed, the bit is held back, or masked out.

Considering that a single hexadecimal number is equivalent to four bits, a four-digit hexadecimal number (16 bits) could act as a mechanism to control data flow bit by bit.

The Allen-Bradley SLC 500 is an example of a PLC that uses hexadecimal masks within certain data manipulation instructions. In this section we will investigate the SLC 500 masked-move instruction and how a hexadecimal mask is used in conjunction with moving selected data from one memory location through a mask to another.

The move instruction is used to move a copy of information contained in a source, which is typically a data table location, to another user specified data table location. The 16-bit data word in the starting location is called the source word; the target location to which data are being moved is called the destination. An example of moving data from one location to another would be moving a value from an integer table into a timer or counter instruction. The data could specify how long a mixer is to mix or how many counts of a specific product signify a full case. Each product that is manufactured might have different mixing parameters or differing counts signifying a full case. The correct time or count value would need to be moved into the user ladder diagram timer or counter parameters according to which product is being manufactured. In applications such as these, the entire 16-bit word represents a whole number that is a time or count value, and the entire data word must be moved.

In some applications, only selected bits of the 16-bit data word need to be moved, causing them to be separated from adjacent bits. The masked move instruction uses a hexadecimal mask to mask, or filter, undesired bits in a 16-bit data word. The mask restricts data from being moved from the source 16-bit word to the destination.

Hexadecimal masks are used, as it is easier to enter a four-digit hexadecimal value rather than 16 bits of binary information. The instructions on your ladder program will prompt you to enter a hexadecimal value for your mask instruction parameter. When working with hexadecimal masks, think of the actual mask in its binary 1s and 0s state. Each mask binary bit will pair up with its associated binary bit in the source word to determine if the source data will pass. This is important, as the bit value of each of the 16 bits will determine which data are to pass. When a source bit is not allowed to pass, the destination bit remains in the same logical state as it was.

Example 1:

Our starting, or source, 16-bit word is: 1111 1000 1010 1000

Will all bits be allowed to pass? Yes.

Our beginning destination 16-bit word is: 0000 0000 0000 0000

For this example we will replace the mask doors with "Y" for yes and "N" for no. Figure C-1 illustrates source, mask, and destination bits before data are filtered through the mask.

Source word	1111	1000	1010	1000
Bits to pass?	YYYY	YYYY	YYYY	YYYY
Original destination bits	0000	0000	0000	0000

Figure C-1   Determining mask bits.

Figure C-2 shows that because source data bits were allowed to pass, all destination bits have been replaced with their respective source bits.

Source word	1111	1000	1010	1000
Bits to pass?	1111	1111	1111	1111
Original destination bits	1111	1000	1010	1000

Figure C-2   Destination bits after execution of masked move instruction with a mask of 1111 1111 1111 1111 or FFFFh.

Example 2:

Our starting, or source, 16-bit word is: 1111 0001 1010 0000

Will bits be allowed to pass? Upper 8 bits only.

Our beginning destination 16-bit word is: 0000 0000 0000 0000

In this example we will replace the "Y" for yes and "N" for no with 1s and 0s. If the mask bit position is a 1 (the door was open), data will pass. If the bit position is a 0, data will not pass. Figure C-3 illustrates the resulting destination bits. Note that where mask bits were 1s, data were allowed to pass through to the destination. Where mask bits were 0s, data were not allowed to pass (the door was closed); as a result, the associated destination bits were not modified.

Source word	1111	0001	1010	0000
Bits to pass?	1111	1111	0000	0000
Destination bits	1111	0001	0000	0000

Figure C-3 Destination bits after execution of masked move instruction with a mask of 1111 1111 0000 0000 or FF00h.

Example 3:

Our starting, or source, 16-bit word is: 1111 1111 1111 1111

Will bits be allowed to pass? Lower 4 bits only.

Our beginning destination 16-bit word is: 0000 0000 0000 0000

Figure C-4 illustrates the resulting destination bits. Note that where mask bits were 1s, data were allowed to pass through to the destination. Where mask bits were 0s, data were not allowed to pass (the door was closed); as a result, the associated destination bits were not modified.

Source word	1111	1111	1111	1111
Bits to pass?	0000	0000	0000	1111
Destination bits	0000	0000	0000	1111

Figure C-4 Destination bits after execution of masked move instruction with a mask of 0000 0000 0000 1111 or 000Fh.

For the PLC to work with this masking procedure, we must enter the mask values as hexadecimal values when we develop the user program. Here the mask bits from Figure C-4 have been changed to their hexadecimal equivalent.

$$0000\ 0000\ 0000\ 1111_2 = 000F$$

Example 4:

Our starting, or source, 16-bit word is: 1010 1010 1010 1010

Which bits will be allowed to pass? Upper 8 bits only.

Our beginning destination 16-bit word is: 0101 0101 0101 0101

Figure C-5 illustrates the resulting destination bits after execution of moving bits through the designated mask.

Source word	1010	1010	1010	1010
Bits to pass?	1111	1111	0000	0000
Destination bits	1010	1010	0101	0101

Figure C-5 Destination bits after execution of masked move instruction with a mask of FF00h.

For the PLC to work with this masking procedure, we must enter the mask values as hexadecimal values when we develop the user program. The mask bits from Figure C-5 changed to their hexadecimal equivalent FF00h.

An important point to remember for example 4 is the state of the resulting bits. Since the lower eight bits of the mask are 0s, the bits from the source will not be allowed to pass through to the destination. The resulting lower eight destination bits will remain undisturbed.

Example 5:

Our starting, or source, 16-bit word is: 1010 1010 1010 1010

Which bits will be allowed to pass? Only bits 2, 3, 4, and 5 of the lower byte and bits 2, 3, 6, and 7 of the upper byte.

Our beginning destination 16-bit word is: 1111 1101 0111 1101

What will the mask be? (See Figure C-6.)

Source word	1010	1010	1010	1010
Bits to pass?	????	????	????	????
Destination bits	1111	1101	0111	1101

Figure C-6 Determining mask for Example 5.

Figure C-7 illustrates the mask and the destination bits after execution.

Source word	1010	1010	1010	1010
Bits to pass?	1100	1100	0011	1100
Destination bits	1011	1001	0110	1001

Figure C-7 The mask CC3Ch provides the desired destination bits.

The mask, 1100 1100 0011 1100, equals CC3C hex.

GLOSSARY

address The address of an input or saved data is the unique identification of the location where the data is stored. Computer memory is arranged in blocks of locations, much like a grouping of post office boxes, each of which has its own unique address.

analog A voltage or current signal that continuously changes in a smooth gradual progression over a specific range is an analog signal.

analog input module An analog input module converts DC analog incoming signals to digital values that can be manipulated by the CPU.

analog output module Analog output modules convert digital signals from the CPU into analog output voltage or current signals to operate analog output hardware devices.

ASCII The American standard code for information interchange (ASCII) is a 7-bit code used for digital communications. ASCII is a binary code used to represent letters and characters. Seven- bit ASCII can represent 128 different binary combinations. Alphanumeric codes used include the letters, symbols, and decimal numbers found on a standard computer keyboard. There are 26 alphanumeric capital characters and 26 lower case characters, ten numerals (0 through 9), plus punctuation, mathematical and special symbols, and control characters. There is a total of 128 characters in 7-bit ASCII.

attach To attach is to establish communication between a PLC and an outside device. After the user program is developed in a personal computer or handheld programmer, it is electronically attached to the PLC's CPU memory through a communication cable. After communications have been established, the user program is transferred between the personal computer and the PLC (downloaded). If the program is to be transferred between PLC memory and the personal computer, it is uploaded.

backplane The printed circuit board that runs along the back of a modular rack, base, or chassis. This board accepts each I/O module's signal through a connector into which each module plugs. Signals are transferred between the modules and the CPU via the backplane.

battery backup A battery is used on a CPU and some specialized modules as a way of providing power to keep volatile memory chips energized to ensure that memory will be retained even if the main power becomes disconnected. Newer memory chips are more energy-efficient, and as a result, many newer PLCs use internal capacitors in place of batteries for memory retention.

baud rate Baud rate is the speed of bits per second transmitted when using serial communication. PLC processors will typically transmit serial data from 1,200 to 19,200 bits per second.

BCD BCD is a simple, 4-bit code number system developed as an easy method to convert decimal numbers into a binary format (binary-coded decimal). Binary-coded decimal numbers were designed to help humans interface with the computer. BCD is a set of 4-bit binary codes used to represent the decimal numbers zero through nine.

binary number system A number system in which only two digits (1 and 0) are used to represent numerical values. Also known as the base 2 number system.

Bit Bit is the abbreviation for a binary digit. A single 1 or 0 is a bit.

branch A branch is a logical, parallel path within a rung of ladder logic.

byte A byte is eight bits, or binary digits. An example of a byte would be the series of bits, 11001010.

chassis The hardware assembly, sometimes also called a rack or base (depending on the manufacturer), that holds the power supply, processor, and I/O together as a unit. The chassis contains the backplane, which transfers control signals and data between the processor and I/O modules.

checksum An error checking routine used for verifying the validity of transmitted data.

clear To reset a bit, memory location, or entire memory to a zero logical status is to clear the bit or memory location. Each data table location must contain either an ON signal level or an OFF signal level. The ON signal level stored in memory is called a one, while an OFF signal level is associated with a zero. Thus, a data table that has been cleared contains only zeros.

comment A rung comment is text that is included with each PLC ladder rung and used to help individuals understand how the program operates or how the rung interacts with the rest of the program.

CPU The central processing unit is the microprocessor device inside any computer that controls the system activities. In a modular programmable controller, the CPU is the module that contains the microprocessor. The CPU controls the execution of the user program, I/O updates, and associated housekeeping chores. Sometimes the CPU is simply called the processor.

database files The database folder contains all user-entered ladder text documentation. The database files included in the database folder are: address/symbol, instruction comments, rung comments/page titles, address/symbol picker, and symbol groups database. Even though the database folder is a part of the project, database information is never downloaded to the processor.

data bus A group of conductors grouped together used for transmission of data from one location to another. If 16-bit data is being transferred, typically 16 conductors will be grouped together to form a 16-bit-wide data bus.

data files The areas of memory that contain the status of inputs, outputs, the processor, the timers, the counters, and so forth.

DDE Dynamic Data Exchange (DDE) provides the capability to link data from one application, such as a running PLC program, to another DDE-compliant program, such as Microsoft Excel. As an example, the previous day's production data contained in the PLC's data tables can be used to populate an Excel spreadsheet on a scheduled basis before the morning's production meeting.

debounced (input signal) An input signal that has had either intermediate mechanical noise or multiple input signals removed from it as a result of mechanical contact bounce inherent when two contacts are brought together.

default value The starting value, or beginning settings, provided to the user by the software or hardware.

diagnostic I/O modules Newer I/O modules have the ability to send diagnostic information back to the PLC processor. Diagnostic output modules that have electronic fusing can alert the processor that the fuse has tripped. Diagnostic input and output modules can also detect an open output circuit and send a diagnostic bit back to the processor. Diagnostic bits will be interpreted by the ladder logic and the appropriate action taken.

digital Digital signals can be only ON or OFF. In a PLC, information is stored, transferred, or processed in a two-state numerical representation; that is, ON/OFF, open/closed, true/false, or 1/0.

discrete signals Two-state signals, usually ON/ OFF, true/ false, yes/no, or 1/0. A discrete signal is another term used for a digital signal.

DOS Disk Operating System. The operating system that makes a personal computer work.

double integer Two 16-bit words, or thirty-two bits used to represent integer values are referred to as a double integer.

double word A double word is two 16-bit words used together.

download To transfer a program from a computer or handheld programmer's memory into a programmable controller's memory.

drop lines Flexible cables that drop from a tap box in a communication network. The drop line is used to connect a hardware device to the main (trunk) line of a communications network.

dumb terminal Regarding PLCs, a programming device such as a handheld programmer that, when not connected to the PLC and communicating back and forth, cannot stand alone and be programmed. A terminal that has no internal processing capability is a dumb terminal. This terminal, which typically has a keyboard, can only be used to input data into, and receive information back from, an intelligent device.

edit To change a ladder diagram or program.

EEPROM Electrically Eraseable Programmable Read-Only Memory (EEPROM) is a nonvolatile memory chip used in PLCs to store a processor's firmware or the user ladder program in a memory chip that can be read but not written to. The EEPROM can be erased electrically and reprogrammed.

enable Either energize or make able to be energized under proper signal conditions.

END Instruction The END Instruction is preprogrammed on the last rung by the programming software. The End rung cannot be deleted. Logic cannot be programmed on the End rung. When executing the End rung, the processor is directed to return to the prior ladder file, routine, or subroutine. If currently executing logic in the main ladder or routine, the processor is directed to move on to the next step in the program scan.

error message An error message is a visual indication (such as "E 008") on the display of the programmer that alerts the user to an improper software instruction entry attempt, incomplete instruction sequence, or hardware malfunction.

Ethernet network The TCP/IP Ethernet network is a local area network designed for high-speed exchange of information between computers, programmable controllers, and other devices. Ethernet has a high bandwidth of 10 million bits per second (10 Mbps) up to 100 Mbps for high-speed communication between computers and PLCs over vast distances. An Ethernet network is desirable as it allows plant floor (PLC) data to be accessed by office or corporate mainframe databases.

examine if closed (XIC) An input instruction that is logically true when the input status bit associated to its address is a 1 and logically false when it is a 0.

examine if open (XIO) An input instruction that is logically true when the input status bit associated with its address is a 0 and logically false when it is a 1.

false An instruction is false when not ON, true, passing power, or failing to provide a continuous logical path on a ladder rung.

Fieldbus architecture A control architecture that uses serial, digital, multidrop, and two-way communication between intelligent field devices.

file A collection of like information organized into one group. As an example, timer data must be stored somewhere. The area of memory that stores timer data is called a timer file. Think of a file as a file folder storing like data in the file cabinet of the processor memory.

firmware The set of software commands that defines the personality of a system such as a PLC processor or variable-frequency drive main computer board. The firmware in a PLC defines its personality as a PLC and not a variable-frequency drive, or other piece of hardware. Firmware is typically stored on EEPROM or Flash Memory.

fixed I/O A fixed-style PLC's I/O screw terminals are built into the unit and not changeable. A fixed I/O PLC has no removable modules. All I/O points are built-in or in the form of fixed screw terminals that are nonchangeable.

Flash EPROM Flash Erasable Programmable Read-Only Memory combines the versatility of the EEPROM with the security provided by a UVPROM.

floating-point data file A floating-point data file is used to store integers and other numerical values that cannot be stored in an integer file. If your PLC stores integers as a 16-bit signed integer format, your data range for an integer file is whole numbers within the range of $-32,768$ to $+32,767$. Any number outside that range, or any fractional number, must be stored in a floating-point file. Numbers such as .333 or .25 are not whole numbers, so they will be stored in a floating-point file. Likewise, values over $+32,767$ or less than $-32,768$ must also be stored as "float" numbers.

function keys Keys on a personal computer, electronic operator device, or handheld programmer keyboard, which are labeled F1, F2, and so on. The operation of each of these keys is defined by the software. Function keys can be user defined on many electronic operator interface devices.

hardware Hardware includes the physical PLC with its racks, modules, power supply, and CPU.

hexadecimal number system Hexadecimal (Hex) numbers are an extension of the BCD number system. Hex is based on base 16. This number system has sixteen digits: 0, 1, 2, 3, 4, 5, 6, 7, 8, 9, A, B, C, D, E, F. Digits 10 through 15 inclusive are represented by the first six letters of the English alphabet. Hex allows the use of the invalid codes associated with BCD.

HHP Handheld programmer.

HHT Handheld programming terminal.

human machine interface (HMI) Graphical display hardware where machine status, alarms, messages, diagnostics, and data entry are available to the operator in graphical display format. The graphical hardware can be a personal computer or industrial computer running software such as Rockwell Automation's RSView 32, or a panelview display terminal.

IEC 1131 International standard for machine control programming tools such as PLCs and their associated programming software. The standard is comprised of five programming languages with standard commands and data structures.

industrial computer An industrial computer is a personal computer that has been built especially to withstand a harsh industrial environment.

input Incoming signals to the PLC from outside hardware devices such as limit switches, sensors, push buttons, and so forth. These incoming signals are stored in memory locations in the input status file.

input module The input section of a modular PLC. Modular PLCs have removable assemblies called input modules. Each input module typically has 8, 16, or 32 terminal-block screw connections where incoming signal wires are attached. Each input module transforms and isolates incoming signal levels from outside hardware devices into signal levels that the CPU can understand and process.

input status file The input status file is the memory area that holds the ON/OFF status of all available inputs to your PLC. A PLC with 256 inputs will have 256 one-bit memory locations to store the status of each input. To monitor I/O status, these locations are organized into a table for user viewing on a personal computer screen.

instruction A rung of logic contains input contacts or other symbols representing action the processor is to take depending on whether these contacts or symbols are found to be true or false. Each of these symbols represents an instruction. Instructions in the user program direct the processor how to react to an ON or OFF input signal seen during the program scan.

instruction set A list of available instructions that a particular processor will understand and execute when running the user program.

integer A positive or negative whole number such as -2, 1, 0, 1, 2, 3, 4, 5, etc.

intelligent field devices Microprocessor-based devices used to provide process variables, performance, and diagnostic information to the PLC processor. These devices are able to execute their assigned control functions with little interaction, except communication, with their host processor.

intelligent I/O modules I/O modules that have their own microprocessor intelligence to process input values and the ability to decide how to control their associated output devices.

Internet A global collection of industrial, personal, commercial, academic, and government computers connected on one large network, or on a collection of smaller networks, for the exchange of information.

interoperability When a product from one vendor can be substituted for a similar product by another vendor. The Device Net Network allows the user to mix and match products from many vendors on the same network. The user is not locked into the product offerings from a single vendor.

I/O I/O is the abbreviation for input/output.

I/O interface An I/O interface is a hardware device that enables the PLC and external hardware devices to work together. The PLC needs to communicate with, or control, outside devices with incompatible signal levels. The interface device, typically an I/O module, contains circuitry that converts and isolates signals. The interface allows hardware devices to send each other understandable signals.

JSR Instruction   When true, the Jump to Subroutine (JSR) Instruction redirects program flow from the current ladder file to another ladder file or routine called a subroutine. The SLC 500, MicroLogix, and PLC 5 subroutine files start with ladder file 3. The ControlLogix does not have ladder files with numbers. Since ControlLogix is a name-based PLC, all routines will be assigned a name rather than a file number. A ControlLogix subroutine will be any routine other than the main routine or a fault routine.

ladder diagram   A shorthand representation of a circuit where symbols are used to represent the actual ON or OFF status of input hardware devices is a ladder diagram. In addition to input and output symbols, the ladder diagram will contain internal instructions including timer, counter, sequencer, math, and other data manipulation instructions.

LEDs   Semiconductor diodes that emit light when energized. LEDs are used in seven-segment output displays in electronic and electrical equipment.

local I/O   When expansion racks are connected close to the base rack with simple cabling and no special communication link is configured, these racks are called local I/O racks. Local I/O distance from the processor varies from one manufacturer to another. Typically, local I/O will range from between 6 and 36 inches up to 50 feet. SLC 500 modular racks have two local I/O interconnect cables available, 6 inches and 36 inches.

logic   When the CPU is running the user program, the microprocessor is solving the program logic. Logic is the set of rules for interconnecting discrete ladder program instructions to arrive at conclusion. The logic represented by each rung's output instruction is the solution of the rung. In digital electronics, complex problems are solved using logic gates such as AND, OR, and NOT, and with other logical rules. Programmable controllers employ logic in ladder diagrams, which are sometimes called ladder logic.

lower byte   The lower eight bits in a 16-bit word.

lower nibble   The lower four bits of an 8-bit data word is the lower nibble. Bits 0 through 3 make up the lower nibble of the lower byte of a 16-bit word. Bits 8 through 11 make up the lower nibble of the upper byte of a 16-bit word.

mask   A 16-bit data word used to filter, or mask out, selected data bits. The mask stops the selected bits from being transferred from a source word, or location, to a destination word. Often the mask will be represented as a hexadecimal value.

memory   A PLC's memory is where data are stored in an orderly manner. Data are typically stored in a file by address. A file cabinet full of file folders storing information in an orderly manner serves as a metaphor for how the PLC stores and organizes data.

micro PLCs   A term used to identify a new generation of physically smaller and more powerful PLCs. Their small size and increased capabilities are a result of advances in smaller and more powerful microprocessors and in solid-state components.

microprocessor   A single integrated circuit chip that is the central processing unit (CPU), or "brain," in computerized hardware such as a PLC. The microprocessor is sometimes called the processor or simply CPU. A modular PLC will have the processor as a separate modular piece of hardware that is either inserted into a rack or chassis or clipped onto a rack. Fixed PLCs have the processor built into the hardware housing that also contains the power supply and I/O screw terminals.

MMIs   Man-machine interface (MMI) refers to graphic terminals used to display status or alarm information about the process being controlled. Operator display terminals can be touch screen units or keypad units. Keypad units use function keys separate from the display screen for operator input of data. MMI devices allow the operator to enter process parameter information, or view status data on numerous screens. The operator interface device screens are developed using a screen development software package on a personal computer and then downloading into the operator interface device.

modular I/O   Modular PLCs have removable assemblies called I/O modules. Each I/O module typically has 8, 16, or 32 terminal-block screw connections where signal wires are attached. The advantage of a modular PLC I/O system is the flexibility to mix and match module signal levels and input and output designations to suit a particular application.

move   Data are moved from one location to another with a move instruction. Although a move instruction typically places the data in a new location, the original data still reside in their original location. The move instruction is deceiving in that data are really copied to the destination rather than physically moved. When using a move instruction, some PLCs allow the use of a mask to filter out specific data bits from being copied to the destination.

network   Connecting hardware devices through a communication link to enable communication between multiple devices forms a simple network. A network can be used to streamline system operation by sharing available hardware resources. A network can be used to share operator interface data between multiple PLCs. A central personal computer can exchange programs, program data, and monitor any station on the network.

nibble   Four bits make up a nibble. Some older computers use 4-bit words to represent information input or output.

node   When incorporating a network into your PLC system architecture, the main network cable makes up the network trunk line. All devices are connected to the trunk line cable using some type of junction box, link coupler, or station connector. The cable segment from the trunk line connector box to the hardware device being connected is called a drop, or node. Each piece of hardware and its associated node will have a unique address. Typically, only one PLC, programming terminal, or operator interface device will be allowed on a single node.

nonvolatile memory Memory that will not lose its contents after the power is lost. Usually this memory is called nonvolatile ROM.

octal number system The octal number system is base eight. Octal numbering uses the numbers 0–7, 10–17, 20–27, 30–37, 40–47, 50–57, 60–67, 70–77, 100–107, and so on. (There are no 8s or 9s in the octal number system.) Octal numbers are used for addressing on older PLCs. Many newer PLCs use decimal numbering for their addressing assignments.

OEMs Refers to Original Equipment Manufacturers.

off-line When a programming device and its associated PLC are not communicating, the devices are considered to be off-line. A programming terminal must be a smart device to stand alone off-line while the user develops or edits programs.

on-line A terminal, such as a handheld programmer or computer, establishes communication by going on-line with the CPU. Being on-line, the programming device and PLC are able to communicate with each other. Devices that are on-line can exchange data, files, or programs between each other's memory. Data that are transferred from the PLC's CPU to a handheld programmer or computer terminal are said to be uploaded. Conversely, data transferred from the handheld programmer or computer terminal to the PLC's CPU are said to be downloaded.

1/0 Digital signals are represented as a 1 for ON and a 0 for OFF. The binary number system is used to identify the two states.

one's complement The system used to represent negative numbers in a digital computer. The left-most bit is the sign bit, while the remaining bits represent the number itself. If the left-most bit is a one, the number is negative. If the left-most bit is a zero, the value is positive. To negate a number, each 1 is replaced by a zero and each 0 is replaced by a one.

open system An open system is one in which the user has interchangeability and connectivity choices.

OTE (OuTput Energize) The output energize instruction is an Allen-Bradley SLC 500 output instruction. The resulting true or false logical status of all input instructions on a particular rung is reflected in the ON or OFF status of the output instruction, OTE. If the logical resultant of all input instructions is true, the rung will become true and the OTE instruction will go true. With the OTE instruction true, the associated output hardware device will energize. If the rung is evaluated as false, the OTE will be false and output devices will either stay off or turn off.

output The resulting ON or OFF signal from solving the ladder rung instructions is sent out, or output, from the microprocessor and stored in the output status file. During the portion of the scan when the processor updates its outputs, the ON or OFF signals residing in the output status file is sent out, or output, to each output module's screw terminals. Output module screw terminals control hardware devices such as motor starters or pilot lights.

output module An output module is part of the output section of a PLC. The output module isolates and converts the low-voltage output signals from the CPU to the proper voltage or current levels needed to control output circuits.

output status file The output status file contains memory locations where the ON/OFF status of each output is stored. The CPU sends the logical ON/OFF status for each output, which is the result of solving the user program, to reserved memory locations for each output's ON/OFF status. The ON/OFF signals from the output status table ultimately control each corresponding output screw terminal.

parity bit The parity bit is added to a binary array to make the sum of the bits always odd or even. Parity is used for error checking during data transmission.

PC control Sometimes referred to as soft control. A control system where the traditional PLC processor has been replaced with a personal or industrial computer running under Windows NT and software control. In many cases soft PLC control is incorporated into systems requiring high degrees of data collection and processing and/or connectivity to multiple networks.

physical chassis A chassis houses the processor, the input and output modules, and in some cases, the power supply. Some PLC manufacturers use the term chassis, some use rack, while others refer to a base. Although each manufacturer's hardware device may have a different name and look different, all are used to hold together the pieces of a modular PLC.

PID Proportional, integral, derivative control. PID control can be executed, either through an intelligent I/O module or a program instruction, to provide automatic closed-loop operation of process control loops.

processor The central processing unit (CPU). Typically, when working with PLCs, the CPU is simply called the processor.

processor configuration When setting up software to begin developing ladder logic, the software needs to be told what processor will be used to run the program. Typically there is a menu in the software from which the processor to be used in this application is selected.

processor file The Allen-Bradley company refers to the set of program and data files that make up a user program as a processor file. Only one processor file may be stored in the SLC 500 at a time.

processor scan or **processor sweep**   The running cycle in which the processor:

1. evaluates all input instructions
2. stores input conditions in the input status table
3. solves the user program instructions as a result of the ON or OFF conditions found in the input status file
4. updates output instruction status from data found in the output status file
5. updates communication between other PLCs (if on a network)
6. performs housekeeping chores and resets the watchdog timer

program   The instructions entered into PLC memory that direct the CPU on how the user wishes the input conditions to control the output circuits. The term "to program" is also used when a user is developing and entering the program into a PLC or programming device.

program files   The areas of memory within an SLC 500 processor file that contain the ladder logic program are the program files.

programmable logic controller (PLC)   A programmable logic controller, usually called a programmable controller, is a digitally operated, electronic industrial computer. The PLC has a programmable memory for the internal storage of instructions and data. Programmed instructions execute control logic functions such as timing, counting, sequencing, arithmetic, communication, and data manipulation instructions. The PLC is typically used to control various applications through fixed or modular inputs and outputs.

program mode   A PLC is either in run mode, solving the user program, or in program mode. In program mode, the processor scan is stopped. The programmer develops or edits the user program while the CPU is in program mode.

program scan   The program scan is one part of the PLC processor scan operating cycle. During the processor scan, the CPU scans each rung of the user program. This is the scan of the program, or the program scan. During the scan and execution of the ladder program, each instruction is executed. The resulting true or false logical state derived from solving all of a rung's instructions results in the output instruction status. The output instructions' ON or OFF statuses are stored in the output status table.

project file   The project file contains all data associated with the PLC project. A project is comprised of five major pieces: help folder, controller folder, ladder folder, data folder, and database folder. The help folder contains all of the user help screens. Processor configuration and status information are stored in the controller folder. The ladder program is stored in the ladder folder, whereas ladder program information, or data, is stored in the data folder. User-entered ladder program documentation is stored in the database folder.

prompt   A symbol used to inform the user that a response is required.

protocol   The set of rules that defines the format and timing of data between data communications devices. As an example, Allen-Bradley PLC processors communicating via RS-232 serial communications use Allen-Bradley's DF-1 protocol.

rack   The rack is the physical hardware device that holds the processor, I/O modules, and in some PLCs, the power supply. When assembled, all these pieces make up a modular PLC. *See also* chassis.

radix   Radix is another way to describe the base of a number system. The decimal number system is base ten or radix 10, while binary is base two or radix 2.

RAM   RAM is the abbreviation for random access memory. This term is a misnomer. RAM is more accurately described as read/write memory. This means that the CPU can write, or place, data into memory locations. The CPU can also read, or take, data out of a memory location. The "random access" portion of this term simply means that the CPU can access data by going directly to the desired address rather than going through each and every address in a serial fashion.

read   To read data is to acquire a copy of the desired data from a storage area. The storage area could be a hard disk, floppy disk, RAM, or ROM. When read, a copy of the original data is made and transferred to the target memory storage location.

remote I/O   Remote I/O (RIO) is a rack or chassis that is located farther away from the base PLC than local I/O will support. Typically, remote I/O communicates between the base CPU and a remote rack or block via a serial data communication link. The base rack or chassis will communicate to an adapter or expansion module that resides in, and serves as the receiver and transmitter for, the remote I/O rack.

remote run mode   Changing a processor from program to run mode by using the personal computer's keyboard while concurrently running programming software. This action puts the PLC into remote run mode.

retentive data   Retentive data is not lost when power is interrupted to the PLC and its memory. Typically, a battery will be used to keep power on volatile RAM chips after a power interruption or shutdown of the PLC. If there is power to RAM chips, memory will be retained. If the battery is dead or missing, data in the user program will be lost during a power interruption.

Return Instruction   The Return (RET) Instruction can be used in two ways. Within a subroutine the Return rung can be conditioned. When the conditioned Return rung is true, the rung directs the processor to stop executing the current subroutine and return back to the main ladder file, main routine, or subroutine that was being executed before entering the current subroutine. If the conditioned Return rung is false, remaining rungs in the subroutine will be executed until either another conditioned Return rung is found to be true, or the End rung is executed. When using a PLC 5 or ControlLogix family PLC, the Return Instruction is required when using input and return parameters.

ROM Abbreviation for read-only memory. Read-only memory information can only be read. Under most circumstances, the average user cannot write any data to memory that is read only. ROM may be programmed by an original equipment manufacturer (OEM) and may contain the operating system, or personality, of the computer system. Specifically relating to a PLC, an original equipment manufacturer may develop and load the user program into some type of ROM as a way to restrict end users from modifying it.

run mode The running cycle in which the pro-cessor:

1. evaluates all input instructions
2. stores input conditions in the input status table
3. solves the user program instructions as a result of the ON or OFF conditions found in the input status table
4. updates output instruction status from data found in the output status file
5. updates communication between other PLCs if on a network
6. performs housekeeping chores and resets the watchdog timer

SBR Instruction The Subroutine (SBR) Instruction is the first instruction programmed within the subroutine. When using an SLC 500 or MicroLogix PLC, the SBR Instruction is optional. Using an SLC 500 subroutine, the SBR Instruction could be used to identify the current ladder file as a subroutine rather than an I/O interrupt. When using a PLC 5 or ControlLogix family PLC, the SBR Instruction is also optional unless input and return parameters are incorporated into the JSR, SBR, and RET Instructions.

self-test Hardware, through its firmware monitors, tests operation of a device, such as a PLC, to verify proper operation and to detect any faults or errors.

sequential function chart Sequential function chart programming is another programming language available only on certain PLC processors. Sequential function chart, sometimes called SFC, is similar to flowchart-type programming. SFC programming uses step-type boxes containing regular ladder logic rungs and transition steps, and containing regular ladder logic type rungs. Each step is executed until it is completed. When the transition step is true, the preceding step is completed and the flow can proceed to the next step. SFC programming is an alternative to ladder logic if the process can be broken into logical steps.

sign bit The left-most bit in a 16- or 32-bit binary number is the bit representing the positive or negative status of the represented word. If the bit is a one, the value is negative. A zero in the sign bit position represents a positive number.

signed integer A signed whole number such as -2, -1, 0, $+1$, $+2$, $+3$, $+4$, $+5$, etc. Numerical data is represented in binary format within the PLC. A 16-bit signed integer uses the left-most bit as the sign bit and the lower fifteen bits as the numerical value. Fifteen bits can represent the values 0 to 32,767. The range of integers represented by a 16-bit signed integer is $-32,768$ to $+32,767$. Likewise, a 32-bit PLC using 32-bit signed integers can represent data ranging from $-2,147,483,648$ to $+2,147,483,647$.

single integer A single integer is one 16-bit word used to represent integer values.

sinking input A sinking input point switches the negative DC current side of its physical input field device. A sourcing inductive proximity switch is interfaced to a sinking 24-Vdc input module. The sourcing proximity sensor switches the positive side while the input point sinks the current to ground.

sinking output A sinking output module switches the negative DC current side in relation to the physical output field device. A sinking output point is interfaced to a sourcing 24-Vdc output load. The sinking output sinks the current to ground after current has been seen by the load.

smart terminal A smart terminal or programmer has its own onboard microprocessor, which enables it to operate independently from the PLC. A smart terminal or programmer, while being used for off-line program development, need not be connected to a PLC. The program developed in the smart terminal must be downloaded to the PLC's CPU after establishing communication or going on-line. *See also* dumb terminal.

software The program on a disk or CD-ROM purchased for use on a personal computer to create your user program. Software also refers to the program a user develops and stores in the programmable controller's memory.

sourcing input A sourcing input point switches the positive DC current side of its physical input field device. A sinking inductive proximity switch is interfaced to a sourcing 24-Vdc input module. The proximity sensor sinks current to ground, or the negative side of the power supply.

sourcing output A sourcing output module switches the positive DC current side in relation to the physical output field device. A sourcing output point is interfaced to a sinking 24-Vdc output load. The sourcing output point switches the positive side while the output field device sinks the current to ground.

station A PLC, computer, operator interface, or other hardware device connected to a network's trunk line. A station connector is the physical hardware connection box where the drop line is connected to the trunk line. The hardware device used to communicate over the network is connected to the drop line. This hardware device is then called a station, or node. Each node has a unique address on the network.

subroutine A subroutine is a subprogram contained within the main program. Subroutine program logic typically will not be solved during every scan of the PLC. A jump to subroutine instruction will be included at the proper point in the PLC ladder logic to direct the processor to jump out of the main program and scan a separate program file should specific inputs or conditions dictate. Subroutines are used to save scan time. There may be many subroutines contained in a PLC processor file. An example of subroutine usage is alarm logic. Under normal conditions, alarm conditions are not present. Alarm logic can be put into a subroutine as it is unnecessary to scan all alarm logic during every program scan. Incorporating a rung into your program will direct the processor to execute this subroutine, or separate program, only under alarm conditions.

test mode Test mode is a PLC processor operating mode used for testing a PLC program. Test mode is typically used in troubleshooting a system or system installation. When the PLC processor is in test mode, the ladder program operates as normal except that output module outputs are not energized. Test mode is for testing input interaction with the ladder program and observing the resulting output status without physical field outputs being energized.

throughput Throughput is the time it takes for an input signal to be processed and seen at an output point.

truth table A truth table represents how outputs are expected to behave as the result of specific input signals. Input signals, for computer purposes, are usually in digital format.

TTL (Transistor-Transistor Logic) TTL is a semiconductor logic family based on transistor switching as the basic logic element. TTL logic is usually +5 Vdc.

two's complement Two's complement is the system used to represent negative numbers while executing math operations in a digital computer. The left-most bit is the sign bit while the remaining bits represent the number itself. If the left-most bit is a one, the number is negative. If the left-most bit is a zero, the value is positive. Negating a number is a two-step process. First we must one's complement the number. To one's complement the number, each 1 is replaced by a zero and each 0 is replaced by a one. Step two is to take the one's complement result and add

one. As a result of the operation, if a carry is generated, it is discarded.

unsigned integer When using all sixteen bits in a 16-bit word to represent a numerical value with no sign bit, there is no sign; therefore, there are only positive numbers. Sixteen bits can represent a value from 0 to 65,535.

upper byte The upper eight bits in a 16-bit word form the upper byte.

upper nibble The upper four bits of any specific byte form the upper nibble.

user program The ladder program (program file) along with the associated data files stored as multiple files on a computer's hard drive. The SLC 500 refers to the collection of program and data files as a processor file.

UVPROM A UVPROM is an EEPROM that is erasable using ultraviolet light.

volatile memory Memory, usually identified as RAM, is volatile memory. It will not retain its original contents if the power is removed.

watchdog timer A hardware timer used in PLCs to ensure that the program scan is completed in a timely manner. The watchdog timer ensures that the program has not been caught in an endless loop or for some reason become hung up and unable to complete its program scan. The processor resets the watchdog timer at a specific point during each scan to ensure continuous operation.

who A utility used with the Allen-Bradley PLCs that enables viewing which devices are on the DH-485 network.

word The bit format that the computer accepts as its standard representation of data or information. The most common PLC word lengths are eight and sixteen bits.

write To write, or transfer a copy of data, from the originator's memory to the specified storage device. Data may be written to RAM or to a disk. Currently output update data are written to the output status file. Data written to a specified storage area write over current data stored there.

INDEX